Transboundary Animal Diseases in Sahelian Africa and Connected Regions

Moustafa Kardjadj • Adama Diallo •
Renaud Lancelot
Editors

Transboundary Animal Diseases in Sahelian Africa and Connected Regions

Editors
Moustafa Kardjadj
Ecole Supérieure en Science de l'Aliment
et des Industries Agro-Alimentaire
Algiers, Algeria

Laboratoire de Recherche
«Santé et Productions Animales»
Ecole Nationale Supérieure
Vétérinaire d'Alger (ENSV), El-Alia
Algiers, Algeria

Adama Diallo
UMR ASTRE
CIRAD
Montpellier, France

ASTRE
Montpellier University, CIRAD, INRA
Montpellier, France

ISRA-LNERV
Dakar-Hann, Senegal

Renaud Lancelot
UMR ASTRE
CIRAD
Montpellier, France

ASTRE
Montpellier University, CIRAD, INRA
Montpellier, France

ISBN 978-3-030-25387-5 ISBN 978-3-030-25385-1 (eBook)
https://doi.org/10.1007/978-3-030-25385-1

This Springer imprint is published by the registered company Springer Nature Switzerland AG.
The registered company address is: Gewerbestrasse 11, 6330 Cham, Switzerland

This book is dedicated to the memory of my father, Mihoub Kardjadj (1954–2015)

Moustafa Kardjadj

Preface

Sahelian African region covers countries located between the Atlantic Ocean (west and south), the Sahara Desert (north), and sub-humid/humid Africa (south). This region faces many political and socioeconomic challenges. According to the Permanent Inter-State Committee for Drought Control in the Sahel, about 6 million people suffered extreme hunger in the Sahel last year, nearly half of whom are pastoralists and agropastoral livestock herders. In this region, livestock production is the main source of food and outcome of the inhabitants; the production systems are based on animal mobility to optimize the access to water and forage resources. Livestock are herded toward best quality grazes either throughout the year (nomadism) or during specific periods (transhumance). These livestock movements have the double potential effect of exposing healthy animals to new pathogens upon their arrival or introducing infected animals into disease-free areas, thus putting at risk origin and destination countries. Moreover, some of these diseases are of zoonotic nature and may result in pathogen transmission to humans by direct contact with live animals or during/after slaughtering and butchering operations.

According to the Food and Agricultural Organization (FAO), Transboundary Animal Diseases (TADs) are defined as those that are of significant economic, trade, and food security importance for a considerable number of countries. TADs can easily spread to other countries, reach epidemic proportions, and where control, management, or exclusion is required, necessitate the cooperation between several countries. The Sahel countries are vulnerable to several TADs by virtue of its political situation and geographical location. In fact, TADs constitute an important setback in livestock economies in the Sahel; they constantly reduce the region capacity to achieve self-sufficiency in food proteins, to assure livestock owners' welfare, and continue to pose obstacles to national, regional, and international trade in livestock and livestock products. It seems impossible at present to completely control one or more TAD solely at a national level in this region, because many countries of the area are dependent on each other's animal disease status. Therefore, the proper implementation of control measures at the regional level are needed, and these measures should be focused to prevent, control, and/or eradicate the principal epizootic diseases that have a strong impact on the economy of the region.

This book focuses on the Sahel region, shedding light on the epidemiology, socioeconomics, clinical manifestations and control approaches of TADs in this specific region including those of zoonotic nature. Other than the description of TADs in Sahelian Africa and connected regions, several issues regarding the burden of TADs, animal mobility, one health, the role of national/regional/international veterinary organizations in the surveillance process and TADs in the dromedary are also discussed. The book contains 22 chapters and is structured in three parts: (1) general features and commonalities, (2) viral diseases, and (3) bacterial diseases. Each chapter was written by experts specialized in the topic.

The hope is that this book will stimulate increased awareness from research institutions and funding agencies to strengthen the cooperative efforts of all the Sahel nations for controlling/preventing TADs and re-establish the international commitment to stabilize the political situation in the region.

I express my deep appreciation to my co-editors Dr. Adama Diallo and Dr. Renaud Lencelot for their insights, support, and expertise. I am also very thankful to the editorial staff of Springer, in particular Lars Koerner and Ejaz Ahmad, for their organization and editorial expertise. Finally, I am extremely grateful to all the contributing authors for their valuable contributions, cooperation, and patience to this project.

Algiers, Algeria Moustafa Kardjadj

Contents

Part I
General Features and Commonalities

Chapter 1
The African Sahel Region: An Introduction

Moustafa Kardjadj

Abstract Armed conflicts, fast-growing populations, extreme poverty, food insecurity, climate change, and epidemics are still raging in the Sahel region, menacing the lives of populations already living on the edge of disaster. Annually, millions of people continue to need urgent help, nearly half of whom are pastoralists and agro-pastoral herders. Transboundary animal diseases (TADs) constitute an important setback to the Sahel economies. They constantly reduce the region's capacity to achieve food self-sufficiency and pose a significant impediment to national, regional, and international trade in livestock and livestock products. In addition, they prevent the majority of Sahel countries from getting access to highly profitable markets in the connected regions. With increasing globalization, these TADs continue to occur, and therefore, it has become imperative that due attention be given to surveillance leading to the control and eventually eradication of those animal diseases that threaten Africa's livestock development potential for high-quality products to satisfy domestic need and for export.

Keywords Africa · Sahel · Climate change · Transboundary animal diseases · Pastoralists

The African Sahel region (*Sahel* in Arabic means shore, border, or coast) forms the transitional zone between Northern African Sahara to the north and the humid Savanna to the south (OECD 2014). It is an arid/semiarid tropical savanna ecoregion (Fig. 1.1), which stretches almost 5500 km across the south-central latitudes of North Africa between the Atlantic Ocean and the Red Sea in a strip 450 km wide. It is composed of more than 2.5 million km^2 of arid and semiarid grasslands (Mathon et al. 2002; Giannini et al. 2003). The Sahel area lies between 12°N and 20°N

M. Kardjadj (✉)
Ecole Supérieure en Science de l'Aliment et des Industries Agro-Alimentaire, Algiers, Algeria

Laboratoire de Recherche «Santé et Productions Animales», Ecole Nationale Supérieure Vétérinaire (ENSV), El-Alia, Algiers, Algeria
e-mail: drkardjadj@live.fr

M. Kardjadj et al. (eds.), *Transboundary Animal Diseases in Sahelian Africa and Connected Regions*, https://doi.org/10.1007/978-3-030-25385-1_1

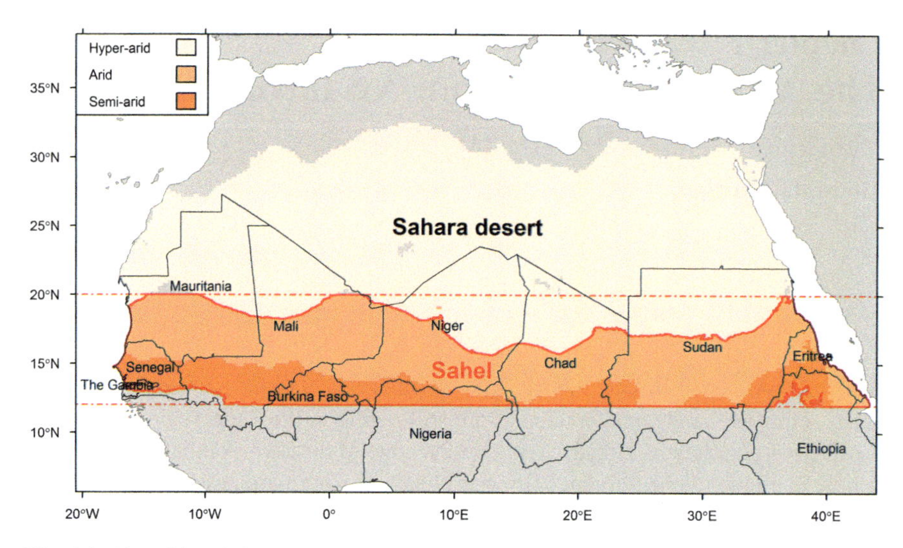

Fig. 1.1 The African Sahel region (personal figure)

(OECD 2014) and covers all or part of 12 countries from the Atlantic coast to the Red Sea: Mauritania, Senegal, the Gambia, Mali, Burkina Faso, Niger, Nigeria, Chad, Sudan, Ethiopia, Eritrea, and Djibouti (Fig. 1.1).

The Sahel countries face numerous political and socioeconomic challenges, such as the undergoing fast demographic alteration, the lack of opportunities for youth (poverty/exclusion undereducation, unemployment), droughts, flooding, and land degradation that are influential drivers for migration (Raleigh 2010; Gonzalez et al. 2012). Current struggles in the Sahel countries have generated massive population movements and demolished the scarce livelihoods of millions (Wehrey and Boukhars 2013).

Currently, as many as 4.5 million people now live in movement throughout the region; it is almost three times more than in 2012. Chaos in Libya, instability/insecurity in Mali, and the escalation of violence by terrorist activities had a devastating impact in the region (UN-ECA 2017). Gradually, these factors are deeply affecting communities and families that are already counted among the world's poorest. Large-scale displacement has exacerbated an already fragile humanitarian situation (Ammour 2013; Wehrey and Boukhars 2013). For instance, around the Chad Lake Basin, food insecurity and acute malnutrition have exceeded the emergency edge in many regions (UN-ECA 2017).

Violence and border closures prevent farmers from accessing their lands and obstruct roads that are crucial for trade and transhumance. Today, criminal and illegal networks use these roads to smuggle drugs, migrants, or illicit products, filling the space left by conflict, weak governance, and lack of cross-border cooperation (Cabot 2017). These illegal activities have stretched the level of danger to governance and social constancy in the Sahel and the connected region (UN-ECA 2017).

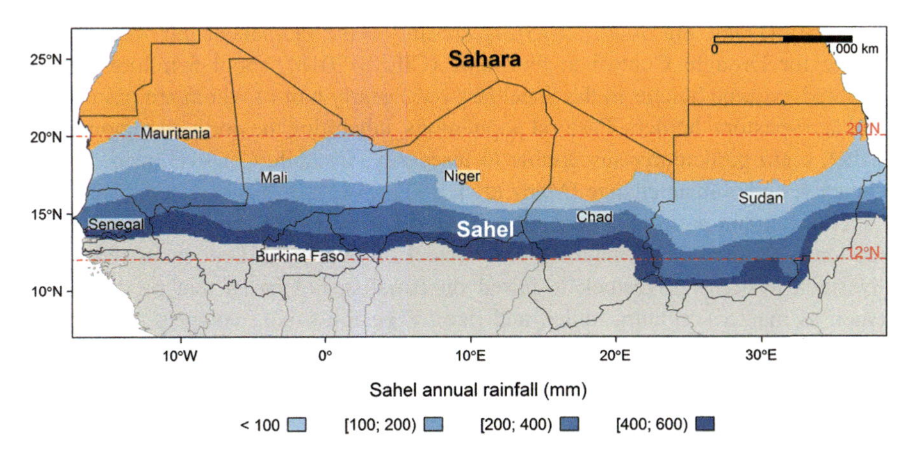

Fig. 1.2 The African Sahel region precipitation quantities (personal figure)

The Sahel is a highly vulnerable region that is facing some of the major problems in the world. It is home to almost 100 million of the world's underprivileged, most forgotten, and poorest population (Cabot 2017). Even though the Sahel has the lowest carbon emissions level, the region faces the most dramatic consequences of human-induced climate change (Panthou et al. 2014; Dong and Sutton 2015).

Studies predict that the Sahel is becoming a "hotspot" of climate change, with unprecedented climates not seen in the rest of the world; Sahel countries have been identified at extreme climate risk (Donat et al. 2016; O'Gorman 2015). Considering the fragility of its economies, fast population growth, and fragile governance, dependence on natural resources and the frequent exposure to dangerous climate risk further worsening of the region's existing vulnerabilities (Gonzalez et al. 2012; Panthou et al. 2014).

The impact of climate change is already being felt in the Sahel region. The Sahel climate tendencies witnessed over the last decades show that overall average temperatures have raised. The annual precipitation quantities (Fig. 1.2) are extremely variable (<400 mm/year), with stretched drought episodes (Donat et al. 2016).

Over the past decades, three droughts have hit the region pushing Sahelians to the brink of humanitarian disasters (Dong and Sutton 2015). Temperature prediction analysis showed that a 1.2–1.9 °C rise (which is under the region estimates) would be enough to increase by 95% the number of malnourished people in the Sahel by 2050 (Dong and Sutton 2015). Some studies have shown that a 3% temperature increase will lead to a 15–25% decrease of food production. To sum up, if the trends do not change, Africa will be able to meet only 13% of its food needs by 2050 (Panthou et al. 2014; Evan et al. 2016).

With such climate shocks occurring, susceptible communities are increasingly less able to survive crises and struggle to recover before the droughts hit again. Various inhabitants have to sell their livestock and make their children drop out of school, which make them more susceptible and vulnerable over time (Dardel et al. 2014; Donat et al. 2016). Nowadays, rainfall is the key factor in determining food insecurity (Dong and Sutton 2015). According to the Permanent Inter-State Committee for Drought Control in the Sahel (CILSS 2018), about 6 million people suffered extreme hunger in the Sahel last year, nearly half of whom are pastoralists and agro-pastoral herders. Pastoral populations, inhabiting the northern region of the Sahel, seem particularly susceptible to insecurity, instability of livelihood, coping capacity, and climate change (Mertz et al. 2009, 2011).

Transhumant pastoralist herds that, in return, provide manure for the fields often use crop residues. The majority of livestock, however, is kept in mixed crop–livestock systems. Livestock is raised on (small-scale) farms that produce crop (such as maize, sorghum, millet, and rice), vegetables (e.g., cowpea, groundnuts, and soybeans), and tubes (e.g., cassava and yams). In these systems, livestock have multiple roles such as producing food, generating income, providing manure, producing power, being financial instruments, and enhancing social status (Mortimore 2010).

The integration of livestock in the Sahel into farming systems increases from North to the more humid South. In the South, vector-borne disease challenges are more prominent than in the dryer northern zones. The urban and peri-urban intensified and intensive livestock keeping systems, where resources and feed are imported and food and wastes exported, can better meet the high demand for locally produced livestock products in the cities, but also enhance the importance of certain infectious diseases, e.g., for poultry and food-borne pathogens and antimicrobial resistances (SWAC-OECD/ECOWAS 2008; OECD 2014).

Since the 1980s, animal population growth has changed the structure of pastoral and agro-pastoral areas in the Sahel region. Human and animal densities have increased and the extension of pastoral areas has reduced due to the expansion of areas devoted to cultivation and irrigated areas around water sources. In agro-pastoral areas, fallows are limited in time; nevertheless, farmers decide to raise animals (Gonzalez et al. 2012). Consequently, the competition for resources is exacerbated and mobility becomes more constrained, in particular in southern regions. Yet this is where, more and more often, transhumant pastoralists look for pastoral resources during the dry season. Moreover, commercial movements have also grown toward capital cities of coastal countries, more and more inhabited and with an increased demand for red meat. Sahelian pastoral areas are becoming everyday more and more interdependent. The shared use of zone and resources becomes more and more complex and necessarily must be considered at regional scale (Donat et al. 2016).

According to Perry et al. (2013), livestock mobility is the third cause of the changed spatial dynamics of animal disease, following climatic change and market concentration. Livestock movements have the dual effect of exposing healthy animals to new viruses upon arrival or introducing infected animals to disease-free areas. Indeed, movements often occur before the vaccination of newborn animals that could be affected by pathogens, either by contact with already infected herds or by entering areas that are natural habitats for the diseases' vectors, such as fields infested with tick vectors of Crimean Congo Hemorrhagic Fever (CCHF); swamps and water courses that are habitats for mosquitoes vectors of RVF; as well as forests and savannas, tsetse fly habitat, and vectors of Trypanosome (FAO 2010).

The Food and Agricultural Organization (FAO 2010) defined transboundary animal diseases (TADs) as those that are of significant economic, trade, and food security importance for a considerable number of countries. TADs can easily spread to other countries and reach epidemic proportions and where control, management, or exclusion requires cooperation between several countries. These diseases are a significant burden on public health and the economic stability of societies. Most developed countries have the systems and resources to cope with TADs, either by preventing them or by responding rapidly and effectively to outbreaks (Rossiter and Al Hammadi 2008). In low-income countries, this burden is deepened by the difficulty to control diseases, due to climatic and host instability, inadequate allocation of resources, and inappropriate surveillance (McMichael et al. 2008). A multitude of factors can destabilize societies, such as economy, internal struggles of power, population growth, environmental changes, and health crises such as an outbreak of Ebola. Recent crises within the Sahel in Mali, Libya, Egypt, and Sudan have intensified levels of violence, insecurity, and poverty and restructured geopolitical and geographical dynamics in the region (OECD 2014). Regardless of the cause, instability has a significant impact on the health systems that service both people and animals, often resulting in the emergence or reemergence of infectious diseases (McMichael et al. 2008).

The transboundary animal diseases constitute an important setback in livestock economies in the Sahel. They constantly reduce the region's capacity to achieve self-sufficiency in food proteins, to assure livestock owners' welfare, and to continue to pose significant impediments to national, regional, and international trade in livestock and livestock products. In addition, they prevent the majority of African countries from getting access to highly profitable markets due to the imposition of non-tariff trade barriers by importing countries. These diseases continue to occur, and therefore, it has become imperative that due attention be given to surveillance leading to the control and eventually eradication of those animals. Any national or international response to TADs must acknowledge these constraints within the Sahelian context if they are to be successful.

Several factors contribute to keep the health risk high in the region. First, the permeability of the borders between countries that facilitates illegal movements and promotes the introduction of pathogens. Second, the lack of harmonized surveillance and control systems between countries: the obligation to vaccinate and the prices of vaccines vary according to the country as well as certain populations that are not sensitized to the problematic; introduction of unvaccinated and/or infectious animals can cause transmission of pathogens and potentially spread at international level. Finally, the absence of a system of identification of animals (or herds) and the knowledge of their movements, which would help predicting the occurrence of epidemics and their burdens.

In this book, we primarily focused on the African Sahel region, shedding new light on the epidemiology, socio-economics, clinical manifestations, and control approaches of these diseases in this specific region. Other than the description of

TADs in Sahelian Africa and connected regions, several issues regarding the burden of TADs, the role of national/regional/international veterinary organizations in the surveillance process, animal mobility, one health, and TADs in the dromedary are also discussed.

References

Ammour LA. Algeria's role in the Sahelian security crisis. Stability: Int J Secur Dev. 2013;2(2):28.

Cabot C. Climate change, security risks and conflict reduction in Africa. A case study of farmer-herder conflicts over natural resources in Cote d'Ivoire, Ghana and Burkina Faso. 2017.

Dardel C, et al. Re-greening Sahel: 30 years of remote sensing data and field observations (Mali, Niger). Remote Sens Environ. 2014;140:350–364. http://africacenter.org/2012/02/regional-security-cooperation-in-the-maghreb-and-sahel/

Donat MG, Lowry AL, Alexander LV, O'Gorman PA, Maher N. More extreme precipitation in the world's dry and wet regions. Nat Clim Chang. 2016;6:508–13.

Dong B, Sutton R. Dominant role of greenhouse-gas forcing in the recovery of Sahel rainfall. Nat Clim Chang. 2015;5:757–60.

Evan AT, Flamant C, Gaetani M, Guichard F. The past, present and future of African dust. Nature. 2016;531:493–5.

Food Agriculture Organization (FAO). La transhumance transfrontalière en Afrique de l'Ouest Proposition de plan d'action. 2010.

Giannini A, Saravanan R, Chang P. Oceanic forcing of Sahel rainfall on interannual to interdecadal time scales. Science. 2003;302:1027–30.

Gonzalez P, Tucker CJ, Sy H. Tree density and species decline in the African Sahel attribute to climate. J Arid Environ. 2012;78:55–64.

Mathon V, Laurent H, Lebel T. Mesoscale convective system rainfall in the Sahel. J Appl Meteorol. 2002;41:1081–92.

Mcmichael AJ, Friel S, Nyong A, Corvalan C. Global environmental change and health: impacts, inequalities, and the health sector. Br Med J. 2008;336:191–4.

Mertz O, Mbow C, Reenberg A, Diouf A. Farmers' perceptions of climate change and agricultural adaptation strategies in rural Sahel. Environ Manage. 2009;43:804–16.

Mertz O, Mbow C, Reenberg A, Genesio L, Lambin EF, D'haen S, Zorom M, Rasmussen K, Diallo D, Barbier B, Moussa IB, Diouf A, Nielsen JO, Sandholt I. Adaptation strategies and climate vulnerability in the Sudano-Sahelian region of West Africa. Atmos Sci Lett. 2011;12:104–8.

Mortimore M. Adapting to drought in the Sahel: lessons for climate change. Wiley Interdiscip Rev Clim Change. 2010;1:134–43.

O'Gorman PA. Precipitation extremes under climate change. Curr Clim Change Rep. 2015;1:49–59.

OECD. An atlas of the Sahara-Sahel: geography, economics and security. Paris: OECD; 2014.

Panthou G, Vischel T, Lebel T. Recent trends in the regime of extreme rainfall in the Central Sahel. Int J Climatol. 2014;34:3998–4006.

Permanent Inter-State Committee for Drought Control in the Sahel (CILSS 2018). The harmonized framework analyzes in the Sahel. http://www.cilss.int/

Perry BD, Grace D, Sones K. Current drivers and future directions of global livestock disease dynamics. Proc Natl Acad Sci. 2013;110:20871–7.

Raleigh C. Political marginalization, climate change, and conflict in African Sahel states. Int Stud Rev. 2010;12:69–8.

Rossiter PB, Al Hammadi N. Living with transboundary animal diseases (TADs). Trop Anim Health Prod. 2008;41:999.

SWAC-OECD/ECOWAS. Livestock and regional market in the Sahel and West Africa: potentials and challenges. 2008.
UN Economic Commission for Africa (UN-ECA). Conflict in the Sahel region and the developmental consequences. 2017. ISBN: 978-99944-68-80-5.
Wehrey F, Boukhars A, editors. Perilous Desert: insecurity in the Sahara. Carnegie Endowment for International Peace. 2013. http://www.jstor.org/stable/j.ctt6wpjcm

Chapter 2
The Burden of Transboundary Animal Diseases and Implications for Health Policy

Mieghan Bruce, Camille Bellet, and Jonathan Rushton

Abstract In Sahelian Africa and connected regions, the burden of transboundary animal diseases is poorly understood. This is due in part to the lack of robust estimates of the distribution and intensity of these diseases within the region. However, the problem is compounded by the complexity of the types of losses attributable to specific diseases, including the impact on human health of zoonotic transboundary diseases such as brucellosis and Rift Valley fever. There is also a balance between disease losses and the cost of our response to the presence or perceived threat of transboundary animal diseases. This chapter presents a framework for measuring the burden of transboundary animal diseases in the Sahel region, explores disease distribution data and collates what information is available on productivity losses and expenditure on disease mitigation, namely surveillance, prevention, control and treatment activities. We highlight the need for standardised data collection processes that capture disease loss estimates as well as expenditure related to our response. Reporting changes in losses and expenditure over time will provide a basis for making informed disease control policies for transboundary animal diseases. The outcome of this will be an evidence-base for mobilising resources in an efficient and effective manner.

Keywords Burden of animal disease · Animal disease losses · Animal health policy

Introduction

Transboundary animal diseases (TADs) cause a range of losses due to mortality and morbidity in animals and, for zoonotic TADs, in humans. To reduce these losses, individuals and institutions respond in various ways, including surveillance,

---

M. Bruce
School of Veterinary Medicine, College of Science, Health, Engineering and Education, Murdoch University, Perth, WA, Australia

C. Bellet · J. Rushton (✉)
Institute of Infection and Global Health, University of Liverpool, Liverpool, UK
e-mail: j.rushton@liverpool.ac.uk

M. Kardjadj et al. (eds.), *Transboundary Animal Diseases in Sahelian Africa and Connected Regions*, https://doi.org/10.1007/978-3-030-25385-1_2

prevention, control and treatment actions (Rushton and Jones 2017). Because of the inherent nature of TADs, the need for international intervention and regional coordination is critical. Quantification of the impact of TADs is an important aspect to determine if current levels of response are appropriate, whether resources need to be reallocated and where additional resources should be assigned. Impact studies should also disaggregate data on public and private responses to the presence or risk of disease, helping to provide information as to whether these reactions are appropriate within current institutional environments.[1]

Transboundary diseases are a significant burden on public health and the economic stability of societies (Nii-Trebi 2017). Most developed countries have the systems and resources to cope with TADs, either by preventing them or by responding rapidly and effectively to outbreaks (Rossiter and Al Hammadi 2008). In low-income countries, this burden is deepened by the difficulty to control diseases, due to climatic and host instability, inadequate allocation of resources and inappropriate surveillance (Mcmichael et al. 2008). A multitude of factors can destabilise societies, such as economy, internal struggles of power, population growth, environmental changes and health crises such as an outbreak of Ebola. Recent crises within the Sahel in Mali, Libya, Egypt and Syria have intensified levels of violence, insecurity and poverty and restructured geopolitical and geographical dynamics in the region (OECD 2014). Regardless of the cause, instability has a significant impact on the health systems that service both people and animals, often resulting in the emergence or re-emergence of infectious diseases (Mcmichael et al. 2008). Any national or international response to TADs must acknowledge these constraints within the Sahelian context if they are to be successful.

To make sound decisions, policy-makers need to understand the true nature of disease challenges and how these challenges are changing over time and in space. This, in turn, allows decisions to be prioritised according to the burden of diseases, shifting the focus of animal health systems to address society needs. This will enable resources to be allocated appropriately, helping to make investments by governments more efficient. In Sahelian Africa and connected regions, the impacts of TADs are, however, still poorly understood (Otte et al. 2004). In this book chapter, first we present a framework for measuring the burden of TADs in this region. Second, we collate what is known about the disease frequency of three TADs in the Sahel, namely brucellosis, foot and mouth disease (FMD) and Rift Valley fever (RVF). Third, estimates on productivity losses attributable to each disease and what is known about expenditure on the prevention, control and treatment of TADs in the Sahel are presented. Finally, we highlight the knowledge gaps and consider future requirements for a systematic way to better capture the burden of TADs. Where data specific to the Sahelian region is not available, data on TADs in general is discussed.

[1]Institutional environment refers to the rules and enforcement that people operate under. This can include official laws and policing, private standards and market incentives and disincentives and cultural norms and traditional management systems.

Measuring the Burden of Transboundary Animal Diseases

Many efforts to control TADs focus on the affected animals or livestock populations, ignoring the fact that livestock owners and society as a whole are affected by the impacts of the diseases and the measures taken to prevent or control them (FAO 2016). Transboundary animal diseases under endemic settings, or their incursion into free regions, affect animal production, reduce the availability of nutritious food in affected communities, limit resilience and prospects for poverty reduction and cause market instability (FAO 2016). Estimates of the burden of TADs provide a basis to examine how resources are allocated currently and identify where these multiple adverse effects can be minimised whilst optimising key outputs from the livestock sector.

The epidemiology and economic impacts of TADs vary widely depending upon the geographical location, host species involved and current policies and regulations. The burden of TADs is complex and dependent upon various factors including: (1) the pathogen itself; (2) whether it is endemic or an exotic incursion to a specific area; (3) the livestock production systems involved; (4) what type of trade agreements a specific country, region or zone has; (5) the political economy, such as the ability of the government to respond, political instability and the level of corruption; and (6) the context within which it occurs, for example seasonal aspects that determine the severity of the disease. Consequently, little has been done to estimate the burden of TADs, particularly in low-income countries, mainly due to the challenges in collating and integrating data on all of these components.

In humans, the global burden of diseases is calculated on the basis of incidence, severity, duration and mortality; it does not take into consideration any costs associated with our efforts to prevent, control or treat these diseases (GBD 2016 DALYs and HALE Collaborators 2017). In contrast, when estimating the impact of livestock diseases, there is a recognition that disease incidence is conditional on the efforts to control it, albeit there has been no attempt to apply such a framework to the burden of disease in a systematic way. In a theoretical manner, it is acknowledged that the true burden of disease is a function of losses attributable to the disease itself (L), and expenditures on disease mitigation (E), which is our reaction to the presence or threat of disease (Mcinerney 1996; Rushton et al. 1999). Examples of direct losses include reduced livestock productivity and increased human illness due to zoonotic TADs. Examples of human reactions to disease include vaccination, restricted animal movement or trade and reduced consumption of specific animal-source foods due to a real or perceived risk to human health (Rushton and Bruce 2017). The relationship between these components can be expressed as: $Cost = Losses + Expenditure$. Intuitively, there is an inverse correlation between L and E; for example, higher expenditure on mitigation usually results in lower losses, and vice versa (Mcinerney et al. 1992). However, the relationship is not linear nor guaranteed; therefore, the multiple dimensions of both losses and expenditure need to be quantified and explicitly incorporated into any burden estimate.

The losses attributable to a specific TAD in livestock are calculated as:

$$L_l = \sum_i \sum_j N_i \times p_i \times b_{i,j} \times e_{i,j} \times c_{i,j}$$

where N_i, is the susceptible population of type i (e.g. pigs, cattle, sheep or goats); p_i is the prevalence of the TAD in population i; $b_{i,j}$ is the value of production parameter j, in non-infected animals of type i (e.g. an annual milk yield in sheep of 100 kg); $e_{i,j}$ is the reduction in production parameter j attributable to the TAD in animals of type i (e.g. 15% reduction in the annual milk yield of infected sheep); and $c_{i,j}$ is the price of j in animals of type i (e.g. $0.60 per litre of sheep milk). Losses attributable to human illness as a result of a specific zoonotic TAD are typically expressed in disability-adjusted life years (DALYs), and occasionally health expenditure and income loss are calculated in monetary terms (Narrod et al. 2012).

Expenditure on disease mitigation[2] can be divided into public and private expenditure. Public mitigation strategies targeting TADs vary in complexity, but a comprehensive strategy will combine surveillance, prevention and control in all affected species, as well as treatment of human cases for zoonotic TADs (Leonard 2004). Individual producers or household are rarely able to directly influence public policy; however, they typically adopt specific biosecurity and behavioural changes to reduce the risk of disease at an individual level. There are situations where producers can organise themselves to manage inputs and output markets and that this may include animal health, and companies of sufficient size can afford major animal health research and support services. Regardless of the organisation of production, the amount livestock producers allocate to disease mitigation depends on many factors, such as the perceived risk to their animals, the cost of the activities and the consequences of contracting the disease (Gilbert and Rushton 2014, 2018). Additionally, the extent to which the government supports the control of a specific TAD will influence private expenditure. As a consequence, the total expenditure on any particular disease will vary between countries and may be vastly different at the individual farm level.

In the next section, data on the incidence and prevalence of three TADs in Sahelian Africa is presented. These diseases were selected to represent the diversity of TADs: brucellosis is a zoonotic TAD which is considered endemic in the Sahel, FMD is an endemic livestock disease that will have different effects depending upon the level of immunity within the livestock population, and RVF is a zoonotic disease with epidemics typically following adverse weather events.

---

[2]Disease mitigation incorporates surveillance, prevention and control activities and treatment of individuals or groups.

Estimating the Occurrence of Transboundary Animal Diseases

The requirement of countries to report outbreaks of TADs enables data on the presence of TADs to be accessed via the World Organisation for Animal Health (OIE) disease reporting systems (WAHIS) and the Food and Agriculture Organization's Global Animal Health Information System (EMPRES-i). However, these data may be unreliable and are frequently missing denominator data. Information on incidence is usually poor and hard to measure in places where TADs are endemic, which is the case for most developing countries. There is little incentive for livestock producers to report outbreaks and only a small part of the information on livestock disease is captured by official information systems (Mcleod et al. 2010). Therefore, official reports often need to be supplemented by published research reporting disease frequency data such as prevalence and incidence.

Official data on the occurrence of brucellosis, FMD and RVF available from the OIE via the World Animal Health Information Database is inconsistent. For example, FMD outbreak data for Burkina Faso reported in their Annual Animal Health Report (OIE 2018a) documented 55 outbreaks of FMD in 2016 and 49 new outbreaks in 2017, including 5002 cases in cattle, of which there were 137 deaths. In contrast, no outbreaks are reported in the Summary of Immediate Notifications and Follow-up reports for either year (OIE 2018b). To supplement OIE data, a literature review was conducted with the aim to collate disease incidence and prevalence data for brucellosis, FMD and RVF from studies published between 1 January 2008 and 31 March 2018 in the Sahel following a similar process to Ducrotoy et al. (2017) and Craighead et al. (2018). Three scientific databases, namely PubMed, Science Direct and Web of Science, were searched for relevant articles using the following search terms: (brucell∗ OR "rift valley fever" OR RVF OR "foot and mouth disease" OR FMD) AND (Sahel OR *individual country names*). Publication dates were limited to the last 10 years as earlier studies are unlikely to represent the current disease situation, due to the dynamic nature of the livestock production systems and rapidly changing demographics of humans and animals.

Brucellosis

Brucellosis is amongst the most economically important zoonotic diseases globally (ILRI 2012; WHO 2006). The disease causes significant illness in people, as well as reproductive failure and reduced milk production in livestock. Government regulations to control brucellosis, such as test and slaughter, can also be devastating to households, especially the rural poor who may be the least economically resilient (ILRI 2012; Perry and Grace 2009). Despite this, brucellosis is rarely a priority for decision-makers within human and animal health systems; thus, it remains a neglected zoonotic disease as recognised by the World Health Organisation (WHO 2015).

As mentioned earlier, for the estimation of the burden of disease, the estimation of the presence of disease is critical. Published papers reporting measures of disease frequency of brucellosis in the Sahelian region were found for 8 of the 13 Sahelian countries: 24 individual studies and 1 meta-analysis (Tadesse 2016) from Ethiopia, 19 from Nigeria, 3 from Sudan and only a single publication each from Burkina Faso, Eritrea, Senegal and the Gambia. There is widespread distribution of antibodies to *Brucella* species in the Sahel, but prevalence estimates at individual level range in terms of the species, production systems and also the type of tests. For example, in Ethiopia, studies report prevalence in cattle ranging from 1.4% to 10.6%, goats 4.7% to 22.8% and camels 0.9% to 15.7%. Cadmus et al. (2011) found that 5.46 % (20/366) and 0.27 % (1/366) of the dogs screened were seropositive to *B. abortus* and *B. canis*, respectively, in Nigeria, indicating that dogs may act as a reservoir for *B. abortus*, complicating control efforts. In contrast, antibodies against brucellosis were not found in any of the 85 bulk samples from smallholder dairy farms in Senegal (Breurec et al. 2010). This immediately creates some difficulties in burden estimates that require modelling approaches that are beyond this chapter to address.

Control of brucellosis in a given area requires an understanding of the *Brucella* species circulating in livestock and humans. However, because of the difficulties intrinsic to *Brucella* isolation and typing, such data are scarce for resource-poor areas, and the majority of studies rely on seroprevalence, interpreting the presence of antibodies as evidence of active infection. As the tests cannot differentiate between vaccinated, infected or previously exposed animals, these results need to be used with caution in any disease burden estimate. Additionally, a positive antibody response is not necessarily correlated with productivity losses in individual animals and may overestimate the burden of disease.

It is not possible to ascertain the infecting *Brucella* species using serological tests, irrespective of the antigen used or host species being tested, which compounds the problems associated with estimating the burden of brucellosis from seroprevalence studies (OIE 2016). In Nigeria, 34 isolates collected between 1976 and 2012 from cattle, sheep and horses were all identified as *Brucella abortus* (Bertu et al. 2015). A review of brucellosis in cattle in West Africa similarly only identified *B. abortus* (Sanogo et al. 2013). The paucity of bacteriological data and the consequent imperfect epidemiological picture are particularly critical for Sahelian countries, where resources utilised for controlling brucellosis need to be allocated appropriately, with knowledge on what livestock species and production systems to target.

Foot and Mouth Disease

The control of FMD in endemic countries is complex and often considered of lower priority, particularly in countries without the potential to benefit from FMD-free export markets (Young et al. 2016). Therefore, FMD outbreaks in endemic countries are generally considered to be greatly underreported (Knight-Jones et al. 2017). A model that integrated livestock density with prevalence indices suggested that the Sahel has one of the highest prevalence in small ruminants and cattle (Sumption et al. 2008). All countries in the Sahelian region reported outbreaks of FMD to the OIE between 2008 and 2017, except Djibouti, Mauritania and the Gambia. Despite this, papers published in the last 10 years reporting FMD frequency data were only found in 4 of the 13 Sahelian countries: Eritrea, Ethiopia, Niger and Nigeria. Seroprevalence studies are complicated in that FMD has different serotypes. This is exemplified in Nigeria, where the spread and co-circulation of serotypes A, O, SAT 1 and SAT 2 are reported (Wungak et al. 2017). This highlights the need to appropriately match multivalent vaccines with circulating strains.

Due to the highly contagious nature of FMD, the disease is distributed widely in endemic countries, as exemplified in Ethiopia, where outbreaks were geographically widespread affecting all major regional states in the country (Jemberu et al. 2016b). In Niger, 791 FMD outbreaks were reported from all regions between 2007 and 2015, 8804 cattle were clinically affected, and amongst these, 247 animals died from the disease (Souley Kouato et al. 2018). Traditionally, the susceptibility of breeds exotic to the Sahel and subsequent production losses were considered to be higher than indigenous breeds (Roeder et al. 1994). In contrast, Mazengia et al. (2010) reported a higher incidence in Fogera cattle indigenous to Ethiopia (15.5%) compared to Holstein Friesian cross-breed cattle (2.5%, $p = 0.02$).[3] Evidently more detailed data on FMD susceptibility, disaggregated by species, breed and production system, is warranted.

Rift Valley Fever

Rift Valley fever is a mosquito-borne viral zoonosis that primarily affects animals but also has the capacity to infect humans. In ruminants, RVF is characterised by abortion, foetal malformation and neonatal mortality, with abortion and mortality rates being highest in sheep and goats. The majority of human infections result from direct or indirect contact with the blood or organs of infected animals, although there are isolated reports of vertical transmission (Adam and Karsany 2008). Herders, farmers, slaughterhouse workers and veterinarians have an increased risk of infection.

Mali, Mauritania, Niger and Senegal reported outbreaks of RVF to the OIE between 2008 and 2017 (OIE 2018b). However, seroprevalence studies conducted in Burkina Faso, Chad, Djibouti, Mali and Sudan in various livestock species, with values ranging from 3.6% to 22.5%, suggest the disease is circulating throughout the region in the inter-epidemic periods. Since 2010, five outbreaks of RVF have been reported by the WHO in the Sahelian region: one in Niger (2016) and four in

---

[3] *p*-value from authors' own calculation.

Mauritania (2015, 2013–2014, 2012 and 2010) (WHO 2018a). The human cases in Mauritania correspond to livestock disease notification by the OIE (2018b). There were 41 human cases confirmed during an RVF outbreak in Mauritania in 2012, including 12 deaths (Sow et al. 2014a). An isolated human case was also reported in Senegal in 2012, which was presumed to be as a result of contact with infected animals imported from Mauritania (Sow et al. 2014b). Notification of an individual case in Gambia was reported in January 2018. Outbreaks are uncommon in Gambia and neighbouring countries, and there was no indication of a risk in Gambia, Senegal or Guinea-Bissau (WHO 2018b). The WHO coordinated a rapid response during the outbreak in Niger, and reported 266 suspected human cases including 32 deaths, concluding that one of the main challenges was under-detection of human cases (WHO 2018a). However, of 196 specimens tested, only 17 were classified as positive PCR, serological identification of IgM antibodies or both. Given that more than 90% tested negative, it is possible that the suspected cases were misclassified, leading to over-reporting. Disease burden estimates of RVF and other zoonotic TADs would clearly benefit from consistent case definition and disease classification systems in humans, livestock and wildlife species.

Estimating Losses Attributable to a Specific Disease

Quantifying the impact of a specific disease on livestock productivity is difficult because their effects are not always obvious, and production parameters can be influenced by other factors such as nutrition, housing and comorbidity with other diseases (Dijkhuizen et al. 1995). Disease is only one of many factors influencing the level of productivity in a production system and cannot be considered in isolation. However, the vast majority of studies reporting differences in production parameters in animals with or without disease do so without considering the system as a whole within its specific geographical and social context.

The severity of the effect on productivity will also vary dependent upon the physiological state of the animal. For example, the effect of brucellosis on milk yield will be dependent upon the stage of pregnancy at which an infected animal aborts; an early abortion is likely to result in greater losses (Godfroid et al. 2004). As the majority of papers reporting economic losses use a cross-sectional design, with notable exceptions described later, only associations between production parameters and seroprevalence can be calculated, negating the ability to measure the dynamic nature of disease severity.

An overlooked component in the socio-economic analysis of animal diseases is the multiplicity of stakeholders that are affected. Animal diseases will directly affect producers, but also service providers within the livestock supply chains such as traders, abattoir workers and retailers. Though harder to assess, there are also lost opportunities where TADs occur. When major diseases threaten livestock, enterprise owners find little incentive to invest in new stock or new systems of husbandry (Rushton 2009). Regular interruptions to sales and other sources of income also reduce hard cash needed for new investments (Otte et al. 2004). Whilst these components of losses attributable to disease are often acknowledged, their complex nature makes quantitative estimates difficult.

There are very few reports of the productivity losses attributable to TADs in the Sahel; therefore, we discuss the impact of the diseases using studies from countries facing similar conditions whilst recognising the inherent limitations of this approach.

Brucellosis

A critical limitation in estimating the burden of brucellosis is the lack of robust evidence of production losses in affected animals. In Nigeria, an absolute reduction in predicted calving rate of 14.6% was found in cattle herds with at least one cow positive in RBT and c-ELISA, with a strong negative association between within-herd *Brucella* seroprevalence and predicted calving rate (Mai et al. 2015). This study was cross-sectional in design; thus, a temporal association could be determined. They also only compared brucellosis-free herds with infected herds defined as having at least one infected animal, with great variation of within-herd prevalence in the infected group. The potential problems associated with assigning productivity losses at herd level to infection at individual level can be overcome using formula described by Putt et al. (1987)

$$A_0 = A_1 - Er$$

where A_0 the production parameter in the population is controlled, A_1 is the observed value in the population in which a proportion, r, of the population is infected by the TAD and E is the disease effect on the production parameter. According to Herrera et al. (2008), a 6-year brucellosis control plan implemented on a Mexican dairy farm with 300 cows resulted in an increase in average milk production of 6 L/day (from 24 L to 30 L) associated with a decrease in prevalence from 8.34% to less than 1%. Using the formula above, if all non-infected cattle produced 30 L/day, then infected cattle would have a decreased milk production of 72 L, which is not biologically possible (30 L − 72 L = −42 L). Therefore, the increase in average milk production reported in this study must be attributed to other factors such as improved nutrition and simultaneous control of other diseases. This highlights the limitations of using unadjusted associations to measure production losses attributable to a single disease.

Comorbidity also poses a challenge in attributing productivity losses to a specific disease. In Nigeria, out of 32 animals that were seropositive for brucellosis, 6 were also demonstrated as being infected with tuberculosis through mycobacterial culture (Cadmus et al. 2008). Although the effects on production for these diseases may be quite distinct, having multiple infectious diseases may exacerbate the clinical picture of any one. Conversely, attributing the production loss to all diseases will result in an overestimate; thus, comorbidity needs to be accounted for and adjusted accordingly (Burstein et al. 2015). Similarly in humans, the presence of co-infection with malaria and brucellosis in humans requires more nuanced approaches to estimate the burden specific to brucellosis (Feleke et al. 2015).

Data on the effect of human brucellosis has been studied in more detail than in livestock, with an average case fatality ratio of 0.5% and 40% of cases resulting in chronic infection and 10% of male cases resulting in orchitis (Dean et al. 2012). Acute brucellosis is given a disability weight of 0.108, whilst the disability weight for chronic brucellosis is 0.079 (WHO 2015). Globally, the median number of food-borne brucellosis illnesses is 393,239 (95% UI: 143,815–9,099,394) (WHO 2015). This large uncertainty interval highlights the lack of robust data on human cases. The report also acknowledges that cases of brucellosis arising from direct contact with infected animals could be just as high as food-borne diseases. The acute and chronic symptoms of the disease in humans can result in a significant loss of workdays, leading to a significant loss of income. Additionally, out-of-pocket healthcare costs can affect the socioeconomic status of infected persons and their families, which is not captured in GBD estimates.

Foot and Mouth Disease

Information on the impact of the disease in FMD endemic countries is poorly characterised, yet essential for the prioritisation of scarce resources for disease control programmes. As demonstrated for brucellosis, impact measurements are limited by study design and data analysis. A study conducted on a dairy farm Ethiopia showed that the milk yield was reduced by half during FMD outbreak (Mazengia et al. 2010). However, diagnosis was made on the basis of clinical signs, and it was not clear how the average milk yield for the individual cows was recorded for 10 days prior and 10 days after infection. Jemberu et al. (2014) reported that overall and calf-specific mortality rates were 2.4% and 9.7% in mixed crop–livestock systems and 0.7% and 2.6% in pastoral systems. The economic losses of foot and mouth disease outbreak due to reduced milk yield, draft power and mortality were on average US$6 per affected herd and US$9.8 per head of cattle in the affected herds in crop–livestock mixed system and US$174 per affected herd and US$5.3 per head of cattle in the affected herds in the pastoral system. However, these data were based on interviews with farmers, collecting retrospective data from past events, thus subject to recall errors and misclassification bias (Rothman et al. 2008). Notable exceptions include a study of FMD on a large-scale dairy farm in Kenya (Lyons et al. 2015a, b). Using daily milk yields from 218 cattle, the authors found the average daily milk yield decreased from 20 to 13 kg/cow, recovering after about 2 months. Cows in parity four or greater and in the first 50 days in milk at the start of the outbreak produced 688.7 kg (95% CI 395.5, 981.8) less milk than predicted, representing a 15% reduction in production. The authors reported increased rate of mastitis in the first month after the onset of the outbreak whilst only weak evidence to support increased culling attributable to FMD (Lyons et al. 2015b). The complexity of the productivity losses due to production stage is evident from these studies, highlighting the need to treat aggregated impact estimates with caution.

A number of economic analyses for FMD have been conducted in Africa with the aim of establishing the potential benefits of control. Randolph et al. (2005) disaggregated the costs and benefits of FMD control by sector and by population groups in Zimbabwe. Their results indicated that the majority of the burden of FMD and consequently the benefits from its control accrue to the commercial sector, such as commercial cattle producers, beef-processing industries and related input industries, rather than the communal sector where the majority of cattle are kept. In Ethiopia, the estimated annual costs of FMD, including production losses, export losses and control costs with no official control programme, were estimated at 1354 million birr (90% CR: 864–2042), with production losses contributing to 94% of the total cost (Jemberu et al. 2016a). Barasa et al. (2008) reported an ex ante benefit–cost ratio for FMD vaccination of 11.5, with 28.2% of the total losses due to chronic disease, indicating that vaccination in South Sudan has considerable impact on improving productivity on infected farms. These estimates could be improved by more robust data on productivity losses directly attributable to FMD, although it is clear that the trade implications have a considerable impact on farms with access to markets.

Rift Valley Fever

Clinical manifestation of RVF in adult animals is characterised by fever of 6–7 days duration, abortions in pregnant animals and decrease in milk production and high mortality in neonates (Kenawy et al. 2018); however, primary research on productivity losses attributable to RVF in livestock is scant. The majority of published studies in the Sahel use questionnaires based on producers self-reporting, which may not reflect morbidity and mortality attributable solely to RVF. In Kenya, case fatality rates in livestock during the 2007 RVF outbreak were estimated at 33% in cattle, 49% in goats, 61% in sheep and 50% in camels (Rich and Wanyoike 2010). There were also wide-ranging impacts on the livestock sector as well as employment losses particularly for casual labour and the total losses were estimated to be over US$32 million. Although these estimates may not be entirely representative, it is clear that RVF causes acute and severe productivity losses.

Human infection with RVF virus is generally asymptomatic, and the majority of those with clinical symptoms present with a short febrile illness and no long-term sequelae. However, severe complications, such as haemorrhagic disease, meningoencephalitis, renal failure and blindness, have been reported (Al-Hazmi et al. 2003; Madani et al. 2003; Adam et al. 2010). It has also been suggested that RVF virus is evolving with increasing severity in humans (Baba et al. 2016). Disability weights used for RVF range from 0.22 and 0.62 (Labeaud et al. 2011); however, standardised values are yet to be established.

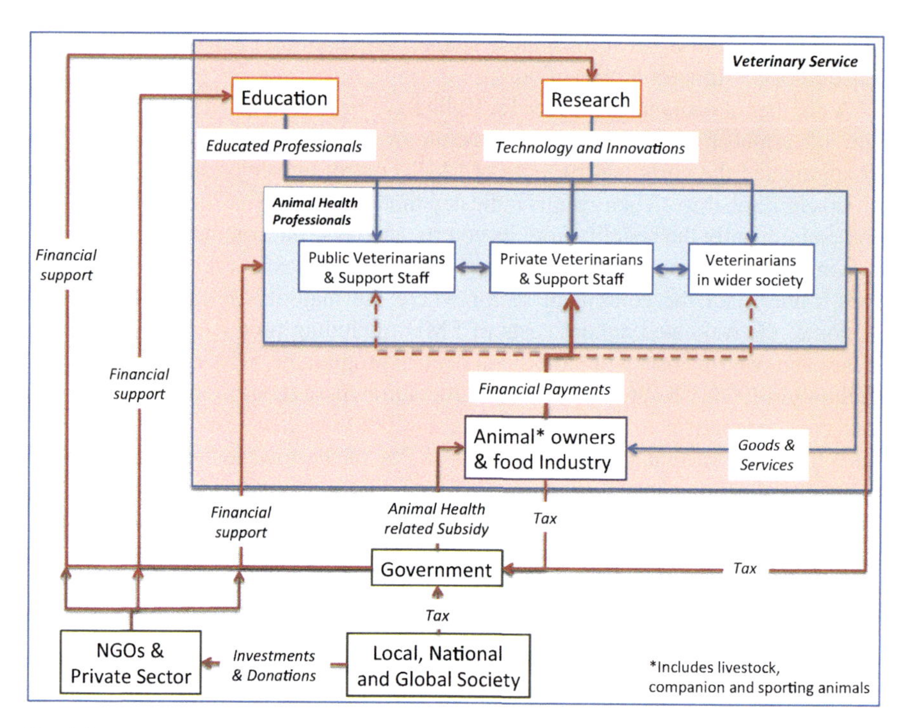

Fig. 2.1 Flows of investment and taxation and critical resources in the veterinary services (from Rushton and Jones 2017)

Calculating Expenditure on Disease Mitigation

Expenditure on disease mitigation is a combination of public and private investments and incorporates fixed costs in core veterinary service activities such as education, research and coordination (Rushton and Jones 2017). Rushton and Jones (2017) separated these flows out into contributions from different parts of society to maintain core animal health activities of research, education and coordination from the day-to-day animal health services for disease and health management (see Fig. 2.1).

Despite some theoretical understanding of the need to separate out the fixed and variable costs of animal health systems and also the flows of resources and money in that system, the data available is limited at a high level. Yet we have some proxies of investments such as employment within the animal health sector (see Table 2.1) and laboratory capabilities.

Table 2.1 Veterinarians and veterinary para-professionals working within the Sahel countries by activity in 2016 (OIE 2018d)

	Burkina Faso	Chad	Djibouti	Eritrea	Ethiopia[a]	Guinea-Bissau	Mali[b]	Mauritania[c]	Niger[c]	Nigeria	Senegal[b]	Sudan[b]	The Gambia[b]
Animal health public sector	62	200	10	7	968	10	385	25	15	1867	33	1571	1
Animal health private sector	38	28	5		160		158	42	17	3259	46	3437	12
Veterinary PH public sector	6	280	5		310	2	145	9	10	799	16	260	1
VPH private sector			1		25		128						
Laboratories public sector		108	4		100	3	81	6	7	228	7	455	
Laboratories private sector										3			
Academic activities and education	15			1	431		38	3	5	2014	6	327	2
Pharmaceutical industry	12	178			179		15			60			
Private clinical practice	38				22	1	461	42	5		116	1819	4
Other				1	150	8							
Animal health and welfare activities	1691	228	35	358	8900	50		35	200	582	287	1055	81
Community animal health workers			70	2	3403	20		44	600			5312	81

(continued)

Table 2.1 (continued)

	Burkina Faso	Chad	Djibouti	Eritrea	Ethiopia[a]	Guinea-Bissau	Mali[b]	Mauritania[c]	Niger[c]	Nigeria	Senegal[b]	Sudan[b]	The Gambia[b]
VPH activities	1328	280	5	32	2159	12		30	86	920	287		81
Other	70				99								
Total	3260	1302	135	401	16,906	106	1411	236	945	9732	798	14,236	263

[a]Data from 2015
[b]Data from 2014
[c]Data from 2013

Brucellosis

Globally, data available on public health expenditure on brucellosis was available from 37 countries with an average annual cost per programme of US$3.0 million (Rushton and Gilbert 2016). However, very little data is available on public expenditure on brucellosis from the individual countries in the Sahel. In terms of public expenditure, only 7 of the 13 countries have laboratories that are equipped to perform diagnostic tests for brucellosis in at least one *Brucella* species, whilst only Mali and Nigeria have laboratories with the ability to perform isolation and culture, which is critical to differentiate the different species (OIE 2018c). Craighead et al. (2018) conducted a workshop on brucellosis in West and Central Africa with the consensus that minimal surveillance and control activities were taking place in that region. The following is a summary of their results from the countries within the Sahel. A written policy is reportedly current in Chad; however, there is no evidence that the policy is implemented. Similarly, Senegal reported historical control programmes that are no longer in existence. In Mali, a private breed improvement programme has taken serum samples, but testing is yet to start due to lack of reagents. In Niger, humans in high-risk occupations are routinely tested; however, testing in livestock populations is not conducted.

Foot and Mouth Disease

Data on expenditure on the control of FMD specific to the Sahelian regions was not available. Using data from 21 FMD outbreaks globally, Rushton and Gilbert (2016) calculated that the average duration of FMD control was 10.9 months with expenditure on control of US$2.1 million per month. Losses due to trade restrictions ranged from US$1306 to US$14.6 million, most likely due to differences in trade restrictions placed on endemic countries versus countries demonstrated free from disease undergoing epidemic situations. During the ban of Ethiopian livestock to Egypt due to FMD, Ethiopia was estimated to have lost more than US$14 million (Ashenafi 2012). According to Ehizibolo et al. (2014), Nigeria does not have a national FMD vaccination programme due to the absence of locally produced FMD vaccines and the lack of permission for vaccine importation, although individual commercial dairy farms may import vaccines. In terms of laboratory capacity, only Burkina Faso, Ethiopia, Mali, Mauritania, Nigeria, Senegal and Sudan have public laboratories capable of diagnosing FMD (OIE 2018c).

Rift Valley Fever

If introduced into non-endemic areas, RVF virus has the potential to spread amongst local mosquito populations, with resultant epidemics causing significant impacts on human and animal health. Therefore, there is significant international support for RVF mitigation strategies. For example, during the 2016 outbreak in Niger, the WHO Country Office provided technical and financial support for surveillance, outbreak investigation, technical guidelines regarding case definition, case management, shipment of samples and risk communication and sent a multisectoral national rapid response team to the affected area (WHO 2018a).

At national level, very little data is available from Sahelian region. In a study of global expenditure on disease control, two countries reported expenditure on RVF control with an average annual cost of US$3.2 million (Rushton and Gilbert 2016). The average cost due to trade restrictions from three outbreaks was US$1.6 million, with an average duration of 37 months. In terms of laboratory capacity, only Ethiopia, Mali, Mauritania, Nigeria, Senegal and Sudan have public laboratories capable of diagnosing RVF in animals (OIE 2018c).

Discussion and Conclusion

The key to optimal resource mobilisation is the collection and collation of data that allows the generation of economic evidence and the rationale for investment. To achieve this, international cooperation is paramount, without which efforts to mitigate TADs are unlikely to be effective or sustained. However the appropriate level and type of intervention, including the optimal balance between private and public, national and international action is difficult to identify and will be context specific. Some disease mitigation efforts can have negative side effects and the decision on surveillance and intervention needs to assess if expenditures are less than the avoidable losses generated.

To make reliable estimates of disease burden, information on basic productivity parameters, such as reproduction and mortality rates, age and weight of animals at sale and offtake rates, is needed. Whilst this information may be available for large-scale commercial production herd recording systems and published industry data, for small-scale and less commercial herds, economists usually rely on a few published reports of variable quality or may need to collect their own primary data (FAO 2016). Currently, there is no systematic way to capture losses associated with livestock diseases, nor are the data on expenditure on disease mitigation analysed in a way that enables comparisons to be made (Rushton et al. 2018). Importantly, there is a need to develop harmonised methods for collecting, collating and monitoring animal health indicators that ensure consistency and comparability of data. Furthermore, due to the dynamic nature of the livestock sector and societal change, there is a need to monitor with different intensities and manage health risks in different ways (Rushton et al. 2012).

In conclusion, there is an obvious need throughout the Sahel to fully enumerate not only the frequency of TADs, but the magnitude of the costs incurred due to the diseases in terms of both productivity losses and control expenditure. Additionally, when developing disease intervention strategies, the distribution of these costs and potential benefits across society needs to be quantified, particularly for strategies enacted through government policies or international agencies. Policy-makers need to weigh the benefits of a regulation that increases epidemiological certainty against the costs of compliance for producers, households and people working within the food system that will be impacted by the policies (Peck and Bruce 2017). The benefits of controlling TADs in livestock sectors that rely on export of livestock and their products are obvious. However, the benefits of control in endemic areas are less well understood, particularly where the disease poses no zoonotic threat. Control strategies for TADs should be realistic and evidence-based, with the need to consider what can be effectively implemented in the countries within the Sahel.

References

Adam I, Karsany MS. Case report: Rift Valley fever with vertical transmission in a pregnant Sudanese woman. J Med Virol. 2008;80:929.

Adam AA, Karsany MS, Adam I. Manifestations of severe Rift Valley fever in Sudan. Int J Infect Dis. 2010;14:e179–80.

Al-Hazmi M, Ayoola EA, Abdurahman M, Banzal S, Ashraf J, El-Bushra A, Hazmi A, Abdullah M, Abbo H, Elamin A, Al-Sammani E-T, Gadour M, Menon C, Hamza M, Rahim I, Hafez M, Jambavalikar M, Arishi H, Aqeel A. Epidemic Rift Valley fever in Saudi Arabia: a clinical study of severe illness in humans. Clin Infect Dis. 2003;36:245–52.

Ashenafi B. Costs and benefits of foot and mouth disease vaccination practices in commercial dairy farms in Central Ethiopia. MSc, Wageningen University. 2012.

Baba M, Masiga DK, Sang R, Villinger J. Has Rift Valley fever virus evolved with increasing severity in human populations in East Africa? Emerg Microb Infect. 2016;5:e58.

Barasa M, Catley A, Machuchu D, Laqua H, Puot E, Tap Kot D, Ikiror D. Foot-and-mouth disease vaccination in South Sudan: benefit-cost analysis and livelihoods impact. Transbound Emerg Dis. 2008;55:339–51.

Bertu WJ, Ducrotoy MJ, Munoz PM, Mick V, Zuniga-Ripa A, Bryssinckx W, Kwaga JK, Kabir J, Welburn SC, Moriyon I, Ocholi RA. Phenotypic and genotypic characterization of Brucella strains isolated from autochthonous livestock reveals the dominance of *B. abortus* biovar 3a in Nigeria. Vet Microbiol. 2015;180:103–8.

Breurec S, Poueme R, Fall C, Tall A, Diawara A, Bada-Alambedji R, Broutin C, Leclercq A, Garin B. Microbiological quality of milk from small processing units in Senegal. Foodborne Pathog Dis. 2010;7:601–4.

Burstein R, Fleming T, Haagsma J, Salomon JA, Vos T, Murray CJ. Estimating distributions of health state severity for the global burden of disease study. Popul Health Metrics. 2015;13:31.

Cadmus SI, Adesokan HK, Stack JA. Co-infection of brucellosis and tuberculosis in slaughtered cattle in Ibadan, Nigeria: a case report. Vet Ital. 2008;44:557–8.

Cadmus SI, Adesokan HK, Ajala OO, Odetokun WO, Perrett LL, Stack JA. Seroprevalence of *Brucella abortus* and *B. canis* in household dogs in southwestern Nigeria: a preliminary report. J S Afr Vet Assoc. 2011;82:56–7.

Craighead L, Meyer A, Chengat B, Musallam I, Akakpo J, Kone P, Guitian J, Häsler B. Brucellosis in West and Central Africa: a review of the current situation in a changing landscape of dairy cattle systems. Acta Trop. 2018;179:96–108.

Dean AS, Crump L, Greter H, Hattendorf J, Schelling E, Zinsstag J. Clinical manifestations of human brucellosis: a systematic review and meta-analysis. PLoS Negl Trop Dis. 2012;6:e1929.

Dijkhuizen AA, Huirne RBM, Jalvingh AW. Economic analysis of animal diseases and their control. Prev Vet Med. 1995;25:135–49.

Ducrotoy M, Bertu WJ, Matope G, Cadmus S, Conde-Alvarez R, Gusi AM, Welburn S, Ocholi R, Blasco JM, Moriyon I. Brucellosis in Sub-Saharan Africa: current challenges for management, diagnosis and control. Acta Trop. 2017;165:179–93.

Ehizibolo DO, Perez AM, Carrillo C, Pauszek S, Alkhamis M, Ajogi I, Umoh JU, Kazeem HM, Ehizibolo PO, Fabian A, Berninger M, Moran K, Rodriguez LL, Metwally SA. Epidemiological analysis, serological prevalence and genotypic analysis of foot-and-mouth disease in Nigeria 2008–2009. Transbound Emerg Dis. 2014;61:500–10.

FAO. Economic analysis of animal diseases. Animal Production and Health Guidelines. No. 18 [Online]. 2016. www.fao.org/3/a-i5512e.pdf. Accessed 12 Dec 2017.

Feleke SM, Animut A, Belay M. Prevalence of malaria among acute febrile patients clinically suspected of having malaria in the Zeway Health Center, Ethiopia. Jpn J Infect Dis. 2015;68:55–9.

GBD 2016 DALYs & HALE Collaborators. Global, regional, and national disability-adjusted life-years (DALYs) for 333 diseases and injuries and healthy life expectancy (HALE) for 195 countries and territories, 1990–2016: a systematic analysis for the Global Burden of Disease Study 2016. Lancet. 2017;390:1260–344.

Gilbert W, Rushton J. Estimating farm-level private expenditure on veterinary medical inputs in England. Vet Rec. 2014;174:276.

Gilbert W, Rushton J. Incentive perception in livestock disease control. J Agric Econ. 2018;69:243–61.

Godfroid J, Bosman PP, Herr S, Bishop GC. Bovine brucellosis. In: Coetzer JAW, Tustin RC, editors. Infectious diseases of livestock, vol. 3. 2nd ed. Oxford, UK: Oxford University Press; 2004. p. 1510–27.

Herrera E, Palomares G, Díaz-Aparicio E. Milk production increase in a dairy farm under a six-year brucellosis control program. Ann N Y Acad Sci. 2008;1149:296–9.

ILRI. Mapping of poverty and likely zoonoses hotspots. Zoonoses Project 4. Report to the Department for International Development, UK. 2012. http://r4d.dfid.gov.uk/Output/190314/Default.aspx.

Jemberu WT, Mourits MC, Woldehanna T, Hogeveen H. Economic impact of foot and mouth disease outbreaks on smallholder farmers in Ethiopia. Prev Vet Med. 2014;116:26–36.

Jemberu WT, Mourits M, Rushton J, Hogeveen H. Cost-benefit analysis of foot and mouth disease control in Ethiopia. Prev Vet Med. 2016a;132:67–82.

Jemberu WT, Mourits MC, Sahle M, Siraw B, Vernooij JC, Hogeveen H. Epidemiology of foot and mouth disease in Ethiopia: a retrospective analysis of district level outbreaks, 2007–2012. Transbound Emerg Dis. 2016b;63:e246–59.

Kenawy MA, Abdel-Hamid YM, Beier JC. Rift Valley fever in Egypt and other African countries: historical review, recent outbreaks and possibility of disease occurrence in Egypt. Acta Trop. 2018;181:40–9.

Knight-Jones TJD, Mclaws M, Rushton J. Foot-and-mouth disease impact on smallholders – what do we know, what don't we know and how can we find out more? Transbound Emerg Dis. 2017;64:1079–94.

Labeaud A, Bashir F, King CH. Measuring the burden of arboviral diseases: the spectrum of morbidity and mortality from four prevalent infections. Popul Health Metrics. 2011;9:1.

Leonard D. Tools from the new institutional economics for reforming the delivery of veterinary services. Rev Sci Tech. 2004;23:47–58.

Lyons NA, Alexander N, Stark KD, Dulu TD, Sumption KJ, James AD, Rushton J, Fine PE. Impact of foot-and-mouth disease on milk production on a large-scale dairy farm in Kenya. Prev Vet Med. 2015a;120:177–86.

Lyons NA, Alexander N, Stärk KD, Dulu TD, Rushton J, Fine PE. Impact of foot-and-mouth disease on mastitis and culling on a large-scale dairy farm in Kenya. Vet Res. 2015b;46:41.

Madani TA, Al-Mazrou YY, Al-Jeffri MH, Mishkhas AA, Al-Rabeah AM, Turkistani AM, Al-Sayed MO, Abodahish AA, Khan AS, Ksiazek TG. Rift Valley fever epidemic in Saudi Arabia: epidemiological, clinical, and laboratory characteristics. Clin Infect Dis. 2003;37:1084–92.

Mai HM, Irons PC, Thompson PN. Brucellosis, genital campylobacteriosis and other factors affecting calving rate of cattle in three states of Northern Nigeria. BMC Vet Res. 2015;11:7.
Mazengia H, Taye M, Negussie H, Alemu S, Tassew A. Incidence of foot and mouth disease and its effect on milk yield in dairy cattle at Andassa dairy farm, North West Ethiopia. Agric Biol J N Am. 2010;1:969–73.
Mcinerney J. Old economics for new problems – livestock disease: Presidential Address. J Agric Econ. 1996;47:295–314.
Mcinerney JP, Howe KS, Schepers JA. A framework for the economic analysis of disease in farm livestock. Prev Vet Med. 1992;13:137–54.
Mcleod A, Honhold N, Steinfeld H. Responses on emerging livestock diseases. In: Steinfeld H, Mooney H, Schneider F, Neville L, editors. Livestock in a changing landscape, Drivers, consequences and responses, vol. 1. Washington, DC: Island Press; 2010.
Mcmichael AJ, Friel S, Nyong A, Corvalan C. Global environmental change and health: impacts, inequalities, and the health sector. Br Med J. 2008;336:191–4.
Narrod C, Zinsstag J, Tiongco M. A one health framework for estimating the economic costs of zoonotic diseases on society. EcoHealth. 2012;9:150–62.
Nii-Trebi NI. Emerging and neglected infectious diseases: insights, advances, and challenges. Biomed Res Int. 2017;2017:15.
OECD. An atlas of the Sahara-Sahel: geography, economics and security. Paris: OECD; 2014.
OIE. Chapter 2.1.4. Brucellosis (*Brucella abortus*, *B. melitensis* and *B. suis*) (infection with *B. abortus*, *B. melitensis* and *B. suis*). Manual of diagnostic tests and vaccines for terrestrial animals. Paris: World Organisation of Animal Health; 2016.
OIE. Burkina Faso: Annual Animal Health Report 2017 [Online]. Paris: World Organisation of Animal Health; 2018a. http://www.oie.int/wahis_2/public/wahid.php/Reviewreport/semestrial/review?year=2017&semester=0&wild=0&country=BFA&this_country_code=BFA&detailed=1. Accessed 23 Mar 2018.
OIE. Summary of immediate notifications and follow-ups [Online]. Paris: World Organisation of Animal Health; 2018b. http://www.oie.int/wahis_2/public/wahid.php/Diseaseinformation/Immsummary. Accessed 23 Mar 2018.
OIE. WAHIS: Laboratory capability [Online]. Paris: World Organisation for Animal Health; 2018c. http://www.oie.int/wahis_2/public/wahid.php/Countryinformation/Countrylaboratoris.
OIE. WAHIS: Veterinarians and veterinary para-professionals [Online]. Paris: World Organisation for Animal Health; 2018d. http://www.oie.int/wahis_2/public/wahid.php/Countryinformation/Veterinarians.
Otte MJ, Nugent R, Mcleod A. Transboundary animal diseases: assessment of socio-economic impacts and institutional responses. Livestock Policy Discussion Paper No. 9. Rome: Food and Agriculture Organization; 2004.
Peck D, Bruce M. The economic efficiency and equity of government policies on brucellosis: comparative insights from Albania and the United States of America. Rev Sci Tech. 2017;36:291–302.
Perry B, Grace D. The impacts of livestock diseases and their control on growth and development processes that are pro-poor. Philos Trans R Soc Lond B Biol Sci. 2009;364:2643–55.
Putt SNH, Shaw APM, Woods AJ, Tyler L, James AD. Veterinary epidemiology and economics in Africa: a manual for use in design and appraisal of livestock health policy. Addis Ababa: International Livestock Centre for Africa; 1987.
Randolph TF, Morrison JA, Poulton C. Evaluating equity impacts of animal disease control: the case of foot and mouth disease in Zimbabwe. Rev Agric Econ. 2005;27:465–72.
Rich KM, Wanyoike F. An assessment of the regional and national socio-economic impacts of the 2007 Rift Valley fever outbreak in Kenya. Am J Trop Med Hyg. 2010;83:52–7.
Roeder PL, Abraham G, Mebratu GY, Kitching RP. Foot-and-mouth disease in Ethiopia from 1988 to 1991. Trop Anim Health Prod. 1994;26:163–7.
Rossiter PB, Al Hammadi N. Living with transboundary animal diseases (TADs). Trop Anim Health Prod. 2008;41:999.

Rothman KJ, Greenland S, Lash TL. Modern epidemiology. Philadelphia: Lippincott Williams and Wilkins; 2008.
Rushton J. The economics of animal health and production. Cambridge, MA: CABI; 2009.
Rushton J, Bruce M. Using a One Health approach to assess the impact of parasitic disease in livestock: how does it add value? Parasitology. 2017;144:15–25.
Rushton J, Gilbert W. The economics of animal health: direct and indirect costs of animal disease outbreaks. Technical item presented at the 84th General Session of the World Assembly of OIE Delegates, 22–27 May, Paris. 2016. www.oie.int/fileadmin/home/eng/Media_Center/docs/pdf/SG2016/A_84SG_9.pdf. Accessed 22 Apr 2017.
Rushton J, Jones D. Sustainability and economic investments in animal health systems. Rev Sci Tech. 2017;36:721–30.
Rushton J, Thornton PK, Otte MJ. Methods of economic impact assessment. Rev Sci Tech. 1999;18:315–42.
Rushton J, Hasler B, De Haan N, Rushton R. Economic benefits or drivers of a 'One Health' approach: why should anyone invest? Onderstepoort J Vet Res. 2012;79:461.
Rushton J, Bruce M, Bellet C, Torgerson P, Shaw APM, Marsh T, Pigott D, Stone M, Pinto J, Mesenhowski S, Wood P, Workshop Participants. Initiation of the Global Burden of Animal Diseases (GBADs). Lancet. 2018;392:538–40.
Sanogo M, Abatih E, Thys E, Fretin D, Berkvens D, Saegerman C. Importance of identification and typing of Brucellae from West African cattle: a review. Vet Microbiol. 2013;164:202–11.
Souley Kouato B, Thys E, Renault V, Abatih E, Marichatou H, Issa S, Saegerman C. Spatio-temporal patterns of foot-and-mouth disease transmission in cattle between 2007 and 2015 and quantitative assessment of the economic impact of the disease in Niger. Transbound Emerg Dis. 2018;65:1049–66.
Sow A, Faye O, Ba Y, Ba H, Diallo D, Faye O, Loucoubar C, Boushab M, Barry Y, Diallo M, Sall AA. Rift Valley fever outbreak, southern Mauritania, 2012. Emerg Infect Dis. 2014a;20:296–9.
Sow A, Faye O, Faye O, Diallo D, Sadio BD, Weaver SC, Diallo M, Sall AA. Rift Valley fever in Kedougou, southeastern Senegal, 2012. Emerg Infect Dis. 2014b;20:504–6.
Sumption K, Rweyemamu M, Wint W. Incidence and distribution of foot-and-mouth disease in Asia, Africa and South America; combining expert opinion, official disease information and livestock populations to assist risk assessment. Transbound Emerg Dis. 2008;55:5–13.
Tadesse G. Brucellosis seropositivity in animals and humans in Ethiopia: a meta-analysis. PLoS Negl Trop Dis. 2016;10:e0005006.
WHO. The control of neglected zoonotic diseases: a route to poverty alleviation. 2006. www.who.int/zoonoses/Report_Sept06.pdf. Accessed 14 Mar 2014.
WHO. WHO estimates of the global burden of foodborne diseases: foodborne disease burden epidemiology reference group 2007–2015. 2015. http://www.who.int/foodsafety/publications/foodborne_disease/fergreport/en/. Accessed 21 Mar 2018.
WHO. Rift Valley Fever [Online]. Geneva: World Health Organization; 2018a. http://www.who.int/news-room/fact-sheets/detail/rift-valley-fever. Accessed 14 Apr 2018.
WHO. Rift Valley Fever – Gambia. Disease outbreak news. [Online]. Geneva: World Health Organization; 2018b. http://www.who.int/csr/don/26-february-2018-rift-valley-fever-gambia/en/. Accessed 14 Apr 2018.
Wungak YS, Ishola OO, Olugasa BO, Lazarus DD, Ehizibolo DO, Ularamu HG. Spatial pattern of foot-and-mouth disease virus serotypes in North Central Nigeria. Vet World. 2017;10:450–6.
Young JR, Suon S, Rast L, Nampanya S, Windsor PA, Bush RD. Benefit-cost analysis of foot and mouth disease control in large ruminants in Cambodia. Transbound Emerg Dis. 2016;63:508–22.

Chapter 3
Livestock Mobility in West Africa and Sahel and Transboundary Animal Diseases

Andrea Apolloni, Christian Corniaux, Caroline Coste, Renaud Lancelot, and Ibra Touré

Abstract Human and animal mobility—especially ruminants for the latter—is a key factor in the Sahelian region. Animals are kept mobile to look for better grazing areas and water resources, to be sold at international markets and to escape insecure areas. The demographic growth of the western African coastal countries (and northern Africa for eastern Sahel) is changing the mobility pattern in the area. Old and new problems are faced by pastoralists and traders. As it is, mobility is a complex phenomenon.

Here we describe livestock mobility in the region, showing some of the benefits and challenges that it faces nowadays. In West Africa, mobility is a complex phenomenon. It involves different spatial (from few kilometres to international journeys) and temporal (movements can take few days till several months) scales and it is contributing to local and regional economy in different ways: through livestock trades and the creation of jobs around the commercial circuits.

A major issue is the scarcity of available data. In the absence of a centralised system, volumes of traded animal are mostly guessed: quantitative information is rarely available. A clearer and more quantitative knowledge of the livestock mobility network

In loving memory of Caroline Coste who passed away during the writing of this chapter.

A. Apolloni (✉)
CIRAD, UMR ASTRE, Montpellier, France

ISRA, Laboratoire National Elevage et Recherche Vétérinaire (LNERV), Dakar, Sénégal
e-mail: andrea.apolloni@cirad.fr

C. Corniaux · I. Touré
ISRA, Laboratoire National Elevage et Recherche Vétérinaire (LNERV), Dakar, Sénégal

UMR SELMET Univ. Montpellier, CIRAD, INRA, SupAgro, Montpellier, France
e-mail: christian.corniaux@cirad.fr; ibra.toure@cirad.fr

C. Coste · R. Lancelot
UMR ASTRE, CIRAD, Montpellier, France

ASTRE, Montpellier University, CIRAD, INRA, Montpellier, France
e-mail: caroline.coste@cirad.fr; renaud.lancelot@cirad.fr

M. Kardjadj et al. (eds.), *Transboundary Animal Diseases in Sahelian Africa and Connected Regions*, https://doi.org/10.1007/978-3-030-25385-1_3

could greatly help veterinary officers to improve control and surveillance of animal diseases. In this chapter, we focus on the movements of cattle and small ruminants.

Keywords Transhumance · Commercial movements · Network · Regional market integration · Socioeconomic aspects

Introduction

According to 2017 FAO estimates (Aubauge et al. 2017), the total number of ruminants (cattle, goats and sheep) in West Africa is around 330 million heads (75 million bovines, 255 million small ruminants) as shown in Fig. 3.1.

The largest fraction of the livestock is concentrated in the Northern Countries (Mauritania, Mali, Chad, Niger) where large uninhabited areas, and unsuitable for agriculture, provide enough space to sustain extensive ruminant farming. The large small-ruminant population found in Nigeria should be considered with respect to the human population, by far the highest in the region (Aubauge et al. 2017). Livestock farming is one of the most important sector in West Africa, and particularly in Sahel, thus contributing to the national gross domestic product (GDP): from 9% in Senegal, Ghana and Togo till 15% in other countries (Kamuanga et al. 2008a). In terms of agricultural GDP, the role of livestock sector ranges from 5% in Benin and Togo to 80% in Mauritania. Agriculture, and in particular livestock sector, employs a large

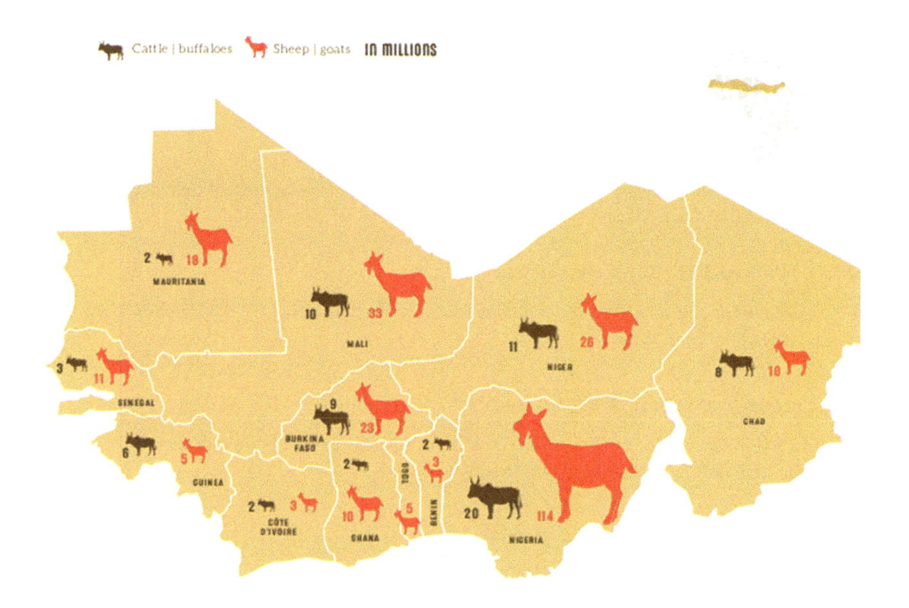

Fig. 3.1 National herds (cattle and small ruminants) in the regions according to FAO estimates, source (Aubauge et al. 2017)

fraction of the active population for an average of 52.5% of the jobs in West Africa (Kamuanga et al. 2008a). In addition, livestock is often a major part of the household capital, and a source of day-to-day income to cover familial needs, or in case of crisis.

Live animals trade is in most of the Sahelian countries the second economical activity after oil-product trade (Various Authors 2018). In general, the consumption and production areas are several hundreds of kilometers apart. Live animals are sold at local markets to traders and then moved to capital or coastal cities where they are slaughtered and butchered. To date (2018), the Sahelian and West African regions are nearly self-sufficient for red meat, unlike poultry whose importation has steadily increased.

Livestock mobility is a key factor in the Sahel and West Africa regions. In the Sahel, livestock mobility is an intrinsic component of the farming systems, aiming at optimising the availability of natural resources (grasslands, surface water) which shows a highly seasonal pattern for a given grazing area. However, the uncontrolled interactions between transhumant herds and their herders and local farmers and their crops have caused many acts of violence along the years: conflicts between Senegalese farmers and Mauritanian herders, causing hundreds of victims, brought the two countries to the brink of war in 1989; in Burkina Faso during the tensions in 1986 and 1995, 10 people were killed in Comoe Province (Salliot 2010), and 7 people died during an incident between herders and farmers on 12 November 2004; in Benin, the same month 2 other people died (Abdi et al. 2014; Abiola et al. 2005), and other 18 deaths were registered between 2007 and 2010; and in Nigeria since 2001 thousands have been killed during conflicts (Cabot 2017). Consequently, governments have taken actions to secure both transhumant herders and local population including border closure (Togo 2016 and 2017) and the creation of transhumant corridors where livestock can pass through (FAO 2010; Kratli et al. 2014).

Livestock mobility in the region is a complex phenomenon involving several temporal (from days to months) and spatial scales (from a few km to reach local markets to international transhumance or trade movements). The mobility drivers include environmental factors (conditioning the availability of natural resources), commercial reasons (demand and market price) (de Jode 2010) and social factors like the Tabaski celebration (during which a young male sheep is slaughtered in each family) (Apolloni et al. 2018). The possibility of providing a reliable picture of livestock mobility in the area is hindered by the fact that few quantitative data are collected. In some cases, like Senegal and Mauritania, a system of movement certificates is implemented, indicating the origin and destination of the examined animals, as well as other data (date, health and vaccination status, etc.). However, this information is often stored at the local level (veterinary post) and thus difficult to retrieve. In other cases, some information can be recovered with ad hoc surveys (Dean et al. 2013) (Motta et al. 2017).

As animals move, so do pathogen agents and possibly their vectors (e.g. ticks). The frequency of the movements, both national and international, could introduce new pathogens in naïve areas and then trigger epidemics: foot and mouth disease is a burning example. Commercial movements covering long distances in few days and transhumant movements have the potential of seeding epidemics in zones far away from the first cases. To assess the risk of disease spreading, understanding the mobility patterns is of utmost importance. Besides the fact that few data are available, disease control is also hindered by the presence of a large fraction of illegal movements. Moreover, control and surveillance actions should take account of the period of the year since mobility can change abruptly along the year.

Pastoral Mobility, Originally a Sahelian Challenge, Has Now Become a Subregional One

Until the 1970–1980 drought waves, pastoral mobility was poorly constrained and in general restricted only to Sahelian countries. Indeed, since 1950, several hydraulic works, built in dry and subhumid areas, have made certain zones accessible to livestock that before couldn't be exploited due to water deficit. After rainy season, herds could move, preferentially, towards humid areas or rivers (Senegal river valley, Niger river delta) and cultivated field after harvesting (exchanging grazing against manure) or towards area put in long fallow.

At the end of the last drought wave in the 1980s, the livestock volume has reduced of the 80% in certain pastoral areas, leaving some areas available to rescued herds. Several herders and animals have migrated further south, reaching from Sahelian areas, regions in coastal countries but also in the north of Nigeria, Benin, Togo, Ghana and Ivory Coast, already inhabited by local communities of herders for the last few decades. At the same time, the decline of trypanosomiasis and the genetic crossings operated by herders have facilitated the displacement of herds of zebus and the hybrid bull/zebus heading south, less susceptible to the disease. At the end of the crisis, most of the herdsmen decided to settle in the hosting area, still keeping some connection with the herders of their origin area and intertwining new ones with local herdsmen.

But since the 1980s, population growth has changed the structure of pastoral and agro-pastoral areas. Human and animal densities have increased and the extension of pastoral areas has reduced due to the expansion of areas devoted to cultivation and irrigated areas around water sources. In agro-pastoral areas, fallows are limited in time; nevertheless, farmers decide to raise animals. Consequently, the competition for resources is exacerbated and mobility becomes more constrained, in particular in southern regions. Yet this is where, more and more often, transhumant pastoralists look for pastoral resources during the dry season. Moreover, commercial movements have also grown towards capital cities of coastal countries, more and more inhabited and with an increased demand for red meat.

Sahelian pastoral areas are becoming everyday more and more interdependent. The shared use of zone and resources becomes more and more complex and necessarily must be considered at regional scale.

Mobility or Mobilities?

Roughly, we can distinguish between two main types of mobility in the area: the commercial and the transhumance ones. The former is related to the economy of the household and to the need of buying products not available locally (Kamuanga et al. 2008a; Wane et al. 2010); the latter is related to the production and the survival of the animals. This is a rough distinction since in some cases it is not possible to make a clear distinction between the two: for example during the transhumance period, some of the herd's animals can be sold to cover some expenses occurred during the journey; in most of the cases, transhumance "finishes" animals that can be sold at markets in transit countries.

Commercial Mobility

Animal product trade represents one of the main economic activities in the region. Due to the lack of slaughterhouses and freezing-storage facilities, the largest fraction of animal trade consists of live animals, one of the first two commercial activities for countries in the area (the other one being the exchange of oil products (Various Authors 2018).

Most of the red meat consumers live in cities on the coastal countries or in rapidly growing Nigeria. The arid and semiarid regions of Sahel provide enough space to raise ruminants that can be sold to provision urban markets. Producers and consumers can be of several hundred kilometres apart. In West Africa, the supply chain consists of 3 types of market (Kamuanga et al. 2008a; de Jode 2010; Wane et al. 2010):

1. *Lumo* or weekly markets. Livestock owners sell their animals to mobile traders to provide for the household income and buy other products (cereal, rice, seeds) not available in their area. Animals are brought to markets walking along paths.
2. Collection market. These are the points where mobile traders meet stockists. These markets are situated in areas with a good route access and sometimes storehouses. These markets are always active, with a peak of activity on a specific day of the week. Animals are sold and sent, most of the time using tracks, to urban markets.
3. Urban markets. They are permanent markets, where transactions occur all days to satisfy the provision of cities' inhabitants. They rely on supply from the collection markets but in some cases also from imported goods which arrived at nearby harbours.

The volume of animal exchanged on the market varies along the year and depends on countries. In countries like Mauritania and Mali, two of the biggest meat exporters in the area, the largest quantity of animal is exchanged in the first half of the year, mainly around April and June, the period indicated as "hunger gap" when the resources accumulated with the last harvest are dwindling, temporary surface water has dried up and grasslands are exhausted or have lost their nutritional value (Bonnet and Guibert 2018; Bouslikhane, 2015). Owners sell sheep and goats offspring, usually born between November and January, to cover household needs. On the other hand, an analysis conducted in Cameroon has shown that trades in the country show a peak of activity between April and August, the rainy season, when animals are more marketable (Motta et al. 2017).

Another important factor affecting animal trades and mobility are religious festivities. In fact, analysis of Cameroon market's data in 2015 (Motta et al. 2017) has shown a consistent peak of animal sales in December to satisfy the demand due to the festivities at the end of the year. Tabaski is a major Muslim feast during which a young male sheep is traditionally slaughtered in many families. The Tabaski date is defined according to the lunar calendar, and therefore, it changes every year. In 2014, this event took place at the beginning of October, and the analysis of mobility data in Mauritania (Apolloni et al. 2018) showed a second, less marked, peak in September–October in small-ruminant trades, involving more than 900,000 heads as a whole. These estimates are comparable with those of the official trade agreement between the Senegalese and Mauritanian veterinary services: this agreement was signed between the two countries to avoid supply disruption on the Senegalese sheep markets. Tabaski is affecting the livestock mobility patterns in many ways, changing the role of the largest part of the mobility network, rerouting animals from export towards national markets, creating new axes of movements, increasing the fraction of animals moved by truck and consequently moving faster, to provision major urban areas.

In West Africa, trades are international by nature, since most of the producers are in arid and semiarid landlocked areas whilst most of the consumers are in the more populated coastal countries in the south or in Nigeria. The annual volume of ruminants traded by Mauritania is around seven million of whom almost the 65% of them were sold on international markets. Similar estimates hold for Mali and Burkina Faso. Historically, international commercial movements were along 3 (vertical) axes (Fig. 3.2) (Bonnet and Guibert 2018; Guibert et al. 2009; Various Authors 2010):

- Western: from Mauritania and Mali towards Senegal Gambia, Guinea Guinea-Bissau and Liberia
- Central: from Mali and Burkina Faso and Niger towards Cote d'Ivoire, Ghana and Togo
- Oriental: from Niger to Nigeria

To these historical circuits, we should add new (horizontal) axes that are getting more and more importance due to the rapid population growth of Nigeria and coastal countries:

- West-East: from South East Burkina towards Central West Nigeria
- Central Africa involving Chad and Central Africa Republic towards Nigeria and Cameroon
- From Central Africa republic towards Congo and RDC

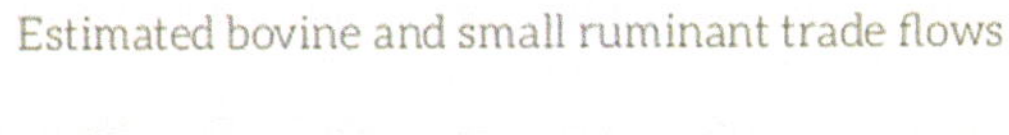

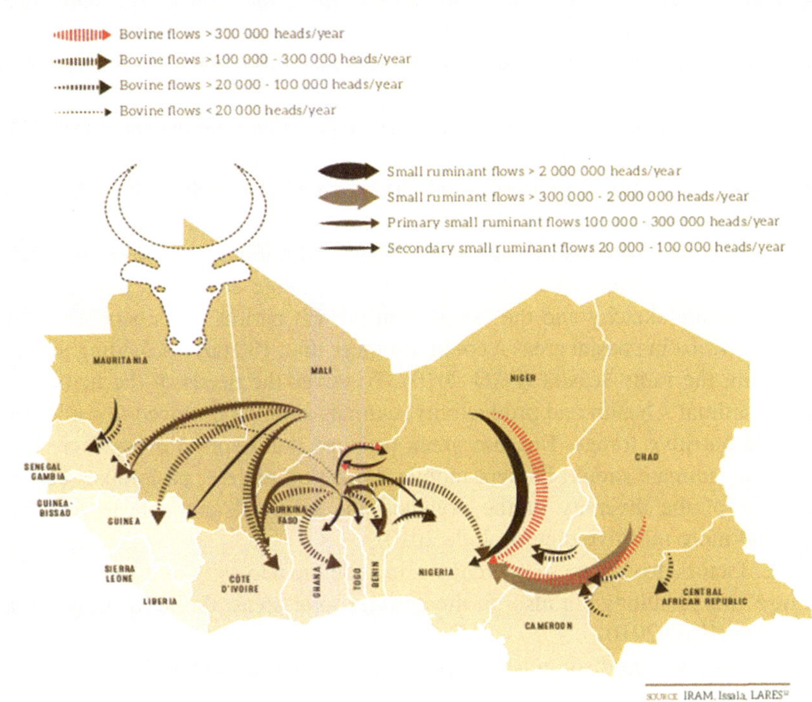

Fig. 3.2 Map of livestock trade mobility in West Africa source (Aubauge et al. 2017)

Most of these movements are done walking, since transportation costs can reach up to 40% of the total costs (Aubauge et al. 2017; Kamuanga et al. 2008b). However, more and more traders are using trucks to deliver livestock to urban markets. Consequently, in few days livestock from Mauritania can reach the market of Guinea-Bissau. This is particularly true, around festivities, when the need to provision cities in a short amount of time is imperative.

Long considered naïve traders, nowadays livestock owners are more opportunistic in their trade. Furthermore, with the advent of cell phones, they can get information about prices of products in different markets, and then decide where to sell and buy products, most of the times not at the same market (de Jode 2010).

Transhumance and Nomadism

In the arid and semiarid regions of Sahel and West Africa, climatic conditions don't allow to have grazing areas all the year round. In these areas, the largest part of the herdsmen and families either adopts a nomadic lifestyle or practises transhumance. In the first case, herds mostly constituted by camels and small ruminants are kept mobile all the year round (Abiola et al. 2005; Bouslikhane, 2015). Transhumance is a system of animal production characterised by seasonal movements of a cyclical nature and to varying extents. These movements take place between complementary ecological zones, under the care of few people, with the rest of the herd remaining sedentary (Kamuanga et al. 2008b). A fraction of cattle varying between 70 and 90% and of small ruminants varying between 30 and 40% of the national numbers is reared using transhumance (Abiola et al. 2005; Bouslikhane 2015).

Transhumance is a form of adaptation to the environmental conditions of Sahel and optimisation of the natural resources. In the arid regions, the presence of good-quality grazing and the access to water sources change along the year. In Sahel, forage areas are sketchy and the "vegetation growth period" lasts between 20 and 180 days, whilst in coastal areas it could last more than 180 days providing nutrients also during the rainy season (FAO 2010). To cover the needs of the herd and to improve animal's health and productivity, animals should be moved towards better and more nutritive forage. Because green pastures don't sprout everywhere at the same time, animals should be kept mobile. By being mobile, pastoralists can take advantage of the diversity of the drylands and livestock are able to feed on a richer and more nutritive diet (de Jode 2010).

Other factors driving these movements include the necessity of escaping drought and insecure situations but also, in the agro-pastoral areas, the need of preserving cultivation (FAO 2010).

In general, we can distinguish 2 types of transhumance (Fig. 3.3):

1. Short transhumance: constituted by small movements, mostly at national level. The aims of this type of movements are to valorise crop residues, access to fresh greener pasture and not ruining cultivation.
2. Long transhumance: this is characterised by long, mainly international movements, lasting from 6 to 9 months per year. During the dry season (from January to May), animals move from Mauritania, Mali and Burkina Faso to Southern countries (Cote d'Ivoire, Guinea, Guinea-Bissau and Togo; see Fig. 3.3). During the rainy season, herds follow the same path in the opposite direction (Aubauge et al. 2017; Kamuanga et al. 2008a; FAO 2010; Bouslikhane, 2015; Corniaux et al. 2016).

Based on data collected by FAO, three axes of transhumance can be identified (FAO 2010). These axes often overlap with commercial ones (Fig. 3.3):

1. Western axis: includes the movements among countries on the Atlantic coast (Mauritania, Senegal, the Gambia, Guinea, Guinea-Bissau, Sierra Leone and Liberia). The zone is characterised by few exchanges towards other countries (except for Mali). A fraction of herds (mainly cattle) from Mauritania passes through Kayes in Mali before entering Senegal at Kidira and then continuing south till Ourossogui and then reaches Guinea. Other herds of cattle from Mauritania enter Senegal through Matam and then continue on the same path of the others. Whilst in Guinea, the herds are dispatched on other directions towards Liberia and coastal countries till Ivory Coast.

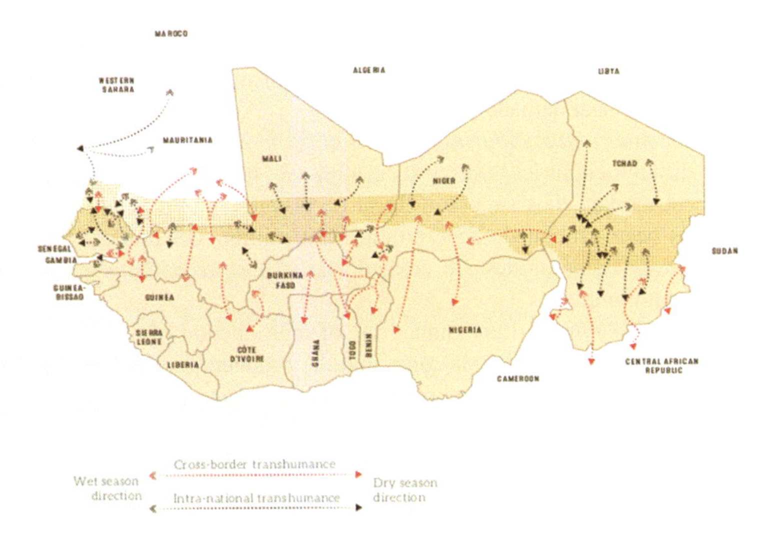

Fig. 3.3 Transhumance movements source (Aubauge et al. 2017)

2. Central axis: includes movements from Mauritania and Burkina Faso to Cote d'Ivoire (mainly), Ghana and Togo.
3. Eastern axis: includes movements from Niger and Burkina Faso towards Nigeria and Benin.

During the dry season, the choice of the path to follow depends on the availability of water and forage along the way. Recently, the introduction of cell phones among transhumant pastoralists has provided new tools to decide the route. Collecting information from other pastoralists about availability of water and grazing areas, state of paths, as well as safety conditions for the herd, herders can decide to change routes in favour of those with better conditions (Aubauge et al. 2017; de Jode 2010).

Constraints and Challenges of Livestock Mobility

For decades, pastoralism practices have been considered archaic and less productive, and the mobility only a conflict-generating cause. Recent studies have, instead, enlightened the prominent role that mobility plays in Sahelian area (Aubauge et al. 2017; de Jode 2010). Some of the benefits of pastoralism are as follows:

1. *It makes possible the sustainable use of drylands:* due to animal mobility, the pressure on grazing area is low and in equilibrium with the environmental

conditions. Moreover, the quantity of greenhouse gases produced by transhumant herds is much less than that by sedentary ones. Mobility is a way to protect herds from natural and human-caused disasters.
2. *Enhance the complementarity between pastoral and agricultural systems:* In this way, exchanges between the two systems are provided. Transhumant herds provide manure and sometimes traction for farmers and the commercial mobility ensures that red meat demand is satisfied.
3. *Generate incomes to all countries involved in the movement:* Pastorals pay taxes to transit through countries. During their journeys, herdsmen sell and buy products at local market, thus supplying local economies. The official cross-border trades have been estimated to be around 150 million US dollars. Social ties among faraway communities are created due to the trades between pastorals and local communities.
4. *The production is efficient:* the annual rate of reproduction of mobile herders is 20% higher than that of sedentary ones and the production of red meat is larger than the sedentary ones. Furthermore, due to the exploitation of natural resources, the meat price per kilo of mobile herd is lower than sedentary one.

Pastoralism provides livelihoods for almost 20 million inhabitants of the Sahelian region who, in turn, provide foods for almost 200 million in the area. Pastoral and agro-pastoral rearing livestock satisfy around 80% of the demand for red meat in the area (FAO 2010; Corniaux et al. 2016). Livestock trades create income for pastoral families from the trade of product, but also employment of different actors (among them traders, stockists, truck drivers) whose economic impact cannot be easily estimated (Grandval 2012).

A recent survey, conducted in the context of the BRACED project, on 386 transhumant families has shown that on average during transhumance period around 1230 million of FCFA (around 2300 dollars) were spent by each respondent family (Aubauge et al. 2017; Thebaud 2017). Table 3.1 shows the repartition of the transhumance costs for each family.

Most of the expenditures occur at local market level, thus benefiting communities of transit countries. Therefore, big markets have appeared at cross-border points. Among the expenditures, taxes paid by herdsmen to access resources generate important fiscal resources for government.

Table 3.1 Distribution of transhumant household expenditure

Reason	Percentage (%)
Livestock feed	44
Food	22
Veterinary products	7
Water costs	4
Other (telephone, taxes to authorities, access to resource, etc.)	23

Source: Aubauge et al. (2017), Kamuanga et al. (2008a), Various Authors (2018)

Furthermore, livestock sector provides services to other sectors. During the transhumance journeys, access to crop residual is exchanged for cash and/or manure (Grandval 2012; Krätli et al. 2013), thus improving the agriculture yields in the area; new financial products, like assurance for drought and diseases, and loans for herdsmen are becoming more and more common.

Finally, pastoralism plays an important role in hostile and marginal zones. The capacity of moving many individuals, either for transhumance or crises, is based on the existence of a community structure among them and a shared set of values and rules. Mobility is also possible due to the existence of a social structure maintaining alliances with sedentary communities. Pastoralism has the effect of stabilising society and provides a valid alternative to violent activities in marginalised area (Grandval 2012; Krätli et al. 2013).

Mobility in Sahel has the potential to integrate markets of the several countries. West Africa comprises two complementary zones, characterised by a variety of productions, and this condition pushes towards the exchange of typical products. As a result, the diet in the area is diversified. Between Niger and Nigeria, there is an intense trade of livestock (95% of national herd in Niger is sent to Nigeria) and cereals (70% of Nigeria's production is sold in Niger). These international trades stimulate domestic production. Except for Nigeria, and maybe Cote d'Ivoire, most of western African countries are less than 15 million inhabitants and their domestic markets are small. An integrated regional market represents a large growth margin for domestic producers. Furthermore, food security in the region is strongly connected to environmental factors that discourages long-term investment and affects price variation in each country. Due to the spatial and temporal dispersion of these events, variations at regional level are mitigated and market is more stable (Blein 2014).

Several factors concur to hinder mobility in the area. Many governments recognise the importance of an integrated regional market; however, the fear of shortage of their own and their neighbour's production has driven towards more country-centred politics including border closure. For example, Cote d'Ivoire aims to become self-sufficient and threatens to close the borders to international movements (Grandval 2012; Blein 2014).

Recent national politics has under-invested in pastoralism instead focusing on the production of crops (cotton, rice, milk to export and or consume) and pushing towards the sedentarisation of livestock rearing. The growing expansion of agriculture, by private farmers and companies, is eroding the grazing areas for transhumant herds (de Jode 2010; Bonnet and Guibert 2018; Bouslikhane, 2015). Governments have seen in sedentarisation a process towards the modernisation of the sector. Furthermore, through sedentarisation, governments can tax a sector always perceived as prosperous, despite the fact that most of livestock owners are low income. Furthermore, the creation of large "modern" factories has created a new elite of rich ranchers, leaving most herdsmen in poor life conditions (Pelon 2015).

Due to the increase of the area reserved for sedentary rearing or cultivation, some of the transhumant corridors have now been invaded and closed to transit. Besides, the water management (both public and private) precludes the access to water sources that is a fundamental factor for animals, like cattle, that should be watered regularly. In the past, these factors gave origin of conflicts among resident farmers and transhumant pastoralists. The situation forced governments of ECOWAS countries to put in place new measures and accords to regulate the transhumance and pastoral movements. In 1998, countries decided to introduce International Certificate of Transhumance (ICT) to regulate transboundary movements, together with a series of public works to secure transhumance movements like securing corridors and water access. Despite this, mobility is still hindered by several administrative factors like the poor level of organisation, the inadequacy of the personnel and the many taxes (legal and illegal ones) to be paid (Bouslikhane, 2015). These factors contribute to incentivise illegal movements across borders more difficult to control and secure.

Livestock Mobility and Sanitarian Risk

According to Perry et al. (2013), livestock mobility is the third cause of the changed spatial dynamics of animal disease, following climatic change and market concentration. Livestock movements have the dual effect of exposing healthy animals to new viruses upon arrival in infected areas or introducing infected animals to disease-free areas. Indeed, movements often occur before the vaccination of newborn animals that could be affected by pathogens, either by contact with already infected herds or by entering areas that are natural habitats for the diseases' vectors, such as fields infested with ticks vectors of Crimean Congo hemorrhagic fever (CCHF); swamps and water courses that are habitats for mosquitoes vectors of RVF; as well as forests and savannas, tsetse fly habitat and vectors of Trypanosome (Abiola et al. 2005).

Whilst in Europe in the last decade the study of movement patterns has allowed us to identify high-risk farms and herds (Volkova et al. 2010; Bajardi et al. 2012), in West Africa this type of study is just beginning to appear (Apolloni et al. 2018; Dean et al. 2013; Motta et al. 2017), limited by the data availability and restricted to certain types of movements. The interplay between the two types of mobilities (commercial and transhumance) is rather complex, and their relations can vary from year to year. Whilst commercial movements are mainly regulated by supply-demand and social dynamics, transhumance movements are more related to environmental and climatic conditions of the origin and destination locations: low rainfall can push herders to anticipate the journey, whilst the harvest time could influence the time of arrival at the destination. Moreover, during the journey herds need to be watered constantly, and the journey could be modified in search of water sources. On top of that we should consider the local mobility, i.e. the daily movements towards water sources and grazing areas of sedentary herds. Around water points or in grazing areas, transhumant, commercial and local herds could get in contact and transmission among herds can occur. The frequency, the duration and the actual number of contacts can vary around seasons, but also depending on the agrosystem (VanderWaal et al. 2017).

Markets, on the other hand, increase the risk of transmission by bringing together animals from faraway regions in close contact, before re-dispatching to other destinations. Moreover, since most of the movements are on foot and the excretion time of the virus could be several days (if not weeks), along the journey more and more animals can get infected. In analogy with human mobility (Balcan et al. 2009; Apolloni et al. 2013), we can suppose that whilst long-range movements (transhumant and commercial cross-border) could be responsible for seeding diseases abroad, local and commercial movements are responsible for spreading among villages and at national level, respectively. Whilst we focus on movements at regional scale, we should consider livestock movements towards other African regions (mainly North Africa and Central) and livestock import from South America as other potential risk factors for the introduction and diffusion of diseases.

Economic losses from animal deaths are important, but difficult to estimate, especially for countries where there is no monitoring or control systems. One of the control measures is often the border closure. This type of measure is not always effective because of the high impact on livestock trade, but also because of illegal movements that could even increase. In addition, diseases such as Foot and Mouth disease and Contagious Bovine Pleuro Pneumonia, which a priori can't show clinical signs, can pass undetected after the reopening of borders and trigger epidemics in the region.

Finally, the existence of few health structures on the transhumance pathways forces herdsmen to resort to inefficient self-medication habits, or to buy drugs of doubtful efficiency at marketplaces. Therefore, an already precarious situation for the herdsmen gets even worse, since treated animals often do not heal and a market, often illegal, for low-quality drugs is promoted (Abiola et al. 2005).

Several factors contribute to keep the health risk high in the region. First, the permeability of the borders between countries that facilitates illegal movements and promotes the introduction of pathogens. Second, the lack of harmonised surveillance and control systems between countries: the obligation to vaccinate and the prices of vaccines vary according to the country as well as certain populations that are not sensitised to the problematic; introduction of unvaccinated and/or infectious animals can cause transmission of pathogens and potentially spread at international level. Finally, the absence of a system of identification of animals (or herds) and the knowledge of their movements would help predicting the occurrence of epidemics and their burdens (FAO 2010). In this context, the International Transhumance Certificates (ITC) are an important tool since a lot of information on the composition of the herd, its vaccination status and the planned route is collected that could be used to inform predictive models. On the other hand, the difficulties and the length of the process to obtain ITC, the harassments at borders, the presence of road blocks and the constraints related to the route to be followed discourage the herdsmen who prefer to take illegal ways.

We conclude this section presenting the case of some diseases in West Africa region, whose patterns of geographical diffusion can be related to animal mobility. This list is far from being exhaustive, and, in some cases, further studies for deepening the relation among mobility and diffusion are still going on. As previously stated, in West Africa, there is no traceability and identification system put in place, and the absence of a centralised surveillance system makes difficult, if not impossible, to reconstruct the patterns of the virus geographical spread. However, the re/appearance of diseases in naïve countries could be related to transboundary animal mobility (Abiola et al. 2005; Bouslikhane, 2015; Lancelot et al. 2011). Contagious Bovine Pleuro Pneumonia (CBPP) is a highly contagious disease affecting cattle caused by *Mycoplasma Mycoides.* Besides an acute phase that could lead to the death of the host, the disease presents a chronic phase during which hosts could still infect at low rate without showing any symptom. The disease is affecting 45% of the countries in West, Central and East Africa (Consultative Group 2015). In Senegal and Gambia, the last outbreaks were reported in 1978. Since then, Senegal has conducted mass vaccination campaigns till 2005. During this period, no cases were recorded. In 2012, the disease made its reappearance in the Tambacounda Region and in Gambia (Mbengue and Sarr 2013; Mbengue et al. 2013), probably related to the introduction of cattle from the nearby countries of Mali and Mauritania (Séry et al. 2015), where CBPP is endemic. Since then, outbreaks of CBPP have been recorded in the regions of Kolda (2013 and 2014), Matam (2014) and Saint Louis (2015) (Consultative Group 2015).

When possible, molecular epidemiology and phylogenetic analysis can help to decipher the trends of virus dispersal at national or even at regional level (Di Nardo et al. 2011). Moreover, the presence of multiple clusters within a country could indicate multiple waves of introduction from neighbouring countries, or recovery of viral population (Padhi and Ma 2014). Foot and mouth disease (FMD) is endemic in sub-Saharan Africa and has expanded in the Middle East and Arabian Peninsula due to livestock exchanges in the area. The virus exists as seven different serotypes, six of which are present in Africa (O, A, C, SAT-1, SAT-2, SAT-3). Several serotypes can co-circulate (Di Nardo et al. 2011). As pointed out by Knowles and Samuel (2001), the virus exhibits a strong correlation between genetic and geographical clusters. Whilst FMD appears to be a disease mostly from sub-Saharan Africa, the disease has done its incursion several times in North Africa. In 1999, serotype O outbreaks were declared in Algeria, Morocco and Tunisia, whose strains were closely related to those isolated in Ghana, Cote d'Ivoire (1999) and Ghana (1993). In Morocco, the outbreaks were in the province of Oujda, appearing few days after Algeria notification. Due to the foci's position, a possible way of introduction was through illegal importation of bullocks from Algeria where the epidemics was already under way. The outbreak investigation conducted in Algeria highlighted the presence of animals imported from Mali and Mauritania and the role of the markers at Alger and Boufarik in disseminating the virus (European Commission for the Control of Foot-and-Mouth and Disease 1999). The SAT 2 strain outbreak of 2003 in Libya, on the other hand, shows a close relationship with the strains of 2000 Cameroon and 2005 Niger outbreaks (Di Nardo et al. 2011). The virus has been circulating most likely due to transhumance movement and through the exchange of milk or dairy products.

Another example of transboundary disease is Peste des Petits Ruminants (PPR). As the name says the main hosts are small ruminants (goats and sheep), whilst the role of wild species is still not clear. The disease, discovered in 1942 in Cote d'Ivoire, has expanded east, during the 1980s has reached the Middle East and Arabian Peninsula and in the first decade of 2000 has reached Far East countries such as China Mongolia and Pakistan. There exist 4 strains of the virus: the lineage I dominating West Africa; lineage II in Central Africa; lineage III in East Africa; and lineage IV in Asia (Albina et al. 2013). Recently, lineage II has moved west replacing lineage I in Senegal (Salami n.d.), and lineage IV is slowly replacing African lineages appearing in Morocco and Niger (Padhi and Ma 2014; Tounkara et al. 2018; Baazizi et al. 2017). The rapid spread of the disease is attributed to the highly contagious nature of the disease, but also to other factors, like the population dynamics and the intensification and extension of livestock movements in the area (Libeau et al. 2014), thus making the eradication problem a regional question.

For an epidemic of vector-borne diseases like Rift Valley fever (RVF) to occur, environmental conditions should be met to ensure the presence of the vectors. Moreover, the virus can survive from a rainy season to another, shedding vector's eggs. The virus, originally from East Africa, made its first appearance in Senegal and Mauritania in 1987 along the River Senegal valley and outbreaks have been recorded in 1988, 1993,1998 and 2003. In 2010, abnormal rainfall in the desert regions of Adrar and Inchiri allowed the creation of temporal ponds, the colonisation by vector species and the growth of forage's herbs. Small ruminants were driven by trucks from southern regions, some of them carrying the virus, and lighting up an epidemic in the region affecting particular camels (El Mamy et al. 2011). A second outbreak of RFV occurred in 2012 in the south-east regions (Hods and Assaba) and central ones (Tagant et Brakna) and continued till 2013 extending to all of the right bank of Senegal river (Ould El Mamy et al. 2014). As in 2010, 2012 was an abnormal year with respect to rainfall, which gave way to a large abundance of vector's species. Herds affected in 2012 epidemics moved to 2013 infected regions, as part of their journey towards Senegal and Mali, most likely disseminating the disease. Concomitantly, an epidemic of Rift Valley has occurred in Northern Senegal reaching southern and more populated areas of Mbour, Thies, Linguere and lately Kedougou (Sow et al. 2016). As pointed out by researchers studying the epidemics in both countries (Ould El Mamy et al. 2014; Sow et al. 2016), the affected area was so widespread, due to animal movements, mainly from Mauritania, for the religious festivities of Tabaski happening during the rainy season (June–October). Phylogenetic analysis has shown at least 5 introductions of RFV from distant African regions (Zimbabwe, Kenya, Madagascar) mainly by animal movements and, in the meantime, the genetic clustering of Mauritanian and Senegalese strains due to the recurrent commercial exchanges (Soumaré et al. 2012; El Mamy et al. 2014).

Assessing the Role of Movements on the Diffusion of Diseases

Network approach has been widely applied in the context of animal diseases to assess the risk of epidemics occurrence depending on the structural characteristics of the underlying network (i.e. the trades network). The trades network is described in terms of nodes and links: nodes correspond to markets, premises

(continued)

and villages; a link exists between two nodes if animals exchanged exist among them. Most of the works in network analysis aim to understand if network structure can facilitate or dampen the spread of virus from a node to the rest of the network. It's in fact known that highly heterogeneous network, where few nodes condense a large fraction of the existing links (called hubs), is more likely to spread the disease than more homogeneous ones (Park and Newman 2004). However, this characteristic of the network can be used to implement surveillance and control measures: vaccinating animals in the hubs has the effect of drastically decreasing the extent of the disease; hubs have higher chances of getting hit during the epidemic, thus monitoring incoming animals in these nodes can help in detecting a virus promptly. Another property to consider is the weight's distribution of the network, where the term weight indicates the number of animals exchanged along a link. Both network's structure and weight's distribution can change along time, thus altering the capacity of the network to transmit epidemics.

Due to the volume of exchanges among West African countries, it would be useful to policymakers and veterinarian services to use some of network tools to support decision on control and surveillance (Natale et al. 2009; Nicolas et al. 2013; Rautureau et al. 2011). However, at this moment data on international exchanges are mostly qualitative, subject to large variations and cross-grained both in space and time resolution. In most of the cases, in fact, those are annual estimates extrapolated from some surveys in specific areas. This is mainly due to the lack of traceability system common to all countries that could permit to reconstruct the movements of the animals. The International Transhumance Certificates, which could help to reconstruct the mobility network, are seldom used and collected. Moreover, these certificates don't include information about each single animal. At national level, some information can be extracted from market ledgers, although they are difficult to collect, and by ad hoc surveys. Mauritania has developed a system of certificates to monitor commercial movements at national level and towards some international destinations. In fact, for certain herd movements, herdsmen should declare at the closest veterinarian office some details of the movements like origin, destination, species and size of the herd. Veterinarian officers deliver a certificate in several copies, among them one for the herdsman to exhibit in case of control and one to send to the general direction. The analysis done on the dataset for the year 2014 has shown a large degree of variability on the network structure and the volume of animal exchanged: in Mauritania, the large part of animals are exchanged in the first half of the year, and then a second peak of activity appears in September October for the occurrence of Tabaski; due to the occurrence of Tabaski, new links are created around that period that account for 3% of total volume; for this religious occurrence,

(continued)

certain movements are done using trucks, instead of walking, to provision urban areas faster.

These data can be used to estimate the threshold parameter (q) (Volkova et al. 2010; Park and Newman 2004). This is a quantity that helps in estimating the minimum value of the transmission probability among animals, so that the virus can spread to all networks. A high value of q indicates that the virus should be strongly virulent to spread around the network; a low value, on the other hand, indicates that the virus can easily spread all over the network; even few animals are circulating. The threshold parameter is related to the network characteristics through the relation:

$$q = \frac{\langle w^{\mathrm{out}} \rangle}{\langle w^{\mathrm{in}} w^{\mathrm{out}} \rangle}$$

where $\langle\rangle$ indicates the average over all network nodes and w^{in}, w^{out} are, respectively, the incoming and outgoing number of animals. Since the network changes along the year we can estimate for each month the threshold parameter and with that assess the risk of a national epizootic. Figure 3.4 shows the monthly mobility networks in Mauritania, as reconstructed from the Health Certificates collected (Apolloni et al. 2018). Due to different causes, some connections can be active for a long period; others can activate just in specific period of the month (like Tabaski) and be active for 1 or 2 months ("occasional ones"). The red lines indicate property related to all the monthly networks and the blue one only to those that are present for several months. In the Mauritanian case, most of the movements are done in the first half of the year, and there are few occasional links, except around Tabaski. A second peak is observed around Tabaski, mainly due to the occasional movements to provision urban areas. The bottom part shows the epidemic threshold estimated for each month and according to the network considered. The low q values indicate that the network is more prone to transmission. Around Tabaski, the risk of epizootic is almost the same for the two countries. In Mauritania, occasional movements poorly affect the transmission of disease since they account for a small fraction of the total volume. Around Tabaski, the risk of an epizootic highly increases, due to the volume of animal traded just for the festivity.

The impact of control measures, like vaccination, can be assessed using the above formula. In fact, vaccination can be modelled through the removing of nodes and all the links connected from the network and re-estimate q. An increase in the value of q corresponds to a decrease of the risk of epizootic. Assessing the variation after the elimination of each node, the highest ones would correspond to nodes that should prioritise in vaccination.

(continued)

A shared information and tracking system built at regional level would allow to extend this approach to the entire region and estimate the risk of the epidemics due to outbreaks happening faraway and estimate the risk of introduction in each country.

A traceability system, where the movement of each animal is recorded, would allow us to perform more sophisticated analysis, like the one of Bajardi et al. (2012) and Bajardi et al. (2011).

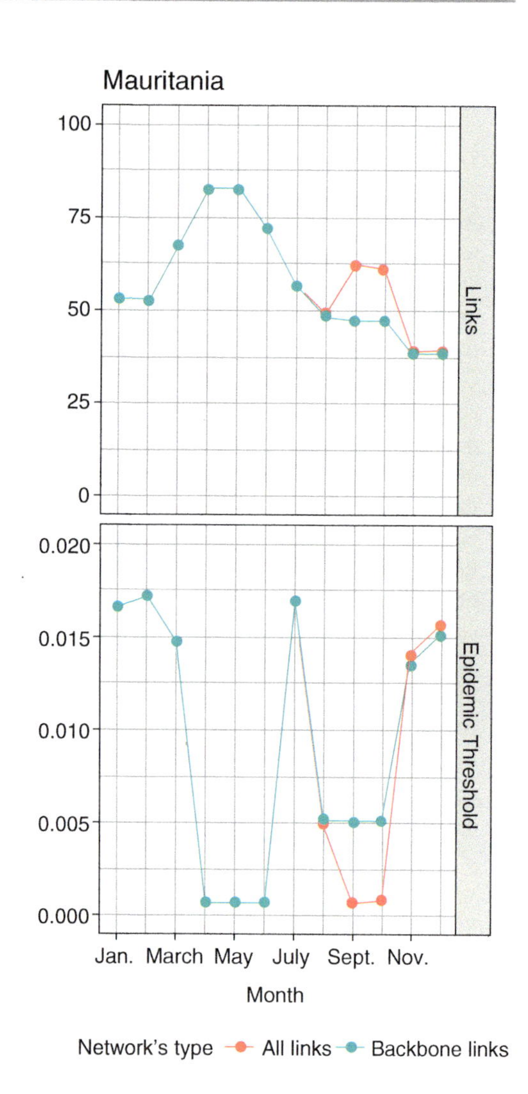

Fig. 3.4 Summary of the mobility network as reconstructed from health certificates from Mauritania. Top shows the number of connections active each month. The red lines considered all the links and the blue line all the links but the occasional ones (appearing only that month). Bottom, the epidemic threshold for each month and for each type of network

Conclusions

Livestock in the region is kept mobile for production, trade and safety reasons. During their movements, animals cover long distances, mainly reaching foreign countries.

Transboundary livestock transhumance is a key factor in West and Sahelian Africa. It represents an efficient and adaptive method of raising animals in a context where natural resources are scarce. Due to international exchanges, nowadays the region has reached quasi-autonomy in red meat consumption, whilst it still depends on importation for poultry meat. However, even though pastoralism is an efficient rearing practice, the offtake is low. Nowadays, the population of West Africa is around 350 million, and it will reach the billion mark around 2050 (Krätli et al. 2013; Holechek et al. 2017). The livestock production is growing but, due to the low carrying capacity of the area, is not growing as fast as population. On the other hand, to improve the offtake of the production would require improving the rangeland productivity and a coordinated framework among countries. The absence of an organisation in the area could allow other countries, namely from South America, to increase their share of imported meat in the areas.

Livestock mobility creates job opportunity for a large fraction of population employed both as primary producers and in the commercial circuits, in all countries crossed by livestock movements. Inhibiting pastoralism and mobility would have huge repercussions both in terms of animal supply and in terms of markets activities, and the consequent unemployment could ignite more turmoils.

Despite the benefits, mobility is hindered by many factors related to the absence of a clear international vision, the lack of good infrastructures and conflicts among herdsmen and farmers. Several efforts have been done in this direction by member countries of ECOWAS.

We have distinguished different types of mobilities present in the area, whose determinant factors are of different nature. To assess their role on the disease diffusion, they must be considered at the same time. The interplay between the different type and scale of mobilities together with the lack of data makes difficult if not impossible to model mobility in this area.

From a sanitary point of view, control and eradication measures in the region will be possible only in terms of a coordinated regional activity. The differences of the control and vaccination systems among countries, together with the lack of identification system, hinder the possibility of controlling disease diffusion in the region. Illegal movements, moreover, play an important role in disease dissemination.

Understanding mobility in the region implies taking account of different factors both social and economical (human and animal population, market presence and religious festivities) and environmental (rainfall and temperature patterns, landscape and forage availability), thus affecting our capabilities of predicting behaviours from one year to another.

Acknowledgements This work was partially funded by INRA grant Meta-programme GISA AMI 2017 and EU grant FP7-613996 VMERGE and is catalogued by the VMERGE Steering Committee as VMERGE000 (http://www.vmerge.eu). The contents of this publication are the sole responsibility of the authors and don't necessarily reflect the views of the European Commission.

References

Abdi AM, Seaquist J, Tenenbaum DE, Eklundh L, Ardo J. The supply and demand of net primary production in the Sahel. Enviromental Res Lett. 2014;9(9):094003.

Abiola FA, Teko-Agbo A, Biaou C Niang, M Socio-economic and animal health impact of transhumance. Conference, OIE 2005;105–119.

Albina E, et al. Peste des petits ruminants, the next eradicated animal disease? Vet Microbiol. 2013;165:38–44.

Apolloni A, Poletto C, Colizza V. Age-specific contacts and travel patterns in the spatial spread of 2009 H1N1 influenza pandemic. BMC Infect Dis. 2013;13:176. https://doi.org/10.1186/1471-2334-13-176.

Apolloni A, et al. Towards the description of livestock mobility in Sahelian Africa: some results from a survey in Mauritania. PLoS One. 2018;13(1):0191565.

Aubauge, S. et al. Pastoral livestock farming in Sahel and West Africa. 2017.

Baazizi R, et al. Peste des petits ruminants (PPR): a neglected tropical disease in Maghreb region of North Africa and its threat to Europe. PLoS One. 2017;12:e0175461.

Bajardi P, Barrat A, Colizza V, Natale F, Savini L. Dynamical patterns of cattle trade movements; 2011. p. 1–51.

Bajardi P, Barrat A, Savini L, Colizza V. Optimizing surveillance for livestock disease spreading through animal movements. J R Soc Interface. 2012;9:2814–25.

Balcan D, et al. Multiscale mobility networks and the spatial spreading of infectious diseases. Proc Natl Acad Sci. 2009;106:21484–9.

Blein R. Taking a regional approach to agricultural and food challrnges in West Africa? Opportunities and difficulties. Food Sovereignity Briefs. 2014;14

Bonnet B, Guibert B. Commerce du bétail en Afrique de l'Ouest, atouts et défis pour les éleveurs. Grain de sel. 2018:37–8.

Bouslikhane M Cross border movements of animals and animal products and their relevance to the epidemiology of animals disease in Africa. (OIE Africa Regional Commision). 2015.

Cabot C Climate change, security risks and conflict reduction in Africa. A case study of farmer -herder conflicts over natural resources in Cote d'Ivoire, Ghana and Burkina Faso. 2017.

Consultative Group. Can contagious bovine pleuropneumonia (CBPP) be eradicated? vol. 82: FAO-OIE; 2015.

Corniaux C, Ancey V, Touré I, Camara AD. Pastoral mobility, from a Sahelian to a sub-regional issue. In: A new emerging rural world - an overview of rural change in Africa: CIRAD/NEPAD; 2016. p. 60–1.

de Jode H. Modern and mobile: the future of livestock production in Africa's drylands. London, UK: IIED and SOS Sahel UK; 2010.

Dean AS, et al. Potential risk of regional disease spread in West Africa through cross-border cattle Trade. PLoS One. 2013;8:e75570.

Di Nardo A, Knowles NJ, Paton DJ. Combining livestock trade patterns with phylogenetics to help understand the spread of foot and mouth disease in sub-Saharan Africa, the Middle East and Southeast Asia. Rev Sci Tech-OIE. 2011;30:63–85.

El Mamy ABO, et al. Unexpected Rift Valley fever outbreak, Northern Mauritania. Emerg Infect Dis. 2011;17:1894–6.

El Mamy AB, et al. Comprehensive phylogenetic reconstructions of Rift Valley fever virus: the 2010 northern Mauritania outbreak in the Camelus dromedarius species. Vector Borne Zoonotic Dis. 2014;14:856–61.
European Commission for the Control of Foot-and-Mouth & Disease. FMD situation in Europe and in other regions: summary of the type O foot-and-mouth disease situation in North Africa as of 29th March, Appendix 3. 1999.
FAO. La transhumance transfrontalière en Afrique de l'Ouest Proposition de plan d'action. 2010.
Grandval F. Pastoralism in Sub-Saharan Africa: know its advantages, understand its challenges, act for its sustainability. Food Sovereignity Briefs. 2012;5
Guibert B, Banzhaf M, Soulé BG Idé G Etude régionale sur les contextes de la commercialisationdu bétail/Accès aux marchés, défis à relever pour améliorer les conditions de vie des communautés pastorales. 2009.
Holechek JL, Cibils AF, Bengaly K, Kinyamario JI. Human population growth, African pastoralism, and rangelands: a perspective. Rangel Ecol Manag. 2017;70:273–80.
Kamuanga MJB, Somda J, Sanon Y Kagoné H. Livestock and regional market in the Sahel and West Africa. 2008a.
Kamuanga M, Somda J, Sanon Y Kagoné H Elevage et marché régional au Sahel et en Afrique de l'Ouest: potentialités et défis. (CSAO/OCDE). 2008b.
Knowles NJ, Samuel AR. Foot-and-mouth disease type O viruses exhibit genetically and geographically distinct evolutionary lineages (topotypes). J Gen Virol. 2001;82:609–21.
Krätli S, Huelsebusch C, Brooks S, Kaufmann B. Pastoralism: a critical asset for food security under global climate change. Anim Front. 2013;3:42–50.
Kratli S, Hesse C, Monimart M, Swift J, Jallo B. Securing the pastoral mobility in Sahel. Quest Dev. 2014;10
Lancelot R, Zundel E, Ducrot C. Spécificités de la santé animale en régions chaudes: le cas des maladies infectieuses majeures en Afrique. INRA Production Animale. 2011;24:65–76.
Libeau G, Diallo A, Parida S. Evolutionary genetics underlying the spread of peste des petits ruminants virus. Anim Front. 2014;4:14–20.
Mbengue MB, Sarr J. Fall Massal Sero-epidemiological studies on contagious bovine pleuropneumonia (CBPP) in Senegal. 2013;10.
Mbengue M, et al. Réémergence de la péripneumonie contagieuse bovine au Sénégal. Bull Soc Pathol Exot. 2013;106:2012–5.
Motta P, et al. Implications of the cattle trade network in Cameroon for regional disease prevention and control. Sci Rep. 2017;7:43932.
Natale F, et al. Network analysis of Italian cattle trade patterns and evaluation of risks for potential disease spread. Prev Vet Med. 2009;92:341–50.
Nicolas G, Durand B, Duboz R, Rakotondravao R, Chevalier V. Description and analysis of the cattle trade network in the Madagascar highlands: potential role in the diffusion of Rift Valley fever virus. Acta Trop. 2013;126:19–27.
Ould El Mamy AB, et al. L'épidémie de fièvre de la Vallée du Rift en Mauritanie en 2012. Rev Afr Santé Prod Anim. 2014;12:169–73.
Padhi A, Ma L. Genetic and epidemiological insights into the emergence of peste des petits ruminants virus (PPRV) across Asia and Africa. Sci Rep. 2014;4(7040)
Park J, Newman MEJ. The statistical mechanics of networks. Phys Rev E. 2004;70:066117.
Pelon V. The Sahel livestock paradox: high stakes, low support. Food Sovereignity Briefs. 2015;16
Perry BD, Grace D, Sones K. Current drivers and future directions of global livestock disease dynamics. Proc Natl Acad Sci. 2013;110:20871–7.
Rautureau S, Dufour B, Durand B. Targeted surveillance of cattle trade using social network analysis tools. Épidémiologie Santé Anim. 2011:58–60.
Salami DH Epidémiologie de la peste des petits ruminants au Sénégal. 67.
Salliot E. A review of past security events in Sahel 1967–2007: Sahel and West Africa Club; 2010.
Séry A, et al. Seroprevalence of contagious bovine pleuropneumonia (CBPP) in Mali. Trop Anim Health Prod. 2015;47:395–402.

Soumaré POL, et al. Phylogeography of Rift Valley fever virus in Africa reveals multiple introductions in Senegal and Mauritania. PLoS One. 2012;7:e35216.
Sow A, et al. Widespread Rift Valley fever emergence in Senegal in 2013–2014. Open Forum Infect Dis. 2016;3(3):ofw149.
Thebaud B. Pastoral and agropastoral resilience in the Sahel. (Nordic Consulting Group). 2017.
Tounkara K, et al. First genetic characterization of Peste des Petits Ruminants from Niger: on the advancing front of the Asian virus lineage. Transbound Emerg Dis. 2018;65(5):1145–51. https://doi.org/10.1111/tbed.12901.
VanderWaal K, Gilbertson M, Okanga S, Allan BF, Craft ME. Seasonality and pathogen transmission in pastoral cattle contact networks. R Soc Open Sci. 2017;4:170808.
Various Authors. Commerce Transfrontalier et Sécurité Alimentaire en Afrique de l'Ouest: Cas du Bassin Ouest. (CILSS, FAO, FEWSNET, WFP, USAID, 2010).
Various Authors. Le pastoralisme a-t-il encore un avenir en Afrique de l'Ouest? 73–74, (Inter reseaux, 2018).
Volkova VV, Howey R, Savill NJ, Woolhouse MEJ. Sheep movement networks and the transmission of infectious diseases. PLoS One. 2010;5:e11185.
Wane A, Ancey V, Touré I. Pastoralisme et recours aux marchés: Cas du Sahel sénégalais (Ferlo). Cah Agric. 2010;19:14–20.

Chapter 4
Transboundary Animal Diseases (TADs) Surveillance and Control (Including National Veterinary Services, Regional Approach, Regional and International Organisations, GF-TAD)

K. Tounkara, Emmanuel Couacy-Hymann, and O. Diall

Abstract The transboundary animal diseases surveillance and control/eradication are key functions of national veterinary services. In designing animal diseases surveillance, the national veterinary services are dealing with major constraints including lack of cooperation of the livestock owners, inadequate resources to control the compliance of established regulations, inadequate communication among national stakeholder institutions and control of livestock movements. The transboundary animal diseases are controlled using one of the two ways in which the chain of transmission of the disease agent can be broken: prevent the infected animals to perform their role of donor of pathogen and the immunisation of susceptible hosts or the combination of the two ways. The immunisation of animals is the most affordable method in the majority of countries in Africa.

The national veterinary services are getting support from the main global and regional initiatives (GF-TAD, EMPRESS, GLEWS, OFFLU, EMC-AH, etc.) and the regional (AU-IBAR, AU-PANVAC) and international organisations (OIE, FAO, IAEA) in their efforts to develop strategies for the control and eradication of transboundary animal diseases.

Keywords Surveillance · Control and eradication · National veterinary services · Regional and international animal health organisations (AU-IBAR, AU-PANVAC, OIE, FAO, IAEA)

K. Tounkara (✉)
OIE Regional Representative for Africa, Bamako, Mali
e-mail: k.tounkara@oie.int

E. Couacy-Hymann
Central Laboratory for Animal Diseases , Bingerville, Côte d'Ivoire

O. Diall
National Committee for Agricultural Research, Bamako, Mali
e-mail: odiall@afribonemali.net

M. Kardjadj et al. (eds.), *Transboundary Animal Diseases in Sahelian Africa and Connected Regions*, https://doi.org/10.1007/978-3-030-25385-1_4

Introduction

The transboundary animal diseases constitute an important setback in livestock sector in Africa. They expose regularly the majority of countries in Africa to deficiency in animal proteins and cause serious obstruction to trade in animal and animal products at national, regional and international levels. As a consequence in Africa animal and animal trade products are not accessing more lucrative markets due to noncompliance with sanitary and phytosanitary (SPS) standards. With increasing globalisation, animal diseases continue to occur, and therefore, it has become imperative that due attention be given to surveillance leading to the control and eventually eradication of those diseases that are reducing drastically livestock possibilities for development in general and for export in particular.

The surveillance and control/eradication of transboundary animal disease are among key functions of national veterinary services. To these functions are also committed extension staff, livestock professional associations, private animal health service providers, livestock owners, regional and international organisations, etc. A functional and effective surveillance system capable of early detecting of transboundary animal diseases (including emerging, reemerging and food-borne diseases) for an early response contributes significantly to improving animal health and productivity worldwide.

Transboundary Animal Diseases (TADS) Surveillance

The internationally accepted overall objective of undertaking surveillance of animal diseases including transboundary animal diseases is provided in Chapter 1.4. of the Terrestrial Animal Health Code (2018) of the World Organisation for Animal Health (OIE). This is to demonstrate the absence of a specific disease or infection, to describe the occurrence or distribution of a specific disease or infection or to detect a specific disease early including emerging and re-emerging diseases. It is a crucial tool for analysing the disease epidemiological parameters and its introduction risk and for decision-making regarding sanitary measures leading to the ultimate goal of controlling and eventually eradicating animal diseases.

There are many types of surveillance: active and passive surveillance, general and pathogen-specific surveillance and clinical, serological or combined surveillance, etc.

The essential prerequisites for the establishment of an effective surveillance system are:

- Defining the precise purpose and objectives
- Specifying disease clinical suspicion and confirmed disease
- Defining the sources, type and frequency of data to be collected
- Clarifying the information distribution list and its channel of dissemination
- Defining the terms of reference of all involved in the surveillance including the chain of command

– Estimating the required budget

During the process of establishing the disease surveillance, the national veterinary services should identify the potential difficulties and address them. These could be among others:

1. Obtain the full cooperation of the livestock owners

The primary source of animal disease information being the livestock owners, their sincere cooperation is vital for the effectiveness of the surveillance. The success of disease control and eradication measures (including ban on trade, control of livestock movement and destruction of livestock) lies in early alert by the livestock owner, which in most cases is requested without payment of any incentive. As a consequence the livestock owners may be not motivated to inform on the disease clinical signs occurring in their farms, which might have an important implication for themselves and all surrounding farmers and also for the entire country. Due attention must be paid to address this constraint before the launching of the surveillance activities.

The national veterinary services could learn lessons from the participatory disease surveillance used during the implementation of the Pan African Rinderpest Campaign (PARC) and the Pan African Programme for the Control of Epizootics (PACE).

2. Inadequate resources to control the compliance of established regulations

The legislation lists the reportable diseases within the OIE notifiable diseases and the punitive measures in case no compliance exists in all countries. However, attempting to take sanctions for not respecting the legislation is not always possible due to inadequate human resources. In the case of imposing sanctions, the relationship between the authorities and the livestock owners can be seriously damaged. An acceptable solution must be found to allow effective reporting.

3. Inadequate communication among national stakeholder institutions. In a globalised world it is essential to ensure good communication between all stakeholders from both public and private sectors, especially for an effective surveillance of zoonotic diseases. The appropriation of "One Health" approach at national level could contribute to the improvement of communication between public and animal health stakeholders.

4. Control of livestock movements. The livestock movements (transhumance, nomadism and commercial movements) may affect the collection of sanitary information for surveillance purpose. For commercial animals, surveillance must be focused on cattle markets and border posts. In the case of transhumance and nomadism, it is crucial to draw up an inventory of all routes used by animals before developing a surveillance strategy. Such surveillance may be external and internal. The external surveillance includes setting up of surveillance posts in areas to be compulsory crossed by animals, animal resting areas and border points. The internal surveillance involves livestock farmers themselves in the surveillance objectives and benefits.

The first step for effective surveillance is to set up the regulatory framework for controlling and monitoring animal movements, which includes:

- Compulsory health certificates for all moving animals
- Clinical inspection of animals, collection and analyses of samples from suspected animals
- Identification of transhumance, nomadism and commercial routes
- Identification of official border checkpoints to be compulsorily used by all moving animals

These measures must be supplemented by:

- In-depth study of official and unofficial animal movements practices
- Communication, training and building trust between animal health workers and livestock owners

Transboundary Animal Diseases (TAD) Control

The occurrence of transboundary animal diseases is maintained endlessly through the passage of the disease pathogen from an infected animal defined as a donor to a susceptible animal defined as recipient within the same epidemiological unit. Consequently, the earlier the transboundary animal disease is detected in its initial epidemiological unit (early warning), the higher the chance to control that disease with a minimum economic loss (early response). It is worldwide accepted that early warning and early response are essential keys for the control and eradication of transboundary animal diseases.

Early warning can be defined as a rapid detection of the introduction or the sudden increase of new cases of any animal disease having the potential for expansion to epizootic proportions or causing important socio-economic losses or public health concerns. It is based on effective surveillance, reporting and analysis of epidemiological data with an ultimate goal to have a better understanding of the disease and to identify the better measures to be implemented for its control or eradication.

Early response can be defined as the prompt implementation of diseases control/ eradication measures required for the containment of the disease outbreak with an ultimate goal of eliminating the disease and infection within the shortest time and with minimum loss.

The disease maintenance chain can be effectively stopped using two ways or their combination: prevent the infected animal to transmit the pathogen and immunise all susceptible animals to prevent them to become infected.

1. Prevent the infected animals to transmit the pathogen.
   - Isolation of infected animals. This is done by creating a physical barrier which does not allow the spread of disease agent between the infected and the susceptible animals.

– Stamping out method.

It can be defined as a method implemented under the authority of veterinary services aiming at eliminating a disease outbreak. It comprises:

1. The killing of all diseased and suspected of being affected animals in the herd or flock and, where appropriate, also those animals exposed to infection in other herds or flocks by direct animal-to-animal contact, or by indirect contact with the causal pathogen.
2. The disposal of carcasses and animal products using internationally accepted methods.
3. The cleaning and disinfection of premises which contained the diseased and exposed animals using internationally accepted methods.

These measures can be successfully applied only with the cooperation of livestock owners, which can be concluded only with the compensation measures.

The application of the stamping out measures is often not feasible in most countries in Africa due to many reasons, e.g. livestock production systems practising transhumance and/or nomadism, rendering disease control strategies involving the restriction of animal movements ineffective. Furthermore, pastoralists tend to be found in inaccessible regions with poor veterinary infrastructure facilities. Armed conflicts and stock thefts lead to social instabilities and uncontrollable displacement of human and livestock populations.

In addition to all these factors, the Veterinary Services in Africa are underfinanced, understaffed and underequipped to deal effectively with the challenges posed by the re-emergence of transboundary animal diseases.

– Modified Stamping out.

This can be envisaged only with animal diseases which are less infectious. It is defined as a method which is not implementing in full the prescriptions of the stamping out complemented with or without vaccination.

– Control of livestock movements. The control of livestock movement is imposed to herds with the main objective to confine the chain of transmission of the disease pathogen exclusively to the infected and exposed herds. The national veterinary authority is empowered (provision in the legislation) with the support of civil administration to quarantine these herds for a specified time.

In many countries in Africa, livestock movement (transhumance and nomadism) is cultural and has been practised for centuries. The amplitude of the movement can be small (within country movement) or wider (between countries). This has to be taken into account when the control of livestock is considered as an option for stopping the spread of transboundary animal diseases.

In order to guarantee the success of the control of livestock movement within a country, the national veterinary services must sensitise the livestock owners and facilitate livestock access to feed and water. For between-countries livestock movement, a regional approach must be considered.

A zoning approach recommended by the OIE and successfully implemented in some countries can be also considered whenever applicable.

2. Immunise all susceptible animals to prevent them to become infected.

The vaccination of susceptible animals is an efficient way of stopping the chain of transmission of disease pathogen. In most of countries in Africa, this is the only viable way to fight transboundary animal diseases. The success of vaccination greatly depends on the use of vaccines of good quality complying with the international standards. A single defective batch of vaccine could have serious consequences on animal health protection and damage public confidence in immunisation programmes. This was one part of the rationale for the establishment of the Pan African Veterinary Vaccine Centre (PANVAC) in support of the rinderpest eradication programmes in Africa. This support was pivotal in achieving rinderpest eradication in Africa as stated in the final evaluation report of the PARC and PACE "The success of the Pan African Rinderpest campaign (PARC) and the Pan African programme for the Control of Epizootics (PACE) clearly demonstrated that no amount of vehicles, syringes, trained personnel, communication materials, would have eliminated Rinderpest if the vaccine batches used were of poor quality. The secondary and independent level of quality control assessment assured by PANVAC played a major role for this success and led, at the same time to a sustained improvement in the quality of vaccines against Rinderpest and Contagious Bovine Pleuropneumonia produced in Africa".

Based on this success, the mandate of PANVAC was expanded to ensure the quality control of all veterinary vaccine batches to be used in Africa and also in producing essential diagnostic reagents. The centre, managed today by the African Union Commission, became an essential tool for all animal diseases control and eradication programmes in Africa.

Role of National Veterinary Services in TAD surveillance and Control

The national veterinary services considered as an international public good and an important instrument of public health constitute the backbone of animal health systems with the main tasks to prevent control and eradicate animal diseases. They must comply with the OIE required standards for quality. Good-quality veterinary services are critical for the successful and sustainable implementation of disease surveillance leading to the control activities.

The failure to detect earlier an animal disease by any veterinary service puts the entire world at risk of introduction of that disease.

The OIE developed a tool entitled "*Performance of Veterinary Services (PVS)*" for the evaluation of current level of performance of veterinary services followed by the identification of gaps in order to make suggestions leading to compliance with OIE international standards.

The Need for a Regional Approach in Surveillance and Control of TADs

The coordination of national disease surveillance systems between African countries is very important for the successful prevention, control and eradication of transboundary animal diseases in Africa.

The past programmes were used and those currently available continue to accompany national veterinary services in their daily efforts aiming at controlling and eradicating transboundary animal diseases in Africa.

Past Programmes

Joint Project 15(JP-15)

The very first major Africa continental programme for the eradication of rinderpest in Africa entitled the Joint Programme (JP–15) was conceived in Kano, Nigeria, in May 1961. Implemented under the auspices of the then Organization of African Unity (OAU), the programme was funded by different international organisations and individual countries and had the overall aim of eradicating rinderpest from Africa using a low-cost rinderpest vaccine for the immunisation of susceptible animals.

The JP15 was implemented in six phases from 1961 to 1976. Unfortunately, the implemented activities did not allow the eradication of the disease and the majority of involved in the campaign countries after 1976 reported the occurrence of many outbreaks. This situation prompted the Head of States of Africa to seek in the early 1980s for a second programme.

The Pan African Rinderpest Campaign (PARC)

The Pan African Rinderpest Campaign (PARC) of the then Organization of African Unity (OAU) was launched in 1986 and coordinated by the Inter-African Bureau for Animal Resources (IBAR) with the objectives to control and eradicate rinderpest and revitalise veterinary services. The strategy adopted was based on carrying out mass vaccination campaigns, serological monitoring and surveillance at a later stage, disease active investigation and control of animal movement.

The PARC programme was implemented for 12 years (1988–1999) and achieved rinderpest disease freedom in Africa. However, there was evidence that the rinderpest virus was not eliminated from the southern part of the then Republic of Sudan, northwest and northeast Kenya and in southern Somalia.

Global Rinderpest Eradication Programme (GREP)

The Global Rinderpest Eradication Programme (GREP) was established in 1994 with the objective to unify the existing regional rinderpest eradication campaigns, unify criteria and assist participating countries in obtaining international accreditation for rinderpest freedom. Its strategy was developed jointly by the OIE, FAO and the participating regional organisations, and its secretariat was established in FAO Rome with the following facilitating terms of reference:

- To propose the required techniques for rinderpest surveillance
- To provide technical support to national veterinary laboratories mandated to organise surveillance programmes
- To assist national veterinary services in using OIE guidelines for disease and infection freedom accreditation
- To develop a strategy for the prevention or response to the reintroduction of rinderpest virus
- To develop guidelines for national and regional emergency plans and action to contain an outbreak
- To promote independent rinderpest vaccine quality control
- To suggest strategy aiming at safeguarding the misuse of rinderpest virus and accidental escape from diagnostic and vaccine producing laboratories
- To assist in the implementation of focused rinderpest vaccination campaigns towards the elimination of endemicity.

GREP was successful in eradicating rinderpest worldwide in 2011.

The Pan African Programme for the Control of Epizootics (PACE)

The Pan African Programme for the Control of Epizootics (PACE) was conceived in 1999 on the success of the PARC and implemented until 2006 in 32 countries in Africa. It was also funded by the European Commission and coordinated by the Inter-African Bureau for Animal Resources of African Union with the main objective to address the problem caused by the circulation of rinderpest virus in the border between Ethiopia, Kenya and Somalia (Somali Ecosystem) and lead the accreditation of rinderpest freedom in all other participating countries.

Somali Ecosystem Rinderpest Eradication Coordination Unit (SERECU)

The Somali Ecosystem Rinderpest Eradication Coordination Unit (SERECU) was established in 2006 in order to coordinate the final eradication of rinderpest virus from the remaining foci in the Somali Ecosystem. This programme was implemented in two phases: from January 2006 to April 2008 for the first phase and from May 2008 to June

2010 for phase II. The expected result for SERECU was to verify scientifically the freedom from rinderpest and obtain OIE accreditation for Somali ecosystem countries: Ethiopia, Kenya and Somalia.

Current Programmes

Animal Diseases Component of the FAO Emergency Prevention System for Transboundary Animal Diseases and Plant Pest and Diseases (EMPRES)

The animal diseases component of the Emergency Prevention System for Transboundary Animal Diseases and Plant Pest and Diseases (EMPRES) was established in 1994 with the objective to strengthen FAO in discharging its role for the prevention and provision of prompt response to the occurrence of transboundary animal diseases which can affect food security, public health and the revenue of livestock owners deriving from the trade of animal and animal products. The operation of GREP is based on early warning for an early reaction, coordination and enabling research.

EMPRES supports surveillance by analysing, disseminating warning messages, setting up surveillance networks, preparing disease prediction models, implementing risk communication and developing systems enabling the improvement of animal health.

EMPRES played an important role in all progammes aimed at controlling/eradicating (globally or regionally) transboundary animal diseases. Its biggest success story is the Global Rinderpest Eradication Programme (GREP).

The Global Early Warning and Response System (GLEWS)

The Global Early Warning and Response System (GLEWS) is a joint OIE, FAO and WHO system based on combining and coordinating the three agencies' warning and response mechanisms to assist animal health stakeholders in addressing animal disease threats, including zoonoses.

It overall objective is to contribute to the improvement of the early warning and early response capacity of veterinary services to animal disease threats. The specific objectives of GLEWS are among others to:

- Strengthen member countries' capacity for an early detection/warning and early response to animal diseases
- Contribute to the improvement of transparency between countries and compliance with OIE requirement for reporting

- Contribute to the improvement of national surveillance and monitoring systems and the strengthening of networks comprising public health, medical and veterinary laboratories involved in the works on zoonotic diseases
- Contribute to the improvement of countries and international preparedness to fight against animal disease outbreaks including zoonotic diseases and provide technical assistance to infected countries
- Contribute to the improvement of the capacity of FAO, OIE and WHO for early detection of new emerging diseases, including zoonoses
- Contribute to the improvement of the integration of public health and animal surveillance for a simultaneous detection of the occurrence of the disease across species

OIE and FAO Network of Expertise on Animal Influenza (OFFLU)

The OIE and FAO Network of Expertise on Animal Influenza (OFFLU), an OIE/FAO joint network of scientific expertise on avian influenza, was launched in 2005. Its mandate was extended to all animal influenza in 2009 with the objective to provide technical support to veterinary services worldwide aiming at reducing risks of transmission of animal influenza viruses to humans and animals.

The main objectives of OFFLUs are to facilitate the exchange of scientific data and biological materials (including virus strains) within the network, provide technical support, strengthen veterinary services capacity in the prevention, detection and containment of animal influenza, and collaborate with the influenza network of the World Health Organization (WHO).

Global Framework for Transboundary Animal Diseases (GF-TADs)

The Global Framework for Transboundary Animal Diseases (GF-TADs), a FAO/OIE joint initiative, was established in 2004 as a mechanism which facilitates the empowerment of national and regional alliances for the control of animal diseases, strengthening of technical capacity and the support for the development of strategies leading to the control of regional targeted transboundary animal diseases.

The overall objective of GF-TADs is to contribute to the reduction of economic losses for livestock owners due to animal diseases and strengthening of national capacities in order to facilitate safe trade of animal and animal products.

The GF-TADs programme is based on four main pillars:

1. Regionally led mechanism, to implement activities for the control of regional priority diseases defined by animal health stakeholders
2. Establishment of regional and global early warning systems for priority animal diseases

3. Molecular and ecological level research on TAD pathogens for more efficient management and control of disease
4. Achieve rinderpest freedom declaration by 2010

GF-TAD stakeholders are national veterinary services, regional organisations and their specialised technical animal health entities. The targeted diseases among others are avian influenza, foot and mouth disease (FMD), rinderpest, African and classical swine fevers, peste des petits ruminants, caprine and bovine contagious pleuropneumonia, Rift Valley fever, Newcastle disease, haemorrhagic septicaemia and sheep and goat poxes.

Global Framework for Transboundary Animal Diseases (GF-TADs) for Africa Region

The GF-TADs for Africa was launched in 2006 with the objective to respond to African region priority diseases in general and rinderpest and highly pathogenic avian influenza in particular. It facilitates the establishment of alliances and partnerships and the development of actions and roadmap for the fight against the following defined region priority diseases: African swine fever (ASF), contagious bovine pleuropneumonia (CBPP), foot and mouth disease (FMD), peste des petits ruminants (PPR), rabies and Rift Valley fever (RVF).

Emergency Centre for Transboundary Animal Diseases Operations (ECTAD)

Emergency Centre for Transboundary Animal Diseases Operations (ECTAD) of FAO was established in 2004 within EMPRES programme as the centre for the implementation of services provided by FAO at field level. It is established in many countries in Africa.

The Progressive Control Pathway for Foot and Mouth Disease (PCP-FMD)

The Progressive Control Pathway for Foot and Mouth Disease (PCP-FMD) was developed by FAO in order to support FMD endemic countries in their effort to reduce progressively the impact of the disease and its virus load. It is a tool used by the infected countries or region to develop their FMD control activities. The FAO and OIE, after consultation, agreed to consider the PCP as their joint tool and the backbone of the Global Strategy for the Control of FMD.

The Peste des Petits Ruminants Global Control and Eradication Strategy (PPR-GCES)

The Peste des Petits Ruminants Global Control and Eradication Strategy (PPR-GCES) was launched in 2015 in Abidjan (Côte d'Ivoire) with three main objectives: to achieve global eradication of PPR by 2030, to strengthen the capacity of veterinary services and to control other priority diseases of small ruminants. The strategy identified four stages for the eradication of the disease: assessment, control, eradication and post-eradication. The stage is determined after assessing the capacity of the country against five key elements: diagnostic system, surveillance system, prevention and control system, legal framework and stakeholder involvement.

The strategy also developed the PPR monitoring and assessment tool (PMAT) to assist the country in evaluating the progress made during the eradication process.

The OIE and FAO established in Rome a Joint global PPR secretariat to lead the control and eradication process of PPR worldwide.

The Emergency Management Centre: Animal Health (EMC-AH)

The Emergency Management Centre—Animal Health (EMC-AH) based at the FAO Headquarters in Rome (Italy) is a joint OIE-FAO mechanism to provide on request rapid technical assistance to countries encountering animal disease outbreaks. The assistance is provided by an EMC-AH multidisciplinary expert team (from OIE and FAO's network of expertise) which is deployed rapidly to the requesting country.

The Regional Networks for Epidemiological Surveillance (RESEPI) and Animal Diseases Laboratory Diagnosis (RESOLAB)

The Regional Networks for Epidemiological Surveillance (RESEPI) and Laboratory Diagnosis (RESOLAB) were established by FAO with the objective to provide support to national epidemiological surveillance systems and also to identify a regional programme for the control of priority animal diseases. The RESOLAB for the control of highly pathogenic avian influenza and other transboundary animal diseases was launched in Bamako (Mali) in December 2007 during the joint FAO/USDA/APHIS workshop on HPAI within the framework of the Regional Animal Health Centre of Bamako.

The immediate objectives assigned to RESOLAB were to increase the efficacy and efficiency of national animal diseases' diagnostic laboratories, improve the communication between the national laboratories and the epidemiological surveillance networks and create conditions for emerging expertise in the diagnosis of avian influenza in the region and subsequently to improve the quality of animal disease diagnosis in general. The scope of the networks was expanded to other transboundary animal diseases such as foot and mouth disease, CBPP and PPR.

The two networks were coordinated by the FAO ECTAD Regional Unit located in Bamako (Mali), and in 2016 the coordination was transferred to the livestock Desk of the Economic Community of West Africa States (ECOWAS) and later to the Regional Animal Health Centre of the ECOWAS.

Role of Regional Organisations

The major regional organisations supporting individual countries in their constant effort to control and ultimately eradicate transboundary animal diseases in Africa are the Inter-African Bureau for Animal Resources of African Union (AU-IBAR), the Pan African Veterinary Vaccine Centre of African Union (AU-PANVAC) and the Regional Animal Health Centres (RAHC) of the Regional Economic Communities.

African Union Inter-African Bureau for Animal Resources (AU-IBAR)

The African Union Inter-African Bureau for Animal Resources (AU-IBAR) was established in 1951 with the primary objective of studying the epidemiological situation of rinderpest and launching the fight against this disease in Africa. Currently, the mandate of AU-IBAR is to contribute to the socio-economic development of African continent by improving animal (livestock, fisheries, wildlife) health situation and productivity. AU-IBAR successfully coordinated the following major continental animal health programme/projects: Joint Project 15, Pan African Rinderpest Campaign (PARC), the African Wildlife Veterinary Project (AWVP: 1998–2000), the Pan African Programme for the Control of Epizootics (PACE), the Somali Ecosystem Rinderpest Eradication Unit—SERECU), the Pastoral Livelihoods Programme (PLP: 2000–2005), the Pastoral Livelihoods Programme HIV/AIDS (PLP HIV/AIDS: 2003–2006), the Community Animal Health and Participatory Epidemiology Project (CAPE: 2000–2004), the Regional Project for Poultry and Milk Production in East Africa Project (1999–2005), the Farming in Tsetse-Controlled Areas Project (FITCA: 1999–2004) and the Regional Programme on Ticks and Tick-borne Disease (RTTDC).

African Union: Pan African Veterinary Vaccine Centre (PANVAC)

In implementing the main recommendation of the FAO Expert Consultation on Rinderpest held in Rome (Italy) requesting the participation of all rinderpest vaccine producing laboratories in Africa in the international and independent vaccine quality programme, the FAO launched the Pan African Veterinary Vaccine Centre (PANVAC) in 1984 to support the Pan African Rinderpest Campaign (PARC). Two Regional Vaccine Quality Control and Training Centres were established in DebreZeit (Ethiopia) for Eastern and Southern Africa and in Dakar (Senegal) for West and Central Africa to provide rinderpest vaccine quality control services to vaccine producing laboratories. The two regional centres merged in 1993 to one: the Pan African Veterinary Vaccine Centre in Debre Zeit (Ethiopia). In 2004, the PANVAC was officially launched as an African Union-specialised office under the management of African Union Commission and its mandate was expended to the quality control of all veterinary vaccines and the production of essential diagnostic reagents.

AU/PANVAC became an OIE Collaborating Centre in vaccine quality control of veterinary vaccines in 2013 and the FAO Reference Centre in 2015.

Regional Animal Health Centres

Four Regional Animal Health Centres of the Regional Economic Communities were established by OIE, FAO and AU/IBAR in Bamako (Mali) for West and Central Africa, Tunis (Tunisia) for Northern Africa, Nairobi (Kenya) for Eastern Africa and Gaborone (Botswana) for Southern Africa following the recommendation of the Second International Conference on Human and Avian Influenza held in Beijing (China) in January 2006. The centres were tasked to function as a multidisciplinary tool at the disposal of member states and regional economic communities.

Currently, only one centre is functional, the Regional Animal Health Centre located in Bamako (Mali) under the management of ECOWAS as of February 2012.

Role of International Organisations

The world internationally recognised organisations, the World Organisation for Animal Health (OIE), the Food and Agriculture Organization of the United Nations (FAO) and the International Atomic Energy Agency (IAEA), are mainly developing animal diseases control and eradication tools, providing technical assistance to individual countries' veterinary services and building and strengthening technical capacity of veterinary services worldwide including Africa.

World Organisation for Animal Health (OIE)

The World Organisation for Animal Health was created in 1924 by 28 countries as the Office International des Epizooties (OIE). In May 2003, the Office International des Epizooties became the World Organisation for Animal Health, keeping its initial acronym OIE. In 2018, the OIE has 182 member countries and signed permanent relations with more than 70 international and regional organisations. The vision of OIE is protecting animals and preserving our future and its activities are carried out on four pillars: standards for international trade of animals and animal products, transparency of the world animal disease situation, expertise for the collection and dissemination of veterinary scientific information and solidarity between countries to strengthen capacities worldwide. The OIE relies on its network of more than 250 reference laboratories and 50 collaborating centres.

Food and Agriculture Organization of the United Nations (FAO) and the Animal Production and Health Section of the Joint FAO/IAEA Division of Nuclear Techniques in Food and Agriculture

The Food and Agriculture Organization of the United Nations (FAO) was created in 1945. It has the mandates from its 194 Member Nations (2018) to contribute to the development of livestock by providing assistance to veterinary services of member countries in their effort to control animal diseases.

The FAO and the World Health Organization jointly established in 1963 the Codex Alimentarius or "Food Code" to develop harmonised international food standards and promote good practices in trade of food commodities.

The Animal Production and Health Section of the Joint FAO/IAEA Division of Nuclear Techniques in Food and Agriculture is mandated mainly to assist its member states to build and strengthen laboratory diagnosis capacity. The section regularly develops protocols, guidelines and standard operating procedures for the application of nuclear and nuclear-related techniques and molecular technologies to be used in the animal disease control and eradication programmes. It also develops guidelines for the use of irradiation techniques to produce animal diseases vaccines.

During the rinderpest eradication process, the section has played a pivotal role through technical support (building or strengthening laboratory diagnostic capacities in individual member states and the establishment of Rinderpest Laboratory Network) using its various Coordinated Research Programmes for the Seromonitoring of Rinderpest and the National and Regional Technical Cooperation Projects.

Bibliography

Bouslikhane M. Cross border movements of animal and animal products and their relevance to the epidemiology of animal diseases in Africa. Proceedings of the Conference, OIE Regional Commission for Africa, Rabat, Morocco. 2015.

Domenech J, Lubroth J, Eddi C, Martin V, Roger F. Regional and international approaches on prevention and control of animal transboundary and emerging diseases. Ann N Y Acad Sci. 2006;1081:90–107.

European Commission for the Control of FMD/OIE/FAO. The progressive control pathway for FMD control (PCP-FMD) principles, stage, descriptions and standards. 2011. 24 p.
Fujita T. Zoning and regulatory for animal disease control. OIE regional representation for Asia and the Pacific Tokyo, Japan, Personal communication.
Halliday J, Daborn C, Auty H, Mtema Z, Lembo T, Bronsvoort BM, Handel I, Knobel D, Hampson K, Cleaveland S. Bringing together emerging and endemic zoonoses surveillance: shared challenges and a common solution. Philos Trans R Soc Lond Ser B Biol Sci. 2012;367(1604):2872–80.
Hendrikx P. Surveillance protocols and the practical establishment of networks. Personal Communication.
New Technologies in the fight against transboundary animal diseases. FAO-Japan cooperative project: collection of information on animal production and health. Phase one: Transboundary animal diseases. 118p. 1999.
OIE/FAO/WHO. Global early warning and response system for major animal diseases, including zoonoses (GLEWS). 2006. 26 p.
OIE/FAO. Global strategy for the control and eradication of PPR. 2015. 83 p.
Regional GF-TADs for Africa. Regional global framework for the progressive control of transboundary animal Diseases (GF-TADS) for Africa: five year action plan for the period 2012–2016, 24 pp. October 2012.
THE OIE/FAO Global Foot and Mouth Disease Control Strategy: Strengthening animal health systems through improved control of major diseases. 2012. 43 p.
Vallat B, Pinto J, Schudel A. International organizations and their role in helping to protect the worldwide community against natural and intentional biological disasters. Rev Sci Tech Off Int Epiz. 2006;25(1):163–172.
Wachida N, Tughgba T, Hambesha P. Participatory disease surveillance of transboundary animal diseases in Gboko, Konshisha and Katsina Ala Local Government areas of Benue State, Nigeria. Epidemiol Rep. ISSN 2054-9911. 2017.
World Organization for Animal Health (OIE). OIE manual of diagnostic tests and vaccines for terrestrial animals. 7th ed. Paris: OIE; 2017a.
World Organization for Animal Health (OIE). OIE terrestrial animal health code. 26th ed. Paris: OIE; 2017b.

Chapter 5
Public and Private Veterinary Services in West and Central Africa: Policy Failures and Opportunities

Mahamat Fayiz Abakar, Vessaly Kallo, Adam Hassan Yacoub, Alhadj Mahamat Souleyman, and Esther Schelling

Abstract The livestock sector in most African countries, in particular in the Sahel region, remains underexploited. It is traditionally managed in pastoralist systems that best guarantee the environmental sustainability of the arid and semi-arid grasslands, which can be hardly used for agriculture. However, pastoralists are vulnerable to exclusion to social services because they are remote to educational and political centres. The majority of livestock, however, are kept in mixed crop–livestock systems in which livestock have multiple roles such as producing food, generating income, providing manure, producing power, being financial instruments and enhancing social status. Livestock breeding faces many challenges and constraints including transboundary animal diseases (TADs) and increasing waves of droughts due to climate change as well as politically and economically instable states. Despite that Sahelian livestock owners have robust empirical methods to protect their basis of livelihood—their livestock—they need and appreciate quality medicines, vaccines and veterinary services.

Operational veterinary services are at the heart of controlling important livestock diseases to reduce impacts on livelihoods. There are effective control measures such as anthrax vaccination of livestock that also safeguard human health. Veterinary services are equally at the heart of early detection of TADs and surveillance and

M. F. Abakar
Institut de Recherche en Elevage pour le Développement (IRED), N'Djamena, Chad

V. Kallo
Direction des Services Vétérinaires, Abidjan, Côte d'Ivoire

Ecole Inter-Etas des Sciences et de Médicine Vétérinaire, Dakar, Senegal

A. H. Yacoub · A. M. Souleyman
Ministère de l'Elevage et des Productions Animales, N'Djamena, Chad

E. Schelling (✉)
Swiss Tropical and Public Health Institute (Swiss TPH), Basel, Switzerland

University of Basel, Basel, Switzerland
e-mail: esther.schelling@swisstph.ch; esther.schelling@vsf-suisse.org

M. Kardjadj et al. (eds.), *Transboundary Animal Diseases in Sahelian Africa and Connected Regions*, https://doi.org/10.1007/978-3-030-25385-1_5

response to epidemic and zoonotic diseases. But how can the services, composed of public and private veterinarians, veterinary technicians, community animal health workers and outreach services, meat inspectors and monitoring/surveillance professionals, better ensure and satisfy the needs of livestock owners, their families and other stakeholders such as public health and rural development? Which roles do international and national policies play?

We review the status of veterinary services in the Sahel over the last 20 years and relate their provided services to overarching policy changes such as the privatisation of veterinary services and external funding schemes and programmes. We conclude on new ways forward such as implementation of intersectoral collaborations of professionals in remote Sahelian zones and needed operational research in optimising services.

Keywords Public and private veterinary services · Livestock systems · International and national policy changes · West and Central Africa

Introduction

Main Livestock Keeping Systems in West and Central Africa

Between 46 and 82% of poor rural households in Asia, Africa and Latin America keep livestock, which generates between 20 and 60% of their incomes (Zezza et al. 2007). Livestock contributes an average of 40% to agricultural gross domestic product (GDP) in West and Central African countries (SWAC-OECD/ECOWAS 2008). The livestock systems in West Africa can be roughly divided into the extensive pastoral systems, mixed crop–livestock systems and the intensified and stall-feeding systems in urban and peri-urban areas (Ly et al. 2010). West Africa carries about 74 million cattle, 270 million sheep and goats and 4.5 million camels; about 13 million pigs and 570 million poultry; and about two million horses and six million donkeys (FAO 2016). Around 18% of all ruminants are kept in pastoral systems. These extensive systems are challenged by social and ecological changes, despite evidence that pastoralism is a viable and sustainable livelihood. They are hotspots of cultural and biological diversity, but need favourable institutional and legal frameworks (Krätli 2016). Among the key ingredients for sustained pastoralism are decentralised governance of natural resources, better locally adapted social services and high flexibility for maintaining mobility. There is still untapped potential to optimise extensive livestock production (Zinsstag et al. 2016). Crop residues are often used by transhumant pastoralist herds that, in return, provide manure for the fields. The majority of livestock, however, are kept in mixed crop–livestock systems. Livestock is raised on (small-scale) farms that produce crop (such as maize, sorghum, millet and rice), vegetables (e.g. cowpea, groundnuts, and soybeans) and tubes (e.g. cassava and yams). In these systems, livestock have multiple roles such as producing food, generating income, providing manure, producing power, being financial instruments and enhancing social status (Randolph et al. 2007). In the Sahel, the integration of livestock into farming systems increases from North to the more humid South. In the South, vector-borne diseases challenges are more prominent than in the dryer northern zones (Ly et al. 2010). The urban and peri-urban intensified and intensive livestock keeping systems, where resources and feed are imported and food and wastes exported, can better provide the high demand on locally produced livestock products in the cities. They can also enhance the importance of certain infectious diseases, e.g. for poultry and food-borne pathogens and antimicrobial resistances. Peri-urban milk production and ruminant, particularly sheep, fattening units are emerging close to the rapidly growing urban centres to support them with locally produced livestock-sourced food (Bonfoh et al. 2010). Urban and peri-urban livestock production importantly includes commercial (up to industrial-scale) poultry production. Animal feed and other production resources are shipped to the production sites, whereas manure and wastes are transported away from the centres, which makes this production more energy-intensive. Livestock owners' demand for veterinary services varies between the different livestock keeping systems.

Training, Education and Career Pathways of Veterinarians

According to the World Organization for Animal Health (OIE)'s Terrestrial Animal Health Code, the concept of "Veterinary Services" refers to the public or private bodies that implement health protection measures in the territory of a country and the welfare of animals as well as other standards and recommendations of the OIE sanitary code for terrestrial animals. This Code says "[....] Veterinary Services are under the direct management and control of the Veterinary Authority. Organizations, veterinarians and veterinary paraprofessionals in the sector are normally approved by the veterinary authority or authorized by it to perform the public service tasks entrusted to them [....]". This implies that a strong central veterinary authority is needed to manage the different organizations and actors.

In the case of Chad for example, the governmental veterinary services within the OIE definition include three main entities: (1) the Directorate of veterinary services—the central veterinary authority, (2) the decentralised services of the Ministry in charge of Livestock (with provincial delegations and veterinary posts) and (3) the central diagnostic laboratory (*Institut de Recherches en Elevage pour le Développement* "IRED") (DGSV 2017).

Veterinary services in West and Central Africa include both the governmental and non-governmental structures. The private sector organisations are composed of veterinarians, veterinary paraprofessionals and aquatic animal health professionals who are commonly accredited or approved by the veterinary authorities of a country to deliver the mandated objectives (OIE 2016). To note is that community animal health workers (CAHW) have not been endorsed by all countries.

The importance of animal welfare, particularly during transport and slaughter of livestock as well as unauthorised keeping of wild animals, yet needs a legal basis—at minimum the welfare regulations according to OIE standards—in most West and Central African countries (Bourzat et al. 2013). In addition, many countries need assistance to ensure a status when they do not endanger the animal health situation of another country, importantly a neighbouring country. According to OIE estimates, despite noteworthy progress, the veterinary services in the least developed countries (LDCs) and many middle-income countries (which represent together a total of 120 countries) need technical and financial assistance to ensure a satisfactory animal health management—one that does not pose a veterinary health risk to other countries (Pradere 2017).

To build veterinary capacity, OIE member countries (180 countries) can apply for the OIE Tool for the Evaluation of the Performance of the Veterinary Services (PVS). The OIE PVS tool is the main resource for improving global animal health, public health and animal welfare (OIE 2006). The necessary budget to enable developing countries to comply with OIE and WHO standards and control zoonotic diseases has been estimated at between 1.9 to 3.4 billion USD per year in 2011 (Bank 2012). These funds, combined with farm-sector reforms, could considerably reduce the economic impact of animal diseases, pandemic risks and the impact of livestock farming on natural resources and the climate. However, structures of veterinary services that can absorb, manage and implement effectively larger funds are scarce in West and Central Africa.

Key Commitments of "Veterinary Services"

Key elements of *good governance* (within national chains of command) of veterinary services include the maintenance of epidemiosurveillance networks, liaising between public and private sectors, offering of quality veterinary services and doing veterinary education and research (AU-IBAR 2010). AU-IBAR (African Union - Interafrican Bureau for Animal Resources) provides technical leadership and advisory services, facilitates the development and harmonisation of policies, coordinates the development of animal resources, articulates common African positions globally, advocates issues relevant for Africa, analyses and disseminates information and provides strategic support to countries in emergency situations (AU-IBAR 2010).

Among the activities covered by the governmental veterinary services in all countries are (1) the establishment of the legislation, (2) the prevention and control of regulated animal diseases, (3) food safety, (4) the establishment of international certificates for the export of animals or products of animal origin, and (5) border control. These activities fall under the concept of "global public good" (OIE 2016). As to economically most important livestock diseases that severely reduce productivity or those diseases poor livestock owners fear most, diseases that can occur at any time and rapidly eliminate their basis of income (Perry et al. 2002), are important zoonoses and public good transboundary diseases. National veterinary services are thus typically responsible for ensuring the protection of "public good" animal health, which includes the safety of food products of animal origin, the control of major animal diseases and the quality control of veterinary pharmaceuticals. The control of zoonoses is considered as a public good in that it protects human and animal public health and thus benefits society as a whole (Zinsstag et al. 2011).

Highly contagious animal diseases and epidemics pose an economic threat to livestock producers and the entire agricultural sector and national economies. Their control and elimination is therefore considered as a public good (The World Bank 2010). The veterinary services lie thus at the heart of the global public good represented by animal health systems. However, they cannot fulfil this mission without the appropriate veterinary legislation and the necessary means to enforce them. Veterinary services have a major role to play in matters of animal health and public health in terms of surveillance, early detection (and notification of) and rapid response to animal disease outbreaks, which can include vaccination, bio-security and bio-containment and compensation of farmers (Johnston 2013).

Excludability principles have been used to group animal health services into private good services, e.g. endemic disease control and prevention, sales of drugs and vaccines and clinical services. Hence, the user captures all benefits and common of public good services like the diagnosis, surveillance, movement control and quarantine services for epidemic or zoonotic disease control, control of food-borne diseases and tsetse control (Ahuja 2004; Riviere-Cinnamond 2005). The public good nature of some services does not necessarily imply that the government must take direct responsibility for their delivery. The government may subcontract these services to private organisations (e.g. non-governmental or research or development organisations) and private veterinarians (Stephen and Waltner-Toews 2015).

Our starting point for this chapter was that animal-health systems have been neglected in many parts of the world in the past two decades, leading to institutional weaknesses and information gaps as well as inadequate investments in animal-health-related public goods (FAO 2009). This is particularly evident in remote and rural zones of West and Central Africa, where international organisations and institutes easily find space to implement livestock protection projects without being forced to ask if they also contribute to the strengthening of veterinary services in the countries where they are active.

Rationale, Objectives and Methods

Governments in West and Central Africa have withdrawn importantly from the provision of veterinary and other input services such as high-quality health services in remote zones. Although more professionals are entering the private veterinary business and are playing active roles in immunisation schemes in partnership with the government and producer organisations—the anticipated emergence of private sector provision for the full range of veterinary and advisory goods and services has not been as successful as hoped for West and Central Africa two decades ago. In the same period, livestock production systems were hindered to develop as hoped as long as there were only low levels of inputs.

Most projects initiated and led by governments failed to create a self-driven development and remained heavily dependent on external funding, despite many good intentions such as stabilising the important livestock production and export markets. The generation of new and continued funding was limited since they had not to seek economic and financial recovery. After the introduction of the World Bank's privatisation policy (De Haan et al. 2001), governmental veterinary services no longer had to make larger investments. Veterinary services may be considered for the past two decades as the most understaffed and under funded services to provide so-called public goods (De Haan et al. 2001). At the same time, poverty reduction was put at the forefront in the design of livestock policies. International policies could easily occupy the vacuum of rigorous national policies.

A thorough assessment of international and national livestock policies in West and Central Africa for the last three decades does not exist, and this chapter does not intend to do so. However, such analyses would help to derive lessons for livestock policy development—and to better guide agreements on regional livestock policies, e.g. within the ECOWAS (Economic Community of West African States) region. It is clear that only regionally harmonised approaches will be successful given the important movements across borders of livestock, people and goods, not to mention animal trade and its role as important factor for diseases transmission in West Africa (Dean et al. 2013).

Low availability and use of veterinary services have allowed the classic endemic infectious diseases to persist. As livestock systems intensified, associated production diseases and syndromes such as mastitis became more important veterinary problems. Because of market segmentation based on food safety concerns, larger volumes of low-quality livestock products passed through informal marketing channels, further compounding the risks of zoonoses and food-borne diseases for low-income consumers (Roesel and Grace 2015). Foot and mouth disease—to name one important transboundary disease—without any veterinary measures such as surveillance, containment and compartmentalisation hinders smallholders to participate in rapidly expanding export markets for livestock products.

Poor access to products, services and information contribute to poor performance, profitability and competiveness and continue to limit the ability of livestock keepers and veterinarians (in a vicious cycle between the two) to address major disease and production constraints. The reduced role of the governments in the provision of veterinary and health services in the context of inability of the private sector to fill the void has led to the resurgence of endemic animal diseases, and reduced livestock productivity in many parts of the Sahel region.

See for example the rapid spread of Ebola in West Africa starting in 2012/13. This was possible because the Ebola epidemic was human to human transmitted (likely after a single index occurrence of a bat-to-human transmission), among others the high mobility of people in the region (also cross-border), but also because of weak health systems which lacked sufficiently trained nurses recognising and reporting abilities of haemorrhagic fever as well as related knowledge on the local traditional customs of washing dead bodies before burying that caused many new cases.

Here we reflect with two case studies on how international and national policies affected negatively and positively the ability of veterinary services in West and Central Africa to become more proactive in reply to old and newer demands of the veterinary services. Also, we want to reflect on how foreign policies and projects with all initial good intentions influenced the livelihoods of livestock owners in West and Central Africa. We present the case studies of Chad and Côte d'Ivoire. These case studies also depict how international policies led to increased self-responsibility of governments who are at the forefront to guarantee quality and good governance of veterinary services.

Given the opportunities and constraints of veterinary services outlined above, we have formulated the following objectives for this chapter:

1. To highlight the recent history of veterinary services performance and operational capacity in West and Central Africa and their role in international disease elimination programmes
2. To give case examples of failure of veterinary services in the past two decades in West and Central Africa, particularly their relations to international livestock policies
3. To outline feasible goals of public and private veterinary services based on the current laws and financial constraints
4. To depict venues and opportunities of services given the current financial and law contexts

The methods we have used are largely empirical (based on our professional experiences) of veterinarians working in West and Central Africa and not systematically (in terms of systematic review or surveys) to describe performance of veterinary services before and after implementation of international laws and programmes.

Results

Privatisation of Veterinary Services in West and Central Africa: The Unfinished Process

Understaffed and Underfunded

Veterinary field (and laboratory) services are too often chronically understaffed. The hundreds of millions of livestock in a variety of livestock production systems are under the main responsibility of few registered veterinarians in West and Central Africa with all the tasks described above. In Central Africa sub-region, the largest number of registered veterinarians in the public and private sectors was in Cameroon with around 1000 registered veterinarians; meanwhile there are only 250 in Chad,

220 in Niger and about 40 in Gabon. In West Africa sub-region, there were around 250 in Senegal and 198 in Côte d'Ivoire (http://www.rr-africa.oie.int/). Veterinary services largely rely on veterinary technicians and on trained CAHW to cover the minimal requirements of operational veterinary services. In Chad for example, the livestock sector counts for more than 50% of agricultural gross domestic product (GDP), yet the sector receives less than 1% of the governmental budget.

A study undertaken just after the Rift Valley fever outbreak 2006/2007 in Kenya showed that the veterinary sector is understaffed to respond adequately to such an epidemic. The public health sector could deploy five times the staff than the veterinary sector, although the latter had more tasks during the outbreak situation. In addition, the infrastructure of veterinarians to reach pastoral zones was insufficient and central veterinary capacity for diagnosis of RVF was neglected during outbreaks and at the beginning of the outbreak hardly operational to manage the high number of blood samples (Schelling and Kimani 2007).

Next to the PVS tool for planning and analyses, and training courses such as field epidemiology courses (spearheaded by the US Centre for Disease Control and Prevention [CDC]) and laboratory capacity courses organised by the Food and Agriculture Organization (FAO), there are few analytical studies on response capacity of veterinary services in face of an event. This is in contrast to assessments on a more regular basis in the human health sector.

In the past two decades, there was first a declining interest from donors and governments to invest in livestock sector, despite that livestock was at the same time depicted as a route out of poverty (Randolph et al. 2007). More recently, more investments in agriculture, in general, and in livestock, more specifically and particularly in pastoral systems within (rural) development programmes, are seen (ILRI 2010). Still, there is a widening technological gap and underinvestment in research targeting the problems of Sahelian communities in terms of livestock keeping (Ly et al. 2010). Efforts to modernise livestock production have focused mainly on the performance of the animals but have neglected rangeland improvement and management—and they largely failed because they did not involve herders themselves (Leonard 2004).

By encouraging privatisation of veterinary services (de Haan 2004) in the 1990s, structural adjustment in the livestock sector was a leading factor in reshaping the supply of veterinary services. Prior to privatisation, governments have been overstretched in servicing major pastoral areas. Financial rationalisation led to disengagement and disinvestment in public veterinary service delivery systems. Private sector development restructured the economy of the livestock subsector, with new schemes to promote the opening of extension of private veterinary practice while maintaining the community animal health workers. Privatisation of veterinary services was initiated in many parts of Africa and Asia as part of a broader effort to improve animal health delivery in the face of falling governmental expenditure and poor public sector performance (Leonard 2000).

A new branch of private veterinary professionals became present in the livestock sector and more and more involved in public contracting of mandatory immunisation (Ly et al. 2010). Numerous incentive schemes were designed to stimulate the privatisation process. Essentially, subcontracted veterinarians were effective in the implementation of vaccination campaigns given that the government is committed to subsidise work in more remote zones (de Haan 2004). However, due to high transaction costs in rural areas and fewer subsidies by governments, nowadays private veterinary services rarely are viable in remote and sparsely populated areas of West and Central Africa.

International Efforts to Eradicate and Eliminate Major Livestock Diseases and Lessons Learnt on Sustainability

Following outbreaks of rinderpest that have ravaged most of the African cattle herds in waves during the early 1980s, African States were committed to deploy together with development partners enormous efforts to eradicate this deadly cattle disease (DSV 2005). More than three decades of immense resources were invested to carry out annual mass vaccination campaigns for cattle, followed by serological surveys to evaluate the immunity levels of vaccinated herds and of the status of potential carriage in wildlife until its official declaration of worldwide eradication in 2011 (DSV 2005). The OIE recognised the following disease-free stages for Chad (1) free from rinderpest disease in the western part of the country in May 2004, (2) freedom from disease for the whole national territory in May 2006 and finally (3) freedom from rinderpest infection (definitive status), in May 2010 (DSV 2005). The European Community, as the largest donor, together with others such as FAO with important technical and financial inputs, financed largely the implementation of the Pan-African Rinderpest Campaign (PARC) through the Inter-Bureau of Animal Resources AU-IBAR for a period of 10 years. One should not, however, forget the hardly documented important investments in terms of personnel, resources and knowledge of the African countries.

To reach this "major success story of veterinary medicine", among other many measures, "*cordons sanitaires*" were established to separate East and West African zones of transmission. This has fostered the exchange of information between countries. Among the results of these shared efforts between national and international funding bodies were for the Chadian veterinary services are, at the subregional level, periodic cross-border meetings between the veterinary authorities of the Central African Republic (CAR), Sudan and Chad were annually organised (with funding from PARC) to discuss mutual interests in animal health. These actions enabled the veterinary authorities of the subregion to set up in 1998 a protection zone to establish the health corridor "cordon sanitaire" located in eastern Chad, bordering Sudan and CAR (DSV 2005). After the cross-border agreement, the Chadian authorities could restrict the obligation of vaccination in this protected zone and move to mass sero-surveillance, which led to the building up of good serology services. However, since the governmental veterinary services only focused on mass serological surveys in most regions during several years, livestock owners started to perceive the veterinary technicians as just arriving to bleed their animals without providing any information and without further services—and they never received a return of information. This triggered among owners of large livestock herds a doubt about the usefulness of veterinarians in general.

As a logical follow-up to the PARC, the Pan African Programme for the Control of Epizootics (PACE), also with important funding of the European Commission, was set up to ensure the control of epizootics with national epidemiological surveillance systems. PACE covered 32 countries in sub-Saharan Africa. One goal put forward was the eradication of Contagious bovine pleuropneumonia (CBPP), which is yet unaccomplished because, among many reasons, a less effective vaccine is available and a cold chain is needed as compared to rinderpest vaccination. Likely also a bit lesser long-term international commitment was available for continued finances of countries (see below). In Chad, the livestock surveillance system, the REPIMAT (*Réseau d'épidémiosurveillance des maladies animales au Tchad*), was set up (AU-IBAR 2010). An effective, sensitive and sustainable animal disease epidemiological surveillance network is the basic element for the management of the animal health in a country and financing of surveillance networks in a sustainable way would be a catalyst for the effectiveness, credibility and conformity to international standards of the national veterinary services (AU-IBAR 2010).

A next large eradication programme to mention is the Strategy for Control and Eradication of PPR (GCES-PPR) developed and piloted by FAO and the OIE Permanent Secretariat on global level, by the AU-IBAR at the regional level and by the Regional Economic Communities (RECs) at subregional level to guide national veterinary services in gradually reducing the prevalence of PPR up to its planned global eradication by 2030. Chad, like other countries in the region, has firmly subscribed to this dynamic—given the promoted worldwide and national importance extrapolated from the rinderpest eradication—and has already adopted its National Strategic Plan for PPR (PNS-PPR) that was validated in August 2017 (Félix 2016).

It should be noted that at the signing of the financing agreement between the European Commission (donor) and the African Union (beneficiary) for the implementation of PACE, it was recommended to the states that the operating costs (from 25% to 50%, and 75%) of African networks should be progressively included in national budgets and that at the end of the programme should be fully covered by national funding (AU-IBAR 2010). Unfortunately, since the closure of PACE at the end of 2006, only very few countries in sub-Saharan Africa draw nowadays from their commitment of increasing funding for surveillance. Most countries struggle to maintain active and passive epidemiological surveillance activities due to lack of means. This critical aspect has been reported by various PVS missions led by the OIE as one of the "weak points" of the national veterinary services in sub-Saharan Africa.

The lessons learned from the eradication of rinderpest and of the sustainability of PACE initiatives should inform policymakers and veterinary authorities of African States to plan well ahead for better control of emerging and re-emerging animal diseases next to the prevailing endemic livestock diseases. Since foot and mouth disease (FMD), *Pest des petits ruminants* (PPR), contagious bovine pleuropneumonia (CBPP), Newcastle disease (ND), Rift valley fever (RVF) and African swine fever (ASF) (DGSV 2017), to name a few, have a substantial economic impact, it seems cost beneficiary to continuously invest in service provision, which makes services also more sustainable once the big programmes stop. Large international programmes should not only recommend to add national budgets but rather insist in terms of promised outputs on documented increasing matching funds because these will foremost strengthen national veterinary services.

To address these challenges, the following policies must be undertaken by states: (1) budgeting and resource mobilisation for surveillance activities (active and passive) and health prophylaxis; (2) prioritisation of actions to be implemented in the short, medium and long terms; (3) synergies of technical actions among countries with common borders; and (4) capacity building of diagnostic laboratories and training of technicians at all levels (DGSV 2017). Making use of synergies between human and veterinary health services in remote rural zones can strengthen both health systems in terms of delivery and surveillance (Schelling et al. 2005).

Unfortunately, in the more recent events of emergency responses for refugee livestock owners returning to Chad, FAO imported to Chad vaccines mainly to ensure vaccination against CBPP in the bordering regions. Also other vaccines were imported—vaccines that are locally produced by the public veterinary services such as anthrax and blackleg vaccines IRED has invested much in quality control of its vaccine production—and production is a central source of income for veterinary services. Import of vaccines cut veterinary service income. In addition, the fact that the imported vaccines were served free of charge to refugee livestock owners but also to local livestock owners has undermined the efforts of veterinary services, after years of struggle, to maintain generally accepted and stable vaccination fees under what is so-called the cost recovery of vaccines (see below).

Case Studies of Public and Private Veterinary Services in Côte d'Ivoire and Chad

Côte d'Ivoire

As part of the implementation of the structural adjustment policy, Côte d'Ivoire has opted for the privatisation of veterinary services. As a result, the law establishing the National Veterinary Order of Côte d'Ivoire and the Veterinary Code of Ethics was adapted by the Ivorian Parliament in 1988. Privatisation was therefore actually implemented with the opening of the first private veterinary clinics in 1995 with the support of the European Union through the PARC project.

The state's withdrawal from the production and marketing of vaccines as well as the refocusing of its mission in the regulatory and control functions has led to the emergence of private veterinary services who became involved in several sectors such as veterinary pharmacies, urban canine clinics, rural veterinary practitioners and prophylaxis in the format of a health mandate.

The development of a private veterinary drug sector for the distribution of veterinary drugs was authorised. Competition in drug selling was encouraged, but their rapid development in a tight market required regulations for the importation and distribution of veterinary medicinal products. In 1996, the law on the veterinary drugs was adopted by the national assembly.

Out of pocket payment for all livestock vaccinations, including those that were compulsory, has been introduced. However, it has been applied gradually and on the basis of cost sharing between the state and the farmers. The introduction of the payment by the farmers also made it possible to carry out the campaigns by governmental staff of the Ministry of Livestock. Since payment was based on numbers of vaccinated animals, the livestock owners became the primary interest group in good-quality services. The state's decreased contribution led to a decree on the remuneration of private veterinarians by the state—in the framework of the implementation of veterinary mandates. The mandatories were then forced to buy vaccines from the state and to pay the benefits to the breeders. This became one of the major constraints for private veterinarians to settle in clinics. The number of established veterinary clinics first remained stable but then slightly decreased from 16 to less than 13 between 1996 and 2010.

The distribution of veterinary drugs became the most important activity of private veterinarians and accounted for nearly 60% of turnover. However, this important source of income was challenged by illegal markets in many parts of the country. Private veterinarians were competing with untrained wholesalers who sold directly to breeders. These transgressions persisted because of the lack of the enforcement existing regulatory frameworks and the lack of an official control of the distribution of veterinary drugs.

Between 2011 and 2017, the number of private veterinary clinics doubled from less than 15 veterinarians in 2010 to 30 veterinarians in 2017. This can be explained by the growth in the poultry and swine industry, but also by the issuing of a new veterinary law (*Ordre des vétérinaires*). Veterinarians became engaged in the fight against the illegal distribution of veterinary drugs. Indeed, more than a dozen complaints were taken to courts and illegal deposits were closed. Also, the Economic Community of West African States (ECOWAS) funded during three successive years campaigns against the illegal distribution of veterinary drugs in West Africa.

The privatisation of the veterinary profession in Côte d'Ivoire can be subdivided into three main phases:

- From 1995 to 2000: the fostering of installation of private veterinarians with important financial support (30,000,000 FCFA, equalling about 60,000 USD at that time) for each veterinarian installed).
- From 2000 to 2010: after a first promising increase, a sharp decline of private veterinary clinics took place.
- From 2011 to 2017: the revival of the instalment of private veterinarians without financial support from the state but with improved governance, particularly in the enforcement of the regulations on illegal sales of veterinary drugs.

In conclusion, the enforcement of national laws was a key factor to establish viable veterinary services in Côte d'Ivoire. Law enforcement led to more sustainable private veterinary services than financial incentives in Côte d'Ivoire. The state must also create an environment conducive to the sustainable evolution of the private veterinary facilities: (1) to enforce health measures; (2) to strengthen the National Order of veterinarians, i.e. extended scope of the sanitary mandate to the control of foodstuffs; and (3) to involve livestock keepers actively in epidemiological surveillance, by promoting and strengthening health defense groups.

Privatisation of Veterinary Services in Chad

In Chad, the process of veterinary privatisation started in 1987 and has been funded by international partners such as the European Commission, the World Bank and others (Arditi and Lainé 1999). In 1988 with the intention to reform the veterinary sector, the Chadian Ministry of Plan and Cooperation signed a letter of intent with the World Bank. These reforms were mainly on the stabilisation of the veterinary workforce by non-replacement of retiring public sector employees and installation of private clinics, the liberalisation of trade for veterinary products; and redefining the roles and functions of the public service sector. In addition, veterinary services were to be offered based on cost recovery. The latter required the training of community animal health workers to deliver basic veterinary services.

To implement all these reforms, legal frameworks were adopted by the signature of several decisions by the Chadian authorities from different sectors such as: (1) Ordinance No. 005/PR/91 establishing the veterinary profession; (2) Decree No. 384/PR/PM/ME/91 regulating the veterinary profession; (3) Decree No. 417/PR/PM/ME/91 regulating the veterinary pharmacy; (4) Decree No. 21/PR/PM/ME/91 laying out the conditions for the assignment of sanitary mandates; (5) Law No. 24/PR/2000 on pharmacies; and (6) Decree of 4 August 2005 relating to veterinary pharmacy in the Republic of Chad.

In 1992, the first eight livestock technicians were endorsed by the National Livestock Project followed in 1993 by granting of an exclusive mandate to eight private veterinarians to carry out the vaccination campaign against rinderpest. "Exclusive" meant that full responsibilities of veterinary zones were assigned to private veterinarians and, in view of excluding competition, the government would not intervene. The following years, the number of veterinarians settled in private clinics increased and reached 28 veterinarians by 2003 who have benefited from financial support as credit up to USD 30,000 (FCFA 15,000,000). This gradual increase in number of veterinarians raised various concerns, which led to the creation of the Union of Private Veterinarians in Chad (UVPT) in 1994 to better help organising the activities of private veterinary services, specifically the prophylactic mandate with vaccinations. Next to the valuable vaccination mandate, the different areas of intervention of private veterinarians were in clinical activities, sales of veterinary products, treatment of animals, training and supervision of breeders and diseases surveillance.

At early implementation, private veterinarians were very interested to fulfil additional tasks of health promotion and increase demand of livestock owners for clinical services in zones where, due to the poor governmental infrastructure and a mandating system, they practically alone had access (Nahar 2000). A contribution from the public health sector to their costs of vaccination delivery first rendered private veterinary services profit-making. With two private veterinarians working in rural districts of Chad, costs of all expenses for cattle vaccination campaigns for the years 1997–2000 were recorded in detail and verified in successive sessions, if possible with accounts. Vaccination in these years was the main activity of the private veterinarians. The average numbers of vaccinated cattle per year were recorded from detailed vaccination records. In parallel, interviews with pastoralists revealed that it was difficult for livestock owners to cope with the situation that vaccine doses first cost 25 FCFA, then raised to 50, 75 and finally to 100 FCFA. To pay 100 FCFA was finally set for the next years. Livestock owners have praised that prices were stable and they could plan accordingly and have sufficient cash before the arrival of the veterinarians for vaccination of livestock. Mirroring increasing costs to livestock owners, subsidies were reduced.

Without subsidies and a payment of 100 FCFA per dose—the marginal income of private veterinarians per vaccinated livestock was very low. The overall costs of the private veterinarians were composed of variable costs (71.4%) and fixed costs (28.6%) (Table 5.1) (own unpublished data). According to the vaccinated livestock number and the number of different vaccines (anthrax, pasteurellosis and blackleg together during one campaign and CBPP during another) per animal, this resulted in costs of 160 FCFA to vaccinate one animal and in 96 FCFA per vaccine. The income for animals with only one vaccine dose was thus only 4 FCFA. This was too tight to save for new investments (e.g. for replacement of a car). Particularly mandated vaccination campaigns were not beneficial during vaccination campaigns CBPP when only one vaccine was available. Variable and fixed costs of transportation represented 20% of all costs. Indeed, reaching the remote herds and livestock keeping families was perceived by private veterinarians the most difficult obstacle. The reimbursement for collecting samples within the epidemiological surveillance programme did not cover the costs and therefore private veterinarians lost interest to be involved in the surveillance network.

Table 5.1 Cost composition of two private veterinarians in Chad who tried to establish mandated rural veterinary clinics. These costs were contrasted to their incomes that were alone from mandated vaccination campaigns in the years 1997–2000

	Private veterinarians	
Variable (recurrent) costs	FCFA	%
Personnel/administration	11,603,750	27.3
Vaccines	14,092,105	33.1
Transportation	3,205,000	7.5
Cold chain	780,000	1.8
Supplies	711,525	1.7
Total variable costs	30,392,380	71.4
Fixed (non-recurrent) costs		
Vaccines	6,039,474	14.2
Transportation	5,215,333	12.3
Cold chain	150,000	0.4
Buildings	750,000	1.8
Total fixed costs	12,154,807	28.6

After an abrupt implementation of privatisation, private veterinarians lack nowadays a backing-up by a clearer legal framework as well as more political and financial commitment of the government. Clinical services of veterinarians remain weak (breeders are willing to pay for drugs and vaccines, but not for proficiency services) and subsidised tasks such as compulsory vaccination, surveillance, meat inspection (to note is that at least 80% of food is traded in informal markets, thus per definition without inspection (Roesel and Grace 2015)), nor training fully occupy private veterinarians. The cessation of programmes such as PACE guaranteeing subsidies (as for vaccination against CBPP) strongly challenged the profitability of private veterinarians in less populated zones.

To date, only one veterinarian still works in a private clinic in Chad—the others went to the government or internationally and periodically funded programmes—showing a failure of the implementation of the sanitary mandate by the government in the last two decades. Among the main reasons of this failure were the non-enforcement of laws and regulations. Unlike in Côte d'Ivoire, veterinary drug sell controls were never established and thus there was a rapid proliferation of street vendors for veterinary drugs. Another reason of failure to set up private practices was the stop of financial support rendering services in remote rural zones no longer viable for a private business. Also, public entities did not easily accept the new situation where the private veterinarians would take over their duties—they were still present, but no longer had a mandate in the huge zones assigned to mandated private veterinarians. Further was the fact that livestock producers were not sufficiently involved in the privatisation process, not to mention the impact of the changing vaccination fees to be paid based on cost recovery. Finally, Chad experienced great numbers of refugee breeders, who have benefited from the assistance of the humanitarian sector and from the hosting country. The refugees have changed the mobility patterns of pastoralists which also negatively impacted on the activities of private veterinarians throughout the national territory.

Given that the privatisation policies have forced the state to retract from mandated zones and the private veterinarians were de facto no longer present, this left for many years huge zones without veterinary services. The government—to protect its national good—nowadays tries to fill the gaps again. However, in the last 20 years, governmental investments in veterinary services such as training of veterinary technicians were minimal, and there is a lack nowadays in veterinary health personnel, infrastructure and funding schemes.

Opportunities

Conventional public health and veterinary strategies often need to be complemented and in some cases replaced by alternative strategies to more effectively reach remote zones, for example, by public contracting of private agents, competent non-governmental organisations (NGOs) and civil society organisations to deliver selected services (Ahuja 2004). Community animal health programmes sought and seek to complement and extend conventional delivery systems to (partly or mostly) fill the gaps in chronically underserved rural areas. These programmes give (short-term) training in basic veterinary techniques to community members. Community animal health workers (CAHWs, sometimes also referred to as paraveterinarians) can have a substantial impact on livestock morbidity and mortality through the treatment or prevention of a limited range of animal health problems (Catley et al. 2002), particularly in remote zones of West and Central Africa and pastoral zones in the Sahel. Essentially, the main success business model is built on the fact that they are within the communities and thus more easily accessible. As shown below—this can be an advantage but also a disadvantage if no budgetary follow-up and quality control mechanisms are established as additional component of such an initial training programme. Their legal support is lacking in several countries (Catley et al. 2005), whereas in some African countries CAHWs have received extra training (sometimes up to certificate level) for 3 months up to 1 year. These cadres are referred to as Animal Health Assistants or Auxiliaries (AHAs), Nomadic Animal Health Auxiliaries (NAHAs) or Veterinary Supervisors (Simpkin 2005).

CAHWs can also be at the forefront of revised veterinary and One Health surveillance (Abakar et al. 2016). There are many opportunities and advantages of CAHWs in pastoral zones mainly because they are capable of moving with livestock herds and travelling to fixed-point outlets for veterinary drugs. Although some pastoral communities have been exposed to free or heavily subsidised veterinary services, they usually acknowledge the rationale for payment of veterinary services (Catley 1999).

In West and Central Africa, thousands of CAHWs have been trained in the past two decades. A key factor for success of continued demand and effectiveness of CAHWs is that community members to be trained are preferably selected by their community for broader acceptance. Other factors of successful CAHW programmes include community involvement in the design and implementation and involvement of the private and governmental sectors to supply and supervise CAHWs (Catley et al. 2005). Supervision is crucial to ensure quality (access and quality of services is commonly a major concern of livestock keepers) but also to follow up on book-keeping. If they do not sell their services also to relatives, income is missing to buy new drugs that guarantees sustained good-quality services. Another important aspect is quality assurance so that customers, who have less knowledge and information than the provider of animal health services, are willing to pay for services that potentially improve their livestock's health (Leonard 2000). Pastoralists in Chad sought that CAHWs can show renewed certificates as a way to distinguish between those with a follow-up in the veterinary system (quality assurance) and those who operate on their own. CAHWs often work also (sometimes on part-time arrangements) in partnership with the private professional segment. The professionals ensure good practice and the arrangement is an alternative model for the development of private sector delivery systems.

In conclusion, any programme training CAHWs must have a plan on how to sustainably follow up on their trainees. Despite that they can fill in gaps of veterinary service delivery and surveillance of TADs with documented good mid-term outcomes, without continued investments for several years after initial training, programmes should be careful to train new CAHWs.

Public–Private Partnerships and the Role of Producer Organisations in Veterinary Service Delivery

As mentioned earlier, almost all West and Central African countries were committed to privatisation of veterinary services in the beginning of the 1990s. However, many difficulties occurred during this process especially with regard to cost recovery of vaccination. In the case of Chad for instance, one can mention among others the arrival of high numbers of refugee breeders from Sudan to eastern Chad at the beginning of the Darfur security crisis in 2003 and then from the Central African Republic (CAR) in 2014 in the southern part of the country. Free of charge vaccination of livestock of the refugees from CAR and of locals started in May 2015 in the Southern regions—implemented by the Emergency Support Project for the improvement of animal health and with support from FAO. This operation made it difficult to recover costs in the provision of veterinary services (private or governmental). In addition, local livestock owners did not necessarily want free vaccines after they have since many years accepted that paying the stable 100 FCFA is also a sort of quality criteria. Importation of vaccines from abroad reduced the income of the Chadian veterinary services who locally produce some vaccines.

Strong *producer organisations* can offer an efficient tool for delivering animal health services, although some attention needs to be paid to the fact that mixing of marketing and service functions may lead to an undesirable outcome of sharing responsibilities between the public and private sectors. In some countries, farmer cooperative structures are still yet in transition from an era of state control to autonomous management. However, where established with backup of good agricultural organisations, they run well for the benefits of families.

Private industry makes available their commodity chains or market networks to producer organisations for provision of services (including financial/micro-credit services) and to establish a pastoralist dialogue platform. This could be supported by programmes such as the World Overview of Pastoral Approaches and Technology (WOPAT) (Bonfoh et al. 2008) and also by other stakeholders seeking to reach the remote agricultural zones such as public health actors expanding the Universal Health Care coverage, which is long sought to be effectively implemented in several countries of West and Central Africa. To include agricultural organisations that are better represented in rural zones when compared to the few professionals—and to show synergies on how to better reach together remote populations with services—is currently being debated.

Intersectoral Cooperation for Service Delivery and Surveillance

OIE, as the "guardian of the Codes", took some time to consider whether embarking on a One Health path was realistic and appropriate (as also did WHO). Once OIE became committed to the concept (2008 onwards), the amplification of One Health approaches across the global network of the veterinary services has been impressive, with the veterinary profession widely promoting One Health to address issues such as food safety, food security, antimicrobial resistance, climate change and the human–animal bond. One Health is still driving strong today—also because—after the avian influenza crisis and the international organisations strongly promoting One Health—it is nowadays no longer perceived as a top-down approach thanks to various well-documented bottom-up projects (such as that joint vaccination services for mobile pastoralists increase the efficiency of both sectors to reach these communities) and evidence on its added value (Zinsstag et al. 2015). The veterinary profession within One Health remains a strong advocate of multidisciplinary approaches to solving the complex challenges of global health and to be in a position to provide decisive leadership (Okello et al. 2015). However, to make One Health more operational for the good of veterinary services, more evidence that integrated human and animal surveillance is more efficient to detect early disease events and that costs can be saved is needed. The potential of the added value of One Health seems to be huge, but we need next to theoretical thinking more evaluated implementation efforts.

The Way Forward

There is a need for more task-sharing between the public and private sectors which, in return, would enhance effectiveness of service delivery to remote zones with shared responsibility. Given the fact that privatisation processes still faces many challenges in West and Central African countries, governments need to invest more in enforcing laws that protect both public and private goods as part of their duties to ensure good quality of veterinary services and augment animal production. Good quality of veterinary services could not be guaranteed without qualified veterinarians in the field supervising the vaccination activities and providing other animal health-related needs of livestock owners at cost recovery and still profitable for private veterinarians.

In countries where the private veterinary sector is not well developed, like Chad for example, public services also carry out animal care activities (consultations, dispensing drugs, surgery, counselling). These activities are not part of a market good and are not included in the OIE definition of public veterinary services. The importance of the concept of "animal welfare", which is often limited to the welfare of animals during transport and slaughter, should be noted. Basic measures of animal welfare according to OIE standards must be taken into account in legislation. However, the public services must withdraw (to avoid any unfair competition) from these activities as soon as a private person mandates for them also if they cannot guarantee their implementation. In view of the foregoing, one of the essential levers for ensuring the sustainability of the veterinary profession, particularly the private clinicians, is the improvement of the politico-legal environment. It is important to establish a favourable legal and regulatory environment. A more systemic review of issued, implemented and missing, but possibly influential, veterinary laws is needed at national and international levels.

Livestock owners must be included in the process of controlling and eliminating animal diseases such as PPR. Their inclusion would allow them to define their needs of clinical veterinary services next to drugs and vaccinations. Making use of synergies between human and veterinary health services in remote rural zones can strengthen both health systems in terms of delivery and surveillance. All larger livestock programmes should be made responsible that they plan for and show how they strengthen national veterinary services next to safeguarding the livestock health. This should be defined in the way also how governments are nudged themselves into more investments in capacity and institution building at all levels and in setting up new public–private partnerships.

References

Abakar MF, et al. Trends in health surveillance and joint service delivery for pastoralists in West and Central Africa. The future of pastoralism (J. Zinsstag, E. Schelling & B. Bonfoh, eds). Rev Sci Tech Off Int Epiz. 2016;35(2):683–91.

Ahuja V. The economic rationale of public and private sector roles in the provision of animal health services. Rev Sci Tech. 2004;23(1):33–45.

Arditi C, Lainé F. Evaluation du processus de privatisation des services de santé animale au Tchad—Rapport de mission. Tchad: N'Djaména; 1999.

AU-IBAR. Rapport final du Programme Panafricaine de Contrôle des Epizooties, PACE, BIRA-UA, Décembre 2010. 2010.

Bank TW. People, pathogens and our planet: the economics of one health. Washington, DC: World Bank; 2012.

Bonfoh, B, et al. Sustainable natural resources management in semi-arid and high land low land contexts: hindering and supporting framework conditions. 2008.

Bonfoh B, et al. Human health hazards associated with livestock production. In: Livestock in a changing landscape, vol. 1; 2010. p. 197–220.

Bourzat, D, Veronique B, Vincent B. Rapport de mission PVS/OIE au Tchad, Novembre 2013. 2013.

Catley A. Methods on the move: a review of veterinary uses of participatory approaches and methods focussing on experiences in dryland Africa. London: International Institute for Environment and Development; 1999. p. 1–97.

Catley A, Blakeway S, Leyland T. Community-based animal healthcare. London: ITDG Publishing; 2002. p. 1–359.

Catley A, Leyland T, Bishop S. Policies, practice and participation in complex emergencies: the case of livestock interventions in South Sudan. 2005. Alan Shawn Feinstein International Famine Centre, Tufts University.

de Haan C. Introduction: the provision of animal health services in a changing world. Rev Sci Tech Off Int Epiz. 2004;23(1):15–9.

De Haan C, et al. Livestock development, implications for rural poverty, the environment and global food security. In: Directions in development. Washington, DC: The World Bank; 2001.
Dean AS, et al. Potential risk of regional disease spread in West Africa through cross-border cattle trade. PLoS One. 2013;8(10):e75570.
DGSV, T. Plan National Stratégique de contrôle et d'éradication de la PPR au Tchad, Direction Générale des Services Vétérinaires. 2017.
DSV, T. Dossier du Tchad adressé à l'OIE pour l'obtention du statut de pays indemne de peste bovine, Direction des Services Vétérinaires, Août 2005. 2005.
FAO. The state of food and agriculture: livestock in the balance. Rome: FAO; 2009.
FAO. FAOSTAT stocks West Africa. 2016.
Félix N. Stratégie mondiale de contrôle et d'éradication de la PPR. In: Secrétariat permanent FAO/OIE pour le GCES-PPR: Exposé du Dr N. Félix à l'atelier de Douala, Juillet 2016. Cameroun: Douala; 2016.
ILRI. Livestock—a pathway out of poverty, ILRI 's strategy to 2010. 2010.
Johnston C. Lessons from medical ethics. In: Wathes CM, et al., editors. Veterinary & animal ethics. Chichester: Wiley-Blackwell; 2013.
Krätli S. Discontinuity in pastoral development: time to update the method. Rev Sci Tech Off Int Epiz. 2016;35(2):485–97.
Leonard DK. Africa's changing markets for health and veterinary services—the new institutional issues. London: MacMillan; 2000.
Leonard DK. Tools from the new institutional economics for reforming the delivery of veterinary services. Rev Sci Tech. 2004;23(1):47–57.
Ly C, Fall A, Okike I. West Africa—the livestock sector in need of regional strategies. In: Gerber P, Mooney HA, Dijkmann J, editors. Livestock in a changing landscape: experiences and regional perspectives. Washington, DC: Island Press; 2010. p. 27–54.
Nahar MT. La santé des pasteurs nomades: une nécessaire collaboration entre vétérinaires privés et services de santé. Sem Ther. 2000;8:108–12.
OIE. Terrestrial Animal Health Code, Chapter 3.2. 2006.
OIE. Code sanitaire pour les animaux terrestres, Organisation Mondiale de la santé Animale (OIE). 25ème édition ed. 2016.
Okello A, Vandersmissen A, Welburn SC. One health into action: integrating global health governance with national priorities in a globalized world. In: Zinsstag J, et al., editors. One health: the theory and practice of integrated health approaches. Oxfordshire: CABI; 2015. p. 283–303.
Perry B, Randolph TF, McDermott J. In: Ilri NK, editor. Investing in animal health research to alleviate poverty; 2002.
Pradere JP. Poor livestock producers, the environment and the paradoxes of development policies. Rev Sci Tech Off Int Epiz. 2017.
Randolph TF, et al. Invited review: role of livestock in human nutrition and health for poverty reduction in developing countries. J Anim Sci. 2007;85(11):2788–800.
Riviere-Cinnamond, A Animal health policy and practice: scaling-up community-based animal health systems, lessons from human health. 2005. Available from http://www.fao.org/ag/againfo/programmes/en/pplpi/docarc/wp22.pdf
Roesel K, Grace D. Food saftety and informal markets—animal products in Sub-Sahran Africa. London: Routledge; 2015.
Schelling E, Kimani T. Human and animal health response capacity and costs: a rapid appraisal of the 2007 rift valley fever outbreak in Kenya. Nairobi: International Livestock Research Institute (ILRI); 2007.
Schelling E, et al. Synergy between public health and veterinary services to deliver human and animal health interventions in rural low income settings. BMJ. 2005:1264–7.
Simpkin SP. Livestock study in the greater horn of Africa. Nairobi Delegation: International Committee of the Red Cross (ICRC); 2005. p. 1–227.

Stephen C, Waltner-Toews D. Non-governmental organizations. In: Zinsstag J, et al., editors. One health: the theory and practice of integrated health approaches. Oxfordshire, London: CABI; 2015. p. 385–96.
SWAC-OECD/ECOWAS. Livestock and regional market in the Sahel and West Africa: potentials and challenges. 2008.
The World Bank. People, pathogens and our planet. Volume 1: towards a one health approach for controlling zoonotic diseases. Washington DC: The World Bank. Agriculture and Rural Development Health, Nutrition and Population; 2010.
Zezza A, et al. Rural household access to assets and agrarian institutions: a cross country comparison. Rome: Agricultural and Development Economics Division, FAO; 2007.
Zinsstag J, et al. From "one medicine" to "one health" and systemic approaches to health and well-being. Prev Vet Med. 2011;101:148–56.
Zinsstag J, Schelling E, Waltner-Toews D, Whittaker M, Tanner M. One health: the added value of integrated health approaches. Oxfordshire: CABI; 2015.
Zinsstag J, et al. A vision for the future of pastoralism. Rev Sci Tech. 2016;35(2):693–9.

Chapter 6
TADs in the Dromedary

Bernard Faye

Abstract The transboundary diseases in camel are mainly linked to the regional camel meat market from Sahelian countries (from Mauritania to Somalia) to the Arabian peninsula and North Africa. Indeed, the camel flow in relationship with this market is based on live animals' export. Because the camel trade can be formal and informal with interconnections between both sectors and despite veterinary controls in the main exporting ports, some diseases such as Rift Valley fever (RVF), PPR-like disease, and MERS-coronavirus can spread from exporting countries to importing ones. However, the epidemiological status of these different diseases is quite variable and the transmission to humans in case of zoonosis (RVF and MERS-Cov) is not necessarily due to transboundary camel trade despite the impact of outbreak on the regional camel market. Globally, dromedary camel is less affected than other ruminants by infectious diseases under transboundary surveillance. But, because camel breeding is concentrated in countries where the disease surveillance systems often lack means, where the frontiers in desert areas are often "porous," and where the herd mobility is difficult to assess, the risk of transboundary diseases' transmission through borders is not negligible. Nowadays, the challenge of TADs control is limited to Rift Valley fever, but special attention must be paid to emerging diseases, including the recent discovery of prion disease in Algeria.

Keywords Camel diseases · Meat market · Live camels' export · Rift Valley fever · MERS-coronavirus

The transboundary flow of living camel in Sahelian countries is linked to two main features: (1) the herd moving for accessing resources under the management of the camel owners located more or less close to one frontier and (2) the market export of living animals for slaughtering or other purposes such as breeding or racing. The

B. Faye (✉)
Centre de coopération internationale en recherche agronomique pour le développement, Montpellier, France
e-mail: bernard.faye@cirad.fr

M. Kardjadj et al. (eds.), *Transboundary Animal Diseases in Sahelian Africa and Connected Regions*, https://doi.org/10.1007/978-3-030-25385-1_6

Photo 6.1 Cohabitation of camels with other species around water point, region of Garissa, Kenya (Photo: B. Faye)

relative importance of these camel flows is not deeply studied, at least from a quantitative point of view.

The Camel Flows in Sahelian Countries

The Camel Herd Mobility

The camel being linked to ecosystems marked by the poverty of their resources (water and pastures), mobility is one of the main strategies for securing farming system (Faye 2013). This mobility can be pendular (transhumance) or random (nomadism) based on short or long distance. For example, the mobility of Afar tribes across the border between Ethiopia and Djibouti is based on short distance (Faye 1994). In reverse, the movement of camel herds in Chad can be of a large amplitude from North Chad up to beyond RCA (Central African Republic) border (Aubague et al. 2011). Thus, the transboundary camel herd moving is not necessary linked to the distance of the cameleers' camp to the border. However, the potential risk for disease spread increases with the covered distance, the cohabitation between animals of different origins (and different species) occurring mainly around water point (Photo 6.1) throughout the transhumance route.

The Camel Market for Export

Contrary to camel milk, essentially limited to local or national market, camel meat is the object of transboundary trade based on living animals. The camel population in the world estimated to 35 million heads (FAOstat 2019) is probably underestimated (Faye and Bonnet 2012) due to the difficulties of census for a mobile species and the absence of official vaccination. So, it is difficult to assess the exact percentage of camels involved in such transboundary trade. The camel flow in relation to the living camel export is consequently also underestimated. In Sahelian region, the camel flow for export is mainly from South to North. Camels in Mauritania, Mali, Niger, Chad, Sudan, and Ethiopia are the main exporting countries of living camels to North Africa (Morocco but overall Libya, Egypt, and Algeria to a lesser extent) and to the Arabian peninsula (Faye et al. 2013). For example, the slaughtering rate which is around 7% at world level reaches 95% in Egypt, 75% in Saudi Arabia, 50% in Emirates, and 31% in Morocco and Libya which shows the importance of imports for the national camel meat markets. In reverse, less than 5% of the camels are slaughtered in Sahelian countries for their local demand. Probably, this percentage is also underestimated due to a large part of slaughtering occurring in the remote pastoral area, not included in the official data (Aklilu 2002). Along the Red Sea, some ports (Port-Sudan, Djibouti, Hargeisa, Mogadishu) are hubs for camel export by boat to Saudi Arabia, Yemen, Oman, Qatar, Bahrain, and Emirates. In Western and Central Africa, the export is achieved by land, sometimes using trucks but overall by foot as the "forty-days road" from El-Obeid (Sudan) to Aswan (Egypt) through the Nubian Desert. If the main flow is South-North (Africa) or South-South (Horn of Africa), recently new flows occurred from the same Sahelian countries to coastal countries like Senegal (coming from Mauritania) or Nigeria (coming from Niger and Chad). Indeed, the recent demand for camel meat for dietetic reason (camel meat is poor in cholesterol and has a high protein value) or even for supposed virtues (it is believed that camel meat is aphrodisiac and cures hypertension) is increasing significantly in these countries (Kurtu 2004; Kadim et al. 2008; Raiymbek et al. 2015).

Formal and Informal Camel Trade

Camel Trade in the Horn of Africa

In a recent analysis of the camel flow in the Horn of Africa, using different sources of data to assess the importance of the camel chain in the economy, a gap was observed between official data (41,000 camels should be exported every year on average from Ethiopia) and the true potential of export. Such a gap is because only a part of the camel flow is passing through official way (Belachew 2005; Alary and Faye 2016). In Somalia, for example, only 10% of the exported animals would use formal trade.

This low part could be explained partly by insecurity along the export routes, especially since the Somalo-Ethiopian war in 1977 and the civil war in Somalia since the 1990s. Indeed, Somalia was the most important camel exporting country all along the twentieth century, and the Horn of Africa represents more than 60% of the estimated camel world population. The pivotal role of Somalia in this regional camel trade started at the end of the nineteenth century; the Somali pastoral area (that covered the eastern part of Ethiopia, northern part of Kenya, and central and northern part of Somalia) established a well-functioning market chain to supply the Brittany garnison established at Aden in 1839. The international trade was extended to include the Persian areas and the boarders of the Indian continent (Djama 1999). The economic boom of the Arabian peninsula with the development of "oil era" stimulated the demand for camel meat and the transboundary camel trade in the region.

But, due to insecurity evocated above, the camel trade is now competing with other countries of the region such as Djibouti, Sudan, Kenya, and Eritrea and even Australia where a feral camel population is problematic for the environment. Globally, the camel farming systems in the Horn of Africa are not well organized to produce high-quality meat. The main consequence of this competitive trade was the variability in the origins of the camels and a more important brewing camel population in the importing countries.

Moreover, the traditional pastoral camel fattening in the Horn of Africa and central Africa is providing mature camels mainly for the trade with Egypt or Libya. In that case, the animals are slaughtered rapidly after their arrival, decreasing the epidemiological risks. In reverse, the demand in the Arabian peninsula is for young camels (1–2 years), fattened for 3–6 months in the importing countries, in special feed-lots where the epidemiological risks could be more important. In addition, a part of the exported camel stock is used for breeding and for racing which is a very popular activity in the Arabian peninsula. In Africa, in reverse, the camel feed-lots are not developed, except in Tunisia to provide the local market with young camels of 250 kg live weight (Khorchani et al. 2005).

Another consequence of the new insecure environment is the disorganization of official services, such as veterinary services, customs, and banking services especially in the export ports. This has consequently favored the official position of Djibouti and Port Sudan in the international market of live animals for the region and the development of various networks of smugglers who export animals via Yemen traders who then re-export animals to Saudi Arabia (Alary and Faye 2016). Except in the peaceful Somaliland, the bargaining power of the traditional traders' networks was weakened (Little et al. 1998).

Regarding the formal market of live camels, the main stakeholders are legal exporters who have their own collecting points where camels are fed, treated, and eventually vaccinated before being exported by rail or truck to the export port. At Djibouti, for example, infrastructure for the transit of the camel herds (Photo 6.2) was implemented including quarantine (usually for 21 days), paddocks, water, and feeding resources for export or re-export from neighboring countries, especially to

Photo 6.2 Camels for export at Nagad park- Djibouti (Photo: Abdallah Barkat)

Saudi Arabia and Emirates. Sanitary controls are certified by international agreement society (Faye 2003).

In the informal market (notably to Yemen), camels are shipped on Yemenite boats without the use of quarantine or veterinary services although controls could be achieved in the native countries (Ethiopia, Sudan, etc.). Despite the pressures from local governments to legalize and harmonize all the sanitary procedures, some stakeholders may use illegal practices along the chain either for the financial/customs procedure to facilitate the convoying of animals or for the veterinary procedures to avoid the legal procedure imposed by importers. These circumventing acts aim to maintain the international market of live animals in the region.

In the region, there is also the trade of live camels for merchandises' transport where the purchasers of animals can be different from the sellers of merchandises. In this system, camels are the main mechanism used to transport illegally the merchandises. The animals used cross the borders and come back to their native country. This smuggling activity is particularly important between Ethiopia, Somalia, and Djibouti.

However, the two chains (formal and informal) are not truly waterproof. A combination of formal and informal circuits can occur, for example, the combination of illegal crossing of the Ethiopia-Somalia-Djibouti border (despite a strengthening control, especially by the Ethiopian government) and legal re-export from Djibouti or Somaliland port to the Arabian peninsula.

Camel Trade in Central Africa

If few data are available in the Horn of Africa, the situation of camel trade in central Africa is less described. In Chad, in a recent report (Mankor and Koussou 2015), three main trade routes are described according to the destination: Libya, Egypt, and Nigeria.

The trade route to Libya is more important (85% of the export). The camels are gathered at Abeche in Chad (the main camel market for export in the country), coming from different regions of the country, and then conveyed to Sebha and Koufra in Libya. The export activity is seasonal and is linked to the transhumance of the cameleers. The peaks of export are August (at the beginning of transhumance) and October (at the return). Twenty-five percent of the export occurs during these 2 months. The moving of camels is ensured by truck or by foot. The duration of the travel is 7–10 days by truck and 50–60 days by foot across the desert. However, this circuit is destabilized since the political insecurity in Libya. From Abeche, a part of the camels is exported to Egypt via Sudan, mainly by foot. The export to Nigeria is achieved from N'Gueli station, close to N'Djamena, but is also disturbed since the Boko-Haram exactions.

Between 2007 and 2015, the official number of exported camels increased from 3000 to 45,000 heads, but this increase could be partly due to a better control of the trade. Moreover, the camel traders complain regularly about illegal taxes imposed by the custom and security forces. Due to this constraint, an important part of the camel export escapes from official control. Similar observations could be done in Niger where the insecurity due to the threat of AQMI attacks is more pronounced. Consequently, the importance of the camel flow is not truly quantified.

Camel Trade in Western Africa

In Western Africa, the most important camel population is encountered in Mauritania where the camel meat (and milk) market for local consumption is the highest of all the regions: indeed, camel meat represents more than 25% of the total consumed red meat in the country. The live camel export is also an important economic activity although the statistics are debatable. Officially, camel is the main species among the animals' export in the country. In the year 2000, the main countries of destination were Morocco and Algeria, but since the Moroccan border is officially closed for sanitary reasons, the camel export market was diversified and oriented to other North-African countries. Although the retail prices are higher in North Africa, a recent trend of export to Senegal as underlined above is observed. This new destination is associated with the implementation of specialized butchers beyond the border, the only purpose of this export being the slaughtering (Faye 2016).

The camels are conveyed mainly by foot, most of the time without official control and with exchange of shepherds when crossing the border. The price differential is important and is the main driver for this export. However, the pressure of the demand is so high that the Ministry of Livestock has forbidden the export of the young females to maintain the national camel herd (Renard 2005). The camel export is in the hands of livestock merchants who buy the animals on the local end markets and export them.

Camel Trade and Transboundary Diseases

The main diseases in camels such as mange, trypanosomiasis, camel pox (the only camel disease controlled by vaccination), gastrointestinal parasites, or calf diarrhea are so endemic and common that the herd mobility has a weak impact on the spread of the diseases. The risk of the live camel flows described above in disease spreading is limited to few emerging diseases that had significant impact on the camel trade. Its concerns mainly Rift Valley fever (RVF), PPR-like disease, and MERS-coronavirus.

The Rift Valley Fever Outbreaks in Dromedary

The Rift Valley fever (RVF) is a zoonotic arboviral disease caused by a Phlebovirus transmitted by mosquitoes. This disease can affect dromedaries. In camels, RVF provokes abortion and over-mortality among the young animals. It is not a specific disease to camels. In addition, the disease can be transmitted to humans and could be lethal. Thus, several outbreaks in Africa largely affected human population in Egypt in 1977 and in West Africa in 1987. In Mauritania, the first severe outbreak occurred in 1987 in Senegal Valley causing the deaths of 28 people. This first outbreak was attributed to some changes in the pluviometry regime marked by late and intensive rains. Later, other outbreaks occurred, in 1998, 2002–2003, and 2010 (in the North of the country) and in 2012, all over the country (El-Mamy et al. 2011, 2014). The last outbreak was probably started by a sick dromedary camel, slaughtered by the owner before it died, but causing the deaths of several contaminated people with intestinal and hemorrhagic symptoms (El-Mamy et al. 2014). The virus being probably introduced through affected camels transported by truck in grazing areas, the risk of spreading all along the export routes could be high. However, thanks to the rapid appropriate control measures (restriction of livestock movement, re-allocation of locust control teams for mass insecticide spraying, risk communication, and public awareness campaigns) by the veterinary and public health authorities, the spread of the virus was limited, although the fall of the livestock's price (up to 40% decrease) created an attractive opportunity for the traders. Yet, the risk of transboundary dissemination to North African countries having similar ecosystem and rainfall conditions is high. Such epidemiological risk encouraged, since 2000, the creation of the Mauritanian Network for the Epidemio-surveillance of Animal Diseases (REMEMA) under the OIE rules and supported by FAO. This network, based on "sentinel herds," has made it possible to control the international exchanges of camels during the outbreaks.

However, on the opposite side of the continent, the epidemiological situation was not identical. The virus is endemic in Egypt and Sudan due to the favorable ecosystems for the vectors (Aradaib et al. 2013; Sayed-Ahmed et al. 2015). Outbreaks were also described in Kenya and Somalia. All species are affected including camel. It is by animal export that the virus reached the Arabian peninsula for the first time in 2000, affecting Saudi Arabia and Yemen. During this outbreak, around 40,000 animals including sheep, goats, cattle, and camels died, whereas about 10,000 of them aborted (Al-Afaleq and Hussein 2011). The main species imported in the Arabian peninsula from the Horn of Africa are sheep and camel. In consequence, restrictions on livestock movement were imposed by Saudi Arabia (1998 to 2000, 2001 to 2004, and 2007) due to health issues in the Horn of Africa and insufficient veterinary control (Faye 2003; Hassan et al. 2014) which landed pastoralists in economic crisis (Pratt et al. 2005) and disturbed the camel trade (Antoine-Moussiaux et al. 2012).

Control and surveillance system was implemented based on the FAO/OIE recommendations: syndromic surveillance (abortion, mortality of the young animals, presence of vectors), participatory surveillance (semi-structured interview of pastoralists), and risk-based surveillance (risk mapping, climatic factors surveillance) in order to identify epidemiological indicators and establish an early warning system. The system is completed by laboratory analyses for confirmation of the disease.

The PPR-Like Disease Outbreak in the Horn of Africa and Other Unidentified Diseases

Since the end of twentieth century, new cases of camel diseases marked by severe symptoms, high mortality, and then impossibility to form a precise diagnosis have emerged in several Sahelian countries. The more emblematic new pathological event occurred in the Horn of Africa in 1995–1996. This disease was characterized by febrile attacks, highly contagious respiratory syndrome with high morbidity (up to 90%), and mortality varying from 5 to 70% depending on the herd and treatment administered (Roger et al. 2000). The high prevalence of peste des petits ruminants (PPR) in the zone as well as symptoms like those of rinderpest suggested the presence of morbillivirus (Roger et al. 2001). Similar observations were published in Sudan (Khalafalla et al. 2005) where several mortal epizooties have been reported in the early 2000s, as well as in Kenya, without clearly identifying the causal agent although PPR virus was suspected. The pastoralists of these countries considered those diseases as new ones and attributed to them vernacular names, e.g., *Firaanfir*, *Laaba*, or *Yudleye* (Khalafalla et al. 2010). The outbreak, however, was apparently confined in the Horn of Africa and has moved throughout the borders of the region in relation to herd mobility, but probably not exported to camel importing countries.

In Central and Western Africa, other unidentified camel diseases occurred. For example, in 2003–2004, several hundred camels died in Mali, Niger, and Chad showing nonspecific symptoms. Some witnesses spoke about a "stunning death." No quantitative data were available because no exhaustive survey was carried out, but the death rates seemed very high. Anthrax was suspected for a moment as well as an acute form of trypanosomiasis, but the results of laboratory analyses did not confirm them (Faye et al. 2012). It seemed that all the dead animals were heavily infested by ticks and hemoparasites such as Anaplasma and Babesia. Still in Niger, many cases of severe xerophthalmia associated with purulent or bilateral sinusitis causing eye loss were observed, without establishing the cause.

More recently in Somalia, Ethiopia, and North Kenya, sudden deaths concerning hundreds of camels were reported and their assumed causes included plant intoxication and mineral deficiencies until the main viral diseases (PPR, blue-tongue, foot-and-mouth disease, Rift Valley fever) were suspected, but at time of writing they have not been confirmed.

Thus, emergence of new diseases in this species, probably linked to the climatic change and geographical expansion of camel breeding, occurred in many places for the last two decades. Unfortunately, investigations have been hindered by the difficult access to sick animals, often located in remote areas. The risk of appearance of these diseases along the trading routes is obviously important.

MERS-Cov

Contrary to former camel emerging diseases described above, MERS-coronavirus is an important zoonotic disease. It was described for the first time in humans in Saudi Arabia in June 2012 and then in many other countries in all the continents (Gossner et al. 2016), but most of the cases occurred in the Middle East, especially in Saudi Arabia. At the world level, at the end of 2017, 2102 human cases of MERS-Cov were confirmed with at least 733 lethal cases. Camels were suspected among some other species at the origin of the outbreak as the main animal reservoir. Positive serology to MERS-Cov in almost 100% of camels in the Middle East testifies the wide circulation of the virus among the camel population, but without clear clinical expression of the disease. Finally, MERS-coronavirus can be regarded as disease for humans and infection only for camels (Younan et al. 2016).

However, if high seroprevalence was also observed in Sahelian countries (up to 100%!), and since more than 30 years, surprisingly, no human case was reported, except by human-to-human transmission. Finally, the absence of infection in human population close to camels in Africa leads to consider the transmission of the virus from camels to humans as unclear and could involve the immunological status of the human population in the Middle East (Al-Osail and Al-Wazzah 2017).

The question of the transboundary transmission of the disease by trade route was investigated. MERS-Cov seropositivity and percentages of MERS-Cov RNA-positive camels were not more important in the two major Saudi camel import entrance ports of Jeddah and Gizan where an increased likelihood of contacts between imported dromedary camels from the Horn of Africa and local Saudi Arabian dromedary camels was expected. Finally, MERS-Cov prevalence appeared higher in local Saudi camels compared to those imported from Somalia or Sudan (Sabir et al. 2016). Recent investigations revealed the presence of two lineages of sub-Saharan MERS-Cov, genetically distinct from virus found in Saudi camels. Moreover, it is stated that the MERS-Cov is poorly transmissible from camels to humans, and clinical human MERS disease is not proportional to potential exposure of humans to virus circulation in camels (Hemida et al. 2014). According to certain authors (Younan et al. 2016), the control measures of Saudi camels should be a priority to those imported from the Horn of Africa. Indeed, the control of an endemic corona-virus infection among nomadic tribes in Africa appears unrealistic. Moreover, the lack of human cases in Egypt or Libya who also import many camels from Sudan or Somalia seems to indicate that control system of the disease based only on the trade restriction is not necessarily efficient in the case of Saudi Arabia. However, it is probably different in camel countries "MERS-Cov free" as in Central Asia or Australia (Miguel et al. 2016). Besides, the World Health Organization does not recommend trade restrictions in relation to MERS but advocates for persons at risk (immune-depressed) to avoid close contact with camels (Mackay and Arden 2015).

However, even without official ban as for RVF, the MERS-Cov crisis has an indirect impact on camel meat consumption in Saudi Arabia and indirectly on the volume of live camel's import. Thus, the decreasing demand for camel meat (despite the lack of risk because the virus does not survive cooking) affected 78% of the urban butchers and 22% of the rural butchers after mediatization of MERS outbreak (Fedoul 2014). The number of camel slaughtering decreased by 10 to 70% per week. In addition to the impact on the camel prices, the number of imported animals decreased by 21% between 2012 and 2013 (after the beginning of MERS outbreak) whereas this number increased regularly since the year 2000 (Fedoul 2014). No similar trend is observed in Sahelian countries.

A New Emerging Disease: Prion Disease?

Prions are simple proteins causing fatal and transmissible neurodegenerative diseases, like Creutzfeldt-Jakob disease in humans, scrapie in small ruminants, or bovine spongiform encephalopathy (BSE) in cattle. The BSE outbreak started in 1996 in the United Kingdom has provoked an important crisis for both public health and cattle meat economy. Since this crisis, scientific community, stakeholders of the meat sector, and public health authorities were highly sensitive to the risks linked to animal prions, especially because of their potential passage to humans. Moreover, the presence of BSE conducted to ban live animals' export in many countries, impacting the livestock economy. It is interesting to note that such a ban imposed by Egypt for European bovines after BSE outbreak is responsible for the high increasing demand for camel meat in the country.

Thus, recently, a camel prion disease was described in Ouargla region, Algeria (Babelhadj et al. 2018). The camels showed symptoms comparable to that of "mad cow disease." The animals found in the desert had difficulty in getting up. At the abattoir, they showed aggressiveness and became nervous when forced to cross an obstacle and showed down- and upwards movements of the head and teeth grinding. These symptoms occurred in 3.1% of dromedaries brought at Ouargla abattoir in 2015–2016. The diagnosis was confirmed by detecting prion protein in brain tissues from three symptomatic animals. Prion detection in lymphoid tissues is suggestive of the infectious nature of the disease. The potential risks for human and animal health are not yet assessed. Obviously, Algeria is not an important camel export country, but the exchanges with the Middle East and Maghreb represent a potential risk of transboundary spreading of the disease, especially because the true occurrence of the disease in Algeria and neighboring regions is unknown. However, a recent case was reported in Tunisia.

Conclusion

Globally, dromedary camel is less affected than other ruminants by infectious diseases under transboundary surveillance. But, because its breeding is concentrated in southern countries (notably in Sahelian ones) where the disease surveillance systems often lack means, where the frontiers in desert areas are often "porous," and where the importance of herd movements is difficult to assess, the risk of disease spreading through the borders is high, especially since the informal trade can be dominant and achieved by foot. For the moment, the challenge of TADs control is limited to Rift Valley fever, but special attention must be paid to emerging diseases, including the recent discovery of prion disease.

References

Aklilu Y. An audit of the livestock marketing status in Kenya, Ethiopia and Sudan- Volume I, Pan African Program for the Control of Epizootics (PACE), Volume II- Issues and proposed measures, Pan African Program for the Control of Epizootics (PACE), Organization of African Unity. 2002.

Al-Afaleq AI, Hussein MF. The status of Rift Valley fever in animals in Saudi Arabia: a mini review. Vector Borne Zoonotic Dis. 2011;11(12):1513–20.

Alary V, Faye B. The camel chains in East Africa- importance of gaps between the data and the apparent reality. J Camelid Sci. 2016;9:1–22.

Al-Osail AM, Al-Wazzah MJ. The history and epidemiology of Middle East respiratory syndrome corona virus. Multidisc Resp Med. 2017;12:20.

Antoine-Moussiaux N, Chevalier V, Peyre M, AbdoSalem Abdullah S, Bonnet P, Roger F. Economic impact of RVF outbreaks on trade within and between East Africa and the Middle East. GF-TADs (FAO/OIE) inter-regional conference on Rift Valley fever in the Middle East and the Horn of Africa: challenges, prevention and control. Mombasa: OIE; 2012.

Aradaib AE, Erickson BR, Elageb RM, Khristova ML, Carroll SA, Elkhidir IM, Karsany ME, Karrar AE, Elbashir MI, Nichol ST. Rift Valley fever, Sudan, 2007 and 2010. Emerg Infect Dis. 2013;19(2):246–53.

Aubague S, Mannany AA, Grimaud P. Difficultés de transhumance des chameliers dans le Tchad central liées aux aléas climatiques. Science et changements planétaires / Sécheresse. 2011;22 (1):25–32.

Babelhadj B, Di Bari M, Pirisinu L, Chiappini B, Gaouar S, Riccardi G, Marcon S, Agrimi U, Nonno R, Vaccari G. Prion disease in dromedary camels, Algeria. Emerg Infect Dis. 2018;24 (6):1029–36.
Belachew H. Assessment of livestock marketing in Ethiopia: constraints and possible intervention measures. Addis Ababa, Ethiopia: Livestock Marketing Department; Ethiopia. 2005.
Djama M. Producteurs pastoraux et commerce international- l'évolution des rapports marchands en pays Nord-Somali. In: Bourgeot A, editor. Horizons nomades en Afrique sahélienne- Sociétés, développement et démocratie. Karthala: Université de Niamey; 1999. p. 339–53.
El-Mamy OB, Baba MO, Barry Y, Isselmou K, Dia ML, Hampate B, Diallo MY, El Kory MOB, Diop M, Lo MM, Thiongane Y, Bengoumi M, Puech L, Plee L, Claes F, De La Rocque S, Doumbia B. Unexpected Rift Valley fever outbreak, Northern Mauritania. Emerg Infect Dis. 2011;17(10):1894–6.
El Mamy OB, Kane Y, El Arbi AS, Barry Y, Bernard C, Lancelot R, Cêtre-Sossah C. L'épidémie de fièvre de la Vallée du Rift en Mauritanie en 2012. Rev Afr Santé Prod Anim. 2014;12 (3–4):169–73.
Faye B. Systèmes pastoraux, agro-pastoraux et agricoles d'Ethiopie. In: Chantal B-P, Jean B, editors. Dynamique des systèmes agraires. A la croisée des parcours. Pasteurs, éleveurs, cultivateurs, vol. 1994. Paris: ORSTOM; 1994. p. 269–87.
Faye B. Surveillance and control procedures for camel diseases. Workshop on the surveillance and control of camels and wildlife diseases in the Middle East. Regional workshop OIE. Sanaa (Yemen). 2003, March:10–2.
Faye B. Camel farming sustainability: the challenges of the camel farming system in the XXIth century. J Sustain Dev. 2013;6(12):74–82.
Faye B. Projet régional d'appui au pastoralisme au sahel (PRAPS)-Analyse de la filière Cameline en Mauritanie. Rapport de mission CIRAD-ES/PRAPS, Montpellier, 45 p. 2016.
Faye B, Bonnet P. Camel sciences and economy in the world: current situation and perspectives. In: Proc. 3rd ISOCARD conference. Keynote presentations. 29th January – 1st February 2012, Mascate (Sultanate of Oman); 2012. p. 2–15.
Faye B, Chaibou M, Vias G. Integrated impact of climate change and socioeconomic development on the evolution of camel farming systems. Br J Environ Clim Change. 2012;2(3):227–44.
Faye B, Abdelhadi O, Raiymbek G, Kadim I. Filière viande de chameau et critère de qualité. Evolution du marché, perspectives de développement et qualité de la viande de chameau. Viandes & Produits carnés. VPC-2013-29-6-2, 1–8. 2013.
Fedoul AA. Analyse de la filière viande cameline en Arabie Saoudite. Mémoire de stage « Ingénierie et Gestion des Territoires (IGT) », Gestion Agricole et territoires. France: CIHEAM/Université de Montpellier; 2014. 67 p.
Gossner C, Danielson N, Gervelmeyer A, Berthe F, Faye B, Kaasik-Aaslav K, Adlhoch C, Zeller H, Penttinen P, Coulombier D. Human–dromedary camel interactions and the risk of acquiring zoonotic Middle East respiratory syndrome coronavirus infection. Zoonose Public Hlth. 2016;63:1–9.
Hassan OA, Ahlm C, Evander M. A need for one health approach-lessons learned from outbreaks of Rift Valley fever in Saudi Arabia and Sudan. Infect Ecol Epidemiol. 2014;4:1–8.
Hemida MG, Chu DKW, Poon LLM, Perera RAPM, Alhammadi MA, Ng H-Y. MERS coronavirus in dromedary camel herd, Saudi Arabia. Emerg Infect Dis. 2014;20:1231–4.
Kadim I, Mahgoub O, Purchas RW. A review of the growth, and of the carcass and meat quality characteristics of the one-humped camel (*Camelus dromedaries*). Meat Sci. 2008;80:555–69.
Khalafalla AI, Intisar KS, Ali YH, Amira MH, Abu Obeida A, Gasim M, Zakia A. Morbillivirus infection of camels in eastern Sudan. New emerging fatal and contagious disease. Proceeding of the International Conference on Infectious Emerging Disease. AlAain, UAE, 26th March–April 1; 2005.
Khalafalla AI, Saeed IK, Ali YH, Abdurrahman MB, Kwiatek O, Libeau G, Obeida AA, Abba Z. An outbreak of peste des petits ruminants (PPR) in camels in the Sudan. Acta Trop. 2010;116 (2):161–5.
Khorchani T, Hammadi M, Moslah M. Artificial nursing of camel calves: an effective technique for calf's safeguard and improving herd productivity. In: Faye B, Esenov P, editors. Proceedings of

the international workshop, "Desertification combat and food safety: the added value of camel producers", Ashkhabad (Turkmenistan), 19–22 April 2004. In: Vol. 362 NATO sciences series, life and behavioral sciences. Amsterdam: IOS; 2005. p. 177–82.

Kurtu MY. An assessment of the productivity for meat and carcass yield of camel (*Camelus dromedarius*) and the consumption of camel meat in the eastern region of Ethiopia. Trop Anim Health Prod. 2004;36:65–76.

Little PD, Teka T, Azeze A. Research methods on cross-border livestock trade in the horn of Africa: further observations. Research report of the broadening access to markets and input systems-collaborative research support program (BASIS-CRSP) and OSSREA project on cross-border trade and food security in the horn of Africa. 1998, Oct. p. 14.

MacKay IM, Arden KE. Middle East respiratory syndrome: an emerging coronavirus infection tracked by the crowd. Virus Res. 2015;202:60–88.

Mankor A, Koussou MO. Etude du sous-secteur camelin du Tchad. Rapport provisoire. N'Djamena, Tchad: FAO; 2015. 101 p.

Miguel E, Perera R, Baubekova A, Chevalier V, Faye B, Akhmetsadykov N, Chun Yin N, Roger F, Peiris M. Absence of Middle East respiratory syndrome coronavirus in camelids, Kazakhstan, 2015. Emerg Infect Dis. 2016;22(3):555–7.

Pratt AN, Bonnet P, Jabbar M, Ehui S, de Haan C. Benefits and costs of compliance of sanitary regulations in livestock markets: the case of Rift Valley fever in the Somali region of Ethiopia. International livestock research institute. Nairobi, Kenya: The World Bank; 2005.

Raiymbek G, Kadim I, Konuspayeva G, Mahgoub O, Serikbayeva A, Faye B. Discriminant amino-acid components of Bactrian (*Camelus bactrianus*) and dromedary (*Camelus dromedarius*) meat. J Food Compos Anal. 2015;41:194–200.

Renard JF. Programme de développement d'une aptitude au commerce. Mission d'identification sur les filières animales (TF/MAU/04/001/11–53), Mauritanie, UNIDO. Rapport de mission CIRAD, Montpellier, 50 p. 2005.

Roger F, Diallo A, Yigezu LM, Hurard C, Libeau G, Mebratu GY, Faye B. Investigations of a new pathological condition of camels in Ethiopia. J Camel Pract Res. 2000;7(2):163–6.

Roger F, Gebre Yesus M, Libeau G, Diallo A, Yigezu L, Yilma T. Detection of antibodies of rinderpest and peste des petits ruminants viruses (Paramyxoviridae, Morbillivirus) during a new epizootic disease in Ethiopian camels (Camelus dromedarius). Rev Vet Med. 2001;152:265–8.

Sabir JSM, Lam TTY, Ahmed MMM, Li L, Shen Y, Abo-Aba SEM, Qureshi MI, Abu-Zeid M, Zhang Y, Khiyami MA. Co-circulation of three camel coronavirus species and recombination of MERS-CoVs in Saudi Arabia. Science. 2016;351(6268):81–4.

Sayed-Ahmed M, Nomier Y, Shoeib SM. Epidemic situation of Rift Valley fever in Egypt and Saudi Arabia. J Dairy Vet Anim Res. 2015;2(3):00034. https://doi.org/10.15406/jdvar.2015.02.00034.

Younan M, Bornstein S, Gluecks I. MERS and the dromedary camel trade between Africa and the Middle East. Trop Anim Health Prod. 2016;48(6):1277–82.

Part II
Viral Diseases

Chapter 7
Rabies

Stephanie Mauti, Monique Léchenne, Céline Mbilo, Louis Nel, and Jakob Zinsstag

Abstract The inevitably fatal outcome of rabies makes it one of the most well-known and feared zoonoses. But rabies is still a neglected zoonotic disease (NZD), despite being one of the oldest diseases known. Dog-mediated rabies is the main cause of human rabies and is globally responsible for approximately 59,000 human deaths per year, nearly all occurring in low- and middle-income countries (LMICs). Most African countries and connected regions are high-risk areas for contracting rabies. Highly effective vaccines and post-exposure prophylaxis (PEP) for humans allow for the disease to be 100% preventable. But PEP is often unavailable or too expensive for affected people in LMICs. Recently, several new and low-tech diagnostic solutions have been developed, which offer opportunities to establish rabies diagnosis in remote areas and decentralise rabies laboratories.

Rabies has gained more global attention in recent years with development of new tools, formation of regional networks and implementation of a realistic global drive to eliminate canine-mediated human rabies by 2030. Canine rabies elimination is biologically feasible due to efficacious, safe and cheap vaccines and the low basic reproductive ratio (R_0) of the disease. It is well known that mass dog vaccination is a cost-effective, sustainable measure to eliminate the disease at its source. Domestic dogs are tied to human populations, so the role of humans in rabies spread needs to be further investigated, for example through anthropogenic landscape features like roads or vaccine corridors, human movements and sociocultural factors. Combining powerful approaches, such as landscape epidemiology and genetics, can facilitate strategic control programmes and, for example, enable appropriate placement of vaccine barriers and surveillance points. It is known that major landscape features, such as oceans, mountains and deserts, can act as natural barriers to disease spread. However,

S. Mauti (✉) · M. Léchenne · C. Mbilo · J. Zinsstag
Swiss Tropical and Public Health Institute, Basel, Switzerland

University of Basel, Basel, Switzerland
e-mail: stephanie.mauti@swisstph.ch; M.S.Lechenne@exeter.ac.uk; celine.mbilo@swisstph.ch; jakob.zinsstag@swisstph.ch

L. Nel
University of Pretoria, Pretoria, South Africa
e-mail: louis.nel@up.ac.za

M. Kardjadj et al. (eds.), *Transboundary Animal Diseases in Sahelian Africa and Connected Regions*, https://doi.org/10.1007/978-3-030-25385-1_7

very little is known about barrier effects at smaller scales. Long-distance transport of infected dogs by humans poses a risk of rabies introduction in novel places. For successful rabies control, it is also important to guarantee access to pre- and post-exposure prophylaxis, to build capacity in disease diagnosis and to conduct educational campaigns. Well-functioning, continuous rabies surveillance systems are crucial to provide reliable data to increase political commitment, which is eminently important for successful, sustainable disease control. For elimination of rabies in Africa and the connected regions, a multidisciplinary, transdisciplinary and regionally well-coordinated approach, with sustained vaccination programmes to maintain sufficient vaccination coverage, across political boundaries is required. To foster reflection on strategic rabies elimination on the African continent, one scenario of a possible spatio-temporal dynamic of dog rabies elimination in West Africa is proposed. Dog rabies elimination might, for example, start in northwestern Mauritania. The approach should be highly coordinated between the involved countries to avoid cross-border transmission and return of disease in rabies-free zones. It is difficult to mobilise large sums initially, but development impact bonds (DIB) offer an alternative funding mechanism. The total cost is estimated at 800 million to one billion Euros for rabies elimination in sub-Saharan West Africa, including Chad and Cameroon.

Keywords Rabies · Dogs · Rabies control · Vaccination campaigns · Zoonoses · PARACON · 'One Health'

History

The inevitably fatal outcome of rabies, a disease caused by a neurotropic virus, makes it one of the most well-known and feared zoonoses. A zoonotic disease is defined as a disease which can be transmitted between humans and vertebrate animals. The majority of rabies cases around the world are caused by the canine-associated classical rabies virus (family *Rhabdoviridae*, genus *Lyssavirus*), but rabies-related viruses of the same genus, mostly circulating in bat species, can also cause rabies (Lagos bat virus, Mokola virus, Duvenhage virus, European bat lyssavirus-1 and -2, Australian bat lyssavirus and others). In general, all mammalian species can be infected by the rabies virus (Banyard and Fooks 2011). In Western and Central Africa, rabies virus mainly belongs to the lineage 'Africa 2' (Talbi et al. 2009).

Geographical Distribution, Economic/Public-Health Impact and Epidemiology

Canine rabies is the main cause of human rabies, being globally responsible for approximately 59,000 human deaths per year, the majority of which are in Asia (60%) and Africa (36%) (Hampson et al. 2015). Almost all human cases occur in low- and middle-income countries (LMICs), with children being the most affected group. In resource-poor countries, the virus is mainly transmitted through dog bites, but transmission is also possible through infected wild animals (Banyard and Fooks 2011; WHO 2006). It is believed that rural areas are more affected by rabies deaths than urban areas (Knobel et al. 2005). Most African countries and connected regions are high-risk areas for contracting rabies (WHO 2014).

Highly effective vaccines and post-exposure prophylaxis (PEP) for humans allow for the disease to be 100% preventable. PEP consists of local wound care, vaccination and immunoglobulin application. Every year, over 15 million people globally are exposed to rabies and should receive PEP, at considerable monetary cost. But PEP is often unavailable or too expensive for affected persons in LMICs. In some contexts, inappropriate recommendation of PEP use by health personnel was observed (Hampson et al. 2015; Mindekem et al. 2017; WHO 2006).

Different factors contribute to the burden of rabies, including mortality and productivity loss due to premature death, morbidity from adverse events following vaccination (nerve tissue vaccines), psychological effects and direct and indirect costs of PEP. In 2015, the global rabies burden was estimated to be 3.7 million disability-adjusted life years (DALYs), with over 95% lost in Africa and Asia (Hampson et al. 2015). Since political will and interest often depend on the economic burden of a disease, the annual economic losses due to premature death, costs of PEP and costs to the veterinary sector were estimated, reaching 8.6 billion USD globally (Hampson et al. 2015). In Africa alone, the costs of rabies were estimated to be 1.3 billion USD annually (Global Alliance for Rabies Control, Partners for Rabies Prevention).

Today, rabies is classified as a neglected zoonotic disease (NZD), despite being one of the oldest diseases known to humankind. In many affected countries, the control of the disease is still hampered by a low level of political commitment, most likely due to lack of reliable data (WHO 2006, 2015a). The fact that many bite victims in LMICs do not visit medical facilities and lack of laboratory confirmation of cases leads to underreporting of the disease contributes to the cycle of neglect (Nel 2013a; WHO 2006). Poor disease surveillance prevents decision-makers and stakeholders from making informed decisions regarding allocation of funding and resources towards better control and intervention strategies. However, because of lack of funding, it is difficult to improve surveillance (Nel 2013b). In Tanzania, a 10- to 100-fold higher incidence of human rabies cases was estimated when incidence was extrapolated from animal bite occurrence by using a probability decision tree modelling method to determine the likelihood of clinical rabies in humans after the bite of a rabid dog, compared to the officially reported human incidence estimated through passive surveillance data (Cleaveland et al. 2002). Ninety-nine per cent of human rabies cases are likely to be unreported in Tanzania (WHO 2010). Knobel et al. (2005) estimated a rate of underreporting of 160 times for Africa in general. Misdiagnosis also adds to underreporting, as seen in Malawi where children with rabies were falsely diagnosed with cerebral malaria (Mallewa et al. 2007).

Symptoms and Lesions

After saliva containing virus comes in contact with peripheral nerve endings through a bite, skin lesion or mucous membrane, the virus affects the central nervous system where it causes fatal encephalitis with dramatic symptoms. Incubation periods last from a few days to several months depending on factors such as the site of virus inoculation and viral load. Symptoms in humans and animals result from brain dysfunction, leading either to a furious (mad) or a dumb (paralytic) form of rabies (Banyard and Fooks 2011; Rupprecht et al. 2002).

Diagnostic

The 'gold standard' for detection of rabies virus antigen in the brain is the direct fluorescent antibody (DFA) test. However, proper application in LMICs remains limited due to inappropriate laboratory facilities, uncooled sample transportation and lack of quality management systems. Because animal rabies diagnosis is typically only conducted at Central Veterinary Laboratories (CVLs) in LMICs, existing surveillance data mainly reflects the rabies situation of urban areas. Several new and low-tech diagnostic solutions have been developed in recent years. A promising one for rabies diagnosis in LMICs is the direct rapid immunohistochemical test (dRIT), which has diagnostic efficacy equal to that of the DFA. Benefits include ease of differentiation between a positive and a negative result, so the test is simple to interpret by inexperienced readers. Unlike the DFA test, the dRIT does not rely on fluorescence, another advantage because fluorescence is difficult to interpret in degraded or archived samples. Since fluorescent microscopes are not needed, in-house calibration is possible and maintenance of equipment is reduced. Furthermore, storage of brain samples in glycerol seems to influence the DFA more than the dRIT, improving diagnosis in archival samples and samples maintained at room temperature. The dRIT requires a smaller initial capital investment and is cheaper to perform (Coetzer et al. 2014; Dürr et al. 2008b; OIE 2012; Scott et al. 2015; Weyer and Blumberg 2007). Another recent test is the rapid immunodiagnostic test (RIDT), a pre-manufactured diagnostic kit based on the lateral flow principle. This test is extremely easy to conduct, does not require a microscope for interpretation and can be stored at ambient temperature. However, currently RIDTs are not considered as reliable as the FAT or the DRIT (Eggerbauer et al. 2016). Léchenne et al. (2016a) showed, on the other hand, higher reliability compared to the FAT when applied in a resource-poor laboratory. Nevertheless, both the dRIT and the RIDT offer opportunity to establish rabies diagnosis in remote areas and therefore decentralise rabies laboratories (Léchenne et al. 2016a).

Application to Prevention and Control/Adopted Surveillance and Control Strategies

Rabies has gained more global attention in recent years, with development of new tools, formation of regional networks and a realistic global drive to eliminate canine-mediated human rabies by 2030, in line with the United Nations Sustainable Development Goals (SDGs) (UN 2016; WHO 2015b). The Blueprint for Rabies Prevention and Control (http://www.rabiesblueprint.org/; Lembo and Partners for Rabies Prevention 2012) was developed to provide guidelines and strategies on rabies control. In 2014, the Pan-African Rabies Control Network (PARACON) was established under the secretariat of the Global Alliance for Rabies Control (GARC). PARACON represents a unified coordinated approach to eliminate canine rabies in sub-Saharan Africa on a regional, national and continental level, with support from global human and animal health organisations such as the World Health Organization (WHO), the World Organisation for Animal Health (OIE), the Food and Agriculture Organization of the United Nations (FAO) and World Animal Protection (WAP). North African countries are incorporated into the Middle East and Eastern Europe Rabies Expert Bureau (MEEREB), but collaborating closely with PARACON (Scott et al. 2015). The 'One Health' concept must be strengthened to result in an added value of closer cooperation between human and animal health. The added value can be defined as health benefits, financial savings or environmental services (Scott et al. 2015; Zinsstag et al. 2005, 2015). To date, interaction and collaboration between the veterinary sector and the public health departments is often non-existent in LMICs even though with zoonotic diseases both sectors are dealing with the same epidemiological complexities. In general, the ministry of health is responsible for prevention of the disease in humans, the ministry of agriculture is in charge of rabies control in animals and the ministry of local government and the ministries of commerce, industry or science and technologies are involved in rabies vaccine production and imports, dog population management and dog immunisation (WHO 2005). A 'One Health' perspective for rabies brings a clear added value by strengthening intersectoral cooperation, for instance, through better disease surveillance and communication. The approach should always be adapted to the local setting and its human–animal relationship which is governed by the cultural and religious context (Léchenne et al. 2015; Zinsstag et al. 2015). The PARACON network encourages collaboration amongst African countries through regular meetings and workshops held in member countries throughout Africa. Different tools are provided through the PARACON network, like capacity building in rabies diagnosis, focusing on the implementation of the dRIT and provision of material, expertise and assistance in rabies control programmes. Additionally, an online African rabies epidemiological bulletin was launched in 2016, which captures high-quality, pan-African data on human and animal rabies cases in a timely manner and assesses the number of vaccine doses administered, vaccination coverage and dog population estimates. The gathered data from the bulletin will be further used for advocacy purposes in order to draw support from governmental authorities and stakeholders for the implementation of

National Control Strategies within countries allowing them to take ownership for the control and elimination of canine-mediated human rabies (Scott et al. 2015, 2017). The Stepwise Approach towards Rabies Elimination (SARE), developed by GARC and the FAO in 2012, helps member countries evaluate their progress towards rabies control and elimination. The SARE tool provides countries with measurable stages to progress from Stage 0 to Stage 5 towards becoming canine rabies free. A country typically begins at Stage 0, with little or no epidemiological understanding of, or control efforts for, rabies in place. The country progresses to the next stage once certain critical and non-critical activities have been achieved, eventually reaching Stage 5—being canine rabies free (Coetzer et al. 2016; FAO and GARC 2012; Scott et al. 2015).

Canine rabies elimination is biologically feasible due to efficacious, safe and cheap vaccines and the low basic reproductive ratio (R_0) of the disease (Hampson et al. 2009; Kayali et al. 2006; Klepac et al. 2013; Morters et al. 2013). It is well known that mass dog vaccination is a cost-effective, sustainable measure to eliminate the disease at its source and prevent humans from exposure (Hampson et al. 2015; Mindekem et al. 2017; Zinsstag et al. 2009). In many African communities, sufficient dogs are accessible for vaccination in order to achieve herd immunity (WHO 2005). PEP alone will never be able to interrupt human exposure, and the simulation of Mindekem et al. (2017) demonstrated in an African city that canine vaccination in combination with PEP becomes more cost-effective after 15 years in comparison to the use of PEP alone. With ideal One Health communication, the cost for dog vaccination and PEP compared to PEP alone breaks even within the timeframe of 10 years. In Western Europe and North America, rabies has been successfully eliminated from domestic dog populations through mass vaccination and legislation (Hampson et al. 2007; Müller et al. 2012, 2015; Slate et al. 2009). Disease control through dog vaccination is even feasible in situations where large wildlife populations are prevalent (Hampson et al. 2009). For efficient rabies control, the World Health Organization (WHO) recommends vaccinating at least 70% of the dog population, which is rarely reached in African countries (Coleman and Dye 1996; WHO 2013). In Bamako, the capital of Mali, a vaccination coverage of only 24% was estimated in the domestic dog population, which is insufficient coverage to interrupt virus transmission (Coleman and Dye 1996; Mauti et al. 2017a). In N'Djamena, the capital of Chad, rabies cases in dogs decreased by more than 90% within 1 year after a free mass vaccination campaign which achieved coverage at the recommended 70% level (Léchenne et al. 2016b). In both the Malian and the Chadian context, an important constraint to vaccination was cost of the vaccination of the dog (Dürr et al. 2008a; Mauti et al. 2017a). In Chad, it was found that vaccination coverage dramatically dropped when dog owners had to pay for the vaccination (Durr et al. 2009). In Latin America and the Caribbean, canine rabies was successfully reduced due to mass dog vaccination. The Pan American Health Organization (PAHO) played an important role in this process, through its constant support and coordination of dog rabies elimination programmes. It is important that not only urban centres are vaccinated, as was the case for Bolivia which prioritised urban areas for vaccination due to financial limitations and the fact that higher levels of rabies vaccination in urban areas are more easily achieved compared to rural areas (Velasco-Villa et al. 2017; Vigilato et al. 2013). Despite some notable efforts and achievements, we are still far from global elimination of canine rabies. Rabies vaccination campaigns risk failure when they are not adequately presented and accepted within different socio-cultural contexts. The vaccine's 'final' effectiveness in a specific setting is determined by several additional parameters, such as availability, accessibility, affordability, adequacy and acceptability (Bardosh et al. 2014; Zinsstag et al. 2011a, b). Prior to attempting rabies control in a target area, it is important to know what the achievable vaccination coverage and the level of community participation is likely to be. Through a deeper qualitative assessment, as in the intervention effectiveness model, an explanatory framework for a specific area can be elaborated. This was recently done by Muthiani et al. (2015) and Mosimann et al. (2017) in Bamako, Mali. Lack of information and inability to handle the dogs were the most frequently stated reasons for not bringing a dog to the vaccination posts. Knowledge generated in specific local contexts, which accounts for cultural practices and social and political realities, is fundamental to design successful effective, sustainable dog rabies control programmes (Léchenne et al. 2015).

Another important factor in canine rabies elimination is better understanding of affected dog populations. It is crucial to know more about dog ecology, size of the dog population in a specific area, dog density, sex ratio, population turnover/growth and roles of dogs in human societies (WHO 1987, 1988). Mass vaccination campaigns should be repeated at short intervals because of the high turnover rate in dog populations in sub-Saharan African countries which leads to a rapid decrease of population-level immunity (Mauti et al. 2017b). In Chad, information on the size of dog populations was used for planning dog rabies elimination on the country level (Anyiam et al. 2016). It is important that dog population reduction through culling and poisoning is not only socially unacceptable but has also been found to be counterproductive in dog rabies elimination campaigns. Such practices should be discontinued (Morters et al. 2013; WHO 2013; Zinsstag et al. 2009). As domestic dogs are tied to human populations, the role of humans in rabies spread needs to be further investigated. This includes for example anthropogenic landscape features like roads or vaccine corridors, human movements and sociocultural factors. Free-roaming dogs in rabies endemic countries need to be considered during implementation of control measures. Genetic work can identify main routes of viral dissemination. Combining landscape epidemiology and genetics approaches would be a powerful method to facilitate strategic control programmes and identify the appropriate placement of vaccine barriers and surveillance points. It is known that major landscape features, such as oceans, mountains or deserts, can act as natural barriers to disease spread. However, very little is known about barrier effects on smaller scales. Long-distance transport of infected dogs by humans poses a risk of rabies reintroduction or introduction to novel places (Bourhy et al. 2016; Brunker et al. 2012). Hampson et al. (2007) and Talbi et al. (2010) demonstrated long-distance translocation of infected dogs by humans in North and sub-Saharan Africa. In rabies-free areas, a focus should, therefore, be put on the risk of rabies re-emergence through dog movement and possible spillover from wild to domesticated animals (Vigilato et al. 2013). In rabies-free areas, reintroduction control needs to be applied as well as controlled dog movement. Transboundary spread of the rabies virus needs to be considered during the disease elimination process, so control programmes should include neighbouring countries (Klepac et al. 2013). In North Africa, there was restricted movement across geopolitical boundaries (Talbi et al. 2010), but the model of Hampson et al. (2007) demonstrated large-scale synchronous cycles of domestic dog rabies, with a period of 3–6 years, in southern and eastern Africa. In their examination, climate did not influence the observed synchronous pattern of rabies outbreaks. Epidemiological analyses from Ghana also demonstrated frequent cross-border incursions (Hayman et al. 2011).

For successful rabies control, it is vital to guarantee access to pre- and post-exposure prophylaxis, to build capacity in disease diagnosis and to conduct educational campaigns. Responsible dog ownership through community education and legislative measures should be promoted (Lembo et al. 2011). Well-functioning continuous rabies surveillance systems are crucial to provide reliable data to increase political commitment, which is necessary for successful and sustainable disease control. Countries with limited financial resources could significantly benefit from international support, including from other countries. One example could be vaccine donations, as the Dominican Republic has provided to Haiti. In addition, NGOs and the public and private sectors need to be included in rabies control programmes (Vigilato et al. 2013). Remote and hard to access areas should not be ignored, as these areas could harbour remaining foci of infection. A sustained focus on communities where rabies incidence has declined is critical, so that decreased investment does not prevent continued control measures which are necessary to achieve complete elimination in those communities (Klepac et al. 2013).

Conclusion

It is evident that rabies elimination in Africa and elsewhere in the dog rabies endemic world requires a multidisciplinary, transdisciplinary and regionally well-coordinated approach with effective, sustained vaccination programmes which span political boundaries. Freedom from rabies should be recognised as a public good. Canine mass vaccinations should be free for dog owners in order to reach sufficiently high coverage. Given the importance of such regional approaches, international organisations like the African Union (AU-IBAR) and the Economic Union of the West African States (ECOWAS) should be involved from the outset, working in close collaboration with PARACON. To illustrate strategic rabies elimination in Africa, a scenario of one possible spatio-temporal dynamic of elimination in West Africa is shown in Fig. 7.1. This proposal begins with dog rabies elimination in north-western Mauritania, continuing simultaneously in a south and east-ward direction. Natural barriers like the Atlantic Ocean and the Sahara Desert are taken into account. Alternatively, elimination efforts could start in Liberia with a simultaneous north, south and east-ward progression. The approach should be highly coordinated

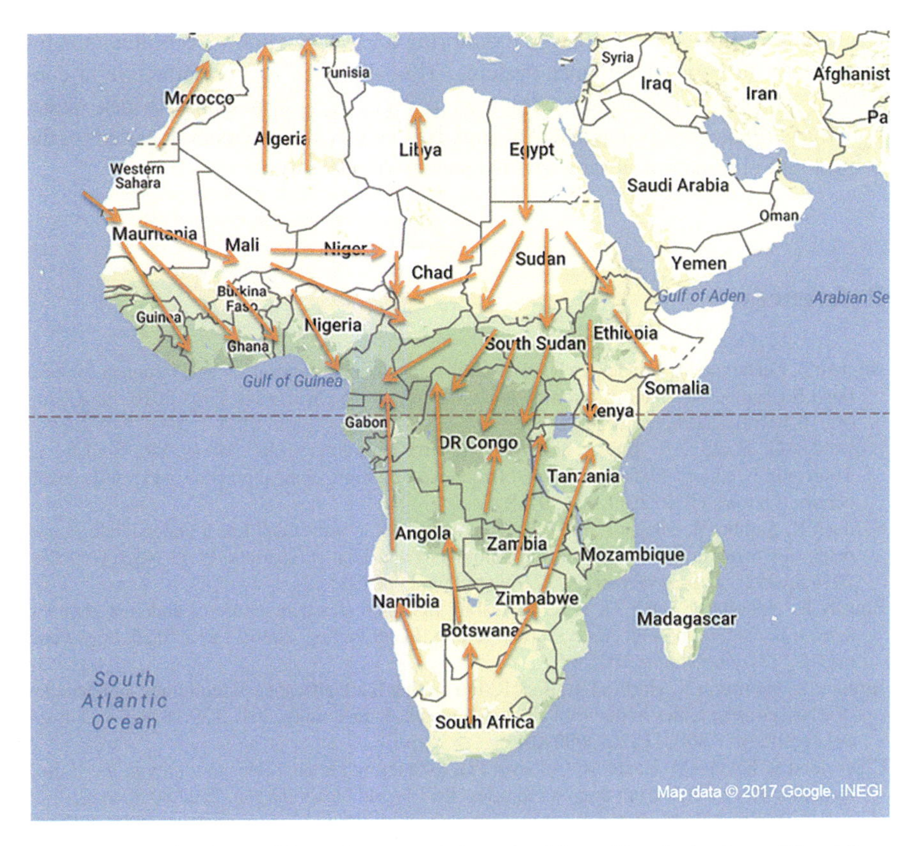

Fig. 7.1 Scenario of a possible spatio-temporal dynamic of dog rabies elimination in West Africa, in the context of a Pan-African campaign (Map data from Google, arrows included by S. Mauti)

between the involved countries to avoid cross-border transmission and return of disease in rabies-free zones. It is difficult to mobilise large sums initially, but development impact bonds (DIB) offer an alternative funding mechanism. Total cost for rabies elimination in sub-Saharan West Africa, including Chad and Cameroon, is estimated at 800 million to one billion Euros.

To identify the optimal operational unit for disease intervention campaigns and optimise existing resources, the role humans play in pathogen spread, natural barriers, virus genetics and host ecology should be assessed for each country. Higher-resolution data from the continent and connected regions would enable a better understanding of spatio-temporal virus dynamics. Such information would allow effective buffer zones for dog rabies to be designed and implemented (Brunker et al. 2012; Hampson et al. 2007). Where control measures deteriorate, rabies epidemics rapidly re-emerge through dispersal from endemic areas. Integrated and effective surveillance systems allow for early disease detection and quick response (Russell et al. 2005). Surveillance is a key element in rabies control, and weak surveillance systems jeopardise disease elimination efforts. Proactive mass dog vaccination followed by 2 years of monitoring and vaccination is recommended for countries with poor surveillance systems. This approach is more effective at controlling rabies in isolated areas compared to vaccinating in response to case detection. Surveillance levels that detect at least 10% of all cases are recommended for the control and elimination of rabies (Townsend et al. 2013).

References

Anyiam F, Lechenne M, Mindekem R, et al. Cost-estimate and proposal for a development impact bond for canine rabies elimination by mass vaccination in Chad. Acta Trop. 2016; https://doi.org/10.1016/j.actatropica.2016.11.005.

Banyard AC, Fooks AR. Rabies and rabies-related lyssaviruses. In: Palmer SR, Soulsby L, Torgerson P, editors. Oxford textbook of zoonoses: biology, clinical practice, and public health control. Oxford: OUP; 2011. p. 650–68.

Bardosh K, Sambo M, Sikana L, et al. Eliminating rabies in Tanzania? Local understandings and responses to mass dog vaccination in Kilombero and Ulanga districts. PLoS Negl Trop Dis. 2014;8:e2935. https://doi.org/10.1371/journal.pntd.0002935.

Bourhy H, Nakouné E, Hall M, et al. Revealing the micro-scale signature of endemic zoonotic disease transmission in an African urban setting. PLoS Pathog. 2016;12:e1005525. https://doi.org/10.1371/journal.ppat.1005525.

Brunker K, Hampson K, Horton DL, Biek R. Integrating the landscape epidemiology and genetics of RNA viruses: rabies in domestic dogs as a model. Parasitology. 2012;139:1899–913. https://doi.org/10.1017/S003118201200090X.

Cleaveland S, Fèvre EM, Kaare M, Coleman PG. Estimating human rabies mortality in the United Republic of Tanzania from dog bite injuries. Bull World Health Organ. 2002;80:304–10.

Coetzer A, Sabeta CT, Markotter W, et al. Comparison of biotinylated monoclonal and polyclonal antibodies in an evaluation of a direct rapid immunohistochemical test for the routine diagnosis of rabies in southern Africa. PLoS Negl Trop Dis. 2014;8:e3189. https://doi.org/10.1371/journal.pntd.0003189.

Coetzer A, Kidane AH, Bekele M, et al (2016) Towards the control and elimination of rabies in Ethiopia: an assessment based on a logical stepwise approach towards rabies elimination.

Coleman PG, Dye C. Immunization coverage required to prevent outbreaks of dog rabies. Vaccine. 1996;14:185–6.

Dürr S, Meltzer MI, Mindekem R, Zinsstag J. Owner valuation of rabies vaccination of dogs, Chad. Emerg Infect Dis. 2008a;14:1650–2. https://doi.org/10.3201/eid1410.071490.

Dürr S, Naïssengar S, Mindekem R, et al. Rabies diagnosis for developing countries. PLoS Negl Trop Dis. 2008b;2:e206. https://doi.org/10.1371/journal.pntd.0000206.

Durr S, Mindekem R, Kaninga Y, et al. Effectiveness of dog rabies vaccination programmes: comparison of owner-charged and free vaccination campaigns. Epidemiol Infect. 2009;137:1558–67. https://doi.org/10.1017/S0950268809002386.

Eggerbauer E, de Benedictis P, Hoffmann B, et al. Evaluation of six commercially available rapid immunochromatographic tests for the diagnosis of rabies in brain material. PLoS Negl Trop Dis. 2016;10:e0004776. https://doi.org/10.1371/journal.pntd.0004776.

FAO and GARC. Developing a stepwise approach for rabies prevention and control. Rome. 2012.

Global Alliance for Rabies Control, Partners for Rabies Prevention Africa rabies costs. https://rabiesalliance.org/media/photo-gallery/africa-rabies-costs. Accessed 30 Aug 2016.

Hampson K, Dushoff J, Bingham J, et al. Synchronous cycles of domestic dog rabies in sub-Saharan Africa and the impact of control efforts. Proc Natl Acad Sci USA. 2007;104:7717–22. https://doi.org/10.1073/pnas.0609122104.

Hampson K, Dushoff J, Cleaveland S, et al. Transmission dynamics and prospects for the elimination of canine rabies. PLoS Biol. 2009;7:e53. https://doi.org/10.1371/journal.pbio.1000053.
Hampson K, Coudeville L, Lembo T, et al. Estimating the global burden of endemic canine rabies. PLoS Negl Trop Dis. 2015;9:e0003709. https://doi.org/10.1371/journal.pntd.0003709.
Hayman DTS, Johnson N, Horton DL, et al. Evolutionary history of rabies in Ghana. PLoS Negl Trop Dis. 2011;5:e1001. https://doi.org/10.1371/journal.pntd.0001001.
Kayali U, Mindekem R, Hutton G, et al. Cost-description of a pilot parenteral vaccination campaign against rabies in dogs in N'Djaména, Chad. Trop Med Int Health TM IH. 2006;11:1058–65. https://doi.org/10.1111/j.1365-3156.2006.01663.x.
Klepac P, Metcalf CJE, McLean AR, Hampson K. Towards the endgame and beyond: complexities and challenges for the elimination of infectious diseases. Philos Trans R Soc Lond Ser B Biol Sci. 2013;368:20120137. https://doi.org/10.1098/rstb.2012.0137.
Knobel DL, Cleaveland S, Coleman PG, et al. Re-evaluating the burden of rabies in Africa and Asia. Bull World Health Organ. 2005;83:360–8. https://doi.org/10.1590/S0042-96862005000500012.
Léchenne M, Miranda ME, Zinsstag J. 16 integrated rabies control. In: Zinsstag J, Schelling E, Waltner-Toews D, et al., editors. One health: the theory and practice of integrated health approaches. Wallingford: CABI; 2015.
Léchenne M, Naïssengar K, Lepelletier A, et al. Validation of a rapid rabies diagnostic tool for field surveillance in developing countries. PLoS Negl Trop Dis. 2016a;10:e0005010. https://doi.org/10.1371/journal.pntd.0005010.
Léchenne M, Oussiguere A, Naissengar K, et al. Operational performance and analysis of two rabies vaccination campaigns in N'Djamena, Chad. Vaccine. 2016b;34:571–7. https://doi.org/10.1016/j.vaccine.2015.11.033.
Lembo T, Partners for Rabies Prevention. The blueprint for rabies prevention and control: a novel operational toolkit for rabies elimination. PLoS Negl Trop Dis. 2012;6:e1388. https://doi.org/10.1371/journal.pntd.0001388.
Lembo T, Attlan M, Bourhy H, et al. Renewed global partnerships and redesigned roadmaps for rabies prevention and control. Vet Med Int. 2011;2011:923149. https://doi.org/10.4061/2011/923149.
Mallewa M, Fooks AR, Banda D, et al. Rabies encephalitis in malaria-endemic area, Malawi, Africa. Emerg Infect Dis. 2007;13:136–9.
Mauti S, Traoré A, Hattendorf J, et al. Factors associated with dog rabies immunisation status in Bamako, Mali. Acta Trop. 2017a;165:194–202. https://doi.org/10.1016/j.actatropica.2015.10.016.
Mauti S, Traoré A, Sery A, et al. First study on domestic dog ecology, demographic structure and dynamics in Bamako, Mali. Prev Vet Med. 2017b;146:44–51. https://doi.org/10.1016/j.prevetmed.2017.07.009.
Mindekem R, Lechenne MS, Naissengar KS, et al. Cost description and comparative cost efficiency of post-exposure prophylaxis and canine mass vaccination against rabies in N'Djamena, Chad. Front Vet Sci. 2017;4:38. https://doi.org/10.3389/fvets.2017.00038.
Morters MK, Restif O, Hampson K, et al. Evidence-based control of canine rabies: a critical review of population density reduction. J Anim Ecol. 2013;82:6–14. https://doi.org/10.1111/j.1365-2656.2012.02033.x.
Mosimann L, Traoré A, Mauti S, et al. A mixed methods approach to assess animal vaccination programmes: the case of rabies control in Bamako, Mali. Acta Trop. 2017;165:203–15. https://doi.org/10.1016/j.actatropica.2016.10.007.
Müller TDP, Moynagh J, Cliquet F, et al. Rabies elimination in Europe—a success story. In: Rabies control—towards sustainable prevention at the source, compendium of the OIE global conference on rabies control. Incheon Seoul: OIE Paris; 2012. p. 31–44.
Müller T, Freuling CM, Wysocki P, et al. Terrestrial rabies control in the European Union: historical achievements and challenges ahead. Vet J Lond Engl. 2015;203:10–7. https://doi.org/10.1016/j.tvjl.2014.10.026.

Muthiani Y, Traoré A, Mauti S, et al. Low coverage of central point vaccination against dog rabies in Bamako. Mali Prev Vet Med doi. 2015; https://doi.org/10.1016/j.prevetmed.2015.04.007.
Nel LH. Discrepancies in data reporting for rabies, Africa. Emerg Infect Dis. 2013a;19:529–33. https://doi.org/10.3201/eid1904.120185.
Nel LH. Factors impacting the control of rabies. Microbiol Spectr. 2013b;1. https://doi.org/10.1128/microbiolspec.OH-0006-2012.
OIE – Manual of diagnostic tests and vaccines for terrestrial animals. Rabies, 2012.
Rupprecht CE, Hanlon CA, Hemachudha T. Rabies re-examined. Lancet Infect Dis. 2002;2:327–43.
Russell CA, Smith DL, Childs JE, Real LA. Predictive spatial dynamics and strategic planning for raccoon rabies emergence in Ohio. PLoS Biol. 2005;3:e88. https://doi.org/10.1371/journal.pbio.0030088.
Scott TP, Coetzer A, de Balogh K, et al. The pan-African rabies control network (PARACON): a unified approach to eliminating canine rabies in Africa. Antivir Res. 2015;124:93–100. https://doi.org/10.1016/j.antiviral.2015.10.002.
Scott TP, Coetzer A, Fahrion AS, Nel LH. Addressing the disconnect between the estimated, reported, and true rabies data: the development of a regional African rabies bulletin. Front Vet Sci. 2017;4:18. https://doi.org/10.3389/fvets.2017.00018.
Slate D, Algeo TP, Nelson KM, et al. Oral rabies vaccination in North America: opportunities, complexities, and challenges. PLoS Negl Trop Dis. 2009;3:e549. https://doi.org/10.1371/journal.pntd.0000549.
Talbi C, Holmes EC, de Benedictis P, et al. Evolutionary history and dynamics of dog rabies virus in western and Central Africa. J Gen Virol. 2009;90:783–91. https://doi.org/10.1099/vir.0.007765-0.
Talbi C, Lemey P, Suchard MA, et al. Phylodynamics and human-mediated dispersal of a zoonotic virus. PLoS Pathog. 2010;6:e1001166. https://doi.org/10.1371/journal.ppat.1001166.
Townsend SE, Lembo T, Cleaveland S, et al. Surveillance guidelines for disease elimination: a case study of canine rabies. Comp Immunol Microbiol Infect Dis. 2013;36:249–61. https://doi.org/10.1016/j.cimid.2012.10.008.
United Nations. Sustainable development goals. Available from http://www.un.org/sustainabledevelopment/sustainable-development-goals/. Accessed 24 Aug 2016.
Velasco-Villa A, Escobar LE, Sanchez A, et al. Successful strategies implemented towards the elimination of canine rabies in the Western Hemisphere. Antivir Res. 2017;143:1–12. https://doi.org/10.1016/j.antiviral.2017.03.023.
Vigilato MAN, Clavijo A, Knobl T, et al. Progress towards eliminating canine rabies: policies and perspectives from Latin America and the Caribbean. Philos Trans R Soc Lond Ser B Biol Sci. 2013;368:20120143. https://doi.org/10.1098/rstb.2012.0143.
Weyer J, Blumberg L. Rabies: challenge of diagnosis in resource poor countries. Infect Dis J Pak Brief Comm. 2007:86–8.
WHO. Guidelines for dog rabies control. WHO document VPH/83.43: Rev. 1. Geneva: World Health Organization; 1987.
WHO. Report of dog ecology studies related to rabies. WHO/Rab.Res/88.25. Geneva: World Health Organization; 1988.
WHO. World Health Organization expert consultation on rabies first report. Technical report series no. 931. Geneva: World Health Organization; 2005.
WHO. The control of neglected zoonotic diseases: a route to poverty alleviation. Geneva: World Health Organization; 2006.
WHO. The control of neglected zoonotic diseases: community-based interventions for prevention and control. Geneva: World Health Organization; 2010.
WHO. World Health Organization expert consultation on rabies. Second report. Technical report series no. 982. Geneva: World Health Organization; 2013.
WHO. Epidemiology and burden of disease. 2014. Available at http://www.who.int/rabies/epidemiology/en/. Accessed 30 Apr 2015.
WHO. Investigating to overcome the global impact of neglected tropical diseases. Third report on neglected tropical diseases. Geneva: World Health Organization; 2015a.

WHO. Global elimination of dog-mediated human rabies – the time is now! Conference report. Geneva: WHO; 2015b.

Zinsstag J, Schelling E, Wyss K, Mahamat MB. Potential of cooperation between human and animal health to strengthen health systems. Lancet. 2005;366:2142–5. https://doi.org/10.1016/S0140-6736(05)67731-8.

Zinsstag J, Dürr S, Penny MA, et al. Transmission dynamics and economics of rabies control in dogs and humans in an African city. Proc Natl Acad Sci USA. 2009;106:14996–5001. https://doi.org/10.1073/pnas.0904740106.

Zinsstag J, Bonfoh B, Cissé G, et al. Towards equity effectiveness in health interventions. In: Wiesmann U, Hurni H, et al., editors. Research for sustainable development: foundations, experiences, and perspectives. Bern: Geographica Bernensia; 2011a. p. 623–40.

Zinsstag J, Schelling E, Waltner-Toews D, Tanner M. From "one medicine" to "one health" and systemic approaches to health and well-being. Prev Vet Med. 2011b;101:148–56. https://doi.org/10.1016/j.prevetmed.2010.07.003.

Zinsstag J, Waltner-Toews D, Tanner M. 2 theoretical issues of one health. In: Zinsstag J, Schelling E, Waltner-Toews D, et al., editors. One health: the theory and practice of integrated health approaches. Wallingford: CABI; 2015.

Chapter 8
Rift Valley Fever: *One Health* at Play?

Renaud Lancelot, Catherine Cêtre-Sossah, Osama Ahmed Hassan, Barry Yahya, Bezeid Ould Elmamy, Assane Gueye Fall, Modou Moustapha Lo, Andrea Apolloni, Elena Arsevska, and Véronique Chevalier

R. Lancelot (✉) · E. Arsevska
UMR ASTRE, CIRAD, Montpellier, France

ASTRE, Montpellier University, CIRAD, INRA, Montpellier, France
e-mail: renaud.lancelot@cirad.fr; elena.arsevska@cirad.fr

C. Cêtre-Sossah
ASTRE, Montpellier University, CIRAD, INRA, Montpellier, France

CIRAD, UMR ASTRE, Sainte-Clothilde, France
e-mail: catherine.cetre-sossah@cirad.fr

O. A. Hassan
Department of Community Medicine and Global Health, Institute of Health and Society, Global Health Center, Faculty of Medicine, Oslo University, Oslo, Norway
e-mail: osama.a.hassan@umu.se

B. Yahya
Service de Parasitologie, Office National de Recherche et de Développement de l'Elevage (ONARDEL), Nouakchott, Mauritania

B. Ould Elmamy
Service de Pathologie Infectieuse, ONARDEL, Nouakchott, Mauritania

A. G. Fall
Service de Bio-Ecologie et Pathologies Parasitaires, ISRA-LNERV, Dakar-Hann, Sénégal

M. M. Lo
Programme Santé Animale, ISRA-LNERV, Dakar-Hann, Sénégal

A. Apolloni
CIRAD, UMR ASTRE, Montpellier, France

Service de Bio-Ecologie et Pathologies Parasitaires, ISRA-LNERV, Dakar-Hann, Sénégal
e-mail: andrea.apolloni@cirad.fr

V. Chevalier
CIRAD, UMR ASTRE, Montpellier, France

ASTRE, Montpellier University, CIRAD, INRA, Montpellier, France

Epidemiology and Public Health Unit, Institut Pasteur du Cambodge, Phnom Penh, Cambodia
e-mail: chevalier@cirad.fr

M. Kardjadj et al. (eds.), *Transboundary Animal Diseases in Sahelian Africa and Connected Regions*, https://doi.org/10.1007/978-3-030-25385-1_8

Abstract Rift Valley fever (RVF) is a mosquito-borne viral infection mostly encountered in Africa. In its acute form, it severely affects domestic and wild ruminants, dromedaries, and humans. It is considered as an emerging disease, with increased frequency in several regions, and a spread potential to many areas under the influence of two main drivers: environmental (including climatic) changes and animal mobility (livestock trade, transhumance). In this chapter, we discuss the peculiarities of RVF epidemiology in Sahelian Africa and we show how the joint influence of these two drivers may trigger RVF epidemics.

The public health impact of RVF can be severe, with tens of thousands of human cases and hundreds of fatalities recorded during large epidemics. Beyond its direct, negative effects on public and animal health, RVF has large economic consequences related to bans on livestock importation from infected countries. Solutions are available to improve surveillance and control of RVF in Sahelian Africa according to well-defined, risk-based strategies. The implementation of coordinated actions between Public Health and Animal Health authorities would represent an important advance in the *One Health* joint approach of RVF for better prevention, early detection, and reaction.

Keywords Rift Valley fever · Zoonotic disease · *One Health* · Climate change · Livestock trade

Introduction

Rift Valley fever (RVF) is a mosquito-borne viral infection mostly met in Africa. In its acute form, it severely affects several species of mammals, in particular, domestic and wild ruminants, dromedaries, and humans (Linthicum et al. 2016). Its occurrence in animals and humans must be notified to the World Animal Health Organization (OIE), and the World Health Organization (WHO), respectively. It is considered as an emerging disease, with increased frequency in several regions, and a spread potential to many areas (Chevalier 2013). Phylogenetic studies revealed its recent emergence probably in the second half of the nineteenth century in eastern or southern Africa, after contact of cattle and sheep of exotic origin with an unknown selvatic cycle (Bird et al. 2007).

RVF is widespread in continental Africa, including Sahelian and northern Africa (Arsevska et al. 2016; Kenawy et al. 2018). Spillovers of large epidemics in the Horn of Africa were reported in the Arabian Peninsula (Balkhy and Memish 2003), the Comoros Archipelago (Lernout et al. 2013, Metras et al. 2017), and Madagascar (Morvan et al. 1992; Andriamandimby et al. 2010) (Fig. 8.1).

The public health impact of RVF can be severe. In Egypt in 1976, 200,000 people were infected, and 600 fatal cases of the hemorrhagic-like disease were recorded (Meegan et al. 1979). The estimated number of deaths was 224 in Mauritania in 1987 (Jouan et al. 1988). In 2007–2008, 738 human cases were officially reported in Sudan, including 230 deaths (Hassan et al. 2011).

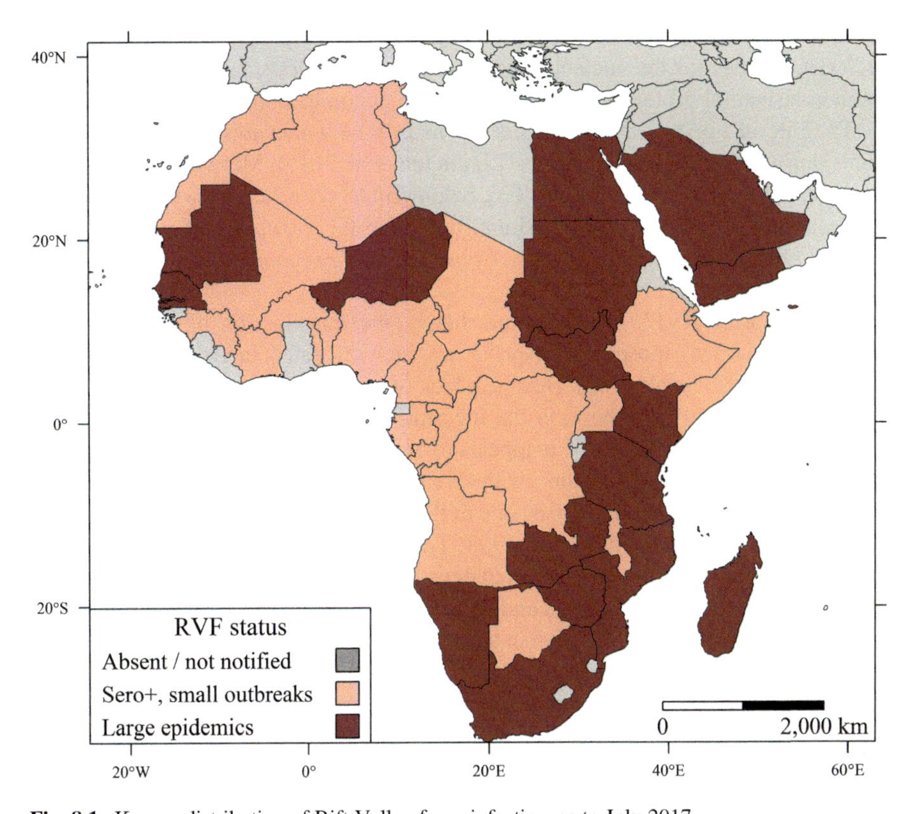

Fig. 8.1 Known distribution of Rift Valley fever infection up to July 2017

Beyond its direct, negative effects on public and animal health, RVF has large economic consequences related to bans on livestock importation from infected countries (Pratt et al. 2005; Peyre et al. 2014).

Epidemiology

Transmission cycle

The Rift Valley fever virus (RVFV) is a member of the *Phlebovirus* genus (Phenuiviridae family of the order Bunyavirales). It is a negative-sense, single-stranded ribonucleic acid (RNA) virus. Only one serotype is recognized but the virulence may vary according to specific RVFV strains. The RVFV genome is organized in three negative-sense, single-stranded segments denominated as large (L), medium (M), and small (S) with a total size of 11.9 kb. The L segment (6.4 kb) encodes the RNA-dependent RNA polymerase; the M segment (3.2 kb) encodes at least four viral proteins in a single open reading frame (ORF): two major envelope surface glycoproteins, the 55 kDa Gn and 58 kDa Gc, plus two accessory proteins, a

14-kDa non-structural NSm and a 78-kDa fusion protein. The S ambisense segment (1.7 kb) encodes for the nucleoprotein in the antigenomic strand (NP; 27 kDa) and the non-structural protein (NSs; 31-kDa) in the genomic strand (Bird et al. 2007).

The bite of infected mosquitoes is the main RVFV transmission route in ruminants during inter-epizootic periods (Linthicum et al. 2016). More than 30 mosquito species were found infected by RVFV, belonging to six genera of which *Aedes* and *Culex* are considered as the most important from the viewpoint of vector competence. Other genera are *Anopheles*, *Coquillettidia*, *Eretmapodite*, *Mansonia,* and *Ochlerotatus*.

Trans-ovarian RVFV transmission, that is, the transmission of the virus from infected females to mosquito offspring was demonstrated in *Aedes mcintoshi* in Kenya (Linthicum et al. 1985). Despite a lack of scientific evidence, this mechanism is still the principal hypothesis in the literature for the survival of RVFV between epizootics (Lumley et al. 2017). It appears to be a likely phenomenon in several other *Aedes* species, including the widespread *Aedes vexans* species complex encompassing *A. vexans* arabiensis, one of the main RVFV vector in Sahelian Africa (Zeller et al. 1997; Fontenille et al. 1998; Traoré-Lamizana et al. 2001). In some of these *Aedes* species, diapaused infected eggs may survive in dried mud during inter-epizootic and/or dry/cold periods and hatch infected imagoes. Thus, *Aedes* mosquitoes are key actors in the primary foci transmission cycle (Fig. 8.2). Following heavy rainfall, temporary ponds (so-called dambos in Kenya) harboring *Aedes* mosquito breeding sites are flooded. In areas where RVF is endemic, some *Aedes* eggs were laid by RVFV-infected females during oviposition, and are thus infected (Fig. 8.2, 1a). Infected eggs hatch and a few days later, emerge as infected imagoes (Fig. 8.2, 2a). They transmit the RVFV to domestic animals during their blood meals. Blood taken from these viremic animals may infect virus-free mosquitoes, thus amplifying the RVFV transmission cycle. Though RVFV transmission by mosquitoes is the dominant infection route, empirical observations indicate that ruminants can also become infected by contact with material containing a virus (e.g., fetus and fetal membranes after abortion) (Nicolas et al. 2013) (Fig 8.2, 3a). The vast majority of human infections result from direct or indirect contact with the blood or organs of infected animals during slaughtering or butchering, assisting with animal parturitions, conducting veterinary cares, or from the disposal of carcasses or fetuses. Fresh and raw meat may be a source of infection for humans, but the virus is destroyed rapidly during meat maturation (Fig. 8.2, 4a). Later in the outbreak progress, human infections may result from the bites of infected *Culex* mosquitoes (Fig. 8.2, 5a).

In addition to these domestic cycles (Figs. 8.2 and 8.3), a selvatic cycle probably allows RVFV persistence in the environment, at least in certain areas where wildlife is abundant and diverse. However, understanding the exact role of wildlife during epizootic or inter-epizoootic periods remain unknown. RVFV was actively searched in wild vertebrate hosts, no strong, and repeated evidence were found (Saluzzo et al. 1987a; Gora et al. 2000; Olive et al. 2012, 2013). Domestic ruminants infected with RVFV at primary transmission sites are introduced (transhumance, trade) into irrigation sites: rice paddies, sugar cane fields, etc. (Fig. 8.3, 1b). *Culex*, *Mansonia*, and *Anopheles* mosquitoes are abundant in these irrigation schemes; they amplify the

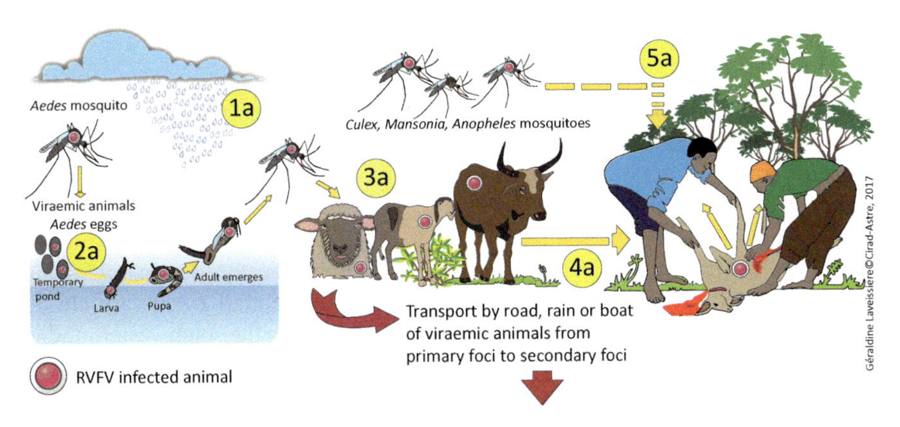

Fig. 8.2 Rift Valley fever transmission cycle: primary foci. Source: Pierre Formenty, Emerging and Dangerous Pathogens Laboratory Network, World Health Organization

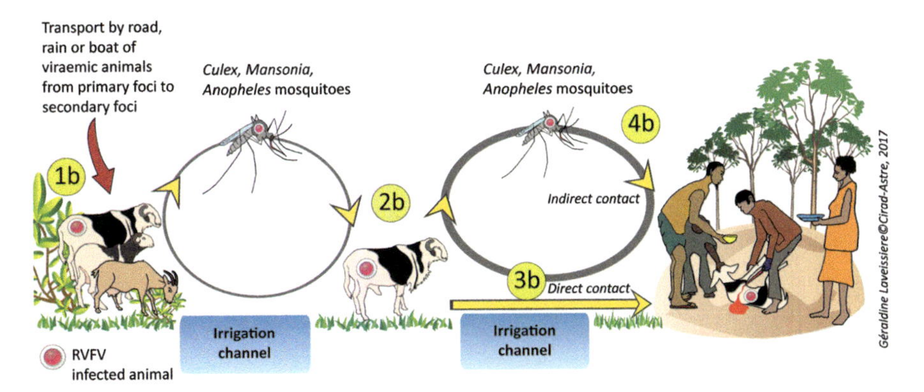

Fig. 8.3 Rift Valley fever transmission cycle: Secondary foci. The cross within a circle indicates RVF-infected sheep. Source: Pierre Formenty, Emerging and Dangerous Pathogens Laboratory Network, World Health Organization

RVF outbreak in animals by biting and infecting them. In Senegal and Mauritania, *Culex poicilipes* is often the dominant *Culex* species and is considered as an important RVFV vector (Diallo et al. 2000, 2005) (Fig. 8.3, 2b). Like for primary foci, human infections may result from direct or indirect contact with the blood or organs of infected animals (Fig. 8.3, 3b). However, many human infections result from the bites of infected *Culex* mosquitoes biting both animals and human. Direct human-to-human transmission has not been reported. RVFV infection may cause miscarriages in pregnant women (Baudin et al. 2016). More generally, trans-placental RVFV transmission may occur in vertebrates. In domestic ruminants, it results in abortion and high newborn mortality rates (Fig. 8.3, 4b). As suggested by several recent serological surveys, wild ruminants, such as African buffaloes may play a role in the epidemiology of RVF in areas where their population density is high (Walsh et al. 2017; Moiane et al. 2017). Rodents and bat species also have been suspected.

Transmission Dynamics

Epizootic and epidemic waves usually last for 2 years (two consecutive rainy seasons). Then, RVFV activity progressively fades out and apparently stops. However, several epidemiological surveys highlighted low-noise RVFV transmission during these inter-epizootic phases (Zeller et al. 1997; Chevalier et al. 2009; Gray et al. 2015; Olive et al. 2017).

In eastern Africa, new RVF epizootics are triggered by heavy autumn rainfall and the subsequent mosquito proliferation. These extreme rainfall events are closely related to the warm phase of El Niño southern oscillation in the southwestern Indian Ocean, that is, high sea surface temperature (Linthicum et al. 1999). Figure 8.4 shows all RVF epidemics occurring in eastern Africa since 1950 were associated with warm phases of El Niño (the reverse is not true). Unfortunately, the situation is more complex in other African regions.

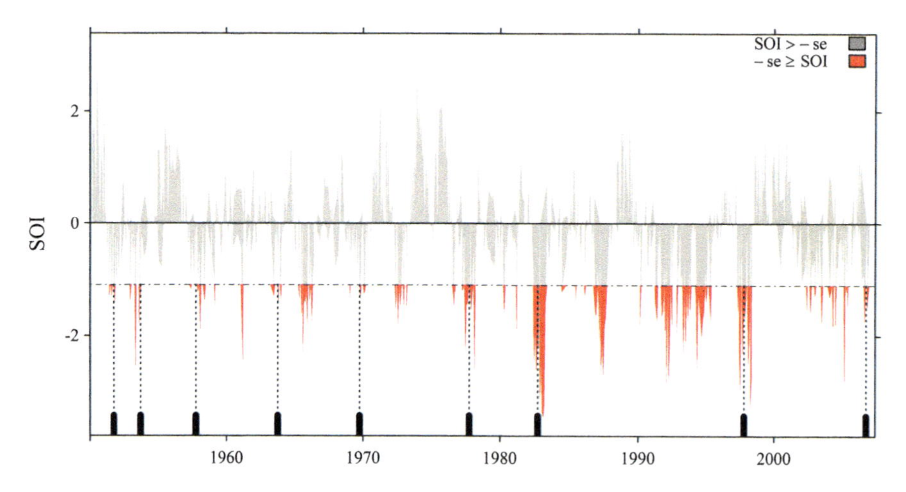

Fig. 8.4 El Niño southern oscillation and RVF epidemics in eastern Africa. The warm phase of El Niño is characterized by negative values of the southern oscillation index (SOI). Such values are encountered when the sea surface temperature is warmer than usual in south-western Indian Ocean. The horizontal, dashed lines show minus standard error (se) of the SOI time series. RVF epidemics are represented by black bars on the *x*-axis. Data source for RVF outbreaks: Greenhalgh (2015); SOI data: NOAA, https://www.esrl.noaa.gov/psd/data/correlation/soi.data (accessed on 24 August 2017)

Clinical Expression

Animals

Clinical expression of RVF is highly variable according to animal species and breed, as well as other factors like age and gravid status (Gerdes 2004; Pépin et al. 2010). During epizootics, mass abortion waves, and high mortality in neonates (especially in newborn lambs) are typically observed in domestic ruminants including dromedaries. When previously RVF-free populations of domestic ruminants are hit by the virus, the abortion rate may be as high as 100% in all species.

In their review, Chevalier et al. (2010) report that (in sheep), "a fever of up to 41–42 °C is observed after a short incubation period. Newborn lambs (and sometimes kids) usually die within 36 to 40 hours after the onset of symptoms, with mortality rates sometimes reaching 95%. Older animals (from 2 weeks to 3 months) either die or develop only a mild infection. In pregnant ewes, abortions are frequent, ranging from 5 to 100%. Twenty percent of the aborting ewes die. Vomiting may be the only clinical sign presented by adult sheep and lambs older than 3 months. However, these animals may experience fever with depression, hemorrhagic diarrhea, bloodstained muco-purulent nasal discharge, and icterus. Case-fatality rates vary between 20 and 30%. Adult goats develop a mild form of the disease, but abortions are frequent (80%). Mortality rates are generally low. In cattle, calves often develop acute illness, with fever, fetid diarrhea, and dyspnea. Mortality rates may vary from 10 to 70%. Abortion is often the only clinical sign and mortality rates are low (10–15%)."

Humans

In humans, asymptomatic forms occur in 50% of infected individuals (Gerdes 2004; Pépin et al. 2010). According to Chevalier et al. (2010), "flu-like syndromes are also very frequent (fever, muscular and articular pains, headaches). However, infected people may experience an undifferentiated, severe, influenza-like syndrome and hepatitis with vomiting and diarrhea. On average, severe—possibly fatal—cases are reported in ca. 1% of infected people. They are manifested in three different clinical syndromes. The most frequent one is a maculo-retinitis, with blurred vision and a loss of visual acuity due to retinal hemorrhage and macular edema. Encephalitis may also occur, accompanied by confusion and coma. This form is rarely fatal but permanent sequels are encountered. The third and most severe form is a hemorrhagic fever, with hepatitis, thrombocytopenia, icterus, and multiple hemorrhages. This form is often fatal. Human case-fatality rates have been lower than 1% in the past, however, an increase has been reported since 1970. In the RVF epidemic in Saudi Arabia in the year 2000, the fatality rate reached 14%."

Diagnostic

RVFV presents a high biohazard for livestock farmers, veterinarians, butchers, slaughterhouse employees, and laboratory staff handling infected biological samples. Biosafety level 3 (BSL3) facilities have to be used for RVFV laboratory diagnostic (Chevalier et al. 2010).

Blood collected on EDTA, as well as serum of infected animals or patients, and organs of dead or euthanized animals such as the liver, brain, spleen, or lymph nodes, are appropriate biological samples for experimental diagnostic. When samples can be quickly sent to a diagnostic laboratory (<2 days), they should be stored at a temperature <+4 °C. Alternatively, samples should be kept at −80 °C (Chevalier et al. 2010).

Virus isolation is the gold standard technique. It is usually performed in suckling or weaned mice by intra-cerebral or intra-peritoneal inoculation or in a variety of mammalian cell cultures: African green monkey kidney Vero, Vero E6, baby hamster kidney BHK21, or mosquito cell lines (*Aedes aegypti* Aag2). Because of its low sensitivity, this technique rarely gives the expected results in due time. Alternative techniques are used, such as the detection of specific RVFV RNA by reverse transcriptase polymerase chain reaction (RT-PCR) and/or by complete genome sequencing performed on RNA extracted directly from biological samples during the acute (febrile) phase of the disease, when high levels of viremia occur in both humans and animals. A range of highly sensitive nucleic acid-based molecular tests has been developed for RVFV including nested RT-PCR methods, quantitative real-time PCR, multiplex PCR-based microarray assay, RT loop-mediated isothermal amplification (RT-LAMP), and recombinase polymerase amplification (RPA). Alternatively, several RT-PCR systems based on one of the three genome segments, or on all of them, have been developed with high levels of sensitivity and specificity. They are the priority tests recommended by the OIE for RVF genome detection, giving results within a few hours (Wilson et al. 2013). The use of RVF RT-LAMP leading to the formation of deoxyribonucleic acid (DNA) precipitate detectable by the naked eye does not require any complex equipment to be read out and are therefore easily applied in field situations. The RVF RPA method utilizing three oligonucleotides and a specific fluorescent probe is as sensitive as LAMP and quantitative real time RT-PCR (qRT-PCR). It has the greater potential to be operated with a hand-held battery device, as a pen-side test.

At the earliest, the detection of antibodies is possible as soon as 4 days following infection or vaccination in animals reacting very early, and 8 days post-vaccination on average. Serological tests to detect antibodies against RVFV include the highly specific virus neutralization test (VNT) considered as the gold standard. However, it requires BSL3 high containment facilities. Commercial enzyme-linked immunosorbent assays (ELISA) are now available for the detection of immunoglobulins (Ig) type G and M. ELISAs are quick, sensitive, and specific. They do not need BSL3 facilities and thus, they progressively replace VNT, allowing serological diagnosis in ruminants and humans. Recently developed ELISA based on the recombinant RVFV N nucleoprotein were evaluated in infected or vaccinated samples of African origin with a sensitivity reaching 91–98% and a specificity of 100% (Kortekaas et al. 2013).

The Emergence of RVF as a Trans-Boundary Disease

Starting in 1910, the Kenyan Veterinary Services reported outbreaks of a sheep disease, so-called enzootic hepatitis of sheep (Bird et al. 2007). The virus was first identified in 1930, during an outbreak in the greater Rift Valley of Kenya (Findlay and Daubney 1931). Large outbreaks also occurred in South Africa, notably between 1950 and 1951, and between 1974 and 1975 (Gerdes 2004). RVFV was then considered as a mild zoonotic agent. During the first large epidemic, reported in Egypt in 1977–1978, over 600 people died of RVF. The epidemic reached the Mediterranean shore but was not reported in neighboring countries. Another large epidemic hit the Horn of Africa in 1997–1998 (Centers for Disease Control and Prevention 1998). In September 2000, RVF was detected for the first time in Saudi Arabia and Yemen (Al-hazmi et al. 2003). By the end of 2006, the disease had re-emerged in Kenya, followed by Tanzania and Somalia.

Another large epidemic hit Sudan in 2007 in the Nile River Valley around Khartoum. In May 2007, RVF was diagnosed in the Comoros Archipelago. The RVFV was probably introduced there by the trade of live ruminants imported from Kenya or Tanzania during the 2006–2007 epidemics. In 2008–2009, RVF epidemic occurred in Madagascar with over 500 human cases.

The Situation in Sahelian Africa

In Sahelian Africa, the first clinical suspicions of RVF in domestic ruminants and humans were reported in 1934 (Curasson 1934). According to this report, the presence of the disease was still older. Indeed, though data were scarce until the late 1980s, several seroepidemiological surveys brought evidence of active RVFV transmission in domestic and wild ruminants, and humans in Sahelian countries (Table 8.1).

The first major RVF epidemic was reported in 1987–1988 in the Senegal River Valley (Fig. 8.5). For the sole area covered by the Rosso hospital (a town located in the Senegal River Delta), the estimated numbers of infected, sick, and deceased humans were more than 10,000, 1200, and 230, respectively (Jouan et al. 1988, 1990). Livestock farmers reported high abortion rates in all ruminant species including dromedaries. Sero-epidemiological surveys reported a large spread of the infection, from far eastern Mauritania to the Senegal River Delta, and from northern Assaba to The Gambia (Ksiazek et al. 1989; Lancelot et al. 1989). Following this

Table 8.1 Reports of RVFV activity in Sahelian Africa before the 1987–1988 epidemics in southern Mauritania

Location	Year	Species	Test	References
Mali	1934	Sheep	Suspicion	Curasson (1934)
Sudan	1936	Human	Serology	Eisa et al. (1980)
Chad, northern Cameroon	1967	Sheep	Serology	Maurice (1967)
Chad, northern Cameroon	1967	Wild ruminants	Serology	Maurice (1967)
Sudan	1973	Sheep, goat, cattle, human	Virology	Eisa et al. (1977a, b)
Sudan	1976	Cattle, humans	Virology	Eisa et al. (1980)
Sudan	1979–1983	Goat	Serology	Eisa (1984)
Sudan	1979–1983	Dromedary	Serology	Eisa (1984)
Senegal	1981–1986	Sheep, goat, cattle	Serology	Saluzzo et al. (1987a)
The Gambia	1981–1986	Cattle	Serology	Saluzzo et al. (1987a)
Burkina Faso	1981–1986	Sheep, goat, cattle	Serology	Saluzzo et al. (1987a)
Niger	1981–1986	Sheep and goat	Serology	Saluzzo et al. (1987a)
Mauritania	1984–1985	Sheep and goat	Serology	Saluzzo et al. (1987a)
Mauritania	1984–1985	Dromedary	Serology	Saluzzo et al. (1987b)
Mauritania	1984–1985	Human	Serology	Saluzzo et al. (1987b)

first epidemic, new outbreaks were recorded in 1993, 1998, and 2003 (Thiongane et al. 1997; Thonnon et al. 1999; Lancelot 2009).

RVFV activity has increased in this region starting in 2010, when 63 human cases—including 13 fatalities, were reported in the hyper-arid region of Adrar, Mauritania, close to the Moroccan border (Fig. 8.5). This RVF event was related to exceptionally high rainfall, which flooded the temporary ponds and allowed the growth of substantial grasslands. The presence of unusually high forage and water resources was attractive for the livestock farmers who decided to bring their animals by truck from the River Senegal Valley. The RVFV was probably introduced together with its hosts on this occasion and found favorable transmission conditions to establish a local cycle (El Mamy et al. 2011).

In 2012, 34 severe cases were reported in humans leading to 17 fatalities in the arid Tagant region (Sow et al. 2014). A new RVF outbreak occurred in 2015 in the same region as well as in central and south-eastern Mauritania, with 57 confirmed cases and 12 fatalities (Bob et al. 2017).

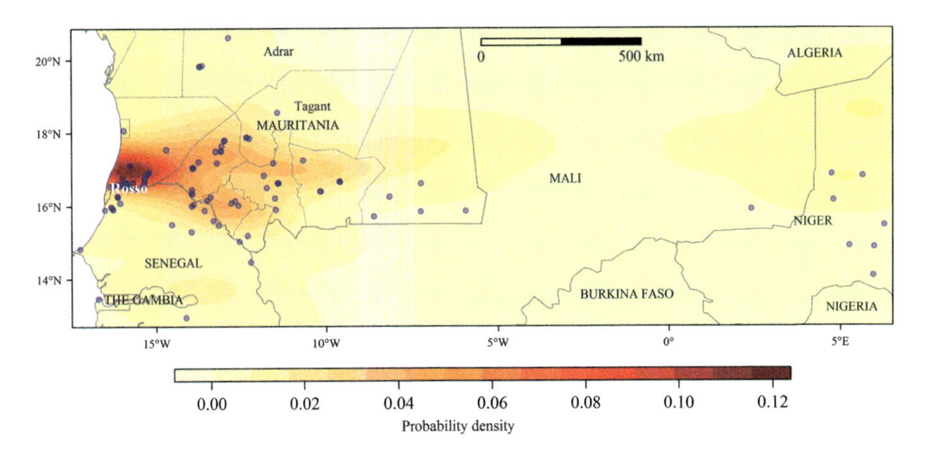

Fig. 8.5 Distribution of reported RVF outbreaks in Mauritania, Senegal, The Gambia, Mali, and Niger, 1987–2016 ($n = 103$ foci). Blue dots represent the foci, and the colored regions show estimates of their probability density. Administrative regions cited in the text are outlined and labeled. Sources of the data: literature review (Arsevska et al. 2016), OIE, WHO, and FAO information systems

In general, there was a 1–2 year delay between the occurrence of RVF in Mauritania and its apparent spread to Senegal, whatever the meteorological situation: rainfall and environmental conditions are not the only drivers of RVF outbreaks in this region. For example, the large outbreak that hit Senegal in 2013 occurred a few months after the 2012 epidemic in Mauritania. This outbreak had a large extent, involving for the first time the densely populated areas of Thiès and Dakar. Animal RVF foci were also reported in 2014 and 2015 (Sow et al. 2016).

In August 2016, an RVF outbreak was reported in Niger, in the region of Tahoua bordering Mali (World Health Organisation 2016). The outbreak also coincided with the annual *Cure Salée* gathering nomadic livestock farmers and their herds in Ingall, Tahoua region. Waves of livestock abortions and deaths were reported in Boni-Bangou while human and animal cases were confirmed in the neighboring region of Menaka in Mali.

Figure 8.5 summarizes the *reported* RVF outbreaks in Mauritania, Senegal, and The Gambia since the 1987 epidemics. The epicenter of reported RVFV activity is on the northern shore of the Senegal River Delta. Most foci were recorded in south-western Mauritania and a small part of northern Senegal.

Drivers of Emergence in Sahelian Africa

In this region, RVF outbreaks do not necessarily occur during wetter rainy seasons. In 1987, the largest ever recorded RVF epidemic in West Africa happened during a drier-than-normal rainy season (Ndione et al. 2005). These authors also rejected the

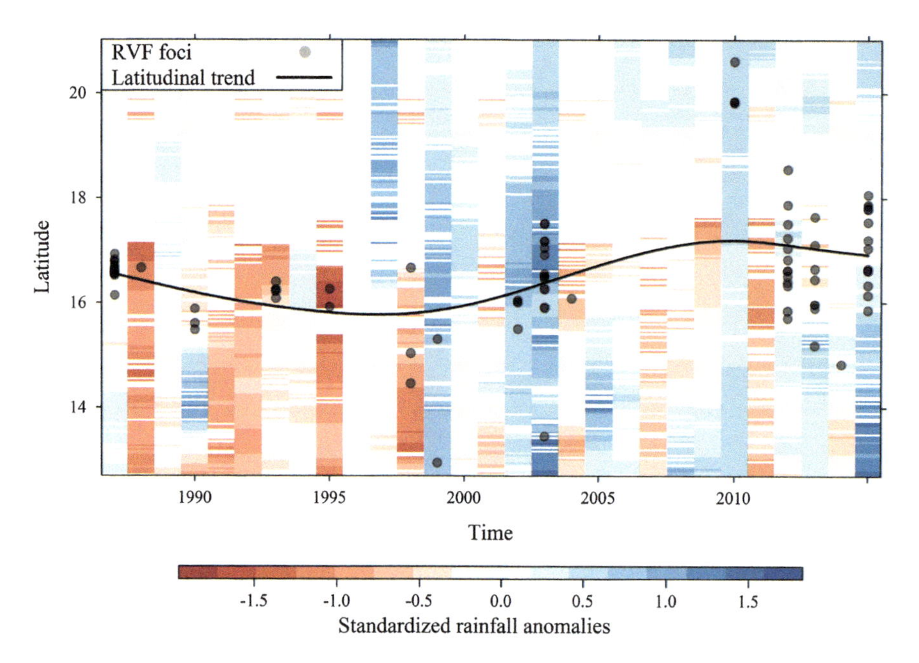

Fig. 8.6 Local rainfall and the occurrence of RVF in Mauritania, Senegal, and The Gambia, 1987–2015. Each pixel shows the standardized rainfall for this latitude over the range of outbreak longitudes. Sources: Literature review and OIE, WHO, and FAO information systems for the outbreak data; TAMSAT for the rainfall data (Maidment et al. 2014; Tarnavsky et al. 2014)

assumption that the epidemic was triggered by the impoundments of the Diama Dam in the River Senegal Delta (Jouan et al. 1988; Ksiazek et al. 1989): the dam had just started operating and the water level in Rosso remained low. Ndione et al. (2005) concluded that the rainfall frequency and intensity might be more important than the overall amount for the population dynamics of floodwater mosquitoes like *Aedes vexans arabiensis*, one of the most important RVFV vectors in the region (Fontenille et al. 1995, 1998; Traoré-Lamizana et al. 2001; Diallo et al. 2005). This assumption was confirmed by fieldwork implemented in the pastoral area of Ferlo, northern Senegal (Mondet et al. 2005). In addition, Fig. 8.6 shows this example is not an exception in the time series of West African RVF outbreaks.

Climatic Drivers

A landscape greening was observed in the Sahelian region during the last decades, as a consequence of increased rainfall since the early 1980s (Fig. 8.7b). However, these rainfall events were more intense and irregular (Anyamba et al. 2014). This change is related to accelerated warming of the northern Atlantic Ocean (NAO) with respect to the average temperature of global tropical oceans, thus suggesting the definition of the subtropical northern Atlantic Ocean index (STNAI) to assess the situation (Giannini et al. 2013). This index is shown in Fig. 8.7a. The

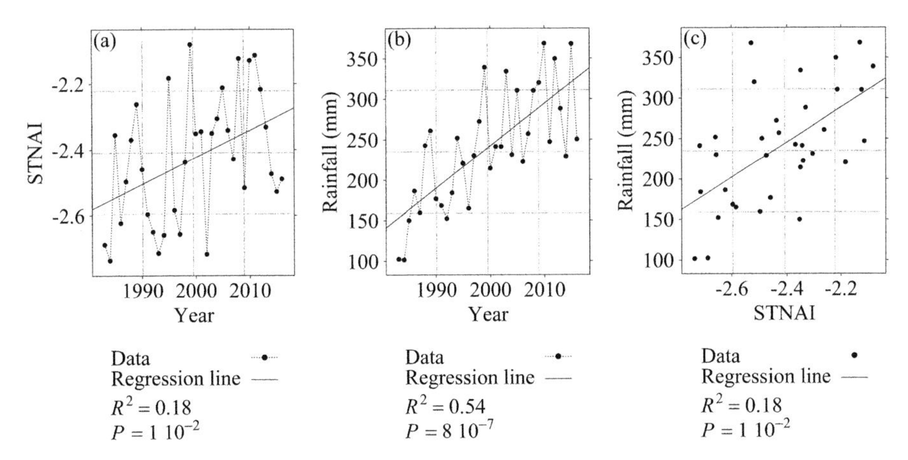

Fig. 8.7 Time trends (**a**) in the subtropical northern Atlantic index (STNAI); (**b**) in annual monsoon rainfall (July–October) in the area covered by the RVF outbreaks in Sahelian Africa, 1987–2016; and (**c**) Relationship between annual monsoon rainfall and STNAI. STNAI was defined by Giannini et al. (2013). Precipitations were estimated from the TAMSAT database (Maidment et al. 2014; Tarnavsky et al. 2014)

underlying mechanism is an increase in solar radiation reaching the sea surface subsequent to the reduction of atmospheric pollution above NAO. A warmer ocean surface temperature means higher water evaporation, and ultimately, more abundant monsoon rainfall (Fig. 8.7c).

As shown by Mondet et al. (2005), sequences of high-intensity rainfall and rainfall pauses are favorable to the population dynamics of *Aedes vexans arabiensis*. On the other hand, the longer persistence of surface water—related to more abundant rainfall—allows the growth of *Culex* populations. In total, the entomological risk is higher. In the meantime, human and animal populations increase, as well as the livestock trade to provide large coastal cities with red meat. In conclusion, ecoclimatic conditions look more favorable today than before for the occurrence and spread of RVF in Sahelian Africa.

Livestock Trade and Transhumance

In general, livestock trade may affect the geographical distribution of RVF and contribute to its introduction into disease-free areas. Thus, RVFV was probably introduced into Egypt from Sudan by infected dromedaries (Hoogstraal et al. 1979). Also, during the outbreak in Saudi Arabia and Yemen in 2000, isolated RVFV strains were genetically close to those isolated in Kenya (1997–1998), suggesting that the virus was probably brought in these two countries from the Horn of Africa by live ruminant trade (Balkhy and Memish 2003; Abdo-salem et al. 2011): 10 to 15 million live small ruminants are imported each year from the Horn of Africa to Saudi Arabia. Similarly, the assumption that RVFV was introduced from Africa mainland to the Comoros Archipelago and to Madagascar, was supported by phylogenetic analyses (Carroll et al. 2011; Maquart et al. 2016). After its introduction into Madagascar, RVFV could quickly spread to the entire island following the cattle trade network (Nicolas et al. 2014; Lancelot et al. 2017).

Major features of Sahelian livestock farming are (1) the strong importance of domestic ruminant populations (and very scarce wild ruminant populations), and (2) the intensity of their mobility, either for transhumance or for trade. In general, domestic ruminants are first reared in Sahelian Africa and later traded to the major cities in the Sahel (Nouakchott, Dakar, Bamako, etc.) and the Guinea Gulf (Abidjan, Lagos, etc.) (Nicolas et al. 2018). Figure 8.8a shows the small ruminant density in Sahelian Africa (colored horizontal strip on the central part of the map). A peculiarity of sheep in Sahelian Africa is their use in religious celebrations. In particular, many Muslim families slaughter a male lamb on the Tabaski day (Aïd el Kebir). Specific commodity chains are implemented to collect lambs in the production areas, bring them to large markets and dispatch them to fattening farms. Also, many

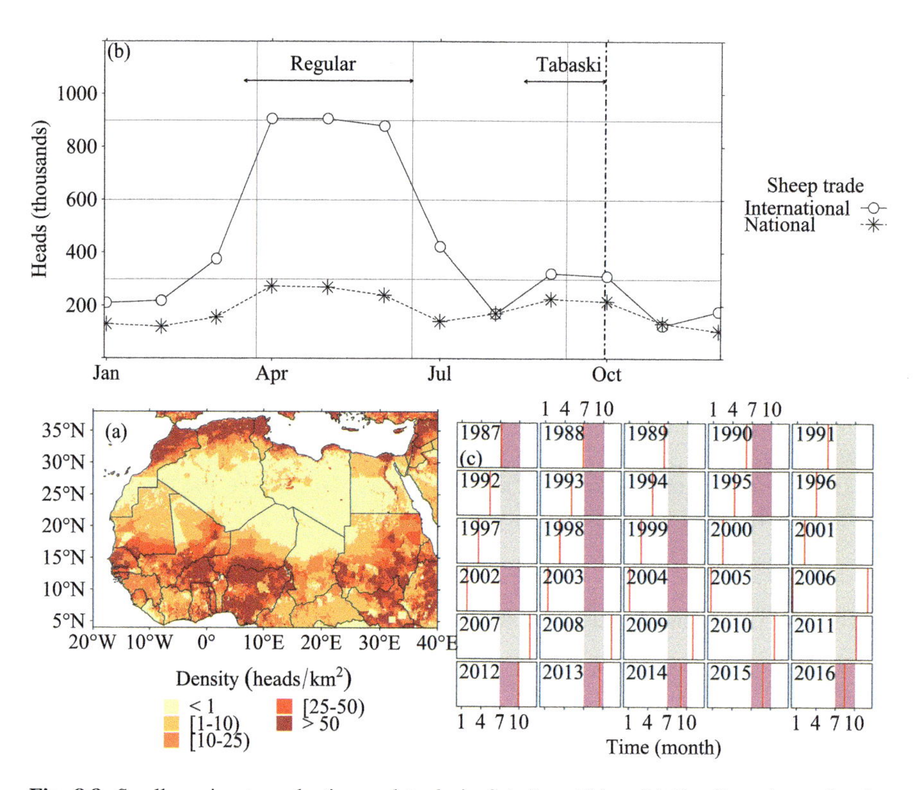

Fig. 8.8 Small ruminant production and trade in Sahelian Africa. (**a**) Small ruminant density (sheep and goats); (**b**) Small ruminant trade flows in Mauritania, 2013. (**c**) Tabaski date (vertical red bar) according to the rainy season (shaded areas). Areas shaded in magenta color indicate the occurrence of Rift Valley fever outbreaks. Data source for small ruminants: Gridded livestock of the world (Nicolas et al. 2016); for small ruminant trade: field surveys implemented by the Mauritanian Veterinary Services and the National Center for Livestock and Veterinary Research (ONARDEL)

families purchase a few sheep and feed them either to cover their own needs or as a speculative activity. As reported by Apolloni et al. (2018), for the sole Mauritania, several million sheep are annually exported to neighboring countries between March and July (Fig. 8.8b), that is, as soon as offspring reach a commercial size. Besides this large exportation peak, a smaller peak can be seen in September–October, just before the Tabaski festival which was celebrated in mid-October when these data were collected (2013). Because Tabaski's date is based on the Lunar calendar (354 days/lunar year), it moves forward by 11 days each year with respect to the Gregorian calendar (365 days/year). Therefore, every 20–30 years, the Tabaski occurs during the rainy season (Fig. 8.8c), that is, when the activity of mosquito populations is high.

Therefore, there is a high risk of spreading the RVFV with livestock trade during this period, and a high risk of human infection if purchased animals are viremic when they are slaughtered. Indeed, the 1987–1988 RVF epidemic, and the outbreaks that have been occurring since 2010, all happened during the rainy season.

Surveillance

Methods and Tools

All the usual disease surveillance methods can be used. We only comment peculiarities of RVF surveillance in Sahelian Africa.

Passive Surveillance

Passive surveillance (PS) is any surveillance activity based on the spontaneous declaration of cases or suspicions of disease under surveillance. Thus, it is based on reporting of RVF suspicions by the field actors: farmers, community animal health workers, veterinary technicians, and veterinarians.

PS is the most widespread surveillance system implemented by the national veterinary services in Sahelian Africa. In countries where RVF is endemic, clinical cases might be difficult to detect and might be confounded with many other diseases (e.g., peste des petits ruminants). Therefore, taking biological samples and sending them to a diagnostic laboratory for confirmation of the suspicion is a necessity. See the section on diagnostic for practical implementation.

PS is known for its low sensitivity (e.g., its ability to detect animal or human cases). This can be improved by training the actors and activating the network in places where, and at time periods when RVF risk is the highest. Therefore, it is best integrated into a risk-based surveillance framework.

Active Surveillance

Active surveillance (AS) is based on data collection implemented according to one or more designed protocols relying on random sampling frames. It aims at detecting the infection and possibly estimating specific indicators such as the RVF incidence rate, or prevalence rate of antibodies, etc. Such indicators may generally not be estimated with passive surveillance. Their repeated estimation allows assessing the change rate of these indicators, which is the basis of post-vaccination evaluation, for instance.

Several kinds of epidemiological surveys may be used for RVF AS:

- **Participatory surveys** are based on the use of participatory techniques (e.g., interviews and focus group discussions) for the collection of qualitative epidemiological intelligence contained within community observations, existing veterinary knowledge, and traditional oral history. It relies on the techniques of participatory rural appraisal, ethno-veterinary surveys and qualitative epidemiology (Mariner and Paskin 2000).
- **Serological surveys** are widely used to assess the spatial distribution of the infection during or after an outbreak or to estimate the seroprevalence rate, following an outbreak or a vaccination round. They are also useful to cross-check and help to interpret the results of participatory surveys.
- **Demographic surveys** based on structured questionnaires may be used to assess the incidence rate of clinical RVF (abortions, neonatal mortality), both at the individual, or at the epidemiological unit levels. For countries where RVF is endemic, demographic surveys are of special interest to detect high levels of abortions and neonatal mortality. Indeed, in Sahelian Africa, many farmers pay little attention to small ruminant health, and results of participatory surveys do not always allow identifying RVFV activity or quantifying RVF incidence at the level of epidemiological units.
- **Sentinel animals** consist in placing receptive animals in a particular environment to detect the presence of an infectious agent, such as the RVFV. They have been widely used for RVF in Sahelian Africa (Thonnon et al. 1999; Thiongane and Martin 2005; Chevalier et al. 2005). In practice, livestock farmers settled in RVF outbreak locations are enrolled in the survey on a voluntary basis. Their animals are individually identified and sampled several times a year—mostly during the rainy season, to detect seroconversions. A major issue regarding this method is the strong spatiotemporal heterogeneity in RVFV activity: a fine mesh of sentinel herds is needed to detect RVFV activity with good sensitivity (Lancelot 2009).

Strategy

These methods can be combined together to build a risk-based RVF surveillance strategy as summarized in Fig. 8.9.

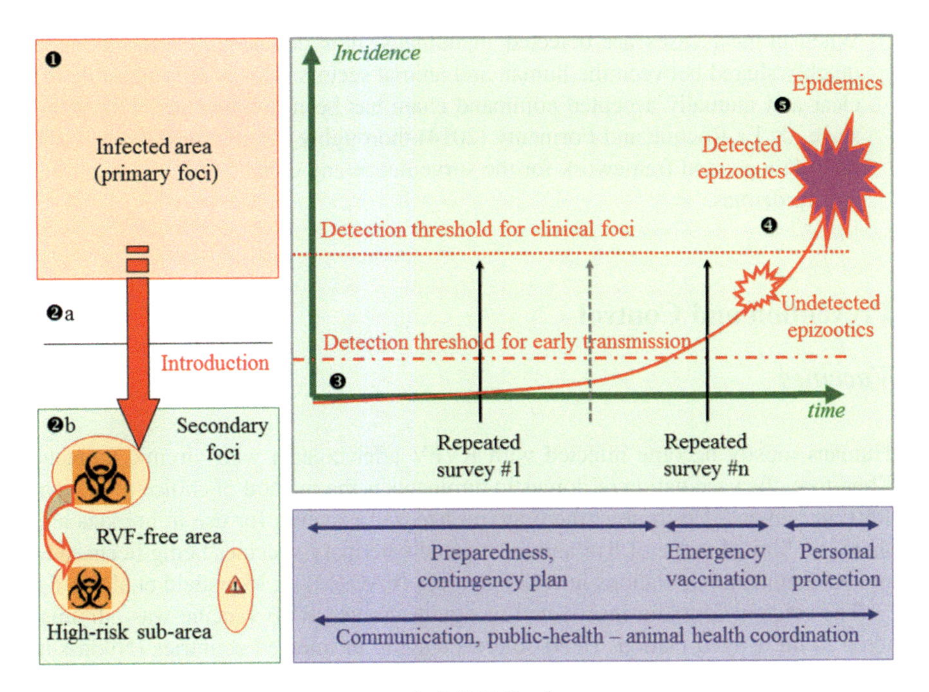

Fig. 8.9 Risk-based surveillance strategy for Rift Valley fever

In the rest of this section, paragraph numbers refer to circled numbers on Fig. 8.9.

1. A regional information system (IS) on RVF outbreaks is needed to gather detailed epidemiological data (hosts, vectors, viruses, locations, dates, etc.). Such an IS does not exist yet. Present systems such as OIE's WAHIS are not appropriate to describe diseases locally considered as enzootic. Anyway, all efforts must be made to draw an accurate picture of RVF situation in the region to be used in the next step.
2. A spatial, qualitative risk analysis (SQRA) should be implemented detailing and combining the risks of introduction (2a) and diffusion (2b), to identify high-risk areas targeted by surveillance activities. A variety of SQRA methods are available (e.g., Arsevska et al. 2015; Tran et al. 2016).
3. In these high-risk areas, implement *repeated* surveys (e.g., once a year) for early detection of RVFV transmission, that is, hopefully before the occurrence of primary clinical foci. A combination of methods can be implemented, according to local features, available resources, and the estimated risk of spread in case of such foci. It looks important to achieve a good spatial coverage, whatever the methodological choices.
4. When clinical outbreaks are detected in animals, surveillance must be quickly reinforced for accurate and timely information of public- and animal-health competent bodies. A key point is coordination between these two bodies.

5. When clinical cases are detected in humans, it is crucial that information is quickly shared between the human and animal sectors. This is only possible if a clear and mutually accepted command chain has been implemented *before* this stage. de La Rocque and Formenty (2014) thoroughly discuss these aspects and provide a general framework for the surveillance and control of RVF epizootics and epidemics.

Prevention and Control

Vaccines

Humans mostly become infected with RVFV after contact with viremic animals. Therefore, the vaccination of domestic ruminants is the method of choice to prevent RVF in humans. Up to now, there are no licensed vaccines for use in humans and there is a limited range of live and inactivated veterinary vaccines being licensed to protect ruminant populations in endemic areas (FAO 2011a; Mansfield et al. 2015).

The original formalin-inactivated mosquito-derived RVF vaccine was administered to the United Nations (UN) soldiers staying in infected countries (Niklasson et al. 1985). It was used by the Israeli veterinary services to prevent RVF introduction into Israel after the 1977–1978 epidemic in Egypt (Shimshony et al. 1981). The Egyptian veterinary services also used it to prevent RVF reintroduction from Sudan in 2007. However, this vaccine requires boost schedules to induce the desired level of protection. This feature makes it inappropriate for use in countries with pastoral farming systems, like in the Sahel.

Live-attenuated Smithburn vaccine is licensed in several African countries. It induces life-long protection in sheep, goats, and cattle after a single administration. However, it has residual pathogenicity and may cause fetal abnormalities and/or abortion in ruminants (Botros et al. 2006). In addition, genetic reassortment between wild and vaccine RVFV was observed in South Africa (Grobbelaar et al. 2011).

The live attenuated MP-12 vaccine is conditionally licensed in the USA. It contains mutations in all three of RVF genome segments. It was promoted as an alternative to the RVF Smithburn vaccine for both human and animal use. However, it has the same potential for reversion to virulence as the Smithburn vaccine. Minimal teratogenic effects were seen in ruminants with suitable immunogenicity in non-human primates following a high dose intravenous or aerosol challenge (Morrill and Peters 2011).

Finally, the clone 13 vaccine is a live attenuated RVFV strain derived from a virus isolated from a moderately ill patient in the Central African Republic. It induces the production of neutralizing antibodies against RVFV in animals (Njenga et al. 2015). The vaccine is well tolerated in West African breeds of sheep and goats—including pregnant animals, without any detectable viremia regardless of physiological status when the recommended dose is administered (Lo et al. 2015). The deletion of the NSs gene offers the potential to develop serological tests to differentiate vaccinated from infected animals (DIVA). This vaccine is licensed in South Africa since 2010 and is widely used there as well as in Kenya for commercial use. It is also registered in Botswana and Namibia. However, the potential for coinfection, reversion of virulence, and reassortment between live-attenuated vaccines and wild-type viruses, suggests that there is still an urgent requirement for a new generation of veterinary RVFV vaccines based on the DIVA strategy, and for a safe and efficacious human vaccine.

New generation vaccines include virus-vectored vaccines based on Poxviridae viruses, Newcastle disease virus, chimpanzee adenovirus, or equine herpesvirus (Mansfield et al. 2015). Importantly, recombinant pox virus-based vaccine preparations offer DIVA potential that is not seen with other vaccines at the exception of clone 13. Other preparations including subunit and DNA vaccines are often poorly immunogenic and require several inoculations to elicit high titer of neutralizing antibodies. Critical progress is the ability to make cDNA copies of each segment of the RVFV genome and rescue recombinant virus-like particles (VLPs) and intact viruses in vitro. In terms of biosafety, the removal of the potential for segment reassortment through reverse genetics has to be considered. Despite progress in VLP vaccine development, high production costs still prevent its use as a veterinary vaccine.

Insecticide Treatments

According to Chevalier et al. (2010), larvicide treatments may be used where mosquito breeding sites are known and restricted to small areas. Both methoprene/pyroxyprofene (hormonal larval growth inhibitors), and *Bacillus thuringiensis israeliensis* (BTI) preparations (a microbial larvicide), are commercially available and can be used to treat temporary ponds and other breeding sites. Unfortunately, most Sahelian regions harboring high densities of mosquito breeding sites are remote with respect to national insect control capacities and difficult to access during the rainy season. In practice, larvicide treatments are not used to control RVF outbreaks.

Adulticide treatments (e.g., pyrethroids) are expensive and difficult to implement. Moreover, adverse environmental and ecological consequences may be important. In practice, they are not used to control RVF outbreaks. Nevertheless, in the line of previous works to improve tick, tsetse, or malaria control, an adaptation of the live-bait trap technique might be used to reduce virus transmission in RVF outbreaks. It consists of treating cattle with a remnant insecticide: mosquitoes are killed when taking their blood meal (Diallo et al. 2008; Poché et al. 2015). Local treatments are even possible with a footbath, to save insecticide and time (Stachurski and Lancelot 2006; Ndeledje et al. 2013). Because of the versatility of RVF outbreaks (scarcity in space and time), the major obstacle for this control measure would be its appropriation by the livestock farmers (Bouyer et al. 2007). Therefore, it should be combined with tick and/or malaria control, implemented with intensive sociological support, for better efficiency.

Both methods may have environmental consequences that should be assessed before a massive implementation.

Other Measures

Preventive measures should also include restrictions on animal movements and markets, the avoidance or control of the slaughter and butchering of ruminants—including personal protection equipment for slaughterhouse workers, the use of insect repellents and bed nets during outbreaks, information campaigns, and increased and targeted surveillance of animals, humans, and vectors (Chevalier et al. 2010; Lancelot et al. 2017). For the same reasons than in the previous section, these measures are difficult to implement for the sole control of RVF: they should be implemented within a global framework for the control of vector-borne infections, including malaria.

Prevention and Control Strategy

A brief overview of RVF prevention and control actions is provided in Fig. 8.9: blue frame at the bottom of the picture.

RVF occurs at the animal, human, and ecosystem interface. In such interconnected determinants, the One Health approach[1] is the best strategy to tackle the tripartite drivers that simultaneously contribute to the emergence of RVF (Hassan et al. 2014). The financial resources expected to be allocated to confront RVF has to follow a One Health economic approach where resources from different ministries could accelerate RVF control at the root such as environment and livestock before it reaches the humans. The successful curb of the RVF at an early stage will protect humans and the public health system from being overwhelmed by the disease later. This could justify why public health sector resources should be on the frontline to strengthen the veterinarian and environmental capacities when the risk of RVF appears.

The primary RVF cycle (Fig. 8.2) is triggered by heavy rains which provide a suitable habitat for mosquitoes that amplify the virus transmission between animals and from animals to humans. Understanding the types of mosquitoes and their behavior is a key to disrupt their habitats and consequently impede RVFV circulation. In essence, RVF is a climate-sensitive disease, which makes it possible to be predicted. Such prediction could offer a window for availing the necessary resources on the ground to control RVFV prior to cause outbreak. However, given the complexity of RVFV epidemiological cycle, the prediction models as an early warning system need accurate data that includes weather, ecology, mosquitoes' activity, livestock densities, and the history of RVF in both animal and human populations. The multicomponents of such efficient predictive model obviously

[1]According to the FAO (2011b), One Health represents a holistic vision to address complex challenges that threaten human and animal health, food security, poverty, and the environments where diseases flourish. These problems threaten global health and economic well-being, including international trade. Many of the dangers stem from diseases circulating in animals, transmitted by food or carried by vectors.

indicates the necessity for the One Health data to make the model protective for all animal, human, and environment.

RVF affects livestock which could then spread the virus along with their movement particularly in open grazing systems, which are dominant in Sahelian Africa. In addition to this, livestock trade may lead to cross-border threat among countries. Therefore, regional livestock trade would need to be reorganized to comply with OIE regulations. This is a major challenge in the Sahelian context with a general lack of animal identification system, and porous borders.

Humans contract the disease by direct contact with infected livestock, in particular, farmers who are at much risk due to their occupational exposure. In such a situation, educating and raising awareness among farmers and their agricultural communities could help to protect them from RVF (Hassan et al. 2017; Affognon et al. 2017). The other aspect of human role in RVF occurrence, is that livestock owners' behavior has the possibility to contribute to the emergence of RVF if they do not comply with the suggested preventive and control measures: for example, if the farmers could not comply with the vaccination strategy, particularly when RVF outbreaks occur after a long interval, considering the vaccine is not for free. Proposing combination vaccines, for example, RVF-contagious bovine pleuropneumonia in cattle, and RVF-peste des petits ruminants in sheep and goats, that could protect livestock against important diseases including RVF in the region could solve such problem if the vaccine is safe, rapid, cost-effective, and easily administered.

The farmers could also sell infected livestock with subclinical signs through livestock markets, which could enhance the circulation of RVFV to new disease-free areas (Fig. 8.2: vaccination might solve such problems). In addition, farmers and livestock traders could also move their viremic livestock for long distance, including crossing borders, which might spread the virus along these routes when the ecology is favorable. Therefore, without the awareness and engagement of farmers and local communities, it would be difficult to fight RVF. To achieve this, the One Health approach should be implemented at the local level of the community, indicating a bottom-up rather than a top-down only approach (Hassan et al. 2017). However, the sensitivity of RVF as a disease that leads to livestock trade ban and accordingly devastative consequences on the rural, national economy as well as disruption of regional and international livestock trade makes the notification quite challenging (Pratt et al. 2005; Hassan et al. 2014, 2017). This sensitivity might impede the farmers and even the countries to notify suspected cases of RVF on livestock if they will be left alone to suffer from the outbreak consequences. Enhancing the early notifications of RVF suspected cases needs reasonable incentives for both farmers and countries which in turn expect to improve the early warning system against RVFV (Hassan et al. 2014).

When RVF has occurred, a fragmented approach to risk communication should be avoided. A disease such as RVF that contains animal, human, and ecosystem interdependence clearly needs a One Health risk communication plan, which includes multiple stakeholders and inevitably local communities, as well as a well-defined and strong coordination at the local, national, and regional levels (Hassan et al. 2014; de La Rocque and Formenty 2014).

Conclusion

More collaboration with sociologists and anthropologists are needed to decipher farmers' perceptions of RVF, and to assess the social acceptability of prevention, surveillance, and control measures, such as cattle, small ruminants, and dromedary vaccination to protect people (should enough vaccine be available), or animal movement restrictions to avoid RVFV spread through hubs in the livestock trade networks (Goutard et al. 2015).

If primary foci continue to occur in Sahelian Africa—a likely situation, targeted (risk based) vaccination campaigns of ruminants might be organized to protect human populations. Individual protection measures, such as vaccination when the human vaccine becomes available, or wearing personal protective equipment such as gowns, gloves, safety glasses, and masks when slaughtering ruminants, would also be important to implement in the most exposed categories of people (Zeller et al. 1998; Bausch and Senga 2017), together with dissemination and training programs.

This implementation of coordinated actions between Public Health and Animal Health authorities would represent an important advance in the One Health joint approach of RVF for better prevention, early detection, and reaction.

Acknowledgments This work was partially funded by INRA grant Meta-programme GISA AMI 2017, and EU grant FP7-613996 VMERGE and is cataloged by the VMERGE Steering Committee as VMERGE000 (http://www.vmerge.eu). The contents of this publication are the sole responsibility of the authors and do not necessarily reflect the views of the European Commission.

References

Abdo-Salem S, Tran A, Grosbois V, Gerbier G, Al-Qadasi M, et al. Can environmental and socioeconomic factors explain the recent emergence of Rift Valley fever in Yemen, 2000–2001? Vector Borne Zoonotic Dis. 2011;11:773–9.

Affognon H, Mburu P, Hassan OA, Kingori S, Ahlm C, Sang R, Evander M. Ethnic groups' knowledge, attitude and practices and Rift Valley fever exposure in Isiolo County of Kenya. PLoS Negl Trop Dis. 2017;11:e0005405. https://doi.org/10.1371/journal.pntd.0005405.

Al-Hazmi M, Ayoola EA, Abdurahman M, Banzal S, Ashraf J, et al. Epidemic Rift Valley fever in Saudi Arabia: a clinical study of severe illness in humans. Clin Infect Dis. 2003;36:245–52.

Andriamandimby SF, Randrianarivo-Solofoniaina AE, Jeanmaire EM, Ravololomanana L, Razafimanantsoa LT, Rakotojoelinandrasana T, Razainirina J, Hoffmann J, Ravalohery J-P, Rafisandratantsoa J-T, Rollin PE, Reynes J-M. Rift Valley fever during rainy seasons, Madagascar, 2008 and 2009. Emerg Infect Dis. 2010;16:963–70. https://doi.org/10.3201/eid1606.091266.

Anyamba A, Small JL, Tucker CJ, Pak EW. Thirty-two years of Sahelian zone growing season non-stationary NDVI3g patterns and trends. Remote Sens. 2014;6:3101–22. https://doi.org/10.3390/rs6043101.

Apolloni A, Nicolas G, Coste C, El Mamy AB, Yahya B, El Arbi AS, Gueya MB, Baba D, Gilbert M, Lancelot R. Towards the description of livestock mobility in Sahelian Africa: some results from a survey in Mauritania. PLoS One. 2018;13:e0191565.

Arsevska E, Hellal J, Mejri S, Hammami S, Marianneau P, Calavas D, Hénaux V. Identifying areas suitable for the occurrence of Rift Valley fever in North Africa: implications for surveillance. Transbound Emerg Dis. 2015; https://doi.org/10.1111/tbed.12331.

Arsevska E, Lancelot R, El Mamy B, Cêtre-Sossah C. Situation épidémiologique de la fièvre de la Vallée du Rift en Afrique de l'Ouest et du Nord [Epidemiological situation of Rift Valley fever in western and northern Africa]. Bulletin épidémiologique, santé animale et alimentation. 2016;74:25–9.

Balkhy HH, Memish ZA. Rift Valley fever: an uninvited zoonosis in the Arabian peninsula. Int J Antimicrob Agents. 2003;21:153–7.

Baudin M, Jumaa A, Jomma H, Karsany M, Bucht G, Näslund J, Ahlm C, Evander M, Mohamed N. Association of Rift Valley fever virus infection with miscarriage in Sudanese women: a cross-sectional study. Lancet Glob Health. 2016;4:e864–71. https://doi.org/10.1016/S2214-109X(16)30176-0.

Bausch DG, Senga M. International encyclopedia of public health. 2nd ed. Amsterdam: Elsevier; 2017. p. 396–409.

Bird BH, Khristova ML, Rollin PE, Ksiazek TG, Nichol ST. Complete genome analysis of 33 ecologically and biologically diverse Rift Valley fever virus strains reveals widespread virus movement and low genetic diversity due to recent common ancestry. J Virol. 2007;81:2805–16. https://doi.org/10.1128/JVI.02095-06.

Bob NS, Bâ H, Fall G, Ishagh E, Diallo MY, Sow A, Sembene PM, Faye O, El Kouri B, Sidi ML, Sall AA. Detection of the northeastern African Rift Valley fever virus lineage during the 2015 outbreak in Mauritania. Open Forum Infect Dis. 2017;4:ofx087. https://doi.org/10.1093/ofid/ofx087.

Botros B, Omar A, Elian K, Mohamed G, Soliman A, Salib A, Salman D, Saad M, Earhart K. Adverse response of non-indigenous cattle of European breeds to live attenuated Smithburn Rift Valley fever vaccine. J Med Virol. 2006;78:787–91. https://doi.org/10.1002/jmv.20624.

Bouyer J, Stachurski F, Kaboré I, Bauer B, Lancelot R. Tsetse control in cattle from pyrethroid footbaths. Prev Vet Med. 2007;78:223–38. https://doi.org/10.1016/j.prevetmed.2006.10.008.

Carroll SA, Reynes J-M, Khristova ML, Andriamandimby SF, Rollin PE, Nichol ST. Genetic evidence for Rift Valley fever outbreaks in Madagascar resulting from virus introductions from the east African mainland rather than enzootic maintenance. J Virol. 2011;85:6162–7. https://doi.org/10.1128/JVI.00335-11.

Centers for Disease Control and Prevention. Rift Valley fever–East Africa, 1997–1998. MMWR Morb Mortal Wkly Rep. 1998;47:261–4.

Chevalier V. Relevance of Rift Valley fever to public health in the European Union. Clin Microbiol Infect Early view. 2013:1–4. https://doi.org/10.1111/1469-0691.12163.

Chevalier V, Lancelot R, Thiongane Y, Sall B, Mondet B. Incidence of Rift Valley fever in small ruminants in the Ferlo pastoral system (Senegal) during the 2003 rainy season. Emerg Inf Dis. 2005;11:1693–700.

Chevalier V, Thiongane Y, Lancelot R. Endemic transmission of Rift Valley fever in Senegal. Transbound Emerg Dis. 2009;56:372–4. https://doi.org/10.1111/j.1865-1682.2009.01083.x.

Chevalier V, Pépin M, Plée L, Lancelot R. Rift Valley fever – a threat for Europe? Euro Surveill. 2010;15, pii=19506.

Curasson G. La fièvre de la Vallée du Rift existe-elle au Soudan français ? [Does Rift Valley fever exist in French Sudan?]. Bulletin de la Société de Pathologie Exotique. 1934;27:599–602.

de La Rocque S, Formenty P. Applying the one health principles: a trans-sectoral coordination framework for preventing and responding to Rift Valley fever outbreaks. Rev Sci Tech OIE. 2014;33:555–67.

Diallo M, Lochouarn L, Ba K, Sall AA, Mondo M, Girault L, Mathiot C. First isolation of the Rift Valley fever virus from *Culex poicilipes* (Diptera: Culicidae) in nature. Am J Trop Med Hyg. 2000;62:702–4.

Diallo M, Nabeth P, Ba K, Sall AA, Ba Y, Mondo M, Girault L, Abdalah MO, Mathiot C. Mosquito vectors of the 1998–1999 outbreak of Rift Valley fever and other arboviruses (Bagaza, Sanar, Wesselsbron and West Nile) in Mauritania and Senegal. Med Vet Entomol. 2005;19:119–26.

Diallo D, Ba Y, Dia I, Lassana K, Diallo M. Utilisation de bœufs traités aux insecticides dans la lutte contre les vecteurs des virus de la fièvre de la Vallée du Rift et de la fièvre West Nile au Sénégal [Use of insecticide-treated cattle to control Rift Valley fever and West Nile virus vectors in Senegal]. Bull Soc Pathol Exot. 2008;101:410–7.

Eisa M. Preliminary survey of domestic animals of the Sudan for precipitating antibodies to Rift Valley fever virus. J Hyg. 1984;93:629–37.

Eisa M, Obeid H, El-Sawi A. Rift Valley fever in the Sudan. I. Results on field investigations of the first epizootic in Kosti District, 1973. Bull Anim Health Prod Afr. 1977a;24:343–7.

Eisa M, Obeid H, El-Sawi A. Rift Valley fever in the Sudan. II - Isolation and identification of the virus from epizootic in Kosti District. Bull Anim Health Prod Afr. 1977b;24:349–55.

Eisa M, Kheir El Sid E, Meegan J. An outbreak of Rift Valley fever in the Sudan – 1976. Trans R Soc Trop Med Hyg. 1980;74:417–8.

El Mamy A, Baba M, Barry Y, Isselmou K, Dia M, Hampate B, Diallo M, El Kory M, Diop M, Lo M, Thiongane Y, Bengoumi M, Puech L, Plée L, Claes F, La Rocque S de, Doumbia B (2011) Unexpected Rift Valley fever outbreak, northern Mauritania. Emerg Infect Dis 17:1894–1896. doi: https://doi.org/10.3201/eid1710.110397.

FAO. Rift Valley fever vaccine development, progress and constraints. In: Proceedings of the GF-TADs meeting, January 2011, Rome, Italy. Roma: FAO; 2011a.

FAO (2011b) One health: food and agriculture of the United Nation. Strategic Action Plan.

Findlay G, Daubney R. The virus of Rift Valley fever or enzoötic hepatitis. Lancet. 1931;218:1350–1. https://doi.org/10.1016/S0140-6736(00)99624-7.

Fontenille D, Traoré-Lamizana M, Zeller H, Mondo M, Diallo M, Digoutte JP. Short report: Rift Valley fever in Western Africa: isolations from *Aedes* mosquitoes during an interepizootic period. Am J Trop Med Hyg. 1995;52:403–4.

Fontenille D, Traoré-Lamizana M, Diallo M, Thonnon J, Digoutte JP, Zeller HG. New vectors of Rift Valley fever in West Africa. Emerg Infect Dis. 1998;4:289–93.

Gerdes G. Rift Valley fever. Rev Sci Tech OIE. 2004;23:613–23.

Giannini A, Salack S, Lodoun T, Ali A, Gaye A, Ndiaye O. A unifying view of climate change in the Sahel linking intra-seasonal, interannual and longer time scales. Environ Res Lett. 2013;8:024010. https://doi.org/10.1088/1748-9326/8/2/024010.

Gora D, Yaya T, Jocelyn T, Didier F, Maoulouth D, Amadou S, Ruel TD, Gonzalez J-P. The potential role of rodents in the enzootic cycle of Rift Valley fever virus in Senegal. Microbes Infect. 2000;2:343–6. https://doi.org/10.1016/S1286-4579(00)00334-8.

Goutard FL, Binot A, Duboz R, Rasamoelina-Andriamanivo H, Pedrono M, Holl D, Peyre MI, Cappelle J, Chevalier V, Figuié M, Molia S, Roger FL. How to reach the poor? Surveillance in low-income countries, lessons from experiences in Cambodia and Madagascar. Prev Vet Med. 2015;120:12–26. https://doi.org/10.1016/j.prevetmed.2015.02.014.

Gray GC, Anderson BD, LaBeaud AD, Heraud J-M, Fèvre EM, Andriamandimby SF, Cook EAJ, Dahir S, de Glanville WA, Heil GL, Khan SU, Muiruri S, Olive M-M, Thomas LF, Merrill HR, Merrill MLM, Richt JA. Seroepidemiological study of interepidemic Rift Valley fever virus infection among persons with intense ruminant exposure in Madagascar and Kenya. Am J Trop Med Hyg. 2015; https://doi.org/10.4269/ajtmh.15-0383.

Greenhalgh E. El Niño, East Africa, and Rift Valley fever. In: Climate.gov. https://www.climate.gov/news-features/understanding-climate/el-ni%C3%B1o-east-africa-and-rift-valley-fever. 2015.

Grobbelaar A, Weyer J, Leman P, Kemp A, Paweska J, Swanepoel R. Molecular epidemiology of Rift Valley fever virus. Emerg Infect Dis. 2011;17:2270–6. https://doi.org/10.3201/eid1712.111035.

Hassan OA, Ahlm C, Sang R, Evander M. The 2007 Rift Valley fever outbreak in Sudan. PLoS Negl Trop Dis. 2011;5:e1229. https://doi.org/10.1371/journal.pntd.0001229.

Hassan OA, Ahlm C, Evander M. A need for one health approach – lessons learned from outbreaks of Rift Valley fever in Saudi Arabia and Sudan. Infection, Ecology & Epidemiology. 2014; https://doi.org/10.3402/iee.v4.20710.

Hassan O, Affognon H, Rocklöv J, Mburu P, Sang R, Ahlm C, Evander M. The one health approach to identify knowledge, attitudes and practices that affect community involvement in the control

of Rift Valley fever outbreaks. PLoS Negl Trop Dis. 2017;11:e0005383. https://doi.org/10.1371/journal.pntd.0005383.
Hoogstraal H, Meegan JM, Khalil GM, Adham FK. The Rift Valley fever epizootic in Egypt 1977–78. 2. Ecological and entomological studies. Trans R Soc Trop Med Hyg. 1979;73:624–9.
Jouan A, Le Guenno B, Digoutte JP, Philippe B, Riou O, Adam F. An RVF epidemic in southern Mauritania. Ann Inst Pasteur Virol. 1988;139:307–8.
Jouan A, Adam F, Riou O, Philippe B, Merzoug N, Ksiazek T, Leguenno B, Digoutte J. Evaluation des indicateurs de santé dans la région du Trarza lors de l'épidémie de fièvre de la vallée du Rift en 1987 [Descriptive study during Rift Valley fever epidemics in Mauritania -Trarza district – in 1987]. Bull Soc Pathol Exot. 1990;83:621–7.
Kenawy M, Abdel-Hamid Y, Beier J. Rift Valley fever in Egypt and other African countries: historical review, recent outbreaks and possibility of disease occurrence in Egypt. Acta Trop. 2018;181:40–9.
Kortekaas J, Kant J, Vloet R, Cêtre-Sossah C, Marianneau P, Lacote S, Banyard AC, Jeffries C, Eiden M, Groschup M, Jäckel S, Hevia E, Brun A. European ring trial to evaluate ELISAs for the diagnosis of infection with Rift Valley fever virus. J Virol Meth. 2013;187:177–81. https://doi.org/10.1016/j.jviromet.2012.09.016.
Ksiazek TG, Jouan A, Meegan JM, Le Guenno B, Wilson ML, Peters CJ, Digoutte JP, Guillaud M, Merzoug NO, Touray EM. Rift Valley fever among domestic animals in the recent West African outbreak. Res Virol. 1989;140:67–77.
Lancelot R. Animaux sentinelles en milieu tropical: vers un système intégré de surveillance. Epidémiol Santé anim. 2009;56:27–34.
Lancelot R, Gonzalez JP, Guenno BL, Diallo BC, Gandega Y. Épidémiologie descriptive de la fièvre de la Vallée du Rift chez les petits ruminants dans le Sud de la Mauritanie après l'hivernage 1988 [Epidemiological investigation on the Rift Valley fever in sheep and goats in Southern Mauritania after 1988 rainy season]. Revue Elev Méd vét Pays trop. 1989;42:485–91.
Lancelot R, Béral M, Rakotoharinome VM, Andriamandimby S-F, Héraud J-M, Coste C, Apolloni A, Squarzoni-Diaw C, La Rocque S de, Formenty PBH, Bouyer J, Wint GRW, Cardinale E (2017) Drivers of rift valley fever epidemics in Madagascar. Proc Natl Acad Sci U S A 114:938–943. doi: https://doi.org/10.1073/pnas.1607948114.
Lernout T, Cardinale E, Jego M, Desprès P, Collet L, Zumbo B, Tillard E, Girard S, Filleul L. Rift Valley fever in humans and animals in Mayotte, an endemic situation? PLoS One. 2013;8: e74192. https://doi.org/10.1371/journal.pone.0074192.
Linthicum KJ, Davis FG, Kairo A, Bailey CL. Rift Valley fever virus. Isolations from Diptera collected during an inter-epidemic period in Kenya. J Hyg. 1985;95:197–209.
Linthicum K, Anyamba A, Tucker C, Kelley P, Myers M, Peters C. Climate and satellite indicators to forecast Rift Valley fever epidemics in Kenya. Science. 1999;285:397–400. https://doi.org/10.1126/science.285.5426.397.
Linthicum KJ, Britch SC, Anyamba A. Rift Valley fever: an emerging mosquito-borne disease. Annu Rev Entomol. 2016;61:395–415. https://doi.org/10.1146/annurev-ento-010715-023819.
Lo M, Mbao V, Sierra P, Thiongane Y, Diop M, Donadeu M, Dungu B. Safety and immunogenicity of onderstepoort biological products' Rift Valley fever clone 13 vaccine in sheep and goats under field conditions in Senegal. Onderstepoort J Vet Res. 2015;82:857. https://doi.org/10.4102/ojvr.v82i1.857.
Lumley S, Horton D, Hernandez-Triana L, Johnson N, Fooks A, et al. Rift Valley fever virus: strategies for maintenance, survival and vertical transmission in mosquitoes. J Gen Virol. 2017;98:875–87. https://doi.org/10.1099/jgv.0.000765.. Epub 2017 May 30. Review
Maidment RI, Grimes D, Allan RP, Tarnavsky E, Stringer M, Hewison T, Roebeling R, Black E. The 30 year TAMSAT African rainfall climatology and time series (TARCAT) data set. J Geophys Res Atmos. 2014;119(10):619, 644 p–0. https://doi.org/10.1002/2014JD021927.

Mansfield KL, Banyard AC, McElhinney L, Johnson N, Horton DL, Hernandez-Triana LM, Fooks AR. Rift Valley fever virus: a review of diagnosis and vaccination, and implications for emergence in Europe. Vaccine. 2015;33:5520–31. https://doi.org/10.1016/j.vaccine.2015.08.020.

Maquart M, Pascalis H, Abdouroihamane S, Roger M, Abdourahime F, Cardinale E, Cêtre-Sossah C. Phylogeographic reconstructions of a Rift Valley fever virus strain reveals transboundary animal movements from eastern continental Africa to the Union of the Comoros. Transbound Emerg Dis. 2016;63:e281–5. https://doi.org/10.1111/tbed.12267.

Mariner JC, Paskin R. Manual on participatory epidemiology: methods for the collection of action-oriented epidemiological intelligence. Rome: FAO; 2000.

Maurice Y. First serological record on the incidence of Wesselsbronn's disease and Rift Valley fever in sheep and wild ruminants in Chad and Cameroon [first serologic verification of the incidence of Wesselsbronn's disease and Rift Valley fever in sheep and wild ruminants in Chad and Cameroon]. Rev Elev Med Vet Pays Trop. 1967;20:395–405.

Meegan JM, Hoogstraal H, Moussa MI. An epizootic of Rift Valley fever in Egypt in 1977. Vet Rec. 1979;105:124–5.

Métras R, Fournié G, Dommergues L, Camacho A, Cavalerie L, et al. Drivers for Rift Valley fever emergence in Mayotte: a Bayesian modelling approach. PLoS Negl Trop Dis. 2017;11: e0005767.

Moiane B, Mapaco L, Thompson P, Berg M, Albihn A, et al. High seroprevalence of Rift Valley fever phlebovirus in domestic ruminants and African buffaloes in Mozambique shows need for intensified surveillance. Infect Ecol Epidemiol. 2017;7:1416248. https://doi.org/10.1080/20008686.2017.1416248.. eCollection 2017

Mondet B, Diaïté A, Ndione J, Fall A, Chevalier V, Lancelot R, Ndiaye M, Ponçon N. Rainfall patterns and population dynamics of *Aedes* (*Aedimorphus*) *vexans arabiensis*, Patton 1905 (Diptera: Culicidae), a potential vector of Rift Valley fever virus in Senegal. J Vector Ecol. 2005;30:102–6.

Morrill J, Peters CJ. Protection of MP-12-vaccinated rhesus macaques against parenteral and aerosol challenge with virulent rift valley fever virus. J Infect Dis. 2011;204:229–36. https://doi.org/10.1093/infdis/jir249.

Morvan J, Rollin PE, Laventure S, Rakotoarivony I, Roux J. Rift Valley fever epizootic in the central highlands of Madagascar. Res Virol. 1992;143:407–15.

Ndeledje N, Bouyer J, Stachurski F, Grimaud P, Belem AMG, Molélé Mbaïndingatoloum F, Bengaly Z, Oumar Alfaroukh I, Cecchi G, Lancelot R. Treating cattle to protect people? Impact of footbath insecticide treatment on tsetse density in Chad. PLoS One. 2013;8:e67580.

Ndione J-A, Bicout DJ, Mondet B, Lancelot R, Sabatier P, Lacaux J-P, Ndiaye M, Diop C. Conditions environnementales associées à l'émergence de la fièvre de la Vallée du Rift (FVR) dans le delta du fleuve Sénégal en 1987. Environnement, Risques et Santé. 2005;4:10005–10.

Nicolas G, Durand B, Tojofaniiry T, Lacote S, Chevalier V, et al. A three years serological and virological cattle follow-up in Madagascar highlands suggests a non-classical transmission route of Rift Valley fever virus. Am J Trop Med Hyg. 2013;90:265–6.

Nicolas G, Chevalier V, Tantely LM, Fontenille D, Durand B. A spatially explicit metapopulation model and cattle trade analysis suggests key determinants for the recurrent circulation of Rift Valley fever virus in a pilot area of Madagascar highlands. PLoS Negl Trop Dis. 2014;8:e3346. https://doi.org/10.1371/journal.pntd.0003346.

Nicolas G, Robinson T, Wint G, Conchedda G, Cinardi G, Gilbert M. Using random forest to improve the downscaling of global livestock census data. PLoS One. 2016;11:1–16. https://doi.org/10.1371/journal.pone.0150424.

Nicolas G, Apolloni A, Coste C, Wint GRW, Lancelot R, Gilbert M. Predictive gravity models of livestock mobility in Mauritania: the effects of supply, demand and cultural factors. PLoS One. 2018;13:e0199547.

Niklasson B, Peters CJ, Bengtsson E, Norrby E. Rift Valley fever virus vaccine trial: study of neutralizing antibody response in humans. Vaccine. 1985;3:123–7.

Njenga MK, Njagi L, Thumbi SM, Kahariri S, Githinji J, Omondi E, Baden A, Murithi M, Paweska J, Ithondeka PM, others (2015) Randomized controlled field trial to assess the immunogenicity and

safety of Rift Valley fever clone 13 vaccine in livestock. PLoS Negl Trop Dis 9: e0003550. doi: https://doi.org/10.1371/journal.pntd.0003550.
Olive M-M, Goodman SM, Reynes J-M. The role of wild mammals in the maintenance of Rift Valley fever virus. J Wildl Dis. 2012;48:241–66.
Olive M-M, Razafindralambo N, Barivelo TA, Rafisandratantsoa J-T, Soarimalala V, Goodman SM, Rollin PE, Heraud J-M, Reynes J-M. Absence of Rift Valley fever virus in wild small mammals, Madagascar. Emerg Infect Dis. 2013;19:1025–7. https://doi.org/10.3201/eid1906.121074.
Olive M-M, Grosbois V, Tran A, Nomenjanahary LA, Rakotoarinoro M, Andriamandimby S-F, Rogier C, Heraud J-M, Chevalier V. Reconstruction of Rift Valley fever transmission dynamics in Madagascar: estimation of force of infection from seroprevalence surveys using bayesian modelling. Sci Rep. 2017;7:39870. https://doi.org/10.1038/srep39870.
Pépin M, Bouloy M, Bird BH, Kemp A, Paweska J. Rift Valley fever virus (Bunyaviridae: Phlebovirus): an update on pathogenesis, molecular epidemiology, vectors, diagnostics and prevention. Vet Res. 2010;41:61. https://doi.org/10.1051/vetres/2010033.
Peyre M, Chevalier V, Abdo-Salem S, Velthuis A, Antoine-Moussiaux N, et al. A systematic scoping study of the socio-economic impact of Rift Valley fever: research gaps and needs. Zoonoses Public Health. 2014; https://doi.org/10.1111/zph.12153.
Poché RM, Burruss D, Polyakova L, Poché DM, Garlapati RB. Treatment of livestock with systemic insecticides for control of *Anopheles arabiensis* in western Kenya. Malar J. 2015;14:351. https://doi.org/10.1186/s12936-015-0883-0.
Pratt A, Bonnet P, Jabbar M, Ehui S, de Haan C. Benefits and costs of compliance of sanitary regulations in livestock markets: the case of Rift Valley fever in the Somali region of Ethiopia. Nairobi: International Livestock Research Institute (ILRI); 2005.
Saluzzo J, Chartier C, Bada R, Martinez D, Digoutte J. La fièvre de la Vallée du Rift en Afrique de l'Ouest. Revue Elev Méd vét Pays trop. 1987a;40:215–23.
Saluzzo JF, Digoutte JP, Chartier C, Martinez D, Bada R. Focus of Rift Valley fever virus transmission in southern Mauritania. Lancet. 1987b;329:504. https://doi.org/10.1016/S0140-6736(87)92110-6.
Shimshony A, Klopfer-Orgad U, Bali S, Chaimovitz M. The influence of information flow on the veterinary policy of Rift Valley fever prevention in Israel, 1978–1979. Contr Epidem Biostatist. 1981;3:159–71.
Sow A, Faye O, Ba Y, Ba H, Diallo D, Faye O, Loucoubar C, Boushab M, Barry Y, Diallo M, Sall AA. Rift Valley fever outbreak, southern Mauritania, 2012. Emerg Infect Dis. 2014;20:296–9. https://doi.org/10.3201/eid2002.131000.
Sow A, Faye O, Ba Y, Diallo D, Fall G, et al. Widespread Rift Valley fever emergence in Senegal in 2013–2014. Open forum. Infect Dis. 2016;3:ofw149.. eCollection 2016
Stachurski F, Lancelot R. Footbath acaricide treatment to control cattle infestation by the tick Amblyomma variegatum. Med Vet Entomol. 2006;20:402–12.
Tarnavsky E, Grimes D, Maidment R, Black E, Allan RP, Stringer M, Chadwick R, Kayitakire F. Extension of the TAMSAT satellite-based rainfall monitoring over Africa and from 1983 to present. J Appl Meteorol Climatol. 2014;53:2805–22. https://doi.org/10.1175/JAMC-D-14-0016.1.
Thiongane Y, Martin V. Système sous-régional d'alerte et de contrôle de la fièvre de la Vallée du Rift (FVR) en Afrique de l'Ouest: ISRA - FAO; 2005.
Thiongane Y, Thonnon J, Fontenille D, Zeller H, Akakpo J, Gonzalez J, Digoutte J (1997) Épidémiosurveillance de la fièvre de la Vallée du Rift au Sénégal. Épidémiol et santé anim 31–32:02.A.08.
Thonnon J, Picquet M, Thiongane Y, Lo M, Sylla R, Vercruysse J. Rift Valley fever surveillance in the lower Senegal river basin: update 10 years after the epidemic. Tropical Med Int Health. 1999;4:580–5.
Tran A, Trevennec C, Lutwama J, Sserugga J, Gély M, et al. Development and assessment of a geographic knowledge-based model for mapping suitable areas for Rift Valley fever transmission in Eastern Africa. PLoS Negl Trop Dis. 2016;10:e0004999.

Traoré-Lamizana M, Fontenille D, Diallo M, Bâ Y, Zeller HG, Mondo M, Adam F, Thonon J, Maïga A. Arbovirus surveillance from 1990 to 1995 in the Barkedji area (Ferlo) of Senegal, a possible natural focus of Rift Valley fever virus. J Med Entomol. 2001;38:480–92.

Walsh M, Willem de Smalen A, Mor S. Wetlands, wild Bovidae species richness and sheep density delineate risk of Rift Valley fever outbreaks in the African continent and Arabian Peninsula. PLoS Negl Trop Dis. 2017;11:e0005756. https://doi.org/10.1371/journal.pntd.0005756.. eCollection 2017 Jul

Wilson W, Romito M, Jasperson D, Weingartl H, Binepal Y, Maluleke M, Wallace D, van Vuren P, Paweska J. Development of a Rift Valley fever real-time RT-PCR assay that can detect all three genome segments. J Virol Methods. 2013;193:426–31. https://doi.org/10.1016/j.jviromet.2013.07.006.

World Health Organisation. Rift Valley fever in Niger. http://www.who.int/csr/don/24-november-2016-rift-valley-fever-niger/en/. 2016.

Zeller HG, Fontenille D, Traore-Lamizana M, Thiongane Y, Digoutte JP. Enzootic activity of Rift Valley fever virus in Senegal. Am J Trop Med Hyg. 1997;56:265–72.

Zeller H, Rakotoharinadrasana H, Rakoto-Andrianarivelo M. Rift Valley fever in Madagascar: infection risks for the abattoir staff in Antananarivo. Revue Elev Méd vét Pays trop. 1998;51:17–20.

Chapter 9
West Nile Fever: A Challenge in Sahelian Africa

Assane Gueye Fall, Modou Moustapha Lo, Nicolas Djighnoum Diouf, Mamadou Ciss, Biram Bitèye, Mame Thierno Bakhoum, and Momar Talla Seck

Abstract West Nile fever is an arthropod-borne viral disease of public health importance transmitted by mosquitoes from the genus *Culex*. Birds are the main hosts/reservoirs of West Nile virus (WNV; Flaviviridae, *Flavivirus*) and ensure its spread worldwide toward migration. The WNV is continuously spreading across the world with significant health and economic impact in human and animals, especially in newly infected areas where many human and animal losses are reported. However, in most of the African countries where the disease is endemic, no or few human or animal cases are reported in contrast with the high prevalence of the disease and the recurrent circulation of the virus. These gaps could be related to the fact that clinical signs are similar to those of dominant pathologies in human or equine encephalitis in animal health allowing to a misdiagnosis.

Significant efforts will be needed in African countries to improve knowledge of the disease and to promote effective surveillance and control strategies. The West African Health Organization (WAHO/OOAS), with support from technical and financial partners, should bring countries in a One Health approach to address these issues through the funding of regional and subregional projects with the key sectors involving in this disease: public, animal, and environmental health.

Introduction

Bibliometric assessment of infectious diseases allows evaluation of the possible emergence of future diseases and their impact on the world. Thus, a bibliometric analysis focus on publications on West Nile fever (WNF) which were published between 1943 and 2016 shows the low research output among African authors of the

A. G. Fall (✉) · M. M. Lo · M. Ciss · B. Bitèye · M. T. Bakhoum · M. T. Seck
ISRA/LNERV, Dakar-Hann, Senegal
e-mail: assane.fall@isra.sn; moustapha.lo@isra.sn; talla.seck@isra.sn

N. D. Diouf
UFR des Sciences Agronomiques, de l'Aquaculture et des Technologies Alimentaires (S2ATA), Université Gaston Berger, Saint Louis, Sénégal

M. Kardjadj et al. (eds.), *Transboundary Animal Diseases in Sahelian Africa and Connected Regions*, https://doi.org/10.1007/978-3-030-25385-1_9

total of 4729 publications considered (Al-Jabi 2017). This highlights the minimal interest given to this disease, which, however, would have a significant One Health impact in African countries.

By discussing the West Nile virus (WNV) spread in the world, we provide a more balanced view of WNF as a threat to public and animal health in a diverse range of settings, thus, leading to a discussion in the gaps in which our knowledge of WNV transmission should be improved in the Sahel Africa and connected regions. The use of existing knowledge to control WNF is also examined, including reference to existing studies carried out during outbreaks of WNV in Europe and the Americas and their application in Africa.

Disease Description and History

WNF is caused by an arbovirus (Flaviviridae, *Flavivirus*) mostly transmitted by mosquitoes from the genus *Culex* (Kramer et al. 2008). First described in 1937 from a febrile illness case in Uganda, WNV has propagated to a vast region of the globe and is now considered the most important causative agent of viral encephalitis worldwide (Fig. 9.1).

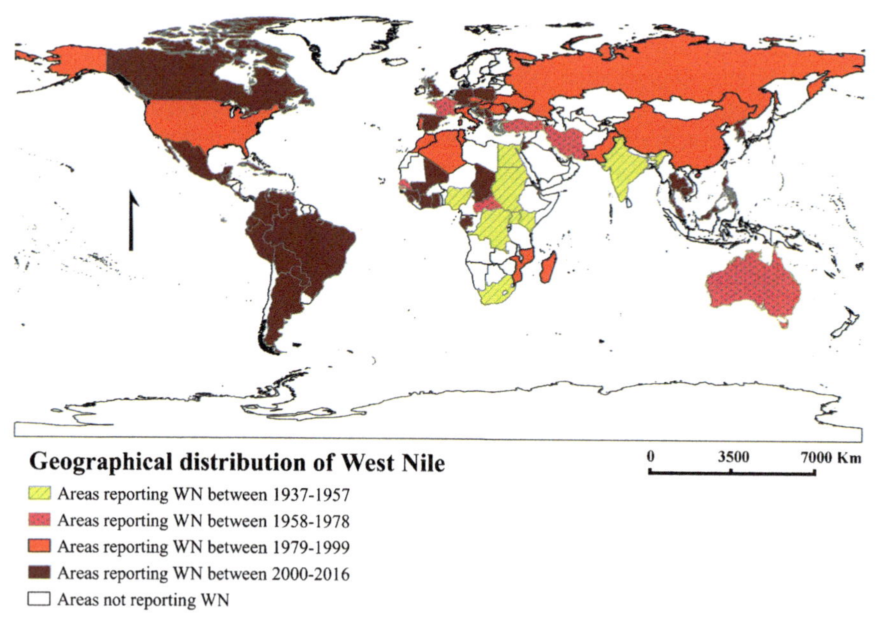

Fig. 9.1 Known geographical distributions of WNV obtained from data of molecular and serology and clinical cases between 1937 and 2016; data were extracted from published literatures and public databases

The WNV is maintained and transmitted through an enzootic cycle involving birds as main amplifying hosts and ornithophilic mosquitoes of the genus *Culex* as main vectors. Indeed, migratory birds are the main hosts of WNV and ensure its long-distance spread via intercontinental migration (Rappole et al. 2000). Humans and horses, although susceptible to WNV, are dead-end hosts (Hubalek and Halouzka 1999).

Despite the fact that the WNV is known to be endemic in many African countries as highlighted by seroepidemiological surveys in animals (Cabre et al. 2006; Chevalier et al. 2006, 2008, 2009; Fall et al. 2013) and humans (Murgue et al. 2002), low number of WNV outbreaks were reported in human African populations. So far in Sahel Africa particularly in Senegal, any clinical case of the disease has been documented to our knowledge in human and animal health. However, clinical signs are similar to those of malaria in human or equine encephalitis in animal health. Thus, many cases may be unnoticed by either a misdiagnosis or either a misunderstanding of the disease. After a silence of several decades from its first record in Europe in 1962 (Murgue et al. 2001), the disease has shown a deadly side in Romania in 1996 (Tsai et al. 1998). This feature was confirmed with the new introduction of WNV in the Americas where many human and animal losses were due to an epidemic that started in 1999 in New York (Lanciotti et al. 1999).

Geographical Distribution

Sub-Sahelian Africa the Birthplace of the Disease

WNF first appeared in tropical Africa, particularly in East Africa, in Uganda in the West Nile district in 1937 (Smithburn et al. 1940). The virus was accidentally isolated during an epidemiological survey focusing on the distribution of yellow fever in Central Africa (Smithburn and Jacobs 1942). South Africa reported WNV circulation for the first time in the 1950s, but since work started on WNV in South Africa, there has been only one large epidemic and one more localized epizootic accompanied by an increase in human infections occurred in the summer of 1973–1974 (Jupp 2001). From 2007 to 2015 WNV infections have been reported routinely in animals (Zaayman and Venter 2012; Venter and Swanepoel 2010; Venter et al. 2017). In Madagascar, WNV was first detected in 1978 from wild birds and the virus is currently distributed across the island, but only one fatal human case of WNV infection was reported in 2011 (Tantely et al. 2016, 2017). Côte d'Ivoire in 2003–2005, the Democratic Republic of Congo in 2004, Djibouti in 2004, and Gabon in 2004 and 2010, recorded the circulation of the virus in human and/or animal (Pourrut et al. 2010; Cabre et al. 2006). Studies in Guinea in 2006 reported that WNV was the most common arbovirus infection in wild rodents and humans (Jentes et al. 2010; Konstantinov et al. 2006). Evidence of WNV infection was recently found in Sierra Leone from sera of patients suspected for Lassa virus infection (Schoepp et al. 2014).

West Nile Virus in the Sahel Region

The WNV antibodies detection was documented as early as 1956 in Sudan (Taylor et al. 1956) followed by several reports in surveys conducted years later. More recently, WNV epidemics occurred in the country with fatal cases reported for the area (Watts et al. 1994; McCarthy et al. 1996; Depoortere et al. 2004). A low incidence of antibodies against WNV was found in sera patients in Nigeria in the early beginning of the 1950s (Macnamara et al. 1959). Thereafter, numerous isolations of WNV from the animal population until the virus was only isolated from human in 1973 (Tomori et al. 1978). Recently high rates of anti-WNV antibody prevalence were observed in horses suggesting that WNV is enzootic in Nigeria (Sule et al. 2015). In Senegal WNV activity has been recorded, in areas where birds are numerous, in human serum samples collected during an epidemiological survey on treponematosis between 1972 and 1975 (Renaudet et al. 1978). Additionally, entomological surveys carried out in the lower Senegal River valley close to the Djoudj National Bird Park (PNOD) showed a large circulation of the virus (Gordon et al. 1992). The very high prevalence rates obtained in different serological surveys in horses (Cabre et al. 2006; Chevalier et al. 2006), migratory, residents, and domesticated birds in the Ferlo region (Chevalier et al. 2008, 2009) and Senegal River Delta (Fall et al. 2013) show the recurrent circulation of the virus in Senegal. More recently, the circulation of WNV on domestic animals has been demonstrated suggesting that WNV is enzootic in the area (Davoust et al. 2016). The virus was also isolated from many mosquito species (Traore-Lamizana et al. 1994; Diallo et al. 2005). In 2003–2004 in Chad, the seroprevalence of WNV was particularly high in horses (97%) (Cabre et al. 2006). Serological evidence of WNV infection was observed recently in Mali in samples collected from suspected yellow fever cases (Safronetz et al. 2016).

Expansion to the Saharo-Arabian Region

Following Uganda, an epidemic of WNF was recorded in Egypt in 1950 during the seroepidemiological surveys in humans, mammals, and birds. In this country, the virus has been isolated also in humans (Melnick et al. 1951). Smithburn et al. (1954) and Taylor et al. (1956) highlighted the endemicity of the disease in this country. Recent studies confirmed active circulation of WNV causing febrile illness in a considerable proportion of human individuals, and virus isolation from both sentinel chickens and mosquitoes (Soliman et al. 2010). WNV was recorded in Algeria for the first time in 1994. Then, no data on WNV circulation was available until 2012 when a fatal human case of WNV neuroinvasive infection occurred (Lafri et al. 2017). Moroccan officials confirmed outbreaks of WNV infection in horses in 1996, 2003, and 2010, with dozens of confirmed cases. Only one case of human infection was reported in 1996, however, virus circulation was evidenced in human in 2012

(El Rhaffouli et al. 2013). Tunisia experienced WNV infection in 1997, 2003, 2007, and 2010–2012 (Benjelloun et al. 2016; Hammami et al. 2017).

West Nile Virus Lineages Distribution in Africa

WNV is a genetically and geographically diverse virus. Nine different lineages (WNV-1 to WNV-9) have been proposed based on phylogenetic analyses of published isolates (Fall et al. 2017; Kemenesi et al. 2014), with WNV-1 and WNV-2 strains have been identified most frequently in human and animal diseases. However, lineage 1 is spread globally and exists in distinct groups: lineage 1a distributed in Europe, Africa, and the Americas; and lineage 1b has been restricted to Oceania (Kemenesi et al. 2014; Fall et al. 2017).

Three distinct WNV genetic lineages (Lineages 1a, 2, and 7) are distributed in Africa (Fig. 9.2). Additionally, a potential new lineage of WNV was isolated from *Culex perfuscus* in Kedougou, Senegal in 1992 (Fall et al. 2017). Lineage 1a is distributed in East Africa (Ethiopia and Kenya), West Africa (Nigeria and Senegal), and Central African Republic, and Saharo-Arabian region (Egypt, Morocco, and Tunisia). Exclusively reported in Africa until 2004, lineage 2 is widely distributed in

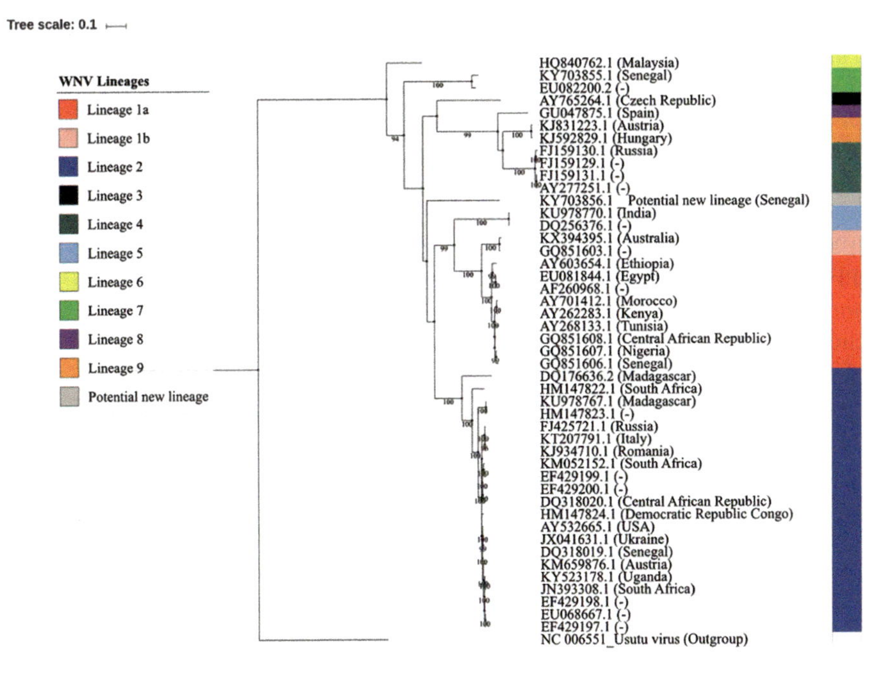

Fig. 9.2 West Nile Virus lineage: Maximum likelihood and Neighbor-Joining trees based on complete genome sequences from Africa. The tree was constructed using PhyML 3.0 and Mega v7 with 1000 bootstrap replications and was rooted using Usutu virus

different African countries: Uganda, Central Africa, Madagascar, Senegal, and South Africa. Lineage 7 has been restricted to Senegal in West Africa (Kemenesi et al. 2014; Fall et al. 2017).

Epidemiology

Birds in WNV Transmission

Wild birds were identified as WNV hosts/reservoirs in the 1950s (Work et al. 1953, 1955). They are also the most affected by the disease. In the United States, 326 bird species are believed to be susceptible to WNV infection (Petersen et al. 2013). Birds can play a role in the transmission cycle only if they develop sufficient viremia to infect vectors that feed on them. Generally, the titers of viremia greater than $10^{6.0}$ plaque-forming units (PFU)/mL of blood are considered infectious for most of *Culex* mosquitoes (Goddard et al. 2002; Sardelis et al. 2001; Turell et al. 2000). A study conducted by Komar et al. (2003) on 25 species of birds belonging to several avian orders showed that viremia titers in birds often exceeded $10^{6.0}$ PFU/mL of blood. The Passeriformes were the most sensitive birds, with viremia titers reaching $10^{12.1}$ PFU/mL. It is generally accepted that Passeriforms play a major role in the transmission of WNV. Serological studies conducted in Senegal identified 13 bird species, including 8 Passeriformes, as potential reservoirs of the WNV (Chevalier et al. 2009). Gallinaceans and Psittacids are less susceptible to WNV infection (Komar et al. 2003). In addition to viremia, the probability that the WNV is transmitted from an infected bird to a mosquito increases with the persistence of the virus in the bird's blood. Such persistence has been demonstrated in experimentally infected gray pigeons (Semenov et al. 1973) in which WNV was isolated from the blood up to 100 days after infection. Komar et al. (2003) detected high virus titers (up to $10^{6.9}$ PFU/0.5 cm^3) in samples of skins taken from dead birds 14 days after experimental inoculation. This persistence of viremia is an important factor in spreading the virus around the world, especially with migratory birds that can travel hundreds or thousands of kilometers with the virus. Serological studies conducted on bird populations in Senegal have identified resident as well as migratory WNV infected birds. Anti-WNV antibodies were detected in 13 bird species of which 5 were migratory (Chevalier et al. 2009). This results in the isolation of very closely related phylogenetic strains in all continents (May et al. 2011; Lanciotti et al. 1999).

Domestic bird (*Gallus gallusdomesticus*) has a short-lived viremia (on average 2 days) (Nemeth and Bowen 2007) and are unlikely to develop clinical disease, but can develop detectable neutralizing antibodies (Langevin et al. 2001). Therefore, its use as a sentinel animal to monitor WNV activity has been recommended by several authors (Barrera et al. 2008; Chaskopoulou et al. 2011; Chevalier et al. 2008; Komar 2001; Komar et al. 2012; Kwan et al. 2010; Langevin et al. 2001; Fall et al. 2013).

Role of Other Vertebrates

Horses and humans, although they may die from the disease, are considered dead-end hosts (Hubalek and Halouzka 1999) because they develop low viremia levels that do not allow arthropods to become infected during blood meals. Serological studies on horses in Senegal have shown high seroprevalence rates varying between 68% and 92% (Cabre et al. 2006; Chevalier et al. 2006; Davoust et al. 2016) without a documented case of mortality due to the disease. Also, domestic animals such as dogs and goats were positive for WNV antibodies but do not play a clear role in WNV epidemiology in Senegal (Davoust et al. 2016). In Nigeria, high prevalence of WNV antibodies was detected in camels, sheep, and goat (Olaleye et al. 1990). Around 30 other vertebrate species are susceptible to WNV infection, but only lemurs, frogs, and hamsters can develop sufficient viremia levels to allow arthropods infection (van der Meulen et al. 2005).

Mosquito Vectors of West Nile Virus

The main biological vectors of WNV are mosquitoes belonging mostly to the genus *Culex*. Several species of mosquitoes belonging to different genera have been found naturally infected with WNV but only a few have the ability to transmit the virus (Hubalek and Halouzka 1999). The vectors capable of transmitting are those whose vector competences are proved in the laboratory or those frequently found associated with the virus in nature.

In Sahelian Africa particularly in Senegal entomological surveys conducted in Barkédji, in the Senegal River Valley and Senegal River Delta, in eastern Senegal and in Kounguel showed that more than 200 WNV strains have been isolated from 14 species of mosquitoes belonging to 5 genera (Digoutte 1995; Gordon et al. 1992; Traore-Lamizana et al. 1994, 2001; Diallo et al. 2005) among them *Culex poicilipes* and *Aedes vexans arabiensis*. These two species have been also found associated with Rift Valley fever virus (Fontenille et al. 1998; Diallo et al. 2005).

The multiple associations of *Aedes vexans arabiensis* with the WNV in the nature (Digoutte 1995; Fontenille et al. 1998; Traore-Lamizana et al. 2001), its high densities at the beginning of the rainy season (Mondet et al. 2005b), its vector competence proven in laboratory (Tiawsirisup et al. 2008; Turell et al. 2005) and its opportunistic trophic behavior make this mosquito one of the main bridge vectors of WNV in the Sahelian region of the Ferlo (Fall et al. 2012). On the other hand, in the areas with permanent rivers notably the Senegal River Delta, *Culex tritaeniorhynchus* and *Culex neavei* are the main vectors of the WNV (Fall et al. 2011). The other species such as *Culex poicilipes*, *Culex perfuscus*, *Aedoemyia africana*, and *Mansonia uniformis* would play minor roles in the epidemiology of the disease in this region.

Transmission Cycles

Like all members of the Japanese encephalitis virus complex, WNV is maintained in nature by transmission cycles in which vectors are arthropods. Mosquitoes are the main vectors, although virus isolates from soft and hard ticks have also been reported (Hubalek and Halouzka 1999). Wild birds are the main hosts of WNF virus. They develop a high level of viremia that can persist for several weeks, allowing vectors' infection (Semenov et al. 1973; Komar et al. 2003). The enzootic or sylvatic cycles of WNV are those occurred between arthropods and birds. The transmission of WNV between arthropods and other vertebrates, including humans and horses, is called epizootic transmission. In addition to transmission by arthropods, direct transmission, horizontal transmission, and vertical transmission can play a role in maintaining the virus in the nature.

Transmission by Arthropods

Two concepts are important in vectorial transmission, vector competence and vector capacity. A vector is considered competent to transmit a virus if it is able, in the laboratory, to infect itself from a viremia source during the blood meal, allowing the multiplication of the pathogen in its organism and transmitting it to a healthy individual during a blood meal. Vector capacity is the ability of a vector given to transmit a virus in a given environment. It depends on the longevity of the vector, the vector density, and the duration of the extrinsic incubation period of the virus. It also depends on intrinsic (genetic) factors that determine the trophic preferences of the vector and other extrinsic factors that influence host/vector contact, such as the density and age of the vector population, aggressiveness, and longevity of vectors and on the host population.

The transmission of WNV by mosquitoes was demonstrated for the first time on *Aedes albopictus* by Philip and Smadel (1943). However, the main mosquitoes involved in the enzootic transmission of the virus belong to the genus *Culex* and are ornithophilic. These are mostly *Cx. univittatus*, *Cx. antennatus*, *Cx. theileri*, and *Cx. pipiens* in Africa, *Cx. pipiens* and *Cx. univittatus* in the Middle East, *Cx. modestus* and *Cx. pipiens* in Europe and Russia and *Cx. tritaeniorhynchus*, *Cx. quinquefasciatus*, and *Cx. vishnui* in Asia (Hubalek and Halouzka 1999). In the United States, *Cx. pipiens*, *Cx. restuans*, *Cx. salinarius*, *Cx. quinquefasciatus*, *Cx. nigripalpus*, and *Cx. tarsalis* are the main vectors (Andreadis et al. 2001; Reisen et al. 2004; Lukacik et al. 2006; Hayes et al. 2005a). In Senegal, *Culex neavei* is highly suspected to play the role of enzootic vector because of the multiple isolates of viruses on this species and its ornithophilic feeding behavior (Fall et al. 2011, 2013). According to numerous studies, several species of mosquitoes belonging to different genera have been found naturally infected with the WNV but do not play an important role in transmission because they do not honor one or more determinants of vector capacity (Fall 2013).

Some mosquitoes that have feeding behavior that is partly ornithophilic, partly mammophilic, or others that have opportunistic feeding behavior, transmit the virus to other vertebrates, including humans and horses. These vectors serve as bridges between the sylvatic cycle (mosquito-bird) and the epizootic cycle and are called "bridge vectors." The repeated isolation of WNV in mammophilic species, such as *Ae. vexans vexans* in the United States (Andreadis et al. 2004) has made it as a potential bridge vector. Similarly in Senegal, where many isolates of the WNV have been realized from the tropical subspecies of *Aedes vexans*, that is, *Aedes vexans arabiensis* (Digoutte 1995; Fontenille et al. 1998; Traore-Lamizana et al. 2001).

Several species of soft ticks (Argasidae) and hard (Ixodidae) ticks have also been found naturally infected (Hubalek and Halouzka 1999) but their role in the transmission of WNV remains unclear. Indeed, experiments carried out have shown that Ixodidae nymphs are capable of infecting and maintaining the virus up to the adult stage but are unable to transmit it (Anderson et al. 2003; Reisen et al. 2007). On the other hand, in some Argasidae, infection is possible in viremic animals and also transmission for several days or several months: *Ornithodoros moubata* (Whitman and Aitken 1960; Lawrie et al. 2004), *Argas hermanni* (Vermeil et al. 1960), *Argas arboreus* (Abbassy et al. 1993), and *Carios capensis* (Hutcheson et al. 2005). These latter thus allow the virus to remain in nature and participate in the endemicity of the disease in a given area.

Mosquito Vectors Population Dynamic Drivers in Sahelian Ecosystems

The influence of climatic conditions on arbovirus mosquito vectors bioecology has been previously documented (Janousek and Kramer 1999; Reeves et al. 1994; Reiter 2001). Many studies have suggested that the first wintering rainfalls and those immediately after long rainless periods have positive influences on *Ae. vexans arabiensis*' abundance (Ba et al. 2005; Janousek and Kramer 1999; Mondet et al. 2005b). However, more recent studies suggested that the abundance of *Ae. vexans arabiensis* is not only influenced by rainfall (Diallo et al. 2011; Biteye et al. 2018). Here (Table 9.1), we highlighted the complex relationship between rainfall, temperature, relative humidity, NDVI, and abundance of WNV mosquito vectors. Relative humidity (min of capture day) and rainfall (mean of 7 days prior the capture event) were significantly related to the abundances of *Ae. vexans arabiensis* populations ($p < 0.001$) while, temperature (min of capture day), rainfall (sum of 7 days prior the capture event), and NDVI (mean of 15 days including the capture day) decreased the abundances of *Ae. vexans arabiensis*. This is further supported by laboratory studies that confirm the influence of temperature range on the development of *Ae. albopictus* (Delatte et al. 2009). On the other hand, temperature (max of capture day), relative humidity (max of capture day), and rainfall (sum of 7 days prior the capture event) increased abundances of *Culex* populations in particular *Cx. neavei*, *Cx. poicilipes*, and *Cx. tritaeniorhynchus* ($p < 0.001$) while, rainfall (mean of 7 days prior the capture event) has negative effect on their abundances ($p < 0.001$). This is further supported by field studies which showed that the dynamics of *Culex* particularly *Cx.*

Table 9.1 Effects of environmental variables on WNV potential mosquito vectors population dynamics in Sahelian ecosystems

GLMM	Regression coefficients	Standard errors	*Z* values	*P* values
Aedes vexans arabiensis				
Intercept	−4.33901	1.43991	−3.01	0.00258
Temperature (min of capture day)	−2.69885	0.10285	−26.24	<2e-16
Humidity (min of capture day)	4.59794	0.15916	28.89	<2e-16
Rainfall (mean of 7 days prior the capture even)	7.87937	0.32144	24.51	<2e-16
Rainfall (sum of 7 days prior the capture even)	−7.95224	0.26214	−30.34	<2e-16
NDVI (mean of 15 days including the capture day)	−3.42669	0.04008	−85.49	<2e-16
Culex neavei				
Intercept	−10.56119	2.03097	−5.200	1.99e-07
Temperature (max of capture day)	0.61937	0.08027	7.716	1.20e-14
Humidity (max of capture day)	3.04573	0.10458	29.123	<2e-16
Rainfall (mean of 7 days prior the capture even)	−14.24671	1.90347	−7.485	7.18e-14
Rainfall (sum of 7 days prior the capture even)	18.95934	2.24661	8.439	<2e-16
NDVI (mean of 15 days including the capture day)	6.58007	0.25927	25.379	<2e-16
Culex poicilipes				
Intercept	−0.78770	0.72264	−1.090	0.275695
Temperature (max of capture day)	−0.48197	0.04286	−11.246	<2e-16
Humidity (max of capture day)	0.82527	0.05924	13.931	<2e-16
Rainfall (mean of 7 days prior the capture even)	−1.10390	0.31507	−3.504	0.000459
Rainfall (sum of 7 days prior the capture even)	1.31127	0.35162	3.729	0.000192
NDVI (mean of 15 days including the capture day)	−0.48581	0.15819	−3.071	0.002133
Culex tritaeniorhynchus				
Intercept	1.36199	0.91337	1.49	0.136
Temperature (max of capture day)	0.13285	0.00606	21.92	<2e-16
Humidity (max of capture day)	1.79438	0.01754	102.28	<2e-16
Rainfall (mean of 7 days prior the capture even)	−1.71089	0.05774	−29.63	<2e-16
Rainfall (sum of 7 days prior the capture even)	1.53809	0.06406	24.01	<2e-16
NDVI (mean of 15 days including the capture day)	0.69073	0.03444	20.06	<2e-16

tritaeniorhynchus were positively associated with temperature and humidity (Fall et al. 2011). NDVI increased abundances of *Cx. neavei* and *Cx. tritaeniorhynchus* but decreased those of *Cx. poicilipes* and *Ae. vexans* ($p \leq 0.002$).

Vertical Transmission

Vertical transmission requires that the virus be transmitted from parents to descendants. For vectors, this mode of transmission has been clearly demonstrated in Culicidae mosquitoes and some ticks of the genera *Argas* (Goddard et al. 2003; Miller et al. 2000; Turell et al. 2001) and *Ixodes* (Anderson et al. 2003) and could allow the virus to cross extreme conditions in regions with unfavorable seasons (Reisen et al. 2006).

Non-vector Transmission

This mode of transmission is rare and is usually done by ingesting substances containing the virus. These substances may be oral or cloacal excretions of birds containing the virus that contaminate other birds (Komar et al. 2003), bird meat containing the virus that contaminates carnivorous birds (Komar et al. 2003), vertebrate meat that contaminates other vertebrates (the case of alligators who consumed horsemeat positive for WNV) (Miller et al. 2003a). Blood transfusion, organ transplantation (Planitzer et al. 2009), and autopsies of dead West Nile animals (Venter and Swanepoel 2010) may also result in the transmission. Intrauterine WNV transmission was documented for the first time in the United States (CDC 2002), and transmission of WNV through breastfeeding is possible but seems to be rare (Hinckley et al. 2007).

Hypothesis of Persistence of WNV in Sahelian Ecosystem

The maintenance of the WNV in the dry season also raises several hypotheses of which the following three seem to be the most plausible: via the eggs of *Aedes* mosquito, via overwintering *Culex* mosquito, or via unknown wild reservoirs. Indeed the eggs of *Aedes* mosquito resist to desiccation and can cross the unfavorable season and hatch to continue their development cycle as soon as they are underwater. Thus, the peaks of *Ae. vexans arabiensis* appear at the beginning of the rainy season with the simultaneous hatching of the stocks of eggs from the previous year (Mondet et al. 2005a, b). If the vertical transmission of WNV exists in *Ae. vexans arabiensis* as demonstrated in *Ae. aegypti* and *Ae. albopictus* (Baqar et al. 1993), eggs could be a route of maintenance of this virus in Sahelian ecosystems. Similarly overwintering *Culex* infected with WNV can cross the unfavorable season in shelters and are not detectable by conventional entomological methods because of their inactivity (Rudolf et al. 2017; Nasci et al. 2001). Wild animals that are carriers of the WNV and dependent on the wetlands (pond and river) may also be involved in the maintenance of the virus. Experimental infection of the lake frogs (*Rana ridibunda*) with a Russian strain of WNV has resulted in high titers of viremia that can even infect a mosquito (Kostiukov et al. 1985) suggesting to consider it as a competent reservoir of this virus. In Sahelian ecosystems, high densities of frogs are observed in wetlands in the rainy season and hibernate during unfavorable periods until the next wintering. Thus, they could be key actors in the maintenance of the virus in such ecosystems. Currently, the role of reptiles and amphibians in the ecology and epidemiology of the WNV is not well known. Rabbit (*Sylvilagus floridanus*), chipmunks (*Tamias striatus*), and Fox squirrels (*Sciurus niger*) are reservoirs of WNV in the United States (Platt et al. 2008). These wild unknown reservoirs should, therefore, be investigated in order to better understand the epidemiology of the disease in Sahelian ecosystems.

Symptoms

In Sahel Africa, WNF is little known in human and animal health services, not because of the absence of cases but rather because of misdiagnosis due to the resemblance of its symptomatology to that of many diseases such as malaria, dengue fever in humans, or African horse sickness in horses. With a few exceptions, the clinical symptoms reported here are therefore those described in Europe and the United States.

Human

Only a small proportion of human infections are symptomatic, with headaches, tiredness, body aches, and hypertrophied lymph nodes typical of many febrile diseases. There is sometimes an abdominal rash. About 1 in 150 people develop one or more signs of neuroinvasive disease: neck stiffness, stupor, disorientation, coma, tremors, convulsions, stiffness, and paralysis. This may occur in people of all ages, but those over 50 years of age are the most at risk (Hayes et al. 2005b). However, an atypical outbreak of WNV that occurred in Sudan in 2002 affected mostly children aged between 6 months and 12 years (median: 36 months) (Depoortere et al. 2004). Similar findings have been observed in Algeria in 1994, where clinically ill patients were children aged between 10 months and 9 years (Le Guenno et al. 1996). Over the past years, more than 4% of laboratory-confirmed cases of clinical infections reported in the United States have been fatal (CDC 2016). In the Sudan 2002 outbreak, the mortality rate reached 13% of the patients (Depoortere et al. 2004). Long-term effects on the kidney have recently been reported in the United States, including the occurrence of chronic kidney disease in 40% of patients who have contracted the disease (Nolan et al. 2012). For the first time, an acute WNV pancreatitis case was documented (Babi et al. 2016).

Horse

Symptomatic infections in horses are also rare and usually benign but can lead to neurological diseases, including fatal encephalomyelitis (Cantile et al. 2000). A higher proportion of infected horses develop encephalitis in relation to human cases. Experimental studies have shown that approximately 10% of infected horses develop clinical disease (Bunning et al. 2002). Clinical signs for WNV include flu-like signs, where the horse seems mildly anorexic and depressed; fine and coarse muscle and skin fasciculation; hyperesthesia (hypersensitivity to touch and sound); changes in mentation (mentality), when horses look like they are daydreaming or "just not with it"; occasional somnolence (drowsiness); propulsive walking (driving or pushing forward, often without control); and "spinal" signs, including asymmetrical weakness. Some horses can show asymmetrical or symmetrical ataxia. There are no specific treatments for WNV; however, supportive care can help horses recover in some cases. Equine mortality rate can be as high as 30–40%. For example, in South Africa 34% of WNV-positive horses died between 2008 and 2015 (Venter et al. 2017).

Birds

WNV-infected birds also exhibit a range of clinical signs ranging from no disease to death. Mortality attributed to WNV infection in North America was reported in 198 bird species in 2002 (Komar 2003). Some species of birds, particularly corvids, are very sensitive to this virus with very often a fatal ending (Komar et al. 2003; McLean et al. 2002). The general signs of infection are lethargy, decubitus, and in some cases, hemorrhages (Komar et al. 2003).

In Europe, bird mortality due to WNV infection is rare (Hubalek and Halouzka 1999). However, high mortalities of birds were noted in the 1998 epidemic in Israel caused by lineage I (Malkinson et al. 2002). Unlike in the United States, where high bird mortality reveals the circulation of the virus (Steele et al. 2000), in Europe the neurological signs in horses are the only indicators of the local circulation of the virus. This contrast in virulence on both sides of the Atlantic is due to the fact that the virus belonging to the same lineage I that was isolated in Israel in 1998 is newly introduced in the United States with a totally naive avian population, whereas in Europe the disease is endemic (Buckley et al. 2003; Lanciotti et al. 1999).

Other Vertebrates

The clinical signs of WNF are not well known in other vertebrates, such as reptiles, amphibians, and other mammals. In North America, captive alligators died of WNV infection between 2001 and 2002 (Miller et al. 2003b). In more than 10,000 captive alligators, 250 died in 2001 and 1000 in 2002, mostly young animals. The dominant symptoms were nervous such as neck spasms and loss of movement control.

Isolated cases were observed in sheep (Tyler et al. 2003), alpacas (Yaeger et al. 2004), dogs, and wolves (Lichtensteiger et al. 2003). In all these cases, the disease was always characterized by progressive neurological signs such as muscle weakness, ataxia, convulsions, paralysis, torticollis, hyperaesthesia, and decubitus. In the wolf, blindness was also observed. In the diseased dog described by Lichtensteiger et al. (2003), numerous other clinical signs were observed, such as polydipsia, nasal and ocular discharge, diarrhea, abdominal pain, ptyalism, and dyspnea.

Lesions

Macroscopic Lesions

In birds, some lesions are evocative of the disease but are not pathognomonic because they are common to several other diseases. These include emaciation, splenomegaly, hepatomegaly, cardiac lesions, and encephalitis. In horses, the brain does not show macroscopic lesions.

Microscopic Lesions

In birds, microscopic lesions vary according to the degree of organ damage. Infiltrating of lymphocytes and heterophils in the capillaries of the brain and meninges are evocative in the case of central nervous system damages (Miller et al. 2003b). In horses, they are characterized by multifocal perivascular infiltrations of neutrophils in the rhombencephalon, spinal cord, or microglia with perivascular hemorrhages (Snook et al. 2001).

Diagnosis

The clinic and epidemiology allow establishing a suspicion that the laboratory diagnosis will then confirm or disprove.

Epidemiological-Clinical Diagnosis

The definition of human suspect cases may vary slightly from one country to another and depending on the epidemiological context. For example, in Romania where WNV infection is routinely detected, all persons aged over 15 years presenting with fever and meningitis or encephalitis or meningoencephalitis and clear cerebrospinal fluid were considered suspected cases and were tested for WNV-specific antibodies during the 2010 outbreak (Sirbu et al. 2011). During the summer of 2003, WNV re-emerged in humans 40 years after the first human clinical WNV infections in France in 1964 (Del Giudice et al. 2004). Thus, a suspected human case of WNF is defined as any hospitalized patient with evocative symptoms (sudden onset fever and neurological manifestation such as meningitis and encephalitis, or sudden onset fever and atypical acute nerve manifestation) who have stayed in areas at risk during a period of risk. In the United States, West Nile fever is categorized into neuroinvasive diseases group that is clinically define as a patient with (1) meningitis, encephalitis, acute flaccid paralysis, or other acute signs of central or peripheral neurologic dysfunction, as documented by a physician, AND (2) in absence of a more likely clinical explanation, other clinically compatible symptoms of arbovirus disease include headache, myalgia, rash, arthralgia, vertigo, vomiting, paresis, and/or nuchal rigidity (Lindsey et al. 2015).

In horses, the onset of acute nerve syndromes with ataxia, muscle weakness and prostration should include WNF in the differential diagnosis. In birds, abnormal behavior, abnormal mortality that cannot be related to an obvious cause and during a period of risk should also lead to a suspicion of WNF, especially when it comes to corvids.

Laboratory Tools

Currently, the laboratory tools for the diagnosis of infection by WNV belong to two main methods, serology and viral detection.

Serological Methods

Laboratory diagnosis can be accomplished by an indirect method using serum or cerebrospinal fluid (CSF) to detect WNV-specific IgM and IgG antibodies. In order to understand the application of serology for WNV diagnosis, it is useful to remember that the mean times from the detection of viral RNA to IgM and IgG seroconversion are approximately 4 and 8 days, respectively, and persist for 30–90 days, but longer persistence has been documented. Specific antibody detection still remains the most widely used approach for the diagnosis of WNV. The main weakness that limits the clinical relevance of serological methods is the broad antigenic cross-reactivity that exists between all flaviviruses: the quite specific viral envelope (E) protein neutralizing antibody response is often combined with less specific tests based on detection of antibodies against the membrane (M) and non-structural (NS) proteins of which the amino acid sequences are more conserved amongst the flaviviruses (Dauphin and Zientara 2007). Furthermore, positive IgM antibodies occasionally may reflect a past infection. If serum is collected within 8 days of illness onset, the absence of detectable virus-specific IgM does not rule out the diagnosis of WNV infection, and the test may need to be repeated on a later sample. The presence of WNV-specific IgM in blood or CSF provides good evidence of recent infection but may also result from cross-reactive antibodies after infection with other flaviviruses or from non-specific reactivity. According to product inserts for commercially available WNV IgM assays, all positive results obtained with these assays should be confirmed by neutralizing antibody testing of acute- and convalescent-phase serum specimens by reference laboratories. Based on this consideration, the principal serological methods can be subdivided into two main groups, the first includes the enzyme-linked immunosorbent assays (ELISAs) and immunofluorescence (IF) based tests; the second includes the Plaque Reduction Neutralization Test that can be carried out using a highly sensitive 50% or less sensitive 90% endpoint (PRNT50 and PRNT90, respectively), both of which require the constant availability of standardized-validated infectious viruses and appropriate cell cultures. The hemagglutination-inhibition test (HIA) is still used to detect pan-flavivirus immune response whereas the complement fixation test (CFT) is rarely used in today's laboratories. The techniques included in the first group are widely used due to their relative applicability in routine laboratory and the ability to automate a part of the workflow but they are less specific as a consequence of their inability to distinguish between WNV-specific and cross-reactive antibody responses. Thus, any positive result identified using these methods must be confirmed by the more specific tests, that is, those that constitute the second group. It is important to emphasize that in either of these analyses, the test should follow the guidelines of the World Health Organization and should include control standardized viruses that are known to be readily neutralizable and to be antigenically closely related but distinct species from WNV. This second group of techniques, particularly PRNT assays are labor intensive and are generally limited to reference or dedicated research laboratories, where appropriate biosafety level could be attained (Monini et al. 2010).

WNV IgG antibodies generally are detected shortly after IgM antibodies and persist for many years following a symptomatic or asymptomatic infection. Therefore, the presence of IgG antibodies alone is the only evidence of previous infection and clinically compatible cases with the presence of IgG, but not IgM, should be evaluated for other etiologic agents.

Virus Detection Methods

As others testing for WNV, direct methods can be applied through virus isolation, molecular methods and immunohistochemistry (Sambri et al. 2013).

Virus Isolation

Although the isolated strains provides the added value of allowing further studies and research on pathogenesis, genetic variation evolution, or epidemiology, WNV is not readily isolatable from tissues, plasma, serum, and CSF samples in cell culture using either mammalian or mosquito-derived cell lines (Rossini et al. 2011; Sudeep et al. 2009; Jayakeerthi et al. 2006). Moreover, WNV isolation procedures must be performed under biosafety level 3 conditions.

Molecular Methods

Detection of virus by molecular methods has been demonstrated to be a useful tool that enables the detection of WNV genomes, even after prolonged times post-infection (Bagnarelli et al. 2011; Murray et al. 2010). WNV genomes in peripheral blood are usually detectable from 2–3 days to 14–18 days post-infection. For the routine detection of WNV RNA using molecular techniques, there are two distinct diagnostic settings, the first involves blood and organ donation screening from subjects living in an area where WNV circulation is known, and the second involves the identification of viral genomes in serum, plasma, and CSF samples from patients presenting with a clinical picture typical of WNV infection. Two molecular methods are usually used to detect WNV: traditional *Reverse Transcription PCR* (RT-PCR) assay whose the sensitivity for WNV detection depends mainly on the target sequence; and TaqMan RT-PCR assay that are generally rapid and reliable and can be used for the detection of WNV on a large variety of samples, including human, animal tissues, and mosquito specimens (Lanciotti et al. 2000). To improve diagnostic knowledge of WNV using RT-PCR, different technical modifications have been proposed in the last 10 years (Jiménez-Clavero et al. 2006; Papin et al. 2004, 2010).

Immunohistochemistry

This method to detect WNV antigen using histochemical protocols in tissues obtained from fatal encephalitis cases has been available for diagnostic purposes for many years. However, this procedure is performed rarely in order to improve the certainty of a clinical diagnosis in cases where laboratory data are minimal (Bhatnagar et al. 2007).

Prevention and Control

There is currently no specific treatment for West Nile. Several studies have demonstrated the efficacy of Ribavirin in inhibiting the in vitro replication of WNV (Morrey et al. 2002; Anderson and Rahal 2002). However, Ribavirin does not appear to be effective against viruses of the genus *Flavivirus* in animal models (Huggins 1989). Similarly, a retrospective study of the efficacy of Ribavirin treatment in the West Nile 2000 epidemic in Israel showed that patients survival was not correlated with treatment with Ribavirin (Chowers et al. 2001). Thus, medical and sanitary prophylaxes are the only means available for the control of the disease.

Prevention

Medical prophylaxis is based on the use of a number of vaccines. There are currently three West Nile vaccines commercially available for horses in the United States and one in Europe. In animals, the adjuvanted vaccines with formalin-inactivated whole virus, named West Nile-Innovator® and Duvaxyn® were marketed by Fort Dodge Animal Health (FDAH) in the United States and Europe, respectively. The West Nile-Innovator® vaccine was used on a large scale in the United States in 2002. Although the number of human cases of WNF did not decline between 2002 and 2003, this mass vaccination may have contributed to the reduction of equine cases during this period (Granwehr et al. 2004). Since then, the following second-generation vaccines have been commercialized: the DNA vaccine called West Nile-Innovator® DNA marketed by Fort Dodge, the Recombinant vaccine based on the canarypox virus called Recombitek® and marketed by Merial (El Garch et al. 2008), and the Chimeric vaccine manufactured from the yellow fever 17D vaccine strain called PreveNile™ and marketed by Intervet (Schering-Plough Animal Health/Merck).

In human, a DNA vaccine was also under development and Phase I clinical trials has shown that it induces the production of neutralizing antibodies (Martin et al. 2007). The Chimeric vaccine technology used to produce the PreveNile™ vaccine for horses was also used to produce a human vaccine called ChimeriVax West Nile. This safe and immunogenic vaccine in humans is currently under clinical trials (Brandler and Tangy 2013). Its marketing is planned during this decade.

Control

Recommended sanitary prophylaxis measures include vectors control, both adults and larval forms. To be effective, vectors control strategies must take into account the biology and ecology of arthropod vectors. Mosquito larvae control consists of the destruction of their breeding sites and/or treatment with chemicals or biological larvicides (Bs, Bti). In Sahelian countries notably in Senegal, larvae control is difficult to implement in the areas where the WNV is routinely detected (Ferlo and the Senegal River region) because of the size, diversity, and number of breeding sites, but also the financial and potential impact of such treatments on the environment and on non-target wildlife. Indeed, in the areas where rivers and ponds constitute the main water sources for human and animal populations, their destruction or treatment with insecticides could cause a real ecological disaster in these ecosystems and poisoning in humans and animals. For the same reasons and also because of very limited effectiveness, the adult mosquito control seems difficult to carry out in the sub-Saharan countries context.

Individual protective measures such as the use of insecticide-treated nets, which have significantly reduced the transmission of malaria in humans, remain difficult to use against the WNV mosquito vectors that are mostly zoophilic and exophagic (Fall et al. 2011, 2012; Ba et al. 2006), and even for some from the genus *Aedes* with diurnal activity. The precautionary measures proposed in the United States and in Europe against West Nile vectors are based on the use of cutaneous repellents and avoiding human and animal exposure to WNV-infected mosquitoes. Adulticide control methods are applied in case of epidemics with limited effectiveness (Lothrop et al. 2008; Carney et al. 2008; Elnaiem et al. 2008). A study conducted in Senegal shows that in the case of West Nile epidemic, insecticide treatment of susceptible domestic animals (horses) could help to control vectors (Diallo et al. 2008). However, the effectiveness of this control method could be hampered by its high cost and difficult implementation in endemic areas. The consequences of chemical control on the environment and non-target wildlife would not be neglected.

To prevent transmission of WNV through blood transfusion, blood donations in WNV-endemic areas should be screened by using sensitive tests such as nucleic acid amplification tests.

Surveillance and Control Strategies in the Sahel Region

According to studies/surveys carried out in the Sahel, WNV can be considered endemic with an epidemiological situation almost homogeneous in the region even epidemiological information are missing in some countries. Efficient surveillance system for WNF in Sahelian region should integrate a “One Health” approach based on a transdisciplinary and trans-sectorial collaboration between institutions involved in public, animal, and environmental health at the national and regional level. This integrated surveillance targets mosquitoes, wild birds, humans, and horses, and aims at early detection of the viral circulation and reducing the risk of infection in the human populations.

Human surveillance aims early detection of WNV infection cases, and identification of affected areas to implement appropriate response measures including vector control and communication to relevant authorities and to the public. To achieve this in Sahelian countries, all human patients with fever and one symptom of neuroinvasive disease described previously should be considered suspect cases of WNF. Samples such as plasma, serum, and cerebrospinal fluid are tested using methods described previously. The surveillance of suspected cases of West Nile neuroinvasive disease should be intensified in the period overlapping with mosquito abundance particularly during the rainy season in Sahelian region.

WNV is maintained in a bird-mosquito-bird transmission cycle. Thus, mosquito and bird surveillance sites should be selected using a risk-based approach, that is, sites with heavy birds and abundant mosquito populations. In protected areas such as bird parks, surveillance must be increased during the period of massive arrival of migratory birds. For mosquito collection, several types of traps could be used altogether or individually, CDC-CO_2 dry ice-baited traps, bird-baited traps, gravid traps, etc. Active surveillance should be carried out in targeted wild bird species and passive surveillance in dead birds. Alive wild birds can be shot or caught using net bird traps. Samples can be blood, brain, spleen, heart, and kidney. Sentinel chickens also are routinely used as a tool for early virus detection. Sentinel chicken seroconversions could be used as an outcome measure in decision support for emergency intervention. One important aspect of the success of this strategy is the location of sentinel chickens. Sentinel chicken approach have been experienced recently in Senegal (Fall et al. 2013). Results showed that the highest overall incidence (7.7%) was observed in chickens located close to the river (<100 m). Sentinel chickens located at 800 m from the river had an overall incidence of 4.6% while the ones who were at 1300 m from the river remained free of WNV infection. This result highlights the importance of installing sentinel chickens near water sources for effective WNV monitoring.

Horse surveillance strategy is more difficult to conduct in Sahel endemic areas where WNV seroprevalence is very high in the equine population (Chevalier et al. 2006; Cabre et al. 2006), clinical signs have been never evidenced or documented and the follow up of animals is not easy because of their agricultural and economic activities. Indeed Diouf (2013) collected blood samples from 570 horses from three different zones in the Senegal River Delta, namely Ross Bethio, Richard Toll, and Saint Louis in order to determine the incidence and circulation of WNV among the equine population. All sera were tested using a competitive ELISA test. Out of 570 horses, 532 (93%) were seropositive for WNV antibodies. The seronegative 38 (7%) horses were selected as sentinel animals and were longitudinally followed up in a period of 11 months with samples collected every 2 months. WNV neutralizing antibodies were detected in 8 horses by the plaque reduction neutralization technique (PRNT) showing a global incidence score of 21%. The study showed high levels of WNV transmission among horses in the Senegal River Delta. Considering that the seroprevalence in the horse is very high in affected countries, it is estimated that surveillance in equids is irrelevant. Additionally, infection in horses may occur at the same time or even later than the identification of the first human cases.

Data sharing on WNV surveillance in human, animal, and vector is a key point for the successfulness of a national and/or regional One Health integrated surveillance system. An integrated collection and analysis of data from human, animal, and vector surveillance is key to obtain a comprehensive understanding of the epidemiological situation of WNV and consequently to implement efficient response measures. Thus, the modalities of the integrated approach should be country-dependent taking into account the local context. However, with the luck of expertise and resources in vector-borne diseases surveillance and control in most of the Sahel countries, and the luck of coordination between the public and veterinary health sectors, promoting in-regional expertise mobility, and the provision of sufficient funding are key to develop adequate surveillance and control strategies.

Conclusion

The main challenge in WNF in southern countries remains to generate knowledge to better understand the epidemiology of the disease. Capacity building of human and animal health workers and facilities in terms of clinical and laboratory diagnostic also is a big need. Thus, the lack of epidemiological and economic impact data makes this disease really neglected in Africa. For instance, few sub-Saharan countries have undertaken sustainable research to resolve this gap. It is up to African scientists to advocate with governments for the treatment of such diseases that make more damage than imagined, and not to limit to external funding.

Regional organizations such as the West African Economic and Monetary Union (WAEMU/UEMOA) and the Economic Community of West African States (ECOWAS/CEDEAO) through the newly created West African Health Organization (WAHO/OOAS) should address these issues through regional and sub-regional funding, as it is done in Europe through the EDEN and EDENext projects (Alexander et al. 2015). To achieve this, WAHO should work with existing networks such as the West African Network of Biomedical Analysis Laboratories (RESAOLAB) and the West and Central Africa Veterinary Laboratory Network for avian influenza and other transboundary diseases (RESOLAB) with support from technical and financial partners such as the World Bank, FAO/OIE, WHO, CDC, and EU. Projects that have a One Health approach should be prioritized to better coordinate actions in the key sectors involving in this disease, public, animal, and environmental health.

References

Abbassy MM, Osman M, Marzouk AS. West Nile virus (Flaviviridae: Flavivirus) in experimentally infected Argas ticks (Acari: Argasidae). Am J Trop Med Hyg. 1993;48(5):726–37.

Alexander N, Allepuz A, Alten B, Bodker R, Bonnet S, Carpenter S, Cetre-Sossah C, Chirouze E, Depaquit J, Dressel K. EDEN & EDENext: the impact of a decade of research (2004–2015) on vector-borne diseases. Montpellier: Cirad; 2015.

Al-Jabi SW. Global research trends in West Nile virus from 1943 to 2016: a bibliometric analysis. Glob Health. 2017;13(1):55.

Anderson JF, Rahal JJ. Efficacy of interferon α-2b and ribavirin against West Nile virus in vitro. Emerg Infect Dis. 2002;8(1):107.

Anderson JF, Main AJ, Andreadis TG, Wikel SK, Vossbrinck CR. Transstadial transfer of West Nile virus by three species of Ixodid ticks (Acari: Ixodidae). J Med Entomol. 2003;40:528–33.

Andreadis TG, Anderson JF, Vossbrinck CR. Mosquito surveillance for West Nile virus in Connecticut, 2000: isolation from *Culex pipiens*, *Cx. restuans*, *Cx. salinarius*, and *Culiseta melanura*. Emerg Infect Dis. 2001;7:670–4.

Andreadis TG, Anderson JF, Vossbrinck CR, Main AJ. Epidemiology of West Nile virus in Connecticut: a five-year analysis of mosquito data 1999–2003. Vector Borne Zoonotic Dis. 2004;4(4):360–78.

Ba Y, Diallo D, Kebe CMF, Dia I, Diallo M. Aspects of bioecology of two Rift Valley fever virus vectors in Senegal (West Africa): Aedes vexans and Culex poicilipes (Diptera: Culicidae). J Med Entomol. 2005;42(5):739–50.

Ba Y, Diallo D, Dia I, Diallo M. Comportement trophique des vecteurs du virus de la fièvre de la vallée du Rift au Sénégal: implications dans l'épidémiologie de la maladie. Bulletin de la Société de Pathologie Exotique. 2006;99:283–9.

Babi M, Waheed W, Wardi S. A first clinical case report of West-Nile viral meningoencephalitis complicated with acute pancreatitis in North America. J Meningitis. 2016;1(104):2.

Bagnarelli P, Marinelli K, Trotta D, Monachetti A, Tavio M, Del Gobbo R, Capobianchi M, Menzo S, Nicoletti L, Magurano F. Human case of autochthonous West Nile virus lineage 2 infection in Italy, September 2011. Eur Secur. 2011;16(43):20002.

Baqar S, Hayes CG, Murphy JR, Watts DM. Vertical transmission of West Nile virus by Culex and Aedes species mosquitoes. Am J Trop Med Hyg. 1993;48:757–62.

Barrera R, Hunsperger E, Muñoz-Jordán JL, Amador M, Diaz A, Smith J, Bessoff K, Beltran M, Vergne E, Verduin M, Lambert A, Sun W. First isolation of West Nile virus in the Caribbean. Am J Trop Med Hyg. 2008;78(4):666–8.

Benjelloun A, El Harrak M, Belkadi B. West Nile disease epidemiology in North-West Africa: bibliographical review. Transbound Emerg Dis. 2016;63(6):e153–9.

Bhatnagar J, Guarner J, Paddock CD, Shieh W-J, Lanciotti RS, Marfin AA, Campbell GL, Zaki SR. Detection of West Nile virus in formalin-fixed, paraffin-embedded human tissues by RT-PCR: a useful adjunct to conventional tissue-based diagnostic methods. J Clin Virol. 2007;38(2):106–11.

Biteye B, Fall AG, Ciss M, Seck MT, Apolloni A, Fall M, Tran A, Gimonneau G. Ecological distribution and population dynamics of Rift Valley fever virus mosquito vectors (Diptera, Culicidae) in Senegal. Parasites Vectors. 2018;11(1):27.

Brandler S, Tangy F. Vaccines in development against West Nile virus. Viruses. 2013;5 (10):2384–409.

Buckley A, Dawson A, Moss SR, Hinsley SA, Bellamy PE, Gould EA. Serological evidence of West Nile virus, Usutu virus and Sindbis virus infection of birds in the UK. J Gen Virol. 2003;84 (10):2807–17.

Bunning ML, Bowen RA, Cropp CB, Sullivan KG, Davis BS, Komar N, Godsey MS, Baker D, Hettler DL, Holmes DA, Biggerstaff BJ, Mitchell CJ. Experimental infection of horses with West Nile virus. Emerg Infect Dis. 2002;8:380–6.

Cabre O, Grandadam M, Marié JL, Gravier P, Prangé A, Santinelli Y, Rous V, Bourry O, Durand JP, Tolou H, Davoust B. West Nile virus in horses, sub-Saharan Africa. Emerg Infect Dis. 2006;12:1958–60.

Cantile C, Di Guardo G, Eleni C, Arispici M. Clinical and neuropathological features of West Nile virus equine encephalomyelitis in Italy. Equine Vet J. 2000;32(1):31–5.

Carney RM, Husted S, Jean S, Glaser C, Kramer V. Efficacy of aerial spraying of mosquito adulticide in reducing incidence of West Nile virus, California, 2005. Emerg Infect Dis. 2008;14(5):747–54.

CDC. Intrauterine West Nile virus infection--New York, 2002. MMWR Morb Mortal Wkly Rep. 2002;51(50):1135.

CDC. West Nile virus disease cases and deaths reported to CDC by year and clinical presentation, 1999–2015. CDC, Atlanta, GA. 2016. https://www.cdc.gov/westnile/resources/pdfs/data/1-WNV-Disease-Cases-by-Year_1999-2015_07072016pdf.

Chaskopoulou A, Dovas CI, Chaintoutis SC, Bouzalas I, Ara G, Papanastassopoulou M. Evidence of enzootic circulation of West Nile virus (Nea Santa-Greece-2010, lineage 2), Greece, May to July 2011. Eur Secur. 2011;16(31):19933.

Chevalier V, Lancelot R, Diaïté A, Mondet B, Sall B, De Lamballerie X. Serological assessment of West Nile fever virus activity in the pastoral system of Ferlo, Senegal. Ann N Y Acad Sci. 2006;1081:216–25.

Chevalier V, Lancelot R, Diaïte A, Mondet B, Lamballerie X. Use of sentinel chickens to study the transmission dynamics of West Nile virus in a Sahelian ecosystem. Epidemiol Infect. 2008;136:525–8.

Chevalier V, Reynaud P, Lefrançois T, Durand B, Baillon F, Balança G, Gaidet N, Mondet B, Lancelot R. Predicting West Nile virus seroprevalence in wild birds in Senegal. Vector Borne Zoonotic Dis. 2009;9(6):589–96.

Chowers MY, Lang R, Nassar F, Ben-David D, Giladi M, Rubinshtein E, Itzhaki A, Mishal J, Siegman-Igra Y, Kitzes R. Clinical characteristics of the West Nile fever outbreak, Israel, 2000. Emerg Infect Dis. 2001;7(4):675.

Dauphin G, Zientara S. West Nile virus: recent trends in diagnosis and vaccine development. Vaccine. 2007;25(30):5563–76.

Davoust B, Maquart M, Roqueplo C, Gravier P, Sambou M, Mediannikov O, Leparc-Goffart I. Serological survey of West Nile virus in domestic animals from Northwest Senegal. Vector Borne Zoonotic Dis. 2016;16(5):359–61.

Del Giudice P, Schuffenecker I, Vandenbos F, Counillon E, Zeller H. Human West Nile virus, France. Emerg Infect Dis. 2004;10(10):1885.

Delatte H, Gimonneau G, Triboire A, Fontenille D. Influence of temperature on immature development, survival, longevity, fecundity, and gonotrophic cycles of *Aedes albopictus*, vector of chikungunya and dengue in the Indian Ocean. J Med Entomol. 2009;46(1):33–41.

Depoortere E, Kavle J, Keus K, Zeller H, Murri S, Legros D. Outbreak of West Nile virus causing severe neurological involvement in children, Nuba Mountains, Sudan, 2002. Trop Med Int Health. 2004;9(6):730–6.

Diallo M, Nabeth P, Ba K, Sall AA, Ba Y, Mondo M, Girault L, Abdalahi MO, Mathiot C. Mosquito vectors of the 1998–1999 outbreak of Rift Valley fever and other arboviruses (Bagaza, Sanar, Wesselsbron and West Nile) in Mauritania and Senegal. Med Vet Entomol. 2005;19:119–26.

Diallo D, Ba Y, Dia I, Konaté L, Diallo M. Utilisation de boeufs traités aux insecticides dans la lutte contre les vecteurs des virus de la fièvre de la vallée du Rift et de la fièvre West Nile au Sénégal. Bulletin de la Société de Pathologie Exotique. 2008;101(5):410–7.

Diallo D, Talla C, Ba Y, Dia I, Sall AA, Diallo M. Temporal distribution and spatial pattern of abundance of the Rift Valley fever and West Nile fever vectors in Barkedji, Senegal. J Vector Ecol. 2011;36(2):426–36.

Digoutte JP. Rapport sur le fonctionnement technique de l'Institut Pasteur de Dakar. Dakar: Institut Pasteur; 1995.

Diouf N. Potentiel rôle épidémiologique du Delta du fleuve Sénégal et impact des arboviroses dans la population équine : cas de la fièvre West Nile et de la peste équine. 2013. These Doctorat, UCAD, Dakar:45.

El Garch H, Minke JM, Rehder J, Richard S, Edlund Toulemonde C, Dinic S, Andreoni C, Audonnet JC, Nordgren R, Juillard V. A West Nile virus (WNV) recombinant canarypox virus vaccine elicits WNV-specific neutralizing antibodies and cell-mediated immune responses in the horse. Vet Immunol Immunopathol. 2008;15(123):230–9.

El Rhaffouli H, Lahlou-Amine I, Loutfi C, Laraqui A, Bajjou T, Fassi-Fihri O, El Harrak M. Serological evidence of West Nile virus infection among humans in the southern provinces of Morocco. J Infect Dev Ctries. 2013;7(12):999–1002.

Elnaiem DE, Kelley K, Wright S, Laffey R, Yoshimura G, Reed M, Goodman G, Thiemann T, Reimer L, Reisen WK, Brown D. Impact of aerial spraying of pyrethrin insecticide on *Culex*

pipiens and *Culex tarsalis* (Diptera: Culicidae) abundance and West Nile virus infection rates in an urban/suburban area of Sacramento County, California. J Med Entomol. 2008;45(4):751–7.
Fall AG. Ecologie des arbovirus au Sénégal: exemple du virus de la fièvre West Nile dans le Delta du fleuve Sénégal et le Ferlo. 2013. These de dcotorat unique UCAD:117.
Fall AG, Diaïté A, Lancelot R, Tran A, Soti V, Etter E, Konaté L, Faye O, Bouyer J. Feeding behaviour of potential vectors of West Nile virus in Senegal. Parasit Vectors. 2011;4(1):99.
Fall AG, Diaïté A, Etter E, Bouyer J, Ndiaye TD, Konaté L. The mosquito *Aedes (Aedimorphus) vexans arabiensis* as a probable vector bridging West Nile virus between birds and horses in Barkedji (Ferlo, Senegal). Med Vet Entomol. 2012;26(1):106–11.
Fall AG, Diaïté A, Seck MT, Bouyer J, Lefrançois T, Vachiéry N, Aprelon R, Faye O, Konaté L, Lancelot R. West Nile virus transmission in sentinel chickens and potential mosquito vectors, Senegal river delta, 2008–2009. Int J Environ Res Public Health. 2013;10(10):4718–27.
Fall G, Di Paola N, Faye M, Dia M, de Melo Freire CC, Loucoubar C, de Andrade Zanotto PM, Faye O. Biological and phylogenetic characteristics of West African lineages of West Nile virus. PLoS Negl Trop Dis. 2017;11(11):e0006078.
Fontenille D, Traore-Lamizana M, Diallo M, Thonnon J, Digoutte JP, Zeller HG. New vectors of Rift Valley fever in West Africa. Emerg Infect Dis. 1998;4(2):289–93.
Goddard LB, Roth AE, Reisen WK, Scott TW. Vector competence of California mosquitoes for West Nile virus. Emerg Infect Dis. 2002;8:1385–91.
Goddard LB, Roth AE, Reisen WK, Scott TW. Vertical transmission of West Nile virus by three *California Culex* (Diptera: Culicidae) species. J Med Entomol. 2003;40(6):743–6.
Gordon SW, Tammariello RF, Linthicum KJ, Dohm DJ, Digoutte JP, Calvo-Wilson MA. Arbovirus isolations from mosquitoes collected during 1988 in the Senegal River basin. Am J Trop Med Hyg. 1992;47(6):742–8.
Granwehr BP, Lillibridge KM, Higgs S, Mason PW, Aronson JF, Campbell GA, Barrett ADT. West Nile virus: where are we now? Lancet Infect Dis. 2004;4:547–56.
Hammami S, Hassine TB, Conte A, Amdouni J, De Massis F, Sghaier S, Hassen SB. West Nile disease in Tunisia: an overview of 60 years. Vet Ital. 2017;53(3):225–34.
Hayes EB, Komar N, Nasci RS, Montgomery SP, O'Leary DR, Campbell GL. Epidemiology and transmission dynamics of West Nile virus disease. Emerg Infect Dis. 2005a;11(8):1167–73.
Hayes EB, Sejvar JJ, Zaki SR, Lanciotti RS, Bode AV, Campbell GL. Virology, pathology, and clinical manifestations of West Nile virus disease. Emerg Infect Dis. 2005b;11(8):1174–9.
Hinckley AF, O'Leary DR, Hayes EB. Transmission of West Nile virus through human breast milk seems to be rare. Pediatrics. 2007;119(3):e666–71.
Hubalek Z, Halouzka J. West Nile virus - a reemerging mosquito-borne viral disease in Europe. Emerg Infect Dis. 1999;5:643–50.
Huggins JW. Prospects for treatment of viral hemorrhagic fevers with ribavirin, a broad-spectrum antiviral drug. Rev Infect Dis. 1989;11(4):S750–61.
Hutcheson HJ, Gorham CH, Machain-Williams C, Loroño-Pino MA, James AM, Marlenee NL, Winn B, Beaty BJ, Blair CD. Experimental transmission of West Nile virus (Flaviviridae: Flavivirus) by *Carios capensis* ticks from North America. Vector Borne Zoonotic Dis. 2005;5 (3):293–5.
Janousek T, Kramer W. Seasonal incidence and geographical variation of Nebraska mosquitoes, 1994–95. J Am Mosq Control Assoc. 1999;15(3):253–62.
Jayakeerthi RS, Potula RV, Srinivasan S, Badrinath S. Shell vial culture assay for the rapid diagnosis of Japanese encephalitis, West Nile and Dengue-2 viral encephalitis. Virol J. 2006;3(1):2.
Jentes ES, Robinson J, Johnson BW, Conde I, Sakouvougui Y, Iverson J, Beecher S, Diakite F, Coulibaly M, Bausch DG. Acute arboviral infections in Guinea, West Africa, 2006. Am J Trop Med Hyg. 2010;83(2):388–94.
Jiménez-Clavero MA, Agüero M, Rojo G, Gómez-Tejedor C. A new fluorogenic real-time RT-PCR assay for detection of lineage 1 and lineage 2 West Nile viruses. J Vet Diagn Investig. 2006;18 (5):459–62.

Jupp PG. The ecology of West Nile virus in South Africa and the occurrence of outbreaks in humans. Ann N Y Acad Sci. 2001;951(1):143–52.

Kemenesi G, Dallos B, Oldal M, Kutas A, Földes F, Németh V, Reiter P, Bakonyi T, Bányai K, Jakab F. Putative novel lineage of West Nile virus in Uranotaeniaunguiculata mosquito, Hungary. Virus. 2014;25(4):500–3.

Komar N. West Nile virus surveillance using sentinel birds. Ann N Y Acad Sci. 2001;951:58–73.

Komar N. West Nile virus: epidemiology and ecology in North America. Adv Virus Res. 2003;61:185–234.

Komar N, Langevin S, Hinten S, Nemeth N, Edwards E, Hettler D, Davis B, Bowen R, Bunning M. Experimental infection of North American birds with the New York 1999 strain of West Nile virus. Emerg Infect Dis. 2003;9(3):311.

Komar N, Bessoff K, Diaz A, Amador M, Young G, Seda R, Perez T, Hunsperger E. Avian hosts of West Nile virus in Puerto Rico. Vector Borne Zoonotic Dis. 2012;12(1):47–54.

Konstantinov O, Diallo S, Inapogi A, Ba A, Kamara S. The mammals of Guinea as reservoirs and carriers of arboviruses. Med Parazitol. 2006;1:34–9.

Kostiukov MA, Gordeeva ZE, Bulychev VP, Nemova NV, Daniiarov OA. The lake frog (*Rana ridibunda*)--one of the food hosts of blood-sucking mosquitoes in Tadzhikistan--a reservoir of the West Nile fever virus. Med Parazitol. 1985;3:49–50.

Kramer LD, Styer LM, Ebel GD. A global perspective on the epidemiology of West Nile virus. Annu Rev Entomol. 2008;53:61–81.

Kwan JL, Kluh S, Madon MB, Nguyen DV, Barker CM, Reisen WK. Sentinel chicken seroconversions track tangential transmission of West Nile virus to humans in the greater Los Angeles area of California. Am J Trop Med Hyg. 2010;83(5):1137–45.

Lafri I, Prat CM, Bitam I, Gravier P, Besbaci M, Zeroual F, Ben-Mahdi MH, Davoust B, Leparc-Goffart I. Seroprevalence of West Nile virus antibodies in equids in the north-east of Algeria and detection of virus circulation in 2014. Comp Immunol Microbiol Infect Dis. 2017;50:8–12.

Lanciotti RS, Roehrig JT, Deubel V, Smith J, Parker M, Steele K, Volpe KE, Crabtree MB, Scherret JH, Hall RA, MacKenzie JS, Cropp CB, Panigrahy B, Ostlund E, Schmitt B, Malkinson M, Banet C, Weissman J, Komar N, Savage HM, Stone W, McNamara T, Gubler DJ. Origin of the West Nile virus responsible for an outbreak of encephalitis in the northeastern United States. Science. 1999;286:2333–7.

Lanciotti RS, Kerst AJ, Nasci RS, Godsey MS, Mitchell CJ, Savage HM, Komar N, Panella NA, Allen BC, Volpe KE. Rapid detection of West Nile virus from human clinical specimens, field-collected mosquitoes, and avian samples by a TaqMan reverse transcriptase-PCR assay. J Clin Microbiol. 2000;38(11):4066–71.

Langevin SA, Bunning M, Davis B, Komar N. Experimental infection of chickens as candidate sentinels for West Nile virus. Emerg Infect Dis. 2001;7:726–9.

Lawrie CH, Uzcátegui NY, Gould EA, Nuttall PA. Ixodid and argasid tick species and west Nile virus. Emerg Infect Dis. 2004;10(4):653–7.

Le Guenno B, Bougermouh A, Azzam T, Bouakaz R. West Nile: a deadly virus? Lancet. 1996;348 (9037):1315.

Lichtensteiger CA, Heinz-Taheny K, Osborne TS, Novak RJ, Lewis BA, Firth ML. West Nile virus encephalitis and myocarditis in wolf and dog. Emerg Infect Dis. 2003;9:1303–6.

Lindsey NP, Lehman JA, Staples JE, Fischer M. West Nile virus and other nationally notifiable arboviral diseases-United States, 2014. MMWR Morb Mortal Wkly Rep. 2015;64(34):929–34.

Lothrop HD, Lothrop BB, Gomsi DE, Reisen WK. Intensive early season adulticide applications decrease arbovirus transmission throughout the Coachella Valley, Riverside County, California. Vector Borne Zoonotic Dis. 2008;8(4):475–89.

Lukacik G, Anand M, Shusas EJ, Howard JJ, Oliver J, Chen H, Backenson PB, Kauffman EB, Bernard KA, Kramer LD, White DJ. West Nile virus surveillance in mosquitoes in New York State, 2000–2004. J Am Mosq Control Assoc. 2006;22(2):264–71.

Macnamara F, Horn D, Porterfield J. Yellow fever and other arthropod-borne viruses: a consideration of two serological surveys made in South Western Nigeria. Trans R Soc Trop Med Hyg. 1959;53(2):202–12.
Malkinson M, Banet C, Weisman Y, Pokamunski S, King R, Drouet MT. Introduction of West Nile virus in the Middle East by migrating white storks. Emerg Infect Dis. 2002;8(4):392–7.
Martin JE, Pierson TC, Hubka S, Rucker S, Gordon IJ, Enama M, Andrews CA, Xu Q, Davis BS, Nason M, Fay M, Koup RA, Roederer M, Bailer RT, Gomez PL, Mascola JR, Chang GJ, Nabel GJ, Graham BS. A West Nile virus DNA vaccine induces neutralizing antibody in healthy adults during a phase 1 clinical trial. J Infect Dis. 2007;196:1732–40.
May FJ, Davis CT, Tesh RB, Barrett AD. Phylogeography of West Nile virus: from the cradle of evolution in Africa to Eurasia, Australia, and the Americas. J Virol. 2011;85(6):2964–74.
McCarthy M, Haberberger R, Salib A, Soliman B, El-Tigani A, Khalid I, Watts D. Evaluation of arthropod-borne viruses and other infectious disease pathogens as the causes of febrile illnesses in the Khartoum Province of Sudan. J Med Virol. 1996;48(2):141–6.
McLean RG, Ubico SR, Bourne D, Komar N. West Nile virus in livestock and wildlife. Curr Top Microbiol Immunol. 2002;267:271–308.
Melnick JL, Paul JR, Riordan JT, Barnett VH, Goldblum N, Zabin E. Isolation from human sera in Egypt of a virus apparently identical to West Nile virus. Proc Soc Exp Biol Med. 1951;77(4):661–5.
Miller BR, Nasci RS, Godsey MS, Savage HM, Lutwama JJ, Lanciotti RS, Peters CJ. First field evidence for natural vertical transmission of West Nile virus in *Culex univittatus* complex mosquitoes from Rift Valley Province, Kenya. Am J Trop Med Hyg. 2000;62:240–6.
Miller DL, Mauel MJ, Baldwin C, Burtle G, Ingram D, Hines ME. West Nile virus in farmed alligators. Emerg Infect Dis. 2003a;9(7):794.
Miller DL, Mauel MJ, Baldwin C, Burtle G, Ingram D, Hines ME, Frazier KS. West Nile virus in farmed alligators. Emerg Infect Dis. 2003b;9(7):794–9.
Mondet B, Diaite A, Fall AG, Chevalier V. Relations entre la pluviométrie et le risque de transmission virale par les moustiques: Cas du virus de la Rift Valley Fever (RVF) dans le Ferlo (Sénégal). Environnement risque et Santé. 2005a;4(2):125–9.
Mondet B, Diaïté A, Ndione J-A, Fall AG, Chevalier V, Lancelot R, Ndiaye M, Ponçon N. Rainfall patterns and population dynamics of Aedes (Aedimorphus) vexans arabiensis Patton, 1905 (Diptera Culicidae), a potential vector of Rift Valley fever in Senegal. J Vector Ecol. 2005b;30(1):102–6.
Monini M, Falcone E, Busani L, Romi R, Ruggeri FM. West Nile virus: characteristics of an African virus adapting to the third millennium world. Open Virol J. 2010;4:42.
Morrey JD, Smee DF, Sidwell RW, Tseng C. Identification of active antiviral compounds against a New York isolate of West Nile virus. Antivir Res. 2002;55(1):107–16.
Murgue B, Murri S, Triki H, Deubel V, Zeller HG. West Nile in the Mediterranean basin: 1950–2000. Ann N Y Acad Sci. 2001;951:117–26.
Murgue B, Zeller H, Deubel V. The ecology and epidemiology of West Nile virus in Africa, Europe and Asia. Curr Top Microbiol Immunol. 2002;267:195–221.
Murray K, Walker C, Herrington E, Lewis JA, McCormick J, Beasley DW, Tesh RB, Fisher-Hoch S. Persistent infection with West Nile virus years after initial infection. J Infect Dis. 2010;201(1):2–4.
Nasci RS, Savage HM, White DJ, Miller JR, Cropp BC, Godsey MS, Kerst AJ, Bennett P, Gottfried K, Lanciotti RS. West Nile virus in overwintering Culex mosquitoes, New York City, 2000. Emerg Infect Dis. 2001;7:742–4.
Nemeth NM, Bowen RA. Dynamics of passive immunity to West Nile virus in domestic chickens (*Gallus gallus domesticus*). Am J Trop Med Hyg. 2007;76(2):310–7.
Nolan MS, Podoll AS, Hause AM, Akers KM, Finkel KW, Murray KO. Prevalence of chronic kidney disease and progression of disease over time among patients enrolled in the Houston West Nile virus cohort. PLoS One. 2012;7(7):e40374.

Olaleye O, Omilabu S, Ilomechina E, Fagbami A. A survey for haemagglutination-inhibiting antibody to West Nile virus in human and animal sera in Nigeria. Comp Immunol Microbiol Infect Dis. 1990;13(1):35–9.

Papin JF, Vahrson W, Dittmer DP. SYBR green-based real-time quantitative PCR assay for detection of West Nile virus circumvents false-negative results due to strain variability. J Clin Microbiol. 2004;42(4):1511–8.

Papin JF, Vahrson W, Larson L, Dittmer DP. Genome-wide real-time PCR for West Nile virus reduces the false-negative rate and facilitates new strain discovery. J Virol Methods. 2010;169 (1):103–11.

Petersen LR, Brault AC, Nasci RS. West Nile virus: review of the literature. JAMA. 2013;310 (3):308–15.

Philip CB, Smadel JE. Transmission of West Nile virus by infected *Aedes albopictus*. Proc Soc Exp Biol Med. 1943;48:537–48.

Planitzer CB, Modrof J, Mei-ying WY, Kreil TR. West Nile virus infection in plasma of blood and plasma donors, United States. Emerg Infect Dis. 2009;15(10):1668.

Platt KB, Tucker BJ, Halbur PG, Blitvich BJ, Fabiosa FG, Mullin K, Parikh GR, Kitikoon P, Bartholomay LC, Rowley WA. Fox squirrels (*Sciurus niger*) develop West Nile virus viremias sufficient for infecting select mosquito species. Vector Borne Zoonotic Dis. 2008;8(2):225–33.

Pourrut X, Nkoghé D, Paweska J, Leroy E. First serological evidence of West Nile virus in human rural populations of Gabon. Virol J. 2010;7(1):132.

Rappole J, Derrickson S, Hubálek Z. Migratory birds and spread of West Nile virus in the western hemisphere. Emerg Infect Dis. 2000;6:319–28.

Reeves WC, Hardy JL, Reisen WK, Milby MM. Potential effect of global warming on mosquito-borne arboviruses. J Med Entomol. 1994;31(3):323–32.

Reisen WK, Lothrop HD, Chiles RE, Madon MB, Cossen C, Woods L, Husted S, Kramer V, Edman JD. West Nile virus in California. Emerg Infect Dis. 2004;10(8):1369–78.

Reisen WK, Fang Y, Lothrop HD, Martinez VM, Wilson J, Oconnor P, Carney R, Cahoon-Young B, Shafii M, Brault A. Overwintering of West Nile virus in Southern California. J Med Entomol. 2006;43(2):344–55.

Reisen WK, Brault AC, Martinez VM, Fang Y, Simmons K, Garcia S, Omi-Olsen E, Lane RS. Ability of transstadially infected *Ixodes pacificus* (Acari: Ixodidae) to transmit West Nile virus to song sparrows or western fence lizards. J Med Entomol. 2007;44(2):320–7.

Reiter P. Climate change and mosquito-borne disease. Environ Health Perspect. 2001;109(Suppl 1):141.

Renaudet J, Jan C, Ridet J, Adam C, Robin Y. A serological survey of arboviruses in the human population of Senegal. Bulletin de la Societe de Pathologie Exotique et de ses Filiales. 1978;71 (2):131–40.

Rossini G, Carletti F, Bordi L, Cavrini F, Gaibani P, Landini MP, Pierro A, Capobianchi MR, Di Caro A, Sambri V. Phylogenetic analysis of West Nile virus isolates, Italy, 2008–2009. Emerg Infect Dis. 2011;17(5):903.

Rudolf I, Betášová L, Blažejová H, Venclíková K, Straková P, Šebesta O, Mendel J, Bakonyi T, Schaffner F, Nowotny N, Hubálek Z. West Nile virus in overwintering mosquitoes, central Europe. Parasit Vectors. 2017;10(1):452. https://doi.org/10.1186/s13071-017-2399-7.

Safronetz D, Sacko M, Sogoba N, Rosenke K, Martellaro C, Traoré S, Cissé I, Maiga O, Boisen M, Nelson D. Vectorborne infections, Mali. Emerg Infect Dis. 2016;22(2):340.

Sambri V, Capobianchi MR, Cavrini F, Charrel R, Donoso-Mantke O, Escadafal C, Franco L, Gaibani P, Gould EA, Niedrig M. Diagnosis of west Nile virus human infections: overview and proposal of diagnostic protocols considering the results of external quality assessment studies. Viruses. 2013;5(10):2329–48.

Sardelis MR, Turell MJ, Dohm DJ, O'Guinn ML. Vector competence of selected North American Culex and Coquillettidia mosquitoes for West Nile virus. Emerg Infect Dis. 2001;7:1018–22.

Schoepp RJ, Rossi CA, Khan SH, Goba A, Fair JN. Undiagnosed acute viral febrile illnesses, Sierra Leone. Emerg Infect Dis. 2014;20(7):1176.

Semenov BF, Chunikhin SP, Karmysheva VY, Takovleva NI. Studies of chronic arbovirus infections in birds. 1. Experiments with West Nile, Sindbis, Bhanja and SFS virus (in Russian). Vestnik Akademii Medicinskikh Nauk SSSR. 1973;2:79–83.

Sirbu A, Ceianu C, Panculescu-Gatej R, Vazquez A, Tenorio A, Rebreanu R, Niedrig M, Nicolescu G, Pistol A. Outbreak of West Nile virus infection in humans. 2011. Romania, July to October 2010.

Smithburn KC, Jacobs HR. Neutralization-tests against neurotropic viruses with sera collected in Central Africa. J Immunol. 1942;43(5):9–23.

Smithburn KC, Hughes TP, Burke AW, Paul JH. A neurotropic virus isolated from the blood of a native of Uganda. Am J Trop Med Hyg. 1940;20:471–92.

Smithburn KC, Taylor RM, Rizk F, Kader A. Immunity to certain arthropod-borne viruses among indigenous residents of Egypt. Am J Trop Med Hyg. 1954;3(1):9–18.

Snook CS, Hyman SS, Del Piero F, Palmer JE, Ostlund EN, Barr BS, Desrochers AM, Reilly LK. West Nile virus encephalomyelitis in eight horses. J Am Vet Med Assoc. 2001;218:1576–9.

Soliman A, Mohareb E, Salman D, Saad M, Salama S, Fayez C, Hanafi H, Medhat I, Labib E, Rakha M. Studies on West Nile virus infection in Egypt. J Infect Public Health. 2010;3(2):54–9.

Steele KE, Linn MJ, Schoepp RJ, Komar N, Geisbert TW, Manduca RM, Calle PP, Raphael BL, Clippinger TL, Larsen T, Smith J, Lanciotti RS, Panella NA, McNamara TS. Pathology of fatal West Nile virus infections in native and exotic birds during the 1999 outbreak in New York City, New York. Vet Pathol. 2000;37(3):208–24.

Sudeep A, Parashar D, Jadi RS, Basu A, Mokashi C, Arankalle VA, Mishra AC. Establishment and characterization of a new *Aedes aegypti* (L.) (Diptera: Culicidae) cell line with special emphasis on virus susceptibility. In Vitro Cell Dev Biol Anim. 2009;45(9):491–5.

Sule WF, Oluwayelu DO, Adedokun RAM, Rufai N, McCracken F, Mansfield KL, Johnson N. High seroprevelance of West Nile virus antibodies observed in horses from southwestern Nigeria. Vector Borne Zoonotic Dis. 2015;15(3):218–20.

Tantely ML, Goodman SM, Rakotondranaivo T, Boyer S. Review of West Nile virus circulation and outbreak risk in Madagascar: entomological and ornithological perspectives. Parasite. 2016;23:49.

Tantely LM, Cêtre-Sossah C, Rakotondranaivo T, Cardinale E, Boyer S. Population dynamics of mosquito species in a West Nile virus endemic area in Madagascar. Parasite. 2017;24:3.

Taylor RM, Work TH, Hurlbut HS, Rizk F. A study of the ecology of West Nile virus in Egypt. Am J Trop Med Hyg. 1956;5:579–620.

Tiawsirisup S, Kinley JR, Tucker BJ, Evans RB, Rowley WA, Platt KB. Vector competence of *Aedes vexans* (Diptera: Culicidae) for West Nile virus and potential as an enzootic vector. J Med Entomol. 2008;45(3):452–7.

Tomori O, Fagbami A, Fabiyi A. Isolations of West Nile virus from man in Nigeria. Trans R Soc Trop Med Hyg. 1978;72(1):103–4.

Traore-Lamizana M, Zeller HG, Mondo M, Hervy JP, Adam F, Digoutte JP. Isolation of West Nile and Bagaza viruses from mosquito (Diptera Culicidae) in Central Senegal (Ferlo). J Med Entomol. 1994;31(6):934–8.

Traore-Lamizana M, Fontenille D, Diallo M, Ba Y, Zeller HG, Mondo M, Adam F, Thonnon J, Maiga A. Arbovirus surveillance from 1990 to 1995 in the Barkedji area (Ferlo) of Senegal, a possible natural focus of Rift Valley fever virus. J Med Entomol. 2001;38:480–92.

Tsai TF, Popovici F, Cernescu C, Campbell GL, Nedelcu NI. West Nile encephalitis epidemic in southeastern Romania. Lancet. 1998;352(9130):767–71.

Turell MJ, O'Guinn M, Oliver J. Potential for New York mosquitoes to transmit West Nile virus. Am J Trop Med Hyg. 2000;62(3):413–4.

Turell MJ, O'Guinn ML, Dohm DJ, Jones JW. Vector competence of North American mosquitoes (Diptera: Culicidae) for West Nile virus. J Med Entomol. 2001;38:130–4.

Turell MJ, Dohm DJ, Sardelis MR, O'Guinn ML, Andreadis TG, Blow JA. An update on the potential of North American mosquitoes (Diptera: Culicidae) to transmit West Nile virus. J Med Entomol. 2005;42:57–62.

Tyler JW, Turnquist SE, David AT, Kleiboeker SB, Middleton JR. West Nile virus encephalomyelitis in a sheep. J Vet Intern Med. 2003;17:242–4.
van der Meulen KM, Pensaert MB, Nauwynck HJ. West Nile virus in the vertebrate world. Arch Virol. 2005;150(4):637–57.
Venter M, Swanepoel R. West Nile virus lineage 2 as a cause of zoonotic neurological disease in humans and horses in southern Africa. Vector Borne Zoonotic Dis. 2010;10(7):659–64.
Venter M, Pretorius M, Fuller JA, Botha E, Rakgotho M, Stivaktas V, Weyer C, Romito M, Williams J. West Nile virus lineage 2 in horses and other animals with neurologic disease, South Africa, 2008–2015. Emerg Infect Dis. 2017;23(12):2060.
Vermeil C, Lavillaureix J, Reeb E. Sur la conservation et la transmission du virus West Nile par des arthropodes. Bulletin de la Société de Pathologie Exotique. 1960;53:273–9.
Watts D, El-Tigani A, Botros B, Salib A, Olson J. Arthropod-borne viral infectious associated with a fever outbreak in the Northern Province of Sudan. 1994. Naval Medical Research Unit No 3, Cairo (Egypt), Department of Medical Zoology.
Whitman L, Aitken TH. Potentiality of *Ornithodoros moubata* Murray (Acarina, Argasidae) as a reservoir vector of West Nile virus. Ann Trop Med Parasitol. 1960;54:192–204.
Work TH, Hurlbut HS, Taylor RM. Isolation of West Nile virus from hooded crow and rock pigeon in the Nile delta. Proc Soc Exp Biol Med. 1953;84(3):719–22.
Work TH, Hurlbut HS, Taylor RM. Indigenous wild birds of the Nile Delta as potential West Nile virus circulating reservoirs. Am J Trop Med Hyg. 1955;4(5):872–88.
Yaeger M, Yoon KJ, Schwartz K, Berkland L. West Nile virus meningoencephalitis in a Suri alpaca and Suffolk ewe. J Vet Diagn Investig. 2004;16:64–6.
Zaayman D, Venter M. West Nile virus neurologic disease in humans, South Africa, September 2008–May 2009. Emerg Infect Dis. 2012;18(12):2051–4.

Chapter 10
Ebola and Other Haemorrhagic Fevers

Mathieu Bourgarel and Florian Liégeois

Abstract Ebola virus disease (EVD) and other viral haemorrhagic fevers (VHF) are mainly acute zoonotic diseases and represent a major threat to public health in Central and West Africa, and worldwide. They are caused by viruses of different families Flaviviridae, Bunyaviridae, Arenaviridae, and Filoviridae. Their circulation is generally restricted to the geographic distribution area of their natural hosts and the viruses emerge or re-emerge continuously where favourable conditions are met. These emergencies are still unpredictable and difficult to control as numerous knowledge gaps in the ecology of the viruses and the transmission routes still need to be filled despite the huge effort of the scientific community. The role of wildlife as a natural host of viruses and the interface/interaction between human/livestock/wildlife yet to be fully appreciated to really understand the ecology, the mechanisms of cross-species spill over and the epidemiology of EVD and other transboundary diseases. The 2014–2016 outbreak of EVD in West Africa, with more than 28,600 reported cases and 11,310 reported deaths, showed the significant epidemic potential, the transboundary nature and the global public health threat of EVD and other VHF in an increasingly interconnected world of intensified travel and trade. Beyond the human loss, the Ebola epidemic impacted the global economy of the African continent and more than USD 3.6 billion were spent to fight the outbreak. Since the first outbreak of EVD in 1976 in South Sudan and RDC, the control of epidemics relied on containment and isolation of the symptomatic patients and dead bodies to stop human-to-human transmission. Today, vaccine and treatment are under trial and show promising first results to fight EVD outbreaks and are currently tested during the 2018 EVD outbreak in RDC.

M. Bourgarel (✉)
CIRAD, UMR ASTRE, Montpellier, France

ASTRE, CIRAD, INRA, Univ. Montpellier, Montpellier, France
e-mail: mathieu.bourgarel@cirad.fr

F. Liégeois
IRD, UMR MIVEGEC, Univ. Montpellier, Montpellier, France
e-mail: florian.liegeois@ird.fr

M. Kardjadj et al. (eds.), *Transboundary Animal Diseases in Sahelian Africa and Connected Regions*, https://doi.org/10.1007/978-3-030-25385-1_10

Keywords Filovirus · VHF · Public-health impact · Economic impact · Transboundary spreading

History

Viral Haemorrhagic Fever: Definitions

According to Centres for Disease Control and Prevention (CDC) '*Viral haemorrhagic fevers refer to a group of illnesses that are caused by several distinct families of viruses. In general, the term "viral haemorrhagic fever" is used to describe a severe multisystem syndrome (multiple organ systems in the body are affected). Characteristically, the overall vascular system is damaged, and the body's ability to regulate itself is impaired. These symptoms are often accompanied by haemorrhage (bleeding); however, the bleeding is itself rarely life-threatening. While some types of haemorrhagic fever viruses can cause relatively mild illnesses, many of these viruses cause severe, life-threatening diseases*' (CDC 2013).

According to World Health Organisation (WHO) '*Viral haemorrhagic fever (VHF) is a general term for a severe illness, sometimes associated with bleeding, that may be caused by a number of viruses. The term is usually applied to diseases caused by Arenaviridae (Lassa fever, Lujo, Guanarito, Machupo, Junin, Sabia, and Chapare viruses), Bunyaviridae (Crimean-Congo haemorrhagic fever, Rift Valley fever, Hantaan haemorrhagic fevers), Filoviridae (Ebola and Marburg haemorrhagic fever) and Flaviviridae (dengue, yellow fever, Omsk haemorrhagic fever, Kyasanur Forest disease, and Alkhurma viruses)*' (Howard 2005; WHO 2018a).

Nonetheless, other pathogens, such as viruses of *Arteviridae*, *Reoviridae*, *Rhabdoviridae*, or *Asfarviridae* families can cause viral haemorrhagic fevers (VHFs). Actually, the use of VHFs term is controversial and pathogens involved in VHFs, as well as the definition of VHFs, need to be updated to regard of scientific progress and new knowledge acquired (Kuhn et al. 2014).

VHFs are generally zoonotic diseases and distributed all over the globe (CDC 2013). As each virus is associated with one or more particular host species, the virus and its related disease are usually observed only in the regions where the host species live(s). Depending on the host species, the virus can be present only in restricted areas (i.e. New World arenaviruses carried by rodent species living only in North and South America) or worldwide (i.e. Seoul virus causing haemorrhagic fever with renal syndrome (HFRS) hosted by the common rat (*Rattus* sp.)).

VHF diseases affect more than 100 million people around the world and more than 60,000 deaths annually (Zapata et al. 2014). Dengue virus is the most prevalent VHF with 50 to 100 million cases of dengue fever reported every year. Yellow fever (YF) virus is the second most globally distributed arthropod-borne disease, causing around 200,000 cases per year. In Africa, we also found Lassa fever virus (LASV), Lujo virus (LUJV), the Rift Valley Fever (RFV), Crimean-Congo haemorrhagic

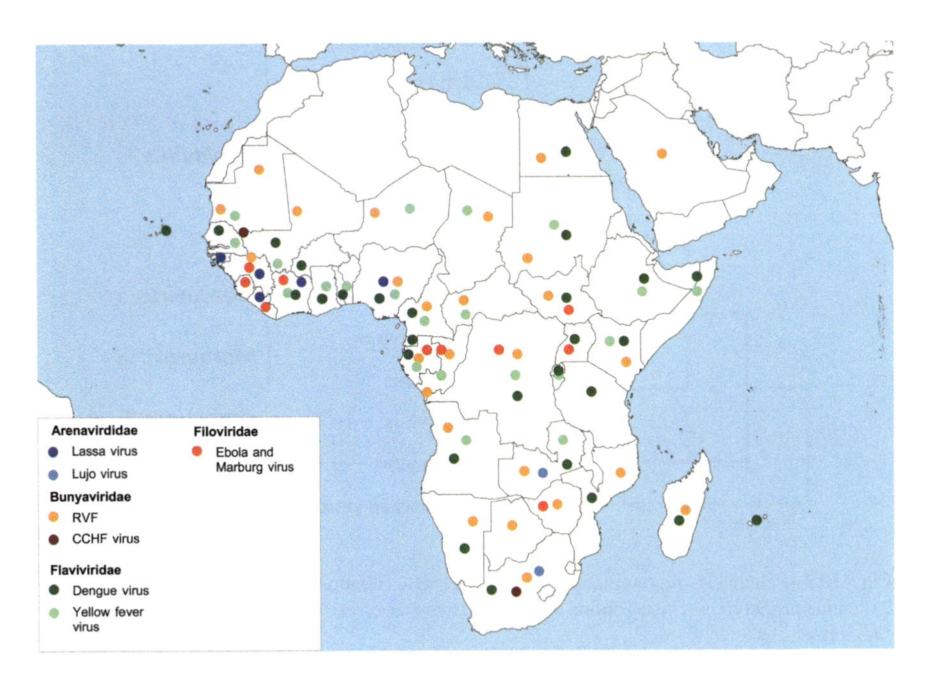

Fig. 10.1 Geographical distribution of the main Viral Haemorrhagic Fevers (VHF) belonging to the virus families Arenaviridae, Bunyaviridae, Flaviviridae and Filoviridae. CCHF stands for Crimean Congo Haemorrhagic Fever (from Were 2012; Zapata et al. 2014).

fever (CCHF), Marburg and Ebola virus diseases (Fig. 10.1), the latter three having the highest mortality rates of all known VHFs (Zapata et al. 2014).

Viral Haemorrhagic Fever Caused by Filovirus

Discovery

Marburg virus was first identified in 1967 following three simultaneous outbreaks in Europe in Germania (Marburg and Frankfurt) and in Serbia (former Yugoslavia) in Belgrade. Laboratory workers were infected after handling blood and tissues from African green monkeys (*Chlorocebus* spp.) imported approximately at the same time from the same location in Uganda.

Ebola virus was identified in 1976 when two successive outbreaks occurred in southern Sudan and northeast Democratic Republic of Congo (former Zaïre). In Sudan, the outbreak began in June 1976 after the infection of a man working in a cotton factory in the township of Nzara. In the Democratic Republic of Congo (DRC), the outbreak occurred near a small river named "Ebola" in the north of Yambuku in September and October 1976. Retrospectively, we found out that there were two distinct viruses: *Sudan ebolavirus* (SUDV) and *Zaire ebolavirus* (ZEBOV).

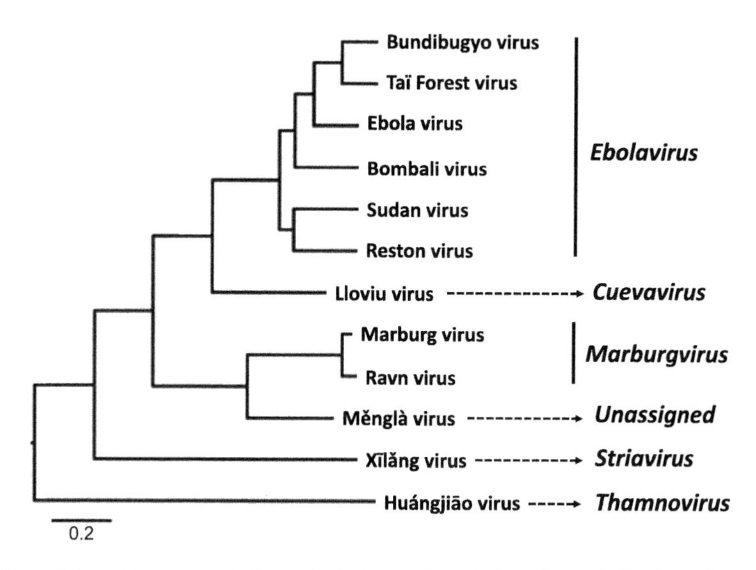

Fig. 10.2 Schematic representation of phylogenetic relationships of filovirus (https://talk.ictvonline.org/ictvreports/ictv_online_report/negative-sense-rna-viruses/mononegavirales/w/filoviridae)

Then, four other species of Ebola virus were found in Ivory Coast in 1994 in the Thai forest; in Bundibugyo District, Western Uganda in 2007; and in Bombali district, Sierra Leonne in 2018. In 2011 and 2019, new filoviruses named Cuevavirus and Měnglà virus, were found in Spain and China, respectively (Negredo et al. 2011; Yang et al. 2019).

Taxonomy of *Filoviridae*'s Family

Filoviridae's family belongs to the Mononegaviruses order (Easton and Pringle 2012). They are enveloped and contain a linear, non-segmented, single-strand RNA genome with the general genomic structure 3′-UTR-core-protein genes-envelope protein genes-polymerase gene-5′-UTR.

According to the International Committee on Taxonomy of Viruses (ICTV, Fig. 10.2), mammals *Filoviridae*'s family includes three genera (Kuhn et al. 2010, 2013; Negredo et al. 2011; Maruyama et al. 2014):

- *Ebola virus* (EBOV) with six distinct species: *Sudan ebolavirus* (SUDV), *Zaire ebolavirus* (ZEBOV), *Taï Forest ebolavirus* (TAFV), *Bundibugyo ebolavirus* (BDBV), *Reston ebolavirus* (RESTV), and *Bombali ebolavirus* (BOMBV). This latter was recently described (Goldstein et al. 2018; Forbes et al. 2019).
- *Marburgvirus* (*Marburg marbugvirus*), with only one species consisting of two viruses, Marburg virus (MARV) and Ravn virus (RAVV).
- *Cuevavirus,* a new genus with one species, *Lloviu virus* (LLOV).

To note, three more genera have been recently proposed with notably two new genera discovered in fishes (Fig. 10.2):

- *Dianlovirus,* with one species, *Měnglà virus* (MLAV), characterized from *Rousettus* bats in China (Yang et al. 2019). However, to date, this new virus has not yet been assigned as new genera by the ICTV.
- *Striavirus*, with one species, *Xilang striavirus,* characterized from striated frogfish (*Antennarius striatus* Shaw) (Shi et al. 2018).
- *Thamnovirus,* with one species, *Huangjiao thamnovirus,* characterized from greenfin horse-face filefish (*Thamnaconus septentrionalis*) (Shi et al. 2018).

Additionally, partial sequences of a putative new filovirus genus were found in bats from China (He et al. 2015; Yang et al. 2017). This taxonomic nomenclature will evolve with the discovery of new species and/or lineages.

Geographical Distribution and Epidemiology

Marburg Haemorrhagic Fever Disease

During the three simultaneous outbreaks in Germania and in Serbia, out of the 31 infected people (23 at Marburg, 6 at Frankfurt, and 2 at Belgrade) and 7 died (Kissling et al. 1970; Martini 1973).

Since this first report of Marburgvirus disease (MVD), sporadic outbreaks occurred mainly in sub-Saharan African countries including Uganda, Zimbabwe, Kenya, DRC, and Angola (Gear et al. 1975; Smith et al. 1982; Johnson et al. 1996; Towner et al. 2006; 2009; Bausch et al. 2006; Adjemian et al. 2011). The latter two countries faced the largest Marburg haemorrhagic fever disease outbreaks. In DRC from 1998 to 2000, 154 people were infected and 128 died from the disease. In Angola between 2004 and 2005, 252 people were infected by the virus and 227 succumbed (Olival and Hayman 2014). Isolated cases were also reported in Russia (former USSR) following laboratory accidents (Nikiforov et al. 1994; Bradfute et al. 2016) and two transboundary cases in South Africa in 1975 and in the Netherlands in 2008, when the tourists were infected after visiting caves in Zimbabwe and Uganda, respectively (Gear et al. 1975; Timen et al. 2009). Finally, a suspected case was reported in the USA in 2008. Interestingly, the patient visited the same cave involved in the Netherlands case. However, the virus was not isolated (January and Health 2009; Bradfute et al. 2016) (Table 10.1; Fig. 10.3).

Although the majority of outbreaks in Africa were linked to bats throughout human mining activities or visiting caves, the identification of *Marburgvirus* host reservoir has been challenging (Johnson et al. 1981, 1982; Towner et al. 2007; Swanepoel et al. 2007). Isolation of *Marburvirus* was first realized in 2009 from Egyptian fruit bats (*Rousettus aegyptiacus*) 40 years after the first reported outbreak (Towner et al. 2009). Additionally, Amman et al. (2012) showed that the circulation of *Marbuvirus* in juvenile Egyptian fruit bats coincided with periods of increased

Table 10.1 History of Marburgvirus diseases outbreaks

Locations	Dates	Sites of infection	Virus	Likely source of infection	Human cases	Case fatality rate (%)
First outbreak						
Germania	1967	Laboratory	MARV	Imported monkeys (Uganda)	29	24
Serbia	1967	Laboratory	MARV	Imported monkeys (Uganda)	2	0
Sub-Saharan Africa outbreaks						
Kenya	1980	Kitum Cave?	MARV	Bats?	2	50
	1987	Kitum Cave	RAVV	Bats?	1	100
DRC	1998–2000	Goroumbwa Cave	MARV/RAVV	Bats?	154	83
Angola	2004–2005	Uige Province	MARV	Unknown	252	90
Uganda	2007	Kitaka Mine	MARV/RAVV	Bats		
Imported cases						
South Africa	1975	Shinoi Cave?	MARV	Bats?	3	33
		Zimbabwe				
Russia	1988	Laboratory	MARV			
Netherland	2008	Piton Cave?	MARV	Bats?	1	100
		Uganda				
USA[a]	2008	Piton Cave?	MARV	Bats?	1	0
		Uganda				

Based on CDC: https://www.cdc.gov/vhf/marburg/outbreaks/chronology.html
[a]Suspected case

risk of human infections. Nonetheless, the question remains concerning other bat species as a potential natural reservoir (Swanepoel et al. 2007; Pourrut et al. 2009; Towner et al. 2009).

Ebola Haemorrhagic Fever Disease

In southern Sudan, the first Ebola haemorrhagic fever disease outbreak began in June 1976 after the infection of a man working in a cotton factory in the township of Nzara. This patient died on the 6th July 1976 and from June through November 1976, 286 people were infected with 53% of case fatalities (Team et al. 1978).

In DRC, the second epidemic of Ebola virus disease (EVD) occurred between September and October 1976. During this outbreak, 280 out of the 318 infected patients succumbed (88%) (Griffiths et al. 2014).

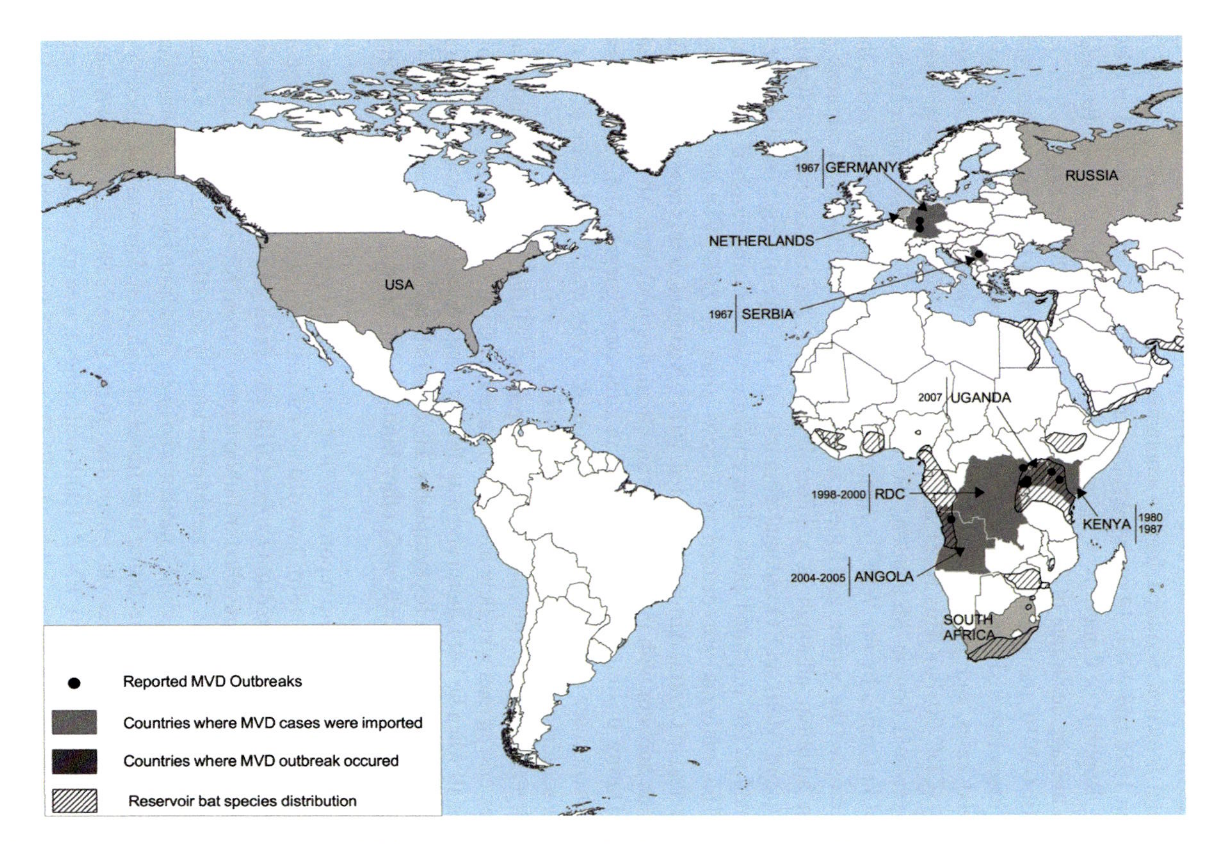

Fig. 10.3 Reported outbreaks or isolated cases of MVD. Countries where outbreaks occurred or where MVD was imported are indicated as well as the known distribution of the *Aegypciacus roussettus*, the bat species identified as the reservoir of Marburg viruses in Africa (from CDC 2012)

Since the discovery of the EBOV, 32 epidemics and/or sporadic cases of infections have been reported from 1976 to 2018 (Table 10.2). To date, seven outbreaks, due to SUDV, occurred in South Sudan (former Sudan) ($n = 3$) and in Uganda ($n = 4$) infecting a total of 761 people with 54.1% (from 41 to 100%) of lethality (CDC 2018a; WHO 2016a). One case was also reported in England after an accidental infection acquired from a contaminated needle (Emond et al. 1977). Between 1976 and 2017, 14 outbreaks, caused by ZEBOV, were recorded in DRC ($n = 7$), Republic of Congo ($n = 3$), and Gabon ($n = 4$) with a total of 1489 cases with an average death rate of 77.2% (from 47 to 100%) (CDC 2018a).

In 2014–2016, the world faced the worse outbreak of EVD known to date with 28,616 cases reported (suspected, probable, and confirmed) with 15,325 laboratory confirmed cases and 11,310 deaths in West Africa (Guinea, Sierra Leone, and Liberia) (CDC 2018a). This EVD stemmed from a single zoonotic spillover event to a 2-year-old boy in Meliandou, Guinea. On March 2014, the outbreak was reported in Guinea by the WHO's African Regional Office and on May 2014, the disease crossed the borders and Liberia and Sierra Leone faced the EVD outbreak as well. In addition, although at a lower level, other African countries (Nigeria, Senegal, and Mali) and Occidental countries (USA, Spain, Italy, and UK) were confronted to few imported cases. This epidemic was declared finished in January 2016. This outbreak was caused by a new strain of ZEBOV distinct of those incriminated in Central Africa (Baize et al. 2014). Nonetheless this new strain shared a common ancestor with the ZEBOV strain circulating in Central Africa (Gire et al. 2014). During this outbreak, three Sahelian countries Senegal, Mali, and Nigeria faced human cases imported through human movements and two of them, Mali and Nigeria, acquired through human transmissions.

In Nigeria, 19 cases (seven lethal) were reported. All these cases were linked to the index case, a man who flew from Morovia in Liberia to Lagos in July 2014. This man was ill when he left Liberia and was directly transported to a private hospital on his arrival in Lagos due to his symptoms. He died five days after his arrival. During his trip, he had 894 contacts with other humans and all of them were monitored (Shuaib et al. 2014).

In Mali, eight cases were reported with seven laboratory confirmed cases. Six of them succumbed (CDC 2018b). Two distinct introductions of EVD were reported in Mali. The first case, reported in October 2014 was a two-year-old girl accompanied by her grandmother who travelled in public transport from Kissidougou (Guinea) to Kayes via Bamako in Mali. The grandmother initially travelled from Mali to Guinea to attend the funeral of the child's mother who likely died from Ebola virus infection. The child was symptomatic during their travel through Mali. She was examined by a health care worker in Kayes who referred her to the Kayes's Hospital. Samples were taken and tested positive for EVD. Health authorities traced more than 400 contacts with the infected child but no additional EVD case was detected. The second EVD introduction was reported when a 70-year-old man travelled from Guinea to Bamako on 25 October 2014. He died on 27 October. Three additional cases occurred in a family visited by the old man before his admission to the clinic. The three patients succumbed and the doctor who treated the old man was also contaminated (WHO 2014a).

Table 10.2 History of human Ebolavirus diseases outbreaks

Locations	Dates	Sites of infection	Virus	Likely source of infection	Human cases	Case fatality rate (%)
First outbreak						
DRC	1976	Yambuku	ZEBOV	Unknown	318	88
South Sudan	1976	Nzara	SUDV	Unknown	284	53
		Maridi				
Sub-Saharan Africa outbreaks						
DRC	1977	Tandala	ZEBOV	Unknown	1	100
	1995	Kikwit	ZEBOV	Unknown	315	81
	2007	Kasai Province	ZEBOV	Bat	264	71
	2008–2009	Kasai Province	ZEBOV	Unknown	32	47
	2012	Isiro	BDBV	Unknown	36[a]	36.1
	2014[a]	Multiple villages	ZEBOV	Monkey	66	74
	2017	Likati	EBOV[b]		8	50
	2018	Bikoro, Equateur Province	ZEBOV	Unknown	54	61
	2018-(ongoing)	Noth Kivu and Ituri Provinces	ZEBOV	Unknown	3133[b]	67[b]
Rep. of Congo	2001–2002	Mbomo district	ZEBOV	Apes?	57	75
		Kelle district				
	2002–2003	Mbomo district	ZEBOV	Apes?	143	89
		Kelle district				
	2003	Mbomo village	ZEBOV	Apes?	35	83
		Mbandza village				
Gabon	1994	Mékouka	ZEBOV	Bats?	52	60
	1996–1997	Boué	ZEBOV	Apes	60	74
	2001–2002	Mékombo	ZEBOV	Apes	65	82
Uganda	2000–2001	Gulu	SUDV		425	53
	2007–2008	Bundibugyo	BDBV		149	29
	2011	Nakisimata	SUDV		1	100
	2012	Kibaale district	SUDV		11[a]	36.4
	2012–2013	Luwero district	SUDV		6[a]	50
	2019	Kasese District	ZEBOV	RDC imported Case	3	75

(continued)

Table 10.2 (continued)

Locations	Dates	Sites of infection	Virus	Likely source of infection	Human cases	Case fatality rate (%)
South Sudan	1979	Nzara	SUDV	Bat?	34	65
	2004	Yambio Maridi	SUDV	Baboon	17	41
Ivory Coast	1994	Tai Forest	TAFV	Apes	1	0
Multicountries	2014–2016		ZEBOV	Bat?		
Sierra Leone		Entire country			14124	28
Liberia		Entire country			10678	45
Guinea		Entire country			3814	66
Nigeria		Lagos			20	40
		Port harcourt				
Senegal		Dakar			1	0
Mali		Bamako			8	75
		Kayes				
Imported cases						
South Africa						
From Gabon	1996	Johannesburg[c]	ZEBOV		2	50
Spain						
From Sierra Leone	2014	Madrid	ZEBOV		2	50
Italy						
From Sierra Leone	2014	Sassari	ZEBOV		1	0
United Kingdom	Laboratory	SUDV			1	0
From Sierra Leone	2014	Glasgow	ZEBOV		1	0
USA						
From Liberia	2014	Dallas[d]	ZEBOV		3	33
From Guinea	2014	New York	ZEBOV		1	0
Uganga						
From DRC	2019	Kasese Distric	ZEBOV		3	75

Based on CDC: https://www.cdc.gov/vhf/marburg/outbreaks/chronology.html

[a]Laboratory confirmed cases

[b]The outbreak is ongoing at the time of chapter writing. The number of cases reported herein are the confirmed cases at the 22 October 2019 (https://apps.who.int/iris/bitstream/handle/10665/311805/SITREP-EVD-DRC-20190407-eng.pdf)

[c]One human-to-human transmission from the index case

[d]Two human-to-human transmissions from the index case

In Senegal, only one case was reported. On the 20 August 2014, a 21-year-old male native from Guinea who was in close contact with a confirmed EVD patient, escaped the Guinean surveillance system and reached Dakar by road. After three days passed with his relatives in the suburbs of Dakar, the young man visited medical care for fever, diarrhoea, and vomiting symptoms received a malaria treatment and was sent back close to his relative before being hospitalized on the 26 August. He survived and no other person was infected (WHO 2014b).

The 9th Ebola outbreak in DRC (ZEBOV) started on the 8 May 2018 in Bikoro, Equateur Province. Fifty-four cases were reported with 33 deaths (61%). The WHO declared the end of this outbreak on 24 July 2018.

The latest ZEBOV outbreak was declared on August 1 North Kivu province. Confirmed cases have been also reported in Ituri province. On April 2019, 1088 confirmed cases were reported with 61% of case fatalities. This outbreak is still ongoing and the number of cases is increasing despite all effort.

Two EVD cases were also reported in Russia (1996 and 2004) and two in South Africa (1996) following laboratory contaminations. These contaminations were linked to the Gabonese outbreak in 1996 (WHO 1996; CDC 2018b). In addition, two other viruses were involved in human Ebola diseases: BDBV who infected 36 persons in DRC (2012) and 149 persons in Uganda (2007–2008) with a fatality rate of 36.1% and 24.8%, respectively (CDC 2018b). TAFV was isolated from a scientist who was infected after conducting an autopsy on a wild chimpanzee in the Taï Forest in Ivory Coast (Le Guenno et al. 1995) (Table 10.2; Fig. 10.4).

The fifth Ebola virus species, the RESTV is the only Asian Ebolavirus species known to date and was never associated with human lethality (Miranda et al. 1999). RESTV was first discovered in Cynomolgus monkeys (*Macaca fascicularis*) exported from the Philippines to Reston, Virginia in the USA in 1989 (Jahrling et al. 1990). Following this first report, RESTV was involved in five other infection episodes among monkeys imported from Philippines (Miranda et al. 1999). In monkeys, the disease showed the clinical signs and pathological lesions of EVDs and was highly fatal. Despite the high viremia observed during the latter clinical stage in infected monkeys (Jahrling et al. 1996), humans exposed to infected monkeys did not develop Ebola-like illness and remained asymptomatic.

The recently reported Bombali ebolavirus species (BOMV) was characterized in two different insectivore bat species, *Chaerephon pumilus* (little free-tailed bats) in Sierra Leone and *Mops condylurus* (Angolan free-tailed bats) in Sierra Leone and Kenya (Goldstein et al. 2018; Forbes et al. 2019). The fact that the BOMV was discovered in two different country from the same bat species suggests a broad distribution of this Ebola virus in view of the Angolan-free tailed bats in habitat distribution. To note, little free-tailed bats have also a broad distribution in Africa, often overlapping with the Angolan-free tailed bats. Although the BOMV has not yet been associated with the EVD, the virus was found in house roof bats in close contact with humans. Besides, Goldstein et al. showed that BOMV is fully competent to mediate viral entry in human cells (Goldstein et al. 2018).

Despite an important research effort made by researchers during the last four decades, the reservoir and the ecology of Ebolavirus remain elusive (Olival and

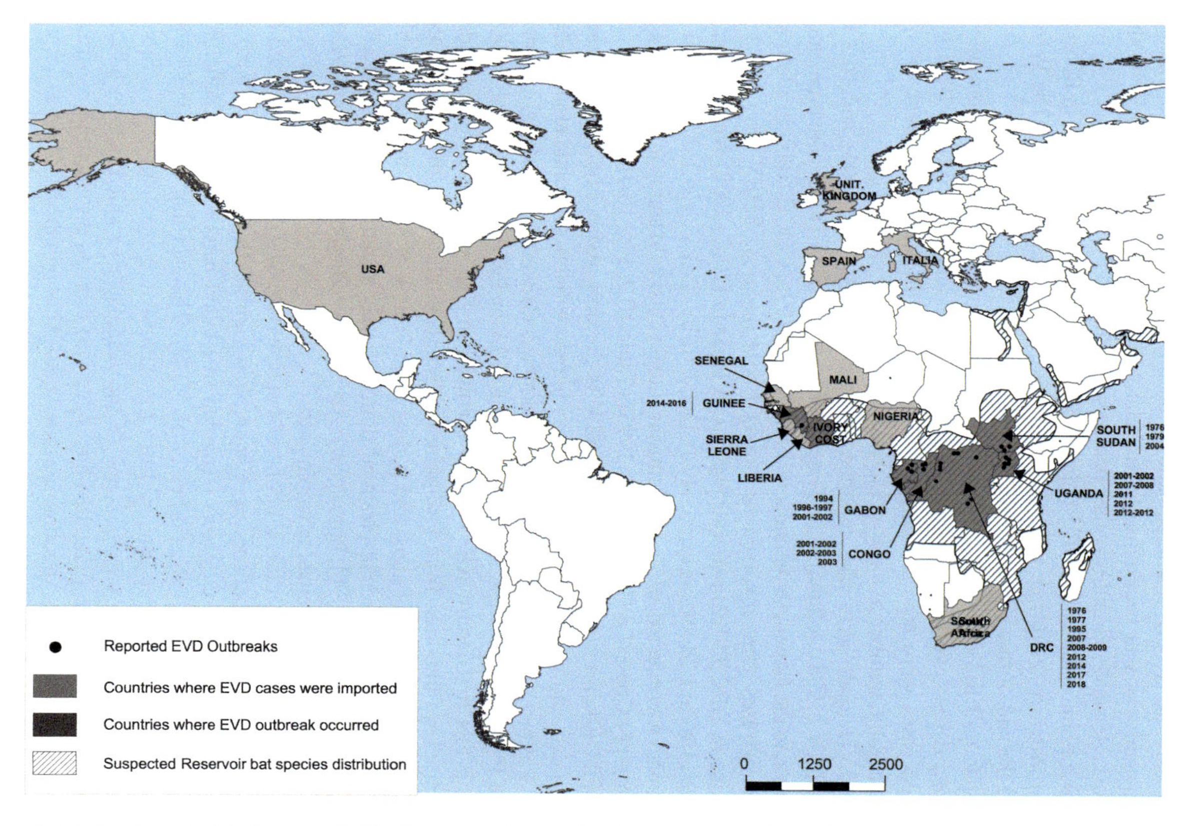

Fig. 10.4 Reported outbreaks or isolated cases of EVD. Countries where outbreaks occurred or where EVD was imported are indicated as well as the known distribution of the bats species suspected to be the reservoir of Ebola viruses in Africa (from Pigott et al. 2016; WHO 2018b)

Hayman 2014). As for Marburgvirus, African Ebola virus strains are suspected to have bats as natural reservoir hosts. Indeed, filoviruses do not appear to be overtly pathogenic in bats (Paweska et al. 2012). Epidemiological investigations show that outbreaks result from a variety of animal sources due to hunting and handling of dead animals such as gorillas, chimpanzees, bats, or ungulates (Leroy et al. 2004). The zoonotic origin was identified in nine out of the 32 known EVD outbreaks: simultaneous EBV outbreaks in chimpanzees and gorillas in Ivory Coast (1994), in Gabon and Congo (1996 and 2003), monkey bushmeat in DRC (2014), and Bat bushmeat in DRC (2007) (Leroy et al. 2004; Pourrut et al. 2005; Pigott et al. 2014). Contamination route from the supposed reservoir to great apes is still unknown. However, the main hypothesis is that gorillas and chimpanzees come to feed fruits in the trees and collect on the ground partially consumed fruits and pulp dropped by the bats feeding on the canopy above. The Great apes ingest them as they are likely to be contaminated from saliva.

Even though anti-Ebolavirus antibodies were detected from several bat species in Africa and in Asia and viral RNA of Ebola virus was amplified from different bat species in Africa (Pourrut et al. 2005; 2009; Leroy et al. 2005; Hayman et al. 2010; 2012; Taniguchi et al. 2011; Yuan et al. 2012; Olival et al. 2013; De Nys et al. 2018; Goldstein et al. 2018; Forbes et al. 2019), to date no Ebolavirus has been yet isolated from bats. As a matter of fact, bats play a role in the ecology of Ebolavirus. Several epidemiological investigations show the link between bats and the Ebola virus. In the 1994–1995 Gabon outbreak, infected people were coming from gold mining camps, which are often used as roosts by the bats. The hunting and butchering of bats for consumption has also been clearly identified as the spillover mechanism in the 2007 DRC outbreak (Leroy et al. 2009). Research must carry on to (1) identify formally the reservoir species of ZEBOV in Central Africa, (2) verify new hypothesis of having other reservoir species than bats involved, and (3) understand the environmental factors that could influence the circulation of the ZEBOV in Africa (Morvan et al. 1999; Leendertz et al. 2015; Leendertz 2016; Buceta and Johnson 2017; Caron et al. 2018).

Concerning the RESTV, this virus was only associated with Asian non-human primates (NHPs) until 2008 when RESTV infection was identified in domestic swine in Philippine. During the survey, 6 out of the 141 humans tested positive for IgG titters to RESTV suggesting a potential transmission from pigs to humans. These six persons were in close contact with pigs throughout their work but they did not become ill (Barrette et al. 2009). Nonetheless, the role of swine in the ecology of RESTV is yet to be determined as all RESTV infection in pigs were associated with porcine reproductive and respiratory syndrome virus (PPRSV) (Barrette et al. 2009; Pan et al. 2014). Finally, in 2015, Jayme et al. detected serological evidence of RESTV infection in different Philippine bat species and showed the presence of partial RESTV RNA in three *Miniopterus* bat species. These bats were captured at less than 40 km from one of the 2008 pig RESTV isolate and RESTV RNA sequences from bats and pigs were closely related indicating a likely link between bat and swine infections (Jayme et al. 2015).

Cuevavirus and Dianlovirus

In addition, the new filovirus genus discovery, Lloviu virus (LLOV), was isolated from *Miniopterus schreibersii* bat carcasses found in Cueva del Lloviu, Spain. However, in contrast with the asymptomatic circulation of MARV and EBOV in bats, LLOV seemed to be pathogenic in *Miniopterus schreibersii* bat species (Negredo et al. 2011).

The *Dianlovirus* (MLAV), characterized recently in China, is phylogenetically close to the *Marburgvirus* genus. Actually, MLAV has been identified from Chinese *Rousettus* bat species that are also the natural reservoirs of *Marburg virus* in Africa (Yang et al. 2019).

Economic/Public Health Impact

The evaluation of the costs of EVD all outbreaks in Africa is very difficult as little information is available.

To stop rapidly the recent EVD Outbreak in RDC, the funds required were evaluated at US$57 million. It increased from US$26 million when the disease spread into an urban area on a major transport route (Mbandaka) (WHO 2018b).

The 2014–2015 EVD outbreak in Sierra Leone, Guinea, and Liberia had a devastating political, economic, sociocultural, environmental effects in West Africa and the rest of Africa. The EVD outbreak had direct and indirect costs for the three countries. The direct human cost remained the worse known to date following an EVD outbreak with 11,316 deaths for more than 28,600 individuals infected. The direct costs for containing and controlling the outbreak were enormous. More than US$3.6 billion were spent to fight the epidemic by the end of the year 2015 (US CDC 2016). In addition, this outbreak impacted durably the mortality and the morbidity in the three countries as their healthcare systems were particularly hit. The healthcare workers have been decimated and the health systems heavily enfeebled. By May 2015, Liberia lost 8% of its doctors, nurses, and midwives due to EVD; Sierra Leone and Guinea lost 7% and 1.5%, respectively, of their healthcare workers. The death of 513 healthcare professionals (doctors, nurses, and midwives) already in limited number, impacted the access of the populations to doctors and other qualified health professionals resulting in an increase of maternal mortality (38% in Guinea, 74% in Sierra Leone, and 111% in Liberia relative to pre-Ebola rates), infants, and less than 5 years mortality rates. Evans et al. (2015) estimated that 4022 women died every year during childbirth as a result of healthcare workers' loss to Ebola. Moreover, the interruption of routine health delivery services stopped the follow-up and the therapeutic management of patients with HIV, tuberculosis, and/or malaria, which tragically increased the severity of the illnesses and the number of deaths (Parpia et al. 2016). Children paid indeed a heavy price: nearly 23,000 children lost one or both parents or their primary caregivers to Ebola and 3508 children died representing more than one in four deaths. Moreover, the percentage of childhood vaccination coverage decreased by 30% when vaccination campaigns were interrupted as funds and means were reallocated to control the EVD epidemics or to avoid public gatherings (UNDP 2014).

Beyond the devastating health effects, the socio-economic impact of the Ebola epidemic in Guinea, Liberia, and Sierra Leone as well as in West Africa is long to resorb. The economic activities and trades between African countries were deeply affected for many reasons linked to the epidemic. Businesses could not operate normally as the movement of goods and people became very problematic. People stopped to work, fled the affected regions, or were put under quarantine, thus impacting countries' productivity (including food production). For example, in Liberia, despite good weather conditions, rice production fell about 25% due to the lack of workers or fear to be contaminated from other workers (FAO 2014). Livestock was also affected by the closure of the borders through lack of feeding and limited restocking. The decrease of the productions and the availability of goods were followed by a rise in prices of basic items like food. Restrictions on movement and international transport affected tourism activities with a fall of about 75% of visitors to West Africa, but also business dependant on foreign skills, as many companies with expatriates reduced or shutdown their activities, send back home non-essential personnel and postponed new investments. Public finances were also deeply affected by EVD epidemic, as slower economic activity had an immediate impact on tax revenues and also because of the huge costs to control the epidemic.

The World Bank estimated that the EBV outbreak impacted the total combined GDP of the three countries of US$2.2 to 2.8 billion (−4.5%), which represents a reduction of the GDP per capita by an average of $125 per person (World Bank Group 2016). The EVD resulted in a decrease of all economic sectors including private and investment, agriculture production, cross-border trade, etc. (US CDC 2016). Even though only 5 of 54 African countries had a positive case of Ebola, the epidemic impacted the entire continent. The forecast for economic growth in sub-Saharan Africa was reduced by 0.5% by the International Monetary Fund because of 'economic spillovers' from the outbreak.

In the other African countries touched by the epidemic (Senegal, Mali, and Nigeria), the impact of EVD has been finally quite low mainly due to the quick health sector response that involved national governments, health, civil and international and national humanitarian actors, the military, the private sectors, as well as NGOs (e.g. doctors without border) with the support of international institutions such as the World Health Organization (WHO) and CDC.

Ebola outbreak also affected indirectly conservation programmes as years of effort and millions of USD engaged to protect endangered wildlife species like great apes (Gorilla—*Gorilla* sp. and Chimpanzee—*Pan* sp.) can be destroyed in a couple of months by one outbreak of EVD as great apes are very sensitive host to EVD (Walsh et al. 2003; Bermejo et al. 2006).

Symptoms and Lesions

Filovirus infections are generally the most severe of the VHFs. Human infection occurs following direct contact with infected individuals or animals (Dowell et al. 1995). The incubation period in human ranges from 2 to 21 days with an average of 8–10 days. Generally, shorter incubation periods are associated with exposure to a larger viral load. Subsequent signs and symptoms indicate multisystem involvement and include systemic (prostration), gastrointestinal (anorexia, nausea, vomiting, abdominal pain, diarrhoea), respiratory (chest pain, shortness of breath, cough), vascular (conjunctival injection, postural hypotension, oedema), and neurologic (headache, confusion, coma) manifestations. Haemorrhagic manifestations appear generally during the peak of the illness and include petechiae, ecchymoses, uncontrolled oozing from venipuncture sites, mucosal haemorrhages, and post-mortem evidence of visceral haemorrhagic effusions. Often a macropapular rash associated with varying degrees of erythema appears by 5–7 days of the illness. Abdominal pain is sometimes associated with hyperamylasaemia and true pancreatitis. In later stages, shock, convulsions, severe metabolic disturbances, and, in more than half of the cases, diffuse coagulopathy supervenes. The WHO and the CDC have established criteria for making a diagnosis of EVD that include the sudden onset of high fever and at least any three of the following: headache, vomiting, loss of appetite, diarrhoea, lethargy, stomach pain, aching muscles or joints, dysphagia, dyspnoea, or hiccupping. The diagnosis is only confirmed with positive serology for Ebola virus.

But, asymptomatic, replicative Ebola virus infection occurs in human populations (Leroy et al. 2000) and Becquart et al. (2010) showed that Gabonese population presented both humoral and cellular immunity to ZEBOV. In this study, seroprevalence for anti-ZEBOV Ig varied from 2.7% to 33.8% according to the area sampled. The seropositivity was higher in forested areas than in urban or semi-rural areas. Becquart et al. (2010) suggested that humans could be infected and developed a natural immunity through handling and ingestion of fruits potentially contaminated by bats.

Diagnostic and Treatment

Laboratory diagnosis of filovirus diseases plays a critical role in outbreak response effort. Early identification of infected patients assists in the provision of rapid access to health care as well as interruption of the chain of transmission by timely isolation of infected patients. However, diagnosing Ebola in a person recently infected is challenging as early symptoms of EVD infection are non-specific (e.g. fever, vomiting, diarrhoea) and often are seen in patients with other diseases such as malaria, typhoid fever, shigellosis, or cholera. Also, a clinical diagnosis of EVD might be based on historical features such as cave or forest exposure, rural travel, contact with sick or dead wild or domestic animals, particularly monkeys and apes, contact with sick persons or treated in a local hospital. For patients with filovirus disease, prostration, lethargy, wasting, and diarrhoea seem to be more severe than what is observed for other VHF infections; the appearance of a characteristic rash is useful in narrowing the differential diagnosis. Diagnosis of a single case is extremely difficult, but the occurrence of clusters of cases with prodromal fever followed by haemorrhagic diatheses and person-to-person transmission are suggestive of VHF and require the implementation of containment procedures. Indeed, the recent 2014–2015 outbreak of EVD in West Africa has highlighted the importance of rapid and accurate diagnosis of the disease. Detection of viral nucleic acid by PCR technology (Real-time RT-PCR) is the recommended technique for laboratory diagnosis of EVD (WHO Ebola Response Team 2014; WHO 2015a). Nonetheless, the implementation of such laboratory techniques is difficult in low-income developing countries. Handling these highly infectious samples require appropriate biosafety BSL3/BSL4 facilities. During previous EBV diseases, blood samples were sent to international reference laboratories. However, some countries have established national reference VHF laboratory thus improving regional diagnostic capacities (Broadhurst et al. 2016). Nonetheless, the important issue during an EVD outbreak is the reduction of the diagnosis delay. To answer this challenge, setting up of field diagnosis capacity throughout the implementation of mobile laboratories have been one of the most appropriate response in order to reduce the time between sample collection and return of results (Broadhurst et al. 2016). Today, 24 laboratories in Guinea, Liberia, and Sierra Leone are able to test for EVD (US CDC 2016).

Several methods have been developed to detect EBOV infection. Virus isolation from serum in cell culture, classically done on Vero E6 African Green Monkey kidney cells, is the traditional gold standard method to confirm the presence of Ebola virus. Although this method is relatively simple and sensitive, it always must proceed in a BSL4 laboratory and thus restricted to research and public health laboratories (Broadhurst et al. 2016). Other methods are based on the detection of host antibodies generated against the virus (serologic tests), the detection of viral proteins (antigen tests), or the detection of viral RNA sequence (molecular tests). The two latter, antigenic and molecular tests are the recommended tests for the detection of Ebola virus in suspected or probable cases (WHO 2015a; Broadhurst et al. 2016).

In September 2014, WHO introduced an emergency procedure under its Prequalification Programme for rapid assessment of Ebola diagnostics for UN procurement to affected countries. The first diagnostic was accepted in November of the same year. In the same month, WHO called on manufacturers to develop rapid and easy to use point-of-care diagnostics that are more adapted to affected countries, where health infrastructures and qualified personnel are insufficient. As a result, seven diagnostics have been approved for Emergency Use Assessment and Listing procedure (EUAL) by WHO (WHO 2015b; Broadhurst et al. 2016).

To date, no specific therapy exists for filovirus infections. The clinical management of EVD consists of supportive treatment and includes oral rehydration salts and intravenous fluids for dehydration, potassium, anti-emetics, and parenteral antibiotics for bacterial infections or comorbidities such as malaria (Murray 2015). Nonetheless, different promising therapeutic approaches and vaccines are underway (Agnandji et al. 2017; Bixler et al. 2017; Geisbert 2017). It goes from monoclonal antibody cocktails (i.e. mAb 114, ZMapp, REGN3470-3471-3479), antiviral drugs (i.e. Remdesivir/GS-5734, Favipiravir), and vaccine (i.e. rVSV-ZEBOV). The vaccine consists of a vesicular stomatitis virus (VSV), an animal virus causing flu-like illness in humans, genetically modified with a protein from the Zaire Ebola virus so that it can provoke an immune response to the Ebola virus (WHO 2018c). It is administrated in one unique dose.

The efficiency of these drugs is still under investigation. However, they were all approved to be used during the 2018 outbreak in RDC by an ethics committee in the DRC to treat Ebola, under the framework of compassionate use/expanded access. This was the first time such treatments were available in the midst of an Ebola outbreak. The vaccine rVSV-ZEBOV was declared to be safe and protect against Ebola virus infection based on several trials involving more than 16,000 volunteers in the USA, Europe, and Africa were done and showed the efficiency of the vaccine and can be utilized as long as the informed consent is obtained from patients and protocols are followed, with close monitoring and reporting of any adverse events (WHO 2016b). The benefit of ZMapp, based on the available data from a randomized controlled trial of ZMapp in patients with EVD, is considered to outweigh the risks of its utilisation. The use of the antiviral drug Remdesivir (GS-5734) (an antiviral drug), was authoriser under 'monitored emergency use of unregistered and experimental interventions' (MEURI), however, additional studies of Remdesivir in appropriate clinical trials are needed to assess its benefits and risks for the treatment of patients with EVD. The monoclonal antibody cocktail REGN3470-3471-3479 seems to be very promising and its utilisation under MEURI is recommended when ZMapp or Remdisivir is not available. However, the effects of REGN3470-3471-3479 must be studied in appropriate clinical trials. For the antiviral drug Favipiravir, there is still uncertainty as to whether it provides benefits for patients with EVD. More appropriate clinical trials are needed to establish whether it provides benefits to patients or not. In 2018 DRC outbreak, MEURI of Favipiravir was considered in select circumstances where the use of ZMapp or Remdesivir or REGN 3470-3471-3479 were not available. The monoclonal antibody mAb114 is in very early stages of development. The early data available looks promising but more information from clinical trials is needed before recommending its use for MEURI (WHO 2016b, 2018d).

For Marburg virus disease (MVD), there is yet no proven treatment available. Only supportive care with rehydration can be proposed in addition to treatments of specific symptoms in order to improve survival. However, potential treatments including blood products, immune therapies, and drug therapies are currently being evaluated (WHO 2017).

Application to Prevention and Control/Adopted Surveillance and Control Strategies

The overall WHO guidance for the prevention and control of haemorrhagic fever disease caused by Marburg and Ebola viruses focuses on strong and efficient surveillance systems and on supporting at-risk countries to prepare and organise their response in case of an outbreak.

For the surveillance of EVD in African countries with no reported cases of EVD (including most of West African countries), the WHO early surveillance prerequisites are to put in place an alert system targeting the major land boundaries with already affected countries, but also capital cities and their ports (airports and seaports), healthcare facilities, especially in major hospitals (WHO 2014c). The health workers must also be trained in case definitions and able to identify quickly the signs and symptoms of the disease, so they can report sick persons crossing the border from a country that has reported cases of EVD. With the alert system, a Rapid Response Team (RRT) is important to start the response sequence (1) case definition, (2) report at the national/international level, and (3) infection prevention and control measures. When a potential case (dead or alive) of EVD is suspected, an RRT must be sent as soon as possible to the site of the reported case to initiate an investigation and put in place initial control measures as required and place the patient(s) in a fully equipped isolation centre with trained staff. If the case meets the WHO definition of 'case under investigation' which is 'a person who has travelled to or stayed in a country that has reported at least one confirmed case of EVD, within a period of 21 days before the onset of symptoms, and who presents with sudden onset of high fever and at least three of the following symptoms: headache, vomiting, diarrhoea, anorexia/loss of appetite, lethargy, stomach pain, aching muscles or joints, difficulty swallowing, breathing difficulties, hiccup, or inexplicable bleeding/haemorrhaging or who died suddenly and inexplicably', then extra action should be taken quickly (WHO 2014c). Clinical samples should then be taken and sent to the pre-identified and WHO recognised laboratory. After isolation of the patient, it is urgent to identify all persons who has been in contact with him and start the medical follow up with their consent. All close contacts should be monitored for 21 days following their last known exposure to the case and must be isolated and receive appropriate care as soon as they show symptoms. It is capital that during the investigation, the RRT works and interacts with the communities respecting social and cultural customs and hierarchies.

Until 2017, for all Ebola or Marburg outbreaks, the control strategy was mainly containment and separation (isolation) of the patients developing clinical symptoms to prevent the further spread at home or in the community, and safely bury the dead without all burial traditional usagės to reduce further spread of the virus through contact with deceased. The year 2018 marks a turning point in the management of EVD outbreaks with vaccines and drug to protect and cure communities contaminated or potentially in contact with the virus. Clinicians working in the treatment centres will make decisions on which drug to use as deemed helpful for their patients, and appropriate for the setting. The treatments can be used as long as informed consent is obtained from patients and protocols are followed, with close monitoring and reporting of any adverse events. The first campaign of vaccination with the rVSV-ZEBOV vaccine in RDC in 2018 targeted persons who have been in contact or contact of contacts with confirmed EVD patients, healthcare and frontline workers (WHO 2018b).

Conclusion

Viral haemorrhagic fever diseases are generally zoonoses. The virus is maintained by a reservoir species and is totally dependent on them for replication and overall survival (CDC 2013). Rodents, bats, and arthropods are the main reservoirs of VHFs, however, the hosts of some viruses remain unknown or uncertain and extra investigations are needed. It is the case of Ebola viruses for which the reservoir hosts are yet to be confirmed. Identifying the reservoir species and understanding their role is important to understand the ecology of the virus, the transboundary nature of the disease, and the epidemiology necessary for effective control of these diseases (Siembieda et al. 2011). In the case of EVD, how an Ebola virus with a most recent common ancestor estimated at 10 years with ZEBOV circulating in Central Africa (Gire et al. 2014) arrived in West Africa and caused the biggest human outbreak ever known, remains unclear and more investigation are needed. The EVD outbreak in West Africa confirmed the transboundary nature of EVD. The spread of the disease from Guinea to Sierra Leone and Liberia was due to human movement (Gire et al. 2014; Dudas et al. 2017). However, the role of wildlife and the international trade of bushmeat (mainly illegal) across borders are clearly identified as the main potential transboundary routes for EVD and other VHFs. And the risk of outbreaks and transboundary spread of VHFs is increased by the environmental changes and natural habitats degradation occurring in Africa and the increased pressure on wildlife to supply the markets in large cities with bushmeat.

Due to the transboundary nature of the VHFs and EVD in particular, a collaboration between countries, regions, and continents is indispensable to control effectively these emerging/re-emerging and transboundary diseases. Systematic data collection, thorough laboratory investigations, and contact tracing during the outbreaks are the key elements to management and the control of the spread of the disease in our interconnected world of intensified travel and trade.

The development of vaccine and drugs to protect and cure communities potentially in contact with EBOV brings new options for the control and the management of EVD outbreaks. However, it is capital to improve during the period between two outbreaks the countries surveillance systems based on cross-sectoral collaboration and information sharing, monitoring of the wildlife/human/livestock interface, and a better understanding of the ecology of the EVD. This will be done through capacity building and training program targeting public health but also animal health professionals together with people involved in wildlife conservation and management, in order to promote a One Health approach.

References

Adjemian J, Farnon EC, Tschioko F, Wamala JF, Byaruhanga E, Bwire GS, Kansiime E, Kagirita A, Ahimbisibwe S, Katunguka F, Jeffs B, Lutwama JJ, Downing R, Tappero JW, Formenty P, Amman B, Manning C, Towner J, Nichol ST, Rollin PE. Outbreak of Marburg hemorrhagic fever among miners in Kamwenge and Ibanda Districts, Uganda, 2007. J Infect Dis. 2011;204:S796–9. https://doi.org/10.1093/infdis/jir312.

Agnandji ST, Fernandes JF, Bache EB, Obiang Mba RM, Brosnahan JS, Kabwende L, Pitzinger P, Staarink P, Massinga-Loembe M, Krahling V, Biedenkopf N, Fehling SK, Strecker T, Clark DJ, Staines HM, Hooper JW, Silvera P, Moorthy V, Kieny M-P, Adegnika AA, Grobusch MP, Becker S, Ramharter M, Mordmuller B, Lell B, Krishna S, Kremsner PG. Safety and immunogenicity of rVSVDeltaG-ZEBOV-GP Ebola vaccine in adults and children in Lambarene, Gabon: A phase I randomised trial. PLoS Med. 2017;14:e1002402. https://doi.org/10.1371/journal.pmed.1002402.

Amman BR, S a C, Reed ZD, Sealy TK, Balinandi S, Swanepoel R, Kemp A, Erickson BR, J a C, Campbell S, Cannon DL, Khristova ML, Atimnedi P, Paddock CD, Crockett RJK, Flietstra TD, Warfield KL, Unfer R, Katongole-Mbidde E, Downing R, Tappero JW, Zaki SR, Rollin PE, Ksiazek TG, Nichol ST, Towner JS. Seasonal pulses of Marburg virus circulation in juvenile *Rousettus aegyptiacus* bats coincide with periods of increased risk of human infection. PLoS Pathog. 2012;8:e1002877. https://doi.org/10.1371/journal.ppat.1002877.

Baize S, Pannetier D, Oestereich L, Rieger T, Koivogui L, Magassouba N, Soropogui B, Sow MS, Keïta S, De Clerck H, Tiffany A, Dominguez G, Loua M, Traoré A, Kolié M, Malano ER, Heleze E, Bocquin A, Mély S, Raoul H, Caro V, Cadar D, Gabriel M, Pahlmann M, Tappe D, Schmidt-Chanasit J, Impouma B, Diallo AK, Formenty P, Van Herp M, Günther S. Emergence of Zaire Ebola Virus disease in Guinea—preliminary report. N Engl J Med. 2014;371:1–8. https://doi.org/10.1056/NEJMoa1404505.

Barrette RW, Metwally S, Rowland JM, Xu L, Zaki SR, Nichol ST, Rollin PE, Towner JS, Shieh W, Batten B, Sealy TK, Carrillo C, Moran KE, Bracht AJ, Mayr G, Sirios-Cruz M, Catbagan DP, Lautner E, Ksiazek TG, White WR, Mcintosh MT. Discovery of swine as a host for the Reston ebolavirus. Science. 2009;325:204–6. https://doi.org/10.1126/science.1172705.

Bausch DG, Nichol ST, Muyembe-Tamfum JJ, Borchert M, Rollin PE, Sleurs H, Campbell P, Tshioko FK, Roth C, Colebunders R, Pirard P, Mardel S, Lemman PA, Zeller H, Tshomba A, Kulidri A, Libande ML, Mulangu S, Formenty P, Grein T, Leirs H, Braack L, Ksiazek T, Zaki S, Bowen MD, Smit SB, P a L, Burt FJ, Kemp A, Swanepoel R. Marburg hemorrhagic fever associated with multiple genetic lineages of virus. N Engl J Med. 2006;355:909–19. https://doi.org/10.1056/NEJMoa051465.

Becquart P, Wauquier N, Mahlakõiv T, Nkoghe D, Padilla C, Souris M, Ollomo B, Gonzalez JP, De Lamballerie X, Kazanji M, Leroy EM. High prevalence of both humoral and cellular immunity to Zaire ebolavirus among rural populations in Gabon. PLoS One. 2010;5:e9126. https://doi.org/10.1371/journal.pone.0009126.

Bermejo M, Rodríguez-Teijeiro JD, Illera G, Barroso A, Vilà C, Walsh PD. Ebola outbreak killed 5000 gorillas. Science. 2006;314:1564. https://doi.org/10.1126/science.1133105.

Bixler SL, Duplantier AJ, Bavari S. Discovering drugs for the treatment of Ebola virus. Curr Treat Options Infect Dis. 2017;9(3):299–317. https://doi.org/10.1007/s40506-017-0130-z.

Bradfute SB, Jahrling PB, Kuhn JH. Filoviruses. Mononegaviruses of veterinary importance. In: Munir M (ed) Molecular epidemiology control. vol. 2; 2016. p. 156–173.

Broadhurst MJ, Brooks TJG, Pollock NR. Diagnosis of Ebola virus disease: past, present, and future. Clin Microbiol Rev. 2016;29:773–93. https://doi.org/10.1128/CMR.00003-16.

Buceta J, Johnson K. Modeling the Ebola zoonotic dynamics: interplay between enviroclimatic factors and bat ecology. PLoS One. 2017;12:1–14. https://doi.org/10.1371/journal.pone.0179559.

Caron A, Bourgarel M, Cappelle J, Liégeois F, De Nys HM, Roger F. Ebola virus maintenance: if not (only) bats, what else? Viruses. 2018;10(10):pii: E549. https://doi.org/10.3390/v10100549.

CDC. Marburg hemorrhagic fever. CDC; 2012. p. 1–6. https://doi.org/10.1056/NEJMp068160.

CDC. Viral hemorrhagic fevers. CDC Fact Sheet; 2013. p. 1–3.

CDC. 40 years of Ebola virus disease around the World. 2018a. https://www.cdc.gov/vhf/ebola/history/chronology.html. Accessed 10 Jun 2018.

CDC. Ebola virus outbreaks by species and size, since 1976. 2018b. https://www.cdc.gov/vhf/ebola/history/distribution-map.html. Accessed 18 Jun 2018.

De Nys HM, Kingebeni PM, Keita AK, Butel C, Thaurignac G, Villabona-Arenas CJ, Lemarcis T, Geraerts M, Vidal N, Esteban A, Bourgarel M, Roger F, Leendertz F, Diallo R, Ndimbo-Kumugo SP, Nsio-Mbeta J, Tagg N, Koivogui L, Toure A, Delaporte E, Ahuka-Mundeke S, Tamfum JM, Mpoudi-Ngole E, Ayouba A, Peeters M. Survey of Ebola viruses in Frugivorous and insectivorous bats in guinea, Cameroon, and the Democratic Republic of the Congo, 2015-2017. Emerg Infect Dis. 2018;24(12):2228–40. https://doi.org/10.3201/eid2412.180740.. Epub 2018 Dec 17

Dowell SF, Mukunu R, Ksiazek TG, Khan AS, Rollin PE, Peters CJ. Transmission of Ebola hemorrhagic fever: a study of risk factors in family members, Kikwit, Democratic Republic of the Congo. J Infect Dis. 1995;179:S87–91.

Dudas G, Carvalho LM, Bedford T, Tatem AJ, Baele G, Faria NR, Park DJ, Ladner JT, Arias A, Asogun D, Bielejec F, Caddy SL, Cotten M, D'Ambrozio J, Dellicour S, Di Caro A, Diclaro JW, Duraffour S, Elmore MJ, Fakoli LS, Faye O, Gilbert ML, Gevao SM, Gire S, Gladden-Young A, Gnirke A, Goba A, Grant DS, Haagmans BL, Hiscox JA, Jah U, Kugelman JR, Liu D, Lu J, Malboeuf CM, Mate S, Matthews DA, Matranga CB, Meredith LW, Qu J, Quick J, Pas SD, Phan MVT, Pollakis G, Reusken CB, Sanchez-Lockhart M, Schaffner SF, Schieffelin JS, Sealfon RS, Simon-Loriere E, Smits SL, Stoecker K, Thorne L, Tobin EA, Vandi MA, Watson SJ, West K, Whitmer S, Wiley MR, Winnicki SM, Wohl S, Wölfel R, Yozwiak NL, Andersen KG, Blyden SO, Bolay F, Carroll MW, Dahn B, Diallo B, Formenty P, Fraser C, Gao GF, Garry RF, Goodfellow I, Günther S, Happi CT, Holmes EC, Kargbo B, Keïta S, Kellam P, Koopmans MPG, Kuhn JH, Loman NJ, Magassouba N, Naidoo D, Nichol ST, Nyenswah T, Palacios G, Pybus OG, Sabeti PC, Sall A, Ströher U, Wurie I, Suchard MA, Lemey P, Rambaut A. Virus genomes reveal factors that spread and sustained the Ebola epidemic. Nature. 2017;544:309–15. https://doi.org/10.1038/nature22040.

Easton A, Pringle C. Mononegavirales. In: AMQ K, Adams MJ, Carstens EB, Lefkowitz EJ, editors. Virus taxonomy: ninth report of the International Committee on Taxonomy of Viruses. London: Academic; 2012. p. 653–7.

Emond RT, Evans B, Bowen ET, Lloyd G. A case of Ebola virus infection. Br Med J. 1977;2:541–4. https://doi.org/10.1136/bmj.2.6086.541.

Evans DK, Goldstein M, Popova A. Health-care worker mortality and the legacy of the Ebola epidemic. Lancet Glob Health. 2015;3:e439–40.

FAO. FAO/WFP crop and food security assessment—Liberia; 2014.

Forbes KM, Webala PW, Jääskeläinen AJ, Abdurahman S, Ogola J, Masika MM, Kivistö I, Alburkat H, Plyusnin I, Levanov L, Korhonen EM, Huhtamo E, Mwaengo D, Smura T, Mirazimi A, Anzala O, Vapalahti O, Sironen T. Bombali virus in *Mops condylurus* bat, Kenya. Emerg Infect Dis. 2019;25(5) https://doi.org/10.3201/eid2505.181666.

Gear JS, Cassel GA, Gear AJ, Trappler B, Clausen L, Meyers a M, Kew MC, Bothwell TH, Sher R, Miller GB, Schneider J, Koornhof HJ, Gomperts ED, Isaäcson M, Gear JH. Outbreak of Marburg virus disease in Johannesburg. Br Med J. 1975;4:489–93. https://doi.org/10.1136/bmj.4.5995.489.

Geisbert TW. First Ebola virus vaccine to protect human beings? Lancet. 2017;389:479–80. https://doi.org/10.1016/S0140-6736(16)32618-6.

Gire SK, Goba A, Andersen KG, Sealfon RSG, Park DJ, Kanneh L, Jalloh S, Momoh M, Fullah M, Dudas G, Wohl S, Moses LM, Yozwiak NL, Winnicki S, Matranga CB, Malboeuf CM, Qu J, Gladden AD, Schaffner SF, Yang X, Jiang P-P, Nekoui M, Colubri A, Coomber MR, Fonnie M, Moigboi A, Gbakie M, Kamara FK, Tucker V, Konuwa E, Saffa S, Sellu J, Jalloh AA, Kovoma A, Koninga J, Mustapha I, Kargbo K, Foday M, Yillah M, Kanneh F, Robert W, Massally JLB, Chapman SB, Bochicchio J, Murphy C, Nusbaum C, Young S, Birren BW, Grant DS, Scheiffelin JS, Lander ES. Genomic surveillance elucidates Ebola virus origin and transmission during the 2014 outbreak. Science. 2014;345:1369–72. https://doi.org/10.1126/science.1259657.

Goldstein T, Anthony SJ, Gbakima A, Bird BH, Bangura J, Tremeau-Bravard A, Belaganahalli MN, Wells HL, Dhanota JK, Liang E, Grodus M, Jangra RK, DeJesus VA, Lasso G, Smith BR, Jambai A, Kamara BO, Kamara S, Bangura W, Monagin C, Shapira S, Johnson CK, Saylors K, Rubin EM, Chandran K, JAK LWIM. The discovery of Bombali virus adds further support for bats as hosts of Ebola viruses. Nat Microbiol. 2018;3(10):1084–9. https://doi.org/10.1038/s41564-018-0227-2.

Griffiths A, Hayhurst A, Davey R, Shtanko O, Carrion RJ, Patterson JL. Ebola virus infection. In: Singh SK, Ruzek D, editors. Viral hemorrhagic fevers. Boca Raton, FL: CRC; 2014. p. 435–56.

Hayman DT, Emmerich P, Yu M, Wang LF, Suu-Ire R, Fooks AR, Cunningham AA, Wood JL. Long-term survival of an urban fruit bat seropositive for Ebola and Lagos bat viruses. PLoS One. 2010;5:e11978. https://doi.org/10.1371/journal.pone.0011978.

Hayman DTS, Yu M, Crameri G, Wang L-F, Suu-Ire R, Wood JLN, Cunningham AA. Ebola virus antibodies in fruit bats, Ghana, West Africa. Emerg Infect Dis. 2012;18:6–8.

He B, Feng Y, Zhang H, Xu L, Yang W, Zhang Y, Li X, Tu C. Filovirus RNA in fruit bats, China. Emerg Infect Dis. 2015;21:1675–7.

Howard C. Viral haemorrhagic fevers. In: Zuckerman AJ, Mushahwar IK, editors. Perspectives in medical virology. Amsterdam: Elsevier; 2005.

Jahrling PB, Geisbert TW, Johnson ED, Peters CJ, Dalgard DW, Hall WC. Preliminary report: isolation of Ebola virus from monkeys imported to USA. Lancet. 1990;335:502–5. https://doi.org/10.1016/0140-6736(90)90737-P.

Jahrling PB, Geisbert TW, Jaax NK, Hanes MA, Ksiazek TG, Peters CJ. Experimental infection of cynomolgus macaques with Ebola-Reston filoviruses from the 1989-1990 U.S. epizootic. Arch Virol Suppl. 1996;11:115–34. https://doi.org/10.1007/978-3-7091-7482-1_11.

January O, Health P. Imported case of Marburg hemorrhagic fever—Colorado, 2008. MMWR Morb Mortal Wkly Rep. 2009;58:1377–81. https://doi.org/10.1097/INF.0b013e3181d467bc.

Jayme SI, Field HE, De Jong C, Olival KJ, Marsh G, Tagtag AM, Hughes T, Bucad AC, Barr J, Azul RR, Retes LM, Foord A, Yu M, Cruz MS, Santos IJ, Lim TMS, Benigno CC, Epstein JH, Wang LF, Daszak P, Newman SH. Molecular evidence of Ebola Reston virus infection in Philippine bats. Virol J. 2015;12:1–8. https://doi.org/10.1186/s12985-015-0331-3.

Johnson BK, Gitau LG, Gichogo A, Tukei PM, Else JG, Suleman MA, Kimani R. Marburg and Ebola virus antibodies in Kenyan primates. Lancet. 1981;317:1420–1.

Johnson BK, Gitau LG, Gichogo A, Tukei PM, Else JG, Suleman MA, Kimani R, Sayer PD. Marburg, Ebola and Rift Valley fever virus antibodies in East African primates. Trans R Soc Trop Med Hyg. 1982;76:307–10.

Johnson E, Johnson B, Silverstein D. Characterization of a new Marburg virus isolated from a 1987 fatal case in Kenya. Arch Virol Suppl. 1996;11:101–14. in Imported Virus Infections edited by Tino F. Sehwarz and Günter Siegl.

Kissling RE, Murphy FA, Henderson BE. Marburg virus. Ann N Y Acad Sci. 1970;174:932–45. https://doi.org/10.1111/j.1749-6632.1970.tb45614.x.

Kuhn JH, Becker S, Ebihara H, Geisbert TW, Johnson KM, Kawaoka Y, Lipkin WI, Negredo AI, Netesov SV, Nichol ST, Palacios G, Peters CJ, Tenorio A, Volchkov VE, Jahrling PB. Proposal for a revised taxonomy of the family Filoviridae: classification, names of taxa and viruses, and virus abbreviations. Arch Virol. 2010;155:2083–103. https://doi.org/10.1007/s00705-010-0814-x.

Kuhn JH, Bao Y, Bavari S, Becker S, Bradfute S, Brister JR, Bukreyev AA, Caì Y, Chandran K, Davey RA, Dolnik O, Dye JM, Enterlein S, Gonzalez JP, Formenty P, Freiberg AN, Hensley LE, Honko AN, Ignatyev GM, Jahrling PB, Johnson KM, Klenk HD, Kobinger G, Lackemeyer MG, Leroy EM, Lever MS, Lofts LL, Mühlberger E, Netesov SV, Olinger GG, Palacios G, Patterson JL, Paweska JT, Pitt L, Radoshitzky SR, Ryabchikova EI, Saphire EO, Shestopalov AM, Smither SJ, Sullivan NJ, Swanepoel R, Takada A, Towner JS, van der Groen G, Volchkov VE, Wahl-Jensen V, Warren TK, Warfield KL, Weidmann M, Nichol ST. Virus nomenclature below the species level: a standardized nomenclature for laboratory animal-adapted strains and variants of viruses assigned to the family Filoviridae. Arch Virol. 2013;158:1425–32. https://doi.org/10.1007/s00705-012-1594-2.

Kuhn JH, Clawson AN, Radoshitzky SR, Wahl-Jensen V, Bavari S, Jahrling PB. Viral hemorrhagic fevers. In: Singh SK, Ruzek D, editors. Viral hemorrhagic fevers. Boca Raton, FL: CRC; 2014. p. 596.

Le Guenno B, Formenty P, Wyers M, Gounon P, Walker F, Boesch C. Isolation and partial characterisation of a new strain of Ebola virus. Lancet. 1995;345:1271–4. https://doi.org/10.1016/S0140-6736(95)90925-7.

Leendertz SAJ. Testing new hypotheses regarding Ebola virus reservoirs. Viruses. 2016;8(2):30. https://doi.org/10.3390/v8020030.

Leendertz SAJ, Gogarten JF, Düx A, Calvignac-Spencer S, Leendertz FH. Assessing the evidence supporting fruit bats as the primary reservoirs for Ebola viruses. EcoHealth. 2015;13(1):18–25. https://doi.org/10.1007/s10393-015-1053-0.

Leroy EM, Baize S, Volchkov VE, Fisher-Hoch SP, Georges-Courbot MC, Lansoud-Soukate J, Capron M, Debré P, McCormick JB, Georges AJ. Human asymptomatic Ebola infection and strong inflammatory response. Lancet. 2000;355:2210–5. https://doi.org/10.1016/S0140-6736(00)02405-3.

Leroy EM, Rouquet P, Formenty P, Souquière S, Kilbourne A, Froment J-M5, Bermejo M, Smit S, Karesh W, Swanepoel R, Zaki SR, Rollin PE. Multiple Ebola virus transmission events and rapid decline of Central African wildlife. Science. 2004;303:387–90. https://doi.org/10.1126/science.1092528.

Leroy EM, Kumulungui B, Pourrut X, Rouquet P, Hassanin A, Yaba P, Délicat A, Paweska JT, Gonzalez J-P, Swanepoel R. Fruit bats as reservoirs of Ebola virus. Nature. 2005;438:575–6. https://doi.org/10.1038/438575a.

Leroy EM, Epelboin A, Mondonge V, Pourrut X, Gonzalez J-P, Muyembe-Tamfum J-J, Formenty P. Human Ebola outbreak resulting from direct exposure to fruit bats in Luebo, Democratic Republic of Congo, 2007. Vector Borne Zoonotic Dis. 2009;9:723–8. https://doi.org/10.1089/vbz.2008.0167.

Martini GA. Marburg virus disease. Postgrad Med J. 1973;49:542–6. https://doi.org/10.1136/pgmj.49.574.542.

Maruyama J, Miyamoto H, Kajihara M, Ogawa H, Maeda K, Sakoda Y, Yoshida R, Takada A. Characterization of the envelope glycoprotein of a novel Filovirus, Lloviu Virus. J Virol. 2014;88:99–109. https://doi.org/10.1128/JVI.02265-13.

Miranda ME, Ksiazek TG, Retuya TJ, Khan AS, Sanchez A, Fulhorst CF, Rollin PE, Calaor AB, Manalo DL, Roces MC, Dayrit MM, Peters CJ. Epidemiology of Ebola (Subtype Reston) virus in the Philippines, 1996. J Infect Dis. 1999;179:S115–9. https://doi.org/10.1086/514314.

Morvan JM, Deubel V, Gounon P, Nakouné E, Barrière P, Murri S, Perpète O, Selekon B, Coudrier D, Gautier-Hion A, Colyn M, Volehkov V. Identification of Ebola virus sequences present as RNA or DNA in organs of terrestrial small mammals of the Central African Republic Ebola virus/RT-PCR/terrestrial small mammals/electron microscopy/DNA form. Microbes Infect. 1999;1:1193–201.

Murray MJ. Ebola virus disease: a review of its past and present. Anesth Analg. 2015;121:798–809. https://doi.org/10.1213/ANE.0000000000000866.

Negredo A, Palacios G, Vázquez-Morón S, González F, Dopazo H, Molero F, Juste J, Quetglas J, Savji N, de la Cruz Martínez M, Herrera JE, Pizarro M, Hutchison SK, Echevarría JE, Lipkin WI, Tenorio A. Discovery of an Ebola virus-like filovirus in Europe. PLoS Pathog. 2011;7:e1002304. https://doi.org/10.1371/journal.ppat.1002304.

Nikiforov VV, IuI T, Kalinin P, Akinfeeva L, Katkova L, Barmin VS, Riabchikova EI, Popkova NI, Shestopalov AM, Nazarov V. Case of Marburg infection in a Russian Laboratory. Z Mikcobiol Epidemiol Immunobiol. 1994;3:104–6.

Olival KJ, Hayman DTS. Filoviruses in bats: current knowledge and future directions. Viruses. 2014;6:1759–88. https://doi.org/10.3390/v6041759.

Olival KJ, Islam A, Yu M, Anthony SJ, Epstein JH, Khan SA, Khan SU, Crameri G, Wang LF, Lipkin WI, Luby SP, Daszak P. Ebola virus antibodies in fruit bats, Bangladesh. Emerg Infect Dis. 2013;19:270–3. https://doi.org/10.3201/eid1902.120524.

Pan Y, Zhang W, Cui L, Hua X, Wang M, Zeng Q. Reston virus in domestic pigs in China. Arch Virol. 2014;159:1129–32. https://doi.org/10.1007/s00705-012-1477-6.

Parpia AS, Ndeffo-Mbah ML, Wenzel NS, Galvani AP. Effects of response to 2014-2015 Ebola outbreak on deaths from malaria, HIV/AIDS, and tuberculosis, West Africa. Emerg Infect Dis. 2016;22:433–41. https://doi.org/10.3201/eid2203.150977.

Paweska JT, Jansen van Vuren P, Masumu J, Leman PA, Grobbelaar AA, Birkhead M, Clift S, Swanepoel R, Kemp A. Virological and serological findings in *Rousettus aegyptiacus* experimentally inoculated with Vero cells-adapted Hogan strain of Marburg virus. PLoS One. 2012;7: e45479. https://doi.org/10.1371/journal.pone.0045479.

Pigott DM, Golding N, Mylne A, Huang Z, Henry AJ, Weiss DJ, Brady OJ, Kraemer MUG, Smith DL, Moyes CL, Bhatt S, Gething PW, Horby PW, Bogoch II, Brownstein JS, Mekaru SR, Tatem AJ, Khan K, Hay SI, Oliver J, Kraemer MUG, Smith DL, Moyes CL, Bhatt S, Gething PW, Horby PW, Bogoch II, Brownstein JS, Mekaru SR, Tatem AJ, Hay SI. Mapping the zoonotic niche of Ebola virus disease in Africa. elife. 2014;3:1–29. https://doi.org/10.7554/eLife.04395.

Pigott DM, Millear AI, Earl L, Morozoff C, Han BA, Shearer FM, Weiss DJ, Brady OJ, Kraemer MUG, Moyes CL, Bhatt S, Gething PW, Golding N, Hay SI. Updates to the zoonotic niche map of Ebola virus disease in Africa. elife. 2016;5:1–13. https://doi.org/10.7554/eLife.16412.

Pourrut X, Kumulungui B, Wittmann T, Moussavou G, Délicat A, Yaba P, Nkoghe D, Gonzalez JP, Leroy EM. The natural history of Ebola virus in Africa. Microbes Infect. 2005;7:1005–14.

Pourrut X, Souris M, Towner JS, Rollin PE, Nichol ST, Gonzalez J-P, Leroy EM. Large serological survey showing cocirculation of Ebola and Marburg viruses in Gabonese bat populations, and a high seroprevalence of both viruses in *Rousettus aegyptiacus*. BMC Infect Dis. 2009;9:159. https://doi.org/10.1186/1471-2334-9-159.

Shi M, Lin XD, Chen X, Tian JH, Chen LJ, Li K, Wang W, Eden JS, Shen JJ, Liu L, Holmes EC, Zhang YZ. The evolutionary history of vertebrate RNA viruses. Nature. 2018;561(7722):E6. https://doi.org/10.1038/s41586-018-0310-0.

Shuaib F, Gunnala R, Musa EO, Mahoney FJ, Oguntimehin O, Nguku PM, Nyanti SB, Knight N, Gwarzo NS, Idigbe O, Nasidi A, Vertefeuille JF. Ebola virus disease outbreak–Nigeria, July-September 2014. MMWR Morb Mortal Wkly Rep. 2014;63:867–72.. mm6339a5 [pii]

Siembieda JL, Kock RA, McCracken TA, Newman SH. The role of wildlife in transboundary animal diseases. Anim Health Res Rev. 2011;12:95–111. https://doi.org/10.1017/S1466252311000041.

Smith DH, Isaacson M, Johnson KM, Bagshawe A, Johnson BK, Swanapoel R, Killey M, Siongok T, Koinange Keruga W. Marburg-virus disease in Kenya. Lancet. 1982;319:816–20. https://doi.org/10.1016/S0140-6736(82)91871-2.

Swanepoel R, Smit SB, Rollin PE, Formenty P, PA L, Kemp A, Burt FJ, AA G, Croft J, Bausch DG, Zeller H, Leirs H, Braack LEO, Libande ML, Zaki S, Nichol ST, Ksiazek TG, Paweska JT. Studies of reservoir hosts for Marburg virus. Emerg Infect Dis. 2007;13:1847–51. https://doi.org/10.3201/eid1312.071115.

Taniguchi S, Watanabe S, Masangkay JS, Omatsu T, Ikegami T, Alviola P, Ueda N, Iha K, Fujii H, Ishii Y, Mizutani T, Fukushi S, Saijo M, Kurane I, Kyuwa S, Akashi H, Yoshikawa Y, Morikawa S. Reston Ebola virus antibodies in bats, the Philippines. Emerg Infect Dis. 2011;17:1559–60.

Team S, Branch SP, Division V, Control D, Eradication S (1978) Ebola haemorrhagic fever in Sudan, 1976. Bull World Health Organ 56:247–270 (WHO 1978a).

Timen A, Koopmans MPG, Vossen ACTM, Van Doornum GJJ, Günther S, Van Den Berkmortel F, Verduin KM, Dittrich S, Emmerich P, Osterhaus ADME, Van Dissel JT, Coutinho RA. Response to imported case of Marburg hemorrhagic fever, The Netherlands. Emerg Infect Dis. 2009;15:1171–5.

Towner JS, Khristova ML, Sealy TK, Vincent MJ, Erickson BR, DA B, Hartman AL, JA C, Zaki SR, Ströher U, Gomes da Silva F, del Castillo F, Rollin PE, Ksiazek TG, Nichol ST. Marburg virus genomics and association with a large hemorrhagic fever outbreak in Angola. J Virol. 2006;80:6497–516. https://doi.org/10.1128/JVI.00069-06.

Towner JS, Pourrut X, Albariño CG, Nkogue CN, Bird BH, Grard G, Ksiazek TG, Gonzalez J-P, Nichol ST, Leroy EM. Marburg virus infection detected in a common African bat. PLoS One. 2007;2:e764. https://doi.org/10.1371/journal.pone.0000764.

Towner JS, Amman BR, Sealy TK, Carroll SAR, JA C, Kemp A, Swanepoel R, Paddock CD, Balinandi S, Khristova ML, Formenty PBH, Albarino CG, Miller DM, Reed ZD, Kayiwa JT, Mills JN, Cannon DL, Greer PW, Byaruhanga E, Farnon EC, Atimnedi P, Okware S, Katongole-Mbidde E, Downing R, Tappero JW, Zaki SR, Ksiazek TG, Nichol ST, Rollin PE. Isolation of genetically diverse Marburg viruses from Egyptian fruit bats. PLoS Pathog. 2009;5:e1000536. https://doi.org/10.1371/journal.ppat.1000536.

UNDP. Assessing the socio-economic impacts of Ebola virus disease in Guinea, Liberia and Sierra Leone: the road to recovery. 2014. http://www.africa.undp.org/content/dam/rba/docs/Reports/EVD Synthesis Report 23Dec2014.pdf. Accessed 1 Jun 2018.

US CDC. Cost of the Ebola epidemic. 2016. Accessed 22 Jan CDC 2014. https://doi.org/10.1016/S2214-109X(15)00065-0.http.

Walsh PD, Abernethy KA, Bermejo M, Beyers R, De WP, Akou ME, Huijbregts B, De Wachter P, Akou ME, Huljbregts B, Mambounga DI, Toham AK, Kilbourn AM, Lahm SA, Latour S, Maisels F, Mbina C, Mihindou Y, Ndong Obiang S, Effa EN, Starkey MP, Teifer P, Thibault M, Tutin CEG, White LJT, Wilkie DS. Catastrophic ape decline in Western equatorial Africa. Nature. 2003;422:611–4. https://doi.org/10.1038/nature01566.

Were F. The dengue situation in Africa. Paediatr Int Child Health. 2012;32:18–21. https://doi.org/10.1179/2046904712Z.00000000048.

WHO. Ebola haemorrhagic fever = Fièvre hémorragique à virus Ebola. Wkly Epidemiol Rec = Relev épidémiologique Hebd. 1996;71:359.

WHO. Mali: Details of the additional cases of Ebola virus disease. Ebola Situation Assessment. 2014a. http://www.who.int/mediacentre/news/ebola/20-november-2014-mali/en/. Accessed 2 May 2018.

WHO. Ebola virus disease update—Senegal. Disease Outbreak News. 2014b. http://www.who.int/csr/don/2014_08_30_ebola/en/. Accessed 10 May 2018.

WHO. Ebola surveillance in countries with no reported cases of Ebola virus disease; 2014c. p. 1–2.

WHO. Laboratory diagnosis of Ebola virus disease: interim guideline. 4; 2015a.

WHO. Selection and use of Ebola in vitro diagnostic (IVD) assays. 2015b. p. 1–9.

WHO. Ebola virus disease—26 May 2016. Situation Report. 2016a. http://apps.who.int/iris/bitstream/10665/206924/1/ebolasitrep_26May2016_eng.pdf. Accessed 31 May 2016.

WHO. Final trial results confirm Ebola vaccine provides high protection against disease. WHO Media Cent. 2016b. http://www.who.int/mediacentre/news/releases/2016/ebola-vaccine-results/en/. Accessed 17 Jun 2018.

WHO. Marburg virus disease; 2017.

WHO. Haemorrhagic fevers, Viral. 2018a. www.who.int/topics/haemorrhagic_fevers_viral/en/. Accessed 2 Jun 2018.

WHO. Ebola virus disease—Democratic Republic of Congo. Health Emergency Information and Risk Assessment. External Situation Report 5–25 May 2018; 2018b.

WHO. Ebola Virus disease. Frequently asked questions on compassionate use of investigational vaccine for the Ebola virus disease outbreak in Democratic Republic of the Congo; 2018c.

WHO. Notes for the record: consultation on monitored emergency use of unregistered and investigational interventions for Ebola Virus Disease (EVD). 2018d.

WHO Ebola Response Team. Ebola Virus Disease in West Africa—the first 9 months of the epidemic and forward projections. N Engl J Med. 2014;371:1481–95. https://doi.org/10.1056/NEJMoa1411100.

World Bank Group. 2014–2015 West Africa Ebola crisis: impact update. World Bank Fisc Rep 4. 2016.

Yang XL, Zhang YZ, Jiang RD, Guo H, Zhang W, Li B, Wang N, Wang L, Waruhiu C, Zhou JH, Li SY, Daszak P, Wang LF, Shi ZL. Genetically diverse Filoviruses in Rousettus and Eonycteris spp. Bats, China, 2009 and 2015. Emerg Infect Dis. 2017;23(3):482–6. https://doi.org/10.3201/eid2303.161119.

Yang XL, Tan CW, Anderson DE, Jiang RD, Li B, Zhang W, Zhu Y, Lim XF, Zhou P, Liu XL, Guan W, Zhang L, Li SY, Zhang YZ, Wang LF, Shi ZL. Characterization of a filovirus (Měnglà virus) from Rousettus bats in China. Nat Microbiol. 2019;4(3):390–5. https://doi.org/10.1038/s41564-018-0328-y.

Yuan J, Zhang YY, Li J, Zhang YY, Wang L-F, Shi Z. Serological evidence of Ebola virus infection in bats, China. Virol J. 2012;9:236. https://doi.org/10.1186/1743-422X-9-236.

Zapata JC, Cox D, Salvato MS. The role of platelets in the pathogenesis of viral hemorrhagic fevers. PLoS Negl Trop Dis. 2014;8:e2858. https://doi.org/10.1371/journal.pntd.0002858.

Chapter 11
Foot-and-Mouth Disease

Tesfaalem Tekleghiorghis Sebhatu

Abstract Foot-and-mouth disease (FMD) is a highly contagious transboundary animal disease, causing worldwide an enormous economic impact and remains a present and continuing severe global threat. When designing FMD control strategies in Africa, most precise information on animal husbandry, trade and wildlife is necessary, in order to coordinate a regional approach based on related ecosystem. Following this objective, FMD control in Africa required mass vaccination using suitable quality conventional vaccine containing the relevant antigens that are specific to the region and could eventually limit primary outbreaks and spread of the virus. Ultimately, monitoring the vaccination coverage in the field is an essential tool for success to control and prevent FMD.

The purpose of this chapter on FMD is to provide an update on the status of FMD virus including serotypes/topotypes distribution, role of animal husbandry system and trade in virus transmission, diagnosis and control strategy in the Sahelian Africa and connected regions (North Africa, Arabian Peninsula and sub-Sahelian Africa north of the equator) and, provide perspectives on better strategies for FMD control. Herein, we argue that the region-based vaccination and zoo-sanitary measures as a control strategy in these endemic settings of Sahelian Africa and connected regions are the most suitable approach of FMD control towards eradication.

Keywords FMD virus · Serotypes · Topotypes · Animal husbandry · Sahelian Africa · Connected regions

T. T. Sebhatu (✉)
Diagnostic Medicine/Pathobiology, College of Veterinary Medicine, International Programs, Kansas State University, Manhattan, KS, USA
e-mail: tesfaalemtsebhatu@vet.k-state.edu

M. Kardjadj et al. (eds.), *Transboundary Animal Diseases in Sahelian Africa and Connected Regions*, https://doi.org/10.1007/978-3-030-25385-1_11

History

Foot-and-mouth disease (FMD) is a highly contagious and transboundary viral disease of domestic and wild cloven-hoofed animals. The disease is endemic in Africa and Asia affecting wide host range (domestic and wildlife) having a rapid spread with a colossal economic impact of growing international concern. Although rarely fatal FMD can be an economically devastating disease in the endemic regions due to serious production loss in the livelihoods directly dependant on livestock. FMD virus host range is extremely wide being capable of infecting nearly 70 species within 20 families of mammals (Hedger 1981). FMD is probably the most important disease constraint to trade in live animals and their products (Kitching 1998; Knight-Jones and Rushton 2013).

The first written description of FMD probably occurred in 1546, when Hieronymus Fracastorius described a similar disease of cattle in Venice, Italy (Fracastorius 1546). In Germany, the first evidence of FMD was reported in 1754, then in Great Britain in 1839, in the USA in 1870 and a year later in South America (Radostits et al. 2000). FMD never occurred in New Zealand (Bachrach 1968) and Australia has been free of it since 1972. The early history of FMD in Africa and Asia is not known except for South Africa where the disease was first officially recorded in 1892 (Thomson 1994).

In 1897, Loeffler and Frosch demonstrated that a filterable agent (i.e. virus) caused FMD (Loeffler and Frosch 1897, 1898) is the first demonstration that a disease of animals was caused by a filterable agent and marked the start of the era of virology. The existence of immunologically distinct serotypes of FMD virus has been shown by Vallée and Carrée in 1922, who demonstrated that there was no cross-immunity between serotypes O and A. These serotypes were named from their areas of origin; O for the department of Oise in France and A for Allemagne (the French word for Germany) (Vallée and Carré 1922). Later serotype C was discovered by Waldmann and Trautwein in Germany in 1926. Subsequently, three new serotypes were identified in samples originating from southern Africa and they were named as Southern African Territories 1, 2 and 3 (SAT-1, SAT-2, SAT-3) (Brooksby 1958). The seventh serotype, Asia1, was initially detected in a sample collected from a water buffalo in Pakistan in 1954 (Brooksby and Rogers 1957).

Afterward it was shown that the agent, FMD virus (FMDV), is a small (27 nm in diameter), non-enveloped, icosahedral, positive-sense, single-stranded RNA virus, with an approximately 8.5 kb genome that encodes for structural proteins (SP) and non-structural proteins (NSP) (Carroll et al. 1984; Forss et al. 1984; Grubman et al. 1984). It belongs to the *Aphthovirus* genus and Picornaviridae family (van Regenmortel et al. 2000). The four main structural proteins of FMDV are VP1, VP2, VP3 and VP4. VP1–3 have surface components and VP4 is internally buried within the capsid (Acharya et al. 1989). These structural proteins are coded respectively by the 1D, 1B, 1C and 1A coding region of the genome. The FMD virus exhibits a high potential for genetic and antigenic variation (Domingo et al. 1990), as shown by the presence of seven immunologically distinct serotypes (A, O, C,

SAT-1, SAT-2, SAT-3 and Asia 1) (Pereira 1977; Domingo et al. 2003). Within these serotypes, more than 65 subtypes have been recognised (Barros et al. 2007; Haydon et al. 2001). The high degree of antigenic variation may be attributed to the high rate of mutation, genetic recombination and the quasispecies nature of the virus. This antigenic variation is the basis for maintenance of FMDV circulation resulting in severe economic loss in livestock productions (Longjam and Tayo 2011; Domingo et al. 1990). FMDV serotypes show genetically and geographically distinct evolutionary lineages (topotypes) based on nucleotide differences of up to 15% for serotype O, A, C and Asia 1. In the case of the SAT serotypes, the level for inclusion within a topotype was raised to 20% since the VP1 coding sequence of these viruses appear to be more inherently variable (Samuel and Knowles 2001).

Also, the NSPs and non-coding elements (NCEs) of FMDV play a critical role in repressing host translation machinery, blocking protein secretion and cellular proteins cleavage associated with signal transduction and the innate immune response to infection (Gao et al. 2016).

Geographic Distribution

Due to underreporting of FMD outbreaks, the current virus serotypes, topotypes and strains circulating in Sahelian Africa and the connected regions are in many cases unknown. The information on FMDV serotypes and topotypes recorded in Sahelian Africa and the connected regions obtained from the various data sources are summarised in Tables 11.1, 11.2, 11.3 and 11.4.

The Sahel Region of Africa

The Sahel belt is roughly situated between the Sahara and the coastal areas of West Africa, which mainly includes (from west to east) Senegal, Mauritania, Mali, Burkina Faso, the extreme south of Algeria, Niger, north of Nigeria, Cameroon, Central African Republic, Chad, Sudan, northern South Sudan, Eritrea and the extreme north of Ethiopia (Fig. 11.1).

Currently, four serotypes of FMD virus circulate (O, A, SAT-1 and SAT-2) in Sahelian Africa. Serotypes O, A and SAT-2 are the most prevalent and responsible for most of the outbreaks in these Sahelian countries. In the Sahelian region of Africa, FMD outbreaks in livestock have been able to spread along east to west and vice versa for many years following the unregulated livestock movement. It has been recorded that these outbreaks were caused by genetically related FMDV serotypes and/or topotypes. The Sahelian Africa region (West, Central and East Africa) is divided into two virus pools (4 and 5) (Paton et al. 2009). There is a considerable co-occurrence in FMD virus serotypes and topotypes within the two pools, for example isolates of serotype O (topotype EA-3) from Niger (2007, 2015) and

Table 11.1 FMDV serotypes and topotype distribution in Sahel region of Africa for the time period of 2003–2017

Serotype	Topotype	Genotype/ strain	Representative country/ countries	References
O	EA-3		Ethiopia (2017), Eritrea (2011), Niger (2007), Nigeria (2015), Sudan (2012),	WRLFMD (2003–2017)
	EA-4		Ethiopia (2016)	Balinda et al. (2010) and WRLFMD (2003–2017)
	WA		Burkina Faso (2002), Cameroon (2005), Mali (2007), Nigeria (2013), Niger (2005), Senegal (2006)	WRLFMD (2003–2017), Ehizibolo et al. (2017) and Gorna et al. (2014)
A	AFRICA	G-II	Ethiopia (2005)	WRLFMD (2003–2017)
	AFRICA	G-III	Ethiopia (2005), Sudan (2007), Cameroon (2005)	Sangula (2010)
	AFRICA	G-IV	Eritrea (2009), Mali (2006), Cameroon (2005), Sudan (2006), Nigeria (2013, 2015)	WRLFMD (2003–2017), Gorna et al. (2014), Ehizibolo et al. (2017) and Gorna et al. (2014)
	AFRICA	G-VI	Mali (2006), Mauritania (2006), Senegal (2006), Benin (2010)	Gorna et al. (2014) and WRLFMD (2003–2017)
	AFRICA	G-VII	Ethiopia (2009)	WRLFMD (2003–2017)
C	AFRICA (II)	Eth-71	Ethiopia (1983)	Sumption et al. (2007)
SAT-1	V		Nigeria (1976), Niger (1976)	Sangaré (2002)
	VI		Nigeria (1981), Sudan (1976)	WRLFMD (2003–2017)
	IX		Ethiopia (2007)	Ayelet et al. (2009)
	X		Nigeria (2015)	Ehizibolo et al. (2017)
SAT-2	VI		Gambia (1979), Senegal (1983)	Sangaré (2002)
	VII		Cameroon (2005), Eritrea (1998), Niger (2005), Nigeria (2008), Senegal (2009), Sudan (2007, 2010)	WRLFMD (2003–2017)
	XIII		Sudan (2008), Ethiopia (2010)	WRLFMD (2003–2017)
	XIV		Ethiopia (1991)	WRLFMD (2003–2017) and Ayelet et al. (2009)

Note: Earlier isolates were included when there is no representative of the genotype in reports after 2003

Table 11.2 FMDV serotypes and topotype distribution in the North Africa region for the time period of 2006–2017

Serotype	Topotype	Genotype/ strain	Representative country/ countries	References
O	EA-3		Libya (2012), Egypt (2012)	WRLFMD (2003–2017)
	ME-SA	Ind-2001d	Egypt (2016), Algeria (2014–15), Tunisia (2014), Morocco (2015), Libya (2013)	WRLFMD (2003–2017) and Bouguedour and Ripani (2016)
	ME-SA	PanAsia-2	Libya (2011), Egypt (2007, 2011)	WRLFMD (2003–2017)
A	AFRICA	G-III	Egypt (2006)	Sangula (2010)
	AFRICA	G-IV	Algeria (2017), Egypt (2014)	WRLFMD (2003–2017) and Gorna et al. (2014)
	AFRICA	G-VII	Egypt (2009)	WRLFMD (2003–2017)
	ASIA	Iran-05^{BAR-08}	Egypt (2010–2014), Libya (2009)	WRLFMD (2003–2017)
SAT-2	VII		Egypt (2016), Libya (2012)	WRLFMD (2003–2017)

Note: Earlier isolates were included when there is no representative of the genotype in reports after 2006

Table 11.3 FMDV serotypes and topotype distribution in the Arabian Peninsula region for the time period of 2008–2017

Serotype	Topotype	Genotype/ strain	Representative country/ countries	References
O	EA-3		Yemen (2009)	WRLFMD (2003–2017)
	ME-SA	Ind-2001d	Saudi Arabia (2016), UAE (2016), Bahrain (2015)	WRLFMD (2003–2017) and Bouguedour and Ripani (2016)
	ME-SA	PanAsia-2^{ANT-10}	Iraq (2010), Kuwait (2016), Bahrain (2014), Jordan (2017), Saudi Arabia (2012, 2016), UAE (2013), PAT (2015)	WRLFMD (2003–2017)
A	ASIA	Iran-05^{BAR-08}	Iraq (2013), Jordan (2006), Bahrain (2009), Kuwait (2009), PAT (2013)	WRLFMD (2003–2017)
	ASIA	G-VII	Saudi Arabia (2016)	WRLFMD (2003–2017)
Asia-1	ASIA	Sindhi-08	Iraq (2013), Bahrain (2011)	WRLFMD (2003–2017)
SAT-2	IV	Ken-09	Bahrain (2012)	WRLFMD (2003–2017)
	VII	Alx-12	Oman (2015)	WRLFMD (2003–2017)

Note: Earlier isolates were included when there is no representative of the genotype in reports after 2008

Table 11.4 FMDV serotypes and topotype distribution in sub-Sahelian Africa region at the North of the equator for the time period of 2003–2017

Serotype	Topotype	Genotype/ strain	Representative country/ countries	References
O	EA-1		Kenya (2009)	WRLFMD (2003–2017)
	EA-2		Kenya (2011), DRC (2011), Uganda (2007)	WRLFMD (2003–2017)
	EA-3		Ethiopia (2017), Somalia (2007), Kenya (1987)	WRLFMD (2003–2017)
	EA-4		Ethiopia (2016), Kenya (2010)	Balinda et al. (2010) and WRLFMD (2003–2017)
	WA		Ghana (2012), Togo (2005), Benin (2010)	WRLFMD (2003–2017), Ehizibolo et al. (2017) and Gorna et al. (2014)
A	AFRICA	G-I	Kenya (2009), Uganda (2002), DRC (2011)	WRLFMD (2003–2017)
	AFRICA	G-II	Ethiopia (2005)	WRLFMD (2003–2017)
	AFRICA	G-III	Kenya (2005), Ethiopia (2005), Uganda (2002)	Sangula (2010)
	AFRICA	G-IV	Togo (2005), Benin (2010)	WRLFMD (2003–2017), Gorna et al. (2014), Ehizibolo et al. (2017) and Gorna et al. (2014)
	AFRICA	G-V	Ghana (1973)	WRLFMD (2003–2017)
	AFRICA	G-VI	Benin (2010)	Gorna et al. (2014) and WRLFMD (2003–2017)
	AFRICA	G-VII	Ethiopia (2009), Kenya (2006)	WRLFMD (2003–2017)
	AFRICA	G-VIII	Kenya (1964)	WRLFMD (2003–2017) and Gorna et al. (2014)
C	AFRICA (I)	Ken-67	Kenya (2004)	Sangula (2010)
	AFRICA (II)	Eth-71	Ethiopia (1983)	Sumption et al. (2007)
SAT-1	I (NWZ)		Kenya (2011)	WRLFMD (2003–2017)
	IV (EA-1)		Uganda (2013, 2007[a])	Ayebazibwe et al. (2010) and Dhikusooka et al. (2016)
	VII (EA-2)		Uganda (1974)	WRLFMD (2003–2017)
	VIII (EA-3)		Uganda (1997[a])	WRLFMD (2003–2017) and Bastos et al. (2001)
	IX		Ethiopia (2007)	Ayelet et al. (2009)

(continued)

Table 11.4 (continued)

Serotype	Topotype	Genotype/ strain	Representative country/ countries	References
SAT-2	I		Kenya (1999)	Bastos et al. (2003a) and WRLFMD (2003–2017)
	II		Ghana (1991)	WRLFMD (2003–2017)
	IV		Kenya (2009)	WRLFMD (2003–2017)
	V		Ghana (1991)	Ayelet et al. (2009)
	IX		Kenya (1996), Uganda (1995)	Ayelet et al. (2009)
	X		Uganda (2007[a])	Ayebazibwe et al. (2010)
	XII		Uganda (1976)	Sahle et al. (2007)
	XIII		Ethiopia (2010)	WRLFMD (2003–2017)
	XIV		Ethiopia (1991)	WRLFMD (2003–2017) and Ayelet et al. (2009)
SAT-3	V		Uganda (1970[a], 1997[a], 2013)	Bastos et al. (2003b) and WRLFMD (2003–2017)

Note: Earlier isolates were included when there is no representative of the genotype in reports after 2003

[a]Foot-and-mouth disease (FMD) virus isolated from African buffalo (*Syncerus caffer caffer*)

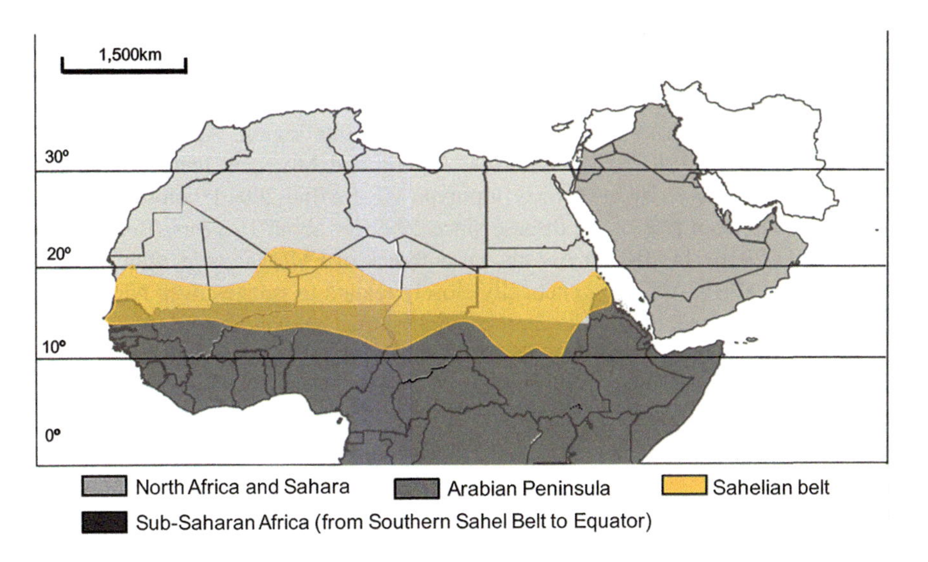

Fig. 11.1 Sahelian region and the connected regions

from Nigeria (2007, 2009 and 2016) in West Africa were genetically related to the serotype and topotype virus found in Eritrea (2004 and 2011), in Ethiopia (2005, 2006, 2008 and 2010–2012) and in Sudan (2005 and 2008–2011) in East Africa (Tekleghiorghis et al. 2014; WRLFMD 2003–2017). For other FMDV topotypes (e.g. A topotype -IV AFRICA (G genotype), and SAT-2 topotype VII, the distribution has increased, and this is most likely linked to the husbandry system and trade in the area where livestock move from west to east and vice versa as well as northward to North Africa and, to Arabian Peninsula from sub-Saharan Africa.

In the last three decades, the occurrence of serotype SAT-1 has been apparently limited to the southern and eastern Africa without any reporting in West and Central Africa (Tekleghiorghis et al. 2014; Vosloo et al. 2002). Recently in 2015 in Nigeria, FMDV serotype SAT-1 was isolated, identified and characterised from an FMD outbreak in cattle, more than 35 years after the last report of FMDV SAT-1 in West Africa in 1981 (Ehizibolo et al. 2017). The VP1 coding sequence of the Nigerian 2015 SAT-1 isolates diverges from reported SAT-1 topotypes resulting in a separate topotype X. In 2014 in Benin (West Africa), the first isolation and molecular characterisation of FMDV serotype O, topotype WA (West Africa) and A topotype G-IV were recorded (Gorna et al. 2014). In West Africa of the Sahel region, it is thought that there is a constant circulation of FMDV, however, very few studies have been published due to very patchy data available and under-reported cases of FMD.

Northern Africa

Northern African countries include Morocco, Algeria, Tunisia, Libya and Egypt (Fig. 11.1) and have been experiencing repeated incursions of FMDV serotypes from sub-Saharan Africa to the Middle East with live animal trade. FMDV serotype O, A and SAT-2 have been recorded in this region. Actually, in cattle, sheep and goats 2014–2015, in Tunisia, Algeria and Morocco, there was an epidemic of FMD caused by serotype O, topotype ME-SA/Ind-2001d; before that, these countries had been free of the disease since 1999 for about 15 years (Bouguedour and Ripani 2016). In 2017 in Algeria, an outbreak of FMD in cattle, with serotype A-topotype AFRICA/G-IV, genetically closely related to isolates from Nigeria and Cameroon was recorded. The unregulated animal movement control system, between Tunisia and Libya, was the potential cause for the emergence of FMDV to Tunisia that later spread to Algeria and Morocco. Since 2009, in Libya, FMDV serotypes A topotype ASIA/Iran-05$^{\text{Bar-08}}$, SAT-2 topotype VII and O topotype ME-SA/Ind-2001d were documented. From 2006 to 2016, Egypt has been endemic for FMDV serotype O topotype EA-3 and ME-SA/PanAsia-2, serotype A topotype AFRICA/G-IV and ASIA/Iran-05$^{\text{Bar-08}}$ and serotype SAT-2 topotype VII were among the isolates characterised at the World Reference Laboratory for FMD (WRLFMD 2003–2017). Libya and Egypt specifically import a considerable number of livestock from FMD endemic countries. These imports could support the emergence of recent outbreaks of FMD from sub-Saharan Africa and the Middle East (Tekleghiorghis et al. 2014).

Arabian Peninsula

Countries in the Arabian Peninsula include Saudi Arabia, Yemen, Oman, Qatar, Kuwait, Bahrain, Iraq, United Arab Emirates (UAE) and Jordan (Fig. 11.1). These countries have been affected by FMD outbreaks mainly introduced from imported live animals from East and North Africa, neighbouring Middle East countries and Asia.

Although export out of Africa is not a risk for FMD transmission within Africa, these exports cause considerable trade within Africa, as not all animals come from the countries that export to the Arabian Peninsula and the Gulf States (Tekleghiorghis et al. 2014). Many are brought in from neighbouring African countries, that is, assembled along the trade routes and finally trucked or trekked to the seaports in the Gulf of Aden and the Red Sea for shipment to the Arabian Peninsula and the Gulf States. These animal movements are of major importance for the dissemination of new strains of FMD (Fevre et al. 2006). For example, serotype SAT-2 topotype VII isolates recorded in Saudi Arabia (SAU/6/00) in 2000 were most closely related to published sequences of isolates from Eritrea (ERI/1/98) in 1998 (Bronsvoort et al. 2004), indicating that the northeast part of Africa was the most likely source of the virus (Bastos et al. 2003a).

FMDV serotypes O, A and Asia-1 are endemic in the Arabian Peninsula region. Serotype Asia-1 has never been reported crossing Africa for unknown reasons while the introduction of serotypes O and A is commonly recorded. Although FMDV SAT serotypes are mainly found in Africa, serotype SAT-1 and SAT-2 incursions have been recorded outside Africa, mainly in the Middle East, very likely through live animal importation from Africa. SAT-1 had spread to Bahrain, Israel, Jordan and Syria in 1962, to Iran in 1964, Turkey in 1965 and Kuwait in 1970 (Aidaros 2002). SAT-2 had been reported in Yemen in 1990 (Ahmed et al. 2012) in Kuwait and Saudi Arabia in 2000 (Knowles and Samuel 2003), in the Palestinian Autonomous Territories (PAT), Bahrain in 2012 (Ahmed et al. 2012) and Oman in 2015 (WRLFMD 2003–2017). The SAT serotypes have been introduced from Eastern and Northern Africa with live animal formal and informal trade practises. Long-distance FMDV movements within Asia and Africa (2009–2016) is common from sub-Saharan Africa (e.g. O/EA-3, SAT-2/VII) to northern Africa and Arabian Peninsula as well as from the Indian subcontinent, and Southeast/East Asia (e.g. O/ME-SA/Ind2001d, O/ME-SA/PanAsia-2) moved into a new geographical locations outside of the endemic pools where they usually circulate in northern Africa and Arabian Peninsula transmission being mainly with live animal movement for trade practises.

Sub-Sahelian Africa Region at the North of Equator

The sub-Sahelian Africa region includes countries between north of the equator and south of the Sahel belt (Fig. 11.1). This region is interconnected with the Sahel belt in the semi-nomadic and transhumance traditional extensive animal husbandry system practises. These allow for increased contact at the livestock/wildlife/human interface, thereby increasing the possibility of FMDV transmission. In countries of West, Central and East Africa, there are well-established livestock trade routes across east to west and vice versa as well as north-south and reverse. Livestock export in the Horn of Africa is important and supplies the Arabian Peninsula and the Gulf States with mostly small ruminants, cattle and the Arabian one hump camel.

In sub-Saharan Africa, two cycles of FMD that impact on livelihoods occur, one in which the virus circulates between wildlife hosts and domestic animals and another in which the virus spreads among domestic animals, without the involvement of wildlife (Vosloo and Thomson 2004). In this region, it is believed that some wild animals play a potential role in the transmission of FMD. Serotype SAT (1–3) viruses have been shown to be constantly evolving in the African buffalo (*Syncerus caffer*) and are a true maintenance host for the SAT serotypes (Vosloo and Thomson 2004; Condy et al. 1985; Bastos et al. 2001, 2003a; Dawe and Flanagan 1994; Hedger 1976; Vosloo et al. 1996; Hedger et al. 1973).

Six of the seven FMDV serotypes have been recorded (O, A, C, SAT-1, SAT-2 and SAT-3) in sub-Sahelian Africa. O, A and SAT-2 are the prevalent serotypes. Serotype C has not been recorded in Africa since 2004, the last being in Kenya (Roeder and Knowles 2008). FMDV serotype SAT-1 and SAT-2 have been isolated from African buffalo in Uganda in 2007 (Ayebazibwe et al. 2010). In this region hitherto, SAT-3 has been recorded only in Uganda in 1970 and 1997 in African buffalo, and in 2013 from cattle (WRLFMD 2003–2017; Dhikusooka et al. 2016) above the equator although it is endemic in southern Africa in both domestic and wildlife. A recent example of unrecognised circulation of SAT-1 (topotype IV) FMDV in cattle herds was found around the Queen Elizabeth National Park in Uganda housing African buffalos (Dhikusooka et al. 2016). African cape buffaloes (*Syncerus caffer*) are known to be carriers of different FMDV SAT strains (Jori et al. 2016) but, unlike in East and southern Africa, African cape buffalos are not present in Sahelian countries.

In East, Central and West Africa including the southern part of the Sahel, there are no sufficient data to substantiate the occurrence of FMDV SAT (1–3) serotypes in wildlife, although there are serological positive titers suggesting that wildlife could be important in the maintenance of FMDV. Nevertheless, there are no reports of SAT serotype virus isolation from buffalo inhabiting regions of East, Central and West Africa except in Uganda (East Africa). Other wild species can also be transiently infected by the SAT serotypes (Thomson et al. 2003) and transmit the disease to susceptible livestock. There are four subspecies of the African buffalo (southern savannah buffalo (*Syncerus caffer caffer*), forest buffalo (*S. c. nanus*), West African savannah buffalo (*S. c. brachyceros*), the Central African savannah buffalo (*S. c.*

aequinoctialis) all found only south of latitude 10° N, where it appears in a variety of habitats, including open savannahs, woodlands, swamps and rainforest. Except for the African southern savannah buffalo (*S. c. caffer*), it is unknown if the other three subspecies of African buffaloes can be persistently infected with FMDV SAT serotypes. The role that subspecies other than the southern savannah buffalo play in the maintenance and transmission of FMDV in Africa is not clear (Tekleghiorghis et al. 2014).

Epidemiology of FMD

The occurrence and distribution of FMD in the Sahelian region and the connected regions have been detailed in the geographic distribution on a regional basis. Transmission of FMD virus is generally effected by direct contact between acutely infected and susceptible animals (Kitching et al. 2005) or, more rarely, indirect exposure of susceptible animals to the excretions and secretions including expired air, saliva, nasal secretions, lachrymal fluid, milk, urine, faeces and semen (Alexandersen et al. 2002; World Organisation for Animal Health (OIE) 2012), products from infected animals (such as meat, milk, or uncooked meat) or inanimate objects contaminated with FMD virus.

The two main ways of FMD transmission in the Sahelian and the connected regions are firstly during the unregulated livestock movement, domestic animals and wildlife mix together on route and congregate for grazing and at watering points with a high potential risk of FMD virus transmission. In this type of traditional extensive animal husbandry system, transmission generally occurred by direct contact between infected and susceptible animals or through virus-contaminated water and grazing pasture with infectious excretions and/or secretions of acutely infected animals. Secondly, live animal and animal products (e.g. meat, meat products, milk and milk products) trade is another cause for the introduction of new FMD virus serotypes/strains from endemic countries through formal and informal livestock trade importation.

Clinical Signs and Lesions

Hosts/Species Affected

FMD affects domestic cloven-hoofed animals, including cattle, pigs, sheep, goats and water buffalo as well as more than 70 species of wild animals, including African buffalo (Fenner et al. 1993), and is characterised by fever, lameness and vesicular lesions on the tongue, feet, snout and teats of lactating animals. Animals affected with FMD show a variety of clinical signs with 2–14 days of the incubation period.

Clinical signs are usually more prominent in cattle and pigs than in sheep and goats. The clinical signs of FMD are indistinguishable from other vesicular diseases.

The Arabian camel (*Camelus dromedarius*) is probably resistant to all strains of FMD virus, while the Central Asian camel or Bactrian camel (*Camelus bactrianus*) is susceptible (Wernery and Kaaden 2004). Other camelid species can be infected under experimental conditions, but usually resist infection in the field. No clinical disease in African buffalo observed but probang sampling has yielded serotype SAT-1, SAT-2 and SAT-3 viruses. However, clinical signs and lesions were observed in impala (*Aepyceros melampus*) when infected with FMDV.

Cattle

Common clinical signs in cattle include pyrexia (fever 40–41 °C), anorexia and reduced milk production, followed by smacking of the lips, grinding of the teeth, drooling of frothy saliva and excess nasal mucous secretions (Fig. 11.2). Lesions from ruptured vesicles on the tongue, gums, buccal, muzzle and nasal mucous membranes and on the mammary glands and teats are very commonly observed

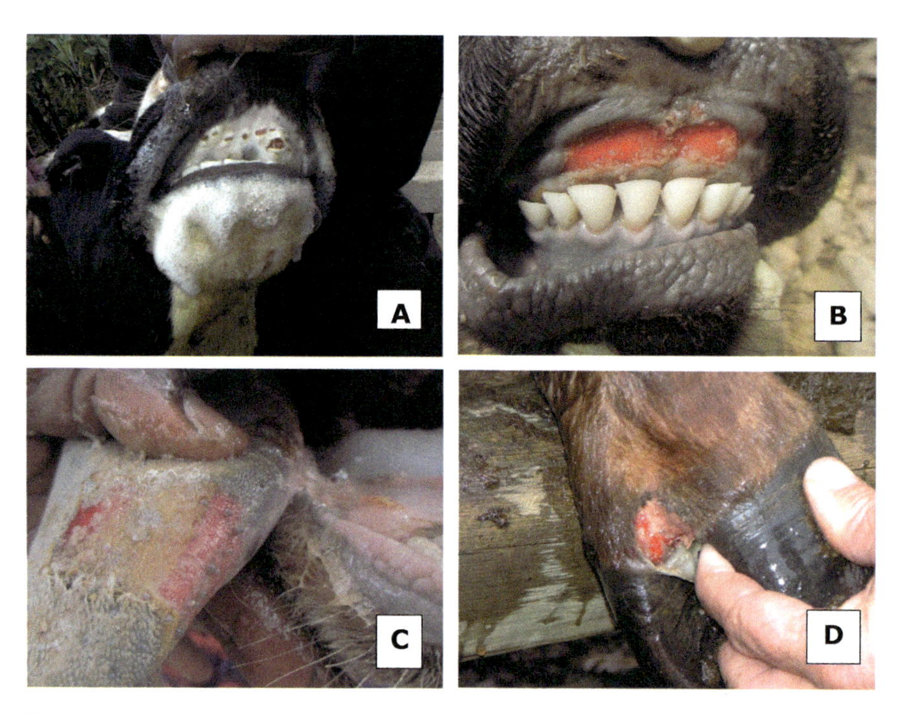

Fig. 11.2 FMD lesions in cattle (from left to right, top to bottom): Drooling of frothy saliva and ruptured dental pad vesicles in cattle (**a**); Necrosis and erosions of the dental pad in cattle (**b**); Necrotic tongue lesions from ruptured vesicle (**c**) and Interdigital lesions (**d**)

when new outbreaks occur. Lameness caused by the ruptured vesicles in the interdigital space, around the claws and coronary band is not uncommon. Abortion in adults and sudden death in young animals due to myocarditis (tiger heart) in FMD-infected animals. Postmortem lesions can be seen in the rumen pillars. The infection usually resolves in 8–15 days unless there is a concurrent secondary bacterial infection.

Although FMD does not result in high mortality in adult animals, the disease has debilitating effect, including weight loss, decrease in milk production and loss of draught power, resulting in a loss in productivity for a considerable time (Rushton and Knight-Jones 2012; Young et al. 2013; Barasa et al. 2008). Mortality, however, can be high in young animals due to myocarditis. Infected zebu cattle in Africa rarely show many disease signs. High-yielding dairy cattle appear very susceptible to severe clinical FMD, frequently leading to secondary complications such as mastitis and chronic lameness.

Pigs

Typical clinical signs of FMD in pigs include pyrexia and blanching of the coronary bands, followed by severe foot lesions sometimes with sloughing of the hoof, severe lameness and pigs do not show drooling of saliva. Lesions in pigs are observed on the snout, muzzle, gums and interdigital spaces. High mortality (up to 45%) in piglets and abortions have been reported in FMD outbreaks with new serotype or strains introduced for the first time.

Sheep and Goats

Clinical signs of FMD in sheep and goats are typically less pronounced and less frequent than in pigs and cattle and may go unrecognised. In sheep and goats, the disease is generally mild and can be difficult to distinguish from other common conditions (Donaldson and Sellers 2000; Geering 1967; Thomson 1994). Mild lameness where there is reddening along the coronary bands is observed, lesions in dental pad, lips, nostrils and teats of sheep and deaths in young animals is not uncommon.

Persistently Infected Animals (Carriers)

Following recovery from the acute stage of infection, the infectious virus disappears with the exception of low levels that may persist in the oropharynx of some ruminants. Animals in which the virus persists in the oropharynx for more than

28 days following infection (Salt 1993) are referred to as carriers. Over 50% of cattle exposed to live virus become carriers. Live virus or viral RNA may continue to be recovered from oropharyngeal fluids and cells collected with a probang cup. In cattle, the virus persists in the basal layer cells of the pharyngeal epithelium, particularly of the dorsal soft palate (Pacheco et al. 2015) and FMD virus has also been shown to persist in a non-replicative form in lymph nodes (Juleff et al. 2008). Van Bekkum et al. (1959) were first to document the presence of infectious FMDV in oropharyngeal fluid (OPF) of asymptomatic cattle several weeks after infection. This was later confirmed by Sutmoller and Gaggero (1965).

Cattle, sheep and goats can become carriers, and cattle can harbour virus for up to 2–3 years (Brooksby 1982). Sheep and goats do not usually carry FMD viruses for more than a few months, whilst pigs are believed to be incapable of establishing persistent infection.

Although a large number of FMD susceptible wildlife species is present in Africa, only the African buffalo is a true persistent carrier for the SAT (1–3) FMDV (Vosloo et al. 1996; Dawe and Flanagan 1994; Hedger 1976; Condy et al. 1985; Bastos et al. 2001). Most healthy buffalo populations maintain SAT viruses and usually become sub-clinically infected (Vosloo et al. 2007). Furthermore, virus has been shown to persist in an individual buffalo for at least 5 years and, in small free-living populations has been maintained for at least 24 years (Condy et al. 1985). Circumstantial evidence indicates, particularly in the African buffalo, that carriers are able, on rare occasions, to transmit the infection to susceptible domestic animals with which they come in close contact, the mechanism involved is unknown.

Differential Diagnoses

Vesicular stomatitis, swine vesicular disease, vesicular exanthema of swine and Seneca virus A (Seneca valley virus) are all clinically indistinguishable from FMD. FMD also has common features with rinderpest, bovine viral diarrhoea, mucosal disease, infectious bovine rhinotracheitis and bluetongue.

Diagnosis

Effective control of FMD needs sensitive, specific and rapid diagnostic tools for each stage of the control strategy. A FMD case is suspected usually after observing the appearance of commonly seen clinical signs of FMD in susceptible species. The clinical signs are usually pronounced when the disease is introduced to naïve cattle or pigs. The highly contagious nature of FMD and high mortality in susceptible animals with the appearance of typical clinical signs and lesions in the oral and nasal mucosa including the interdigital space are indicative to diagnose FMD in the field. Vesicular diseases manifesting similar clinical signs and lesions should not be overlooked, thus, laboratory diagnosis becomes compulsory for confirmation of FMD.

Laboratory diagnosis of FMD is made by virus isolation or by the demonstration of FMD viral antigen or nucleic acid in samples of tissue or fluid. Detection of virus-specific antibody can also be used for diagnosis, and, antibodies to viral non-structural proteins (NSPs) can be used also as indicators of infection, irrespective of vaccination status (targeting structural proteins). The success of the laboratory confirmation depends on the submission of adequate and suitable material. In animals with a history of vesicular disease, the detection of FMDV in samples of vesicular fluid, epithelial tissue, oesophageal–pharyngeal (OP) sample, milk or blood is sufficient to establish a diagnosis.

Epithelial tissue samples should be placed in a transport medium composed of equal amounts of glycerol and 0.04 M phosphate buffer, pH 7.2–7.6, preferably with antibiotics (Penicillin, 1000 International Units [IU], Neomycin sulphate [100 IU], Polymyxin B sulphate [50 IU] and Mycostatin [100 IU]). If 0.04 M phosphate buffer is not available, tissue culture medium or phosphate buffered saline (PBS) can be used instead, but it is important that the final pH of the glycerol/buffer mixture be in the range 7.2–7.6. FMDV is extremely labile in low pH and, buffering of the transport media is critical for successful sample collection. Samples should be kept refrigerated (4 °C) or on wet ice (0 °C) until received by the laboratory.

Virus Isolation and Identification

Virus isolation (VI), antigen enzyme-linked immunosorbent assays (ELISAs), and real-time reverse transcriptase polymerase chain reaction (rRT-PCR) assays are used to detect FMDV-infected animals. Samples of choice for testing include vesicular epithelium, vesicular fluid, epithelial tissues, oesophageal-pharyngeal fluid (OPF), blood, serum and oral and nasal swabs.

(a) Virus Isolation in Cell Cultures Primary cell culture of bovine (Sellers 1955; Snowdon 1966; House and House 1989), ovine and porcine (Bachrach et al. 1955) origin has exhibited susceptibility to FMDV from infected tissues. However, the most sensitive culture system for virus isolation is primary bovine thyroid cells (Snowdon 1966) but cryopreservation of bovine thyroid cells directly after trypsinisation results in the loss of susceptibility to FMDV (House and House 1989).

A suspension of tissue samples (10%) prepared and clarified by centrifuging at 2000 $\times g$ for 10 min. The clarified supernatant is then inoculated into a sensitive cell culture systems, which include primary bovine (calf) thyroid cells and primary pig, calf or lamb kidney cells. Established cell lines, such as BHK-21 (baby hamster kidney) and IB-RS-2 cells (porcine kidney cell line), may also be used but are generally less sensitive than primary cells for detecting low amounts of viable virus infectivity. The sensitivity of any cells used should be tested with standard preparations of FMDV. The use of IB-RS-2 cells aids the differentiation of swine vesicular disease virus (SVDV) from FMDV (as SVDV will only grow in cells of porcine origin) and is often essential for the isolation of porcinophilic strains, such as O Cathay. The cell cultures should be examined for cytopathic effect (CPE) daily for 48 h. If no CPE is detected, the cells should be frozen and thawed, used to inoculate fresh cultures and examined for CPE for another 48 h. However, the cell culture system is laborious, time consuming requiring 4–6 days for test completion, and relatively low sensitivity. It also requires careful handling of specimens and a biosafety laboratory. In the case of OP fluids, pretreatment with an equal volume of chlorofluorocarbons may improve the rate of virus detection by releasing the virus from immune complexes. VI is highly sensitive and specific when used with antigen ELISAs to confirm the presence of FMDV after cytopathic effect is observed.

(b) Antigen ELISA (AgELISA) Detects viral proteins for serotyping (using polyclonal or monoclonal antibodies to FMDV) and is useful for FMD diagnosis in suspect cases. It is also efficient in detecting South African Territories (SATs) serotypes. CPE positive cultures are tested for serotype-specific antigen ELISA following a protocol based on the description of the OIE Manual of Diagnostic Tests and Vaccines for Terrestrial Animals.

Serological Tests

The following serological assays detect FMDV-exposed animals and some help to differentiate vaccinated from infected animals (DIVA).

Structural Protein-Based Assays Virus neutralisation test (VNT), solid-phase competitive ELISA (SPCE) and liquid-phase blocking ELISA (LPBE) are OIE-prescribed tests for trade purposes (World Organisation for Animal Health (OIE) 2012). These are highly sensitive, serotype-specific tests that detect FMDV antibodies. These assays may be used for confirmation of infection (previous or ongoing) and to monitor post-vaccination immunity. ELISA doubtful results must be confirmed by VNT to exclude false-positive results. SPCE and VNT tests do not distinguish infected from vaccinated animals. The VNT confirms the FMDV serotype and a version of this test is used to determine the serotype subtype during vaccine matching.

Non-Structural Protein (NSP)-Based Antibody Assays The FMDV NSP ELISA assay measures antibodies to NSP 3ABC. Commercial ELISAs measure antibodies to 3ABC. This assay is not serotype-specific and it is used as a screening test. The PrioCHECK® FMDV NSP ELISA (Prionics®) (Sørensen et al. 1998) is an ELISA that detects antibodies against the highly conserved NSP of the FMD virus. The test can, therefore, be used for all species. The FMDV NSP ELISA kit contains the FMDV NSP, 3ABC, captured by the coated anti-3ABC monoclonal antibody (mAb). A second mAb labelled with an enzyme that generates a colour signal (the detection antibody) is then added. The reaction between the FMDV-NS antigen and the detection antibody is blocked by antibodies directed against the NSPs present in the sample. The PrioCHECK® FMDV NS ELISA that detects antibodies to NSP 3ABC of FMDV with specificity greater than 97 percent for vaccinated and non-vaccinated cattle, and greater than 99 percent in non-vaccinated sheep and pigs. The sensitivity of PrioCHECK® is 100 percent in non-vaccinated cattle but varies greatly in vaccinated cattle, sheep and pigs depending upon time between infection and testing, clinical signs and carrier status. PrioCHECK® FMDV NS can discriminate vaccinated from infected animals and is best used as a herd test rather than an individual animal test.

Nucleic Acid Recognition Methods

Reverse Transcription-Polymerase Chain Reaction (RT-PCR) PCR was the most widely used nucleic acid-based diagnostic technique since its invention (Mullis and Faloona 1987). With the development of RT-PCR to amplify RNA targets it has been assessed the usefulness of it as a reliable tool for FMD diagnosis. RT-PCR can be used to amplify genome fragments of FMDV in diagnostic materials including epithelium, milk, serum and OP samples.

Agarose Gel-Based RT-PCR Assay A gel-based RT-PCR procedure is described (Reid et al. 2000). The RT-PCR assay consists of the three successive procedures of (1) extraction of template RNA from the test or control sample followed by (2) RT of the extracted RNA, (3) PCR amplification of the RT product and (4) detection of the PCR products by agarose gel electrophoresis. RNA extraction is done with the help of viral nucleic acid extraction kit followed by complementary DNA synthesis (cDNA) using a commercial kit. Subsequently, universal and serotype-specific primer sets are used in simple RT-PCR assays on field samples of epithelium and vesicular fluid to determine for primary diagnosis of all seven serotypes of FMDV. For each primer set, template-free amplification controls are performed in parallel to the RNA samples to monitor for cross-contamination. The specificity of reactions is confirmed by using other vesicular disease viruses (Swine vesicular disease virus, Vesicular stomatitis virus, Seneca Valley virus) parallel with the FMD virus serotypes. Serotype-specific primers are designed to identify the serotypes. Test procedures are performed as prescribed in the OIE Manual of Diagnostic Tests and Vaccines for Terrestrial Animals in Chap. 3.1.8.

Real-Time RT-PCR Assay Real-time RT-PCR (qRT-PCR), also called quantitative real-time polymerase chain reaction (Q-PCR/qPCR) is a laboratory technique based on the PCR, which is used to amplify and simultaneously quantify a targeted DNA/cDNA molecule.

Real-time PCR has given rise to a wider acceptance of PCR due to its improved rapidity and sensitivity (Mackay et al. 2002) overcoming poor precision, low sensitivity, low resolution, absence of automation, only size-based discrimination, absence of expression of results in numbers, poor quantitative performance

(Ethidium bromide for staining is not very quantitative), and post-PCR processing, rendering the conventional PCR not very suitable for accurate diagnosis.

The real-time RT-PCR assay can use the same procedures of extraction of total RNA from the test or control sample followed by RT of the extracted RNA as for the conventional agarose gel-based procedure. Automated extraction of total nucleic acid from samples followed by automated pipetting programmes for the RT and PCR steps (Reid et al. 2003) can be used as an alternative to the manual procedures described above. PCR amplification of the RT product is performed by a different procedure. A more simple one-step method for combining the RT and PCR steps has also been described by Shaw et al. (2007) and is widely used by laboratories. Detection of the PCR products in agarose gels is not required following real-time amplification. The standard setting of the OIE Manual of Diagnostic Tests and Vaccines for Terrestrial Animals (2018) has set the procedure as follows:

1. Take the RT products from the cold store −20 °C
2. Prepare the PCR mix described below for each sample. Again it is recommended to prepare the mix in bulk for the number of samples to be tested plus one extra sample: nuclease-free water (6 μl); PCR reaction master mix, 2× conc. (12.5 μl); real-time PCR forward primer, 10 pmol/μl (2.25 μl); real-time PCR reverse primer, 10 pmol/μl (2.25 μl); labelled probe, 5 pmol/μl (1 μl).
3. Add 24 μl PCR reaction mix to a well of a real-time PCR plate for each sample to be assayed followed by 1 μl of the RT product to give a final reaction volume of 25 μl.
4. Spin the plate for 1 min in a suitable centrifuge to mix the contents of each well.
5. Place the plate in a real-time PCR machine for PCR amplification and run the following programme:

   50 °C for 2 min: 1 cycle;
   95 °C for 10 min: 1 cycle;
   95 °C for 15 s, 60 °C for 1 min: 50 cycles.

Times and temperatures may need to be optimised to the particular enzymes, reagents and PCR equipment used in individual laboratories.

6. Reading the results: Assign a threshold cycle (CT) value to each PCR reaction from the amplification plots (a plot of the fluorescence signal versus cycle number; different cut-off values may be appropriate for different sample types (Parida et al. 2007). The CT values used to assign samples as either FMDV positive or negative should be defined by individual laboratories using appropriate reference material. For example at the OIE Reference Laboratory at Pirbright, negative test samples and negative controls should have a CT value at >50.0. Positive test samples and positive control samples should have a CT value <40. Samples with CT values falling within the range 40–50 are designated "borderline" and can be retested. Strong positive FMD samples have a CT value below 20.0 (Reid et al. 2001).

Sequencing and Phylogenetic Analysis

Strain characterisation by nucleotide sequencing: RT-PCR amplification of the P1 region of FMDV genome or a portion of the P1 region that contains VP1 of the genome, followed by nucleotide sequencing is the preferred method for generating sequence data for strain characterisation. Sequencing and analysis of the VP1 coding region of the FMDV are becoming an increasingly more accessible laboratory tool, providing information that aids our understanding of the spread and global epidemiology of the virus. The molecular epidemiology of FMD is based on the comparison of genetic differences between viruses. Phylogenetic trees showing the genomic relationship between vaccine and field strains for all seven serotypes based on sequences derived from the 1D gene (encoding the VP1 viral protein) have been published (Knowles and Samuel 2003); see also http://www.wrlfmd.org/).

Surveillance and Control Strategies

The epidemiology of FMD in the Sahelian region is not yet studied; although it is assumed that FMD virus spreads among domestic animals, with probably little involvement of wildlife as the African buffalo (*Syncerus caffer caffer*) is not present in this part of Africa. Despite the extensive occurrence of FMD, clinical and laboratory investigation for identification and genotyping of the virus has never been exhaustive and complete because there is no FMD virus network surveillance programme in place at the regional level. Some countries do not report or submit samples to diagnostic and reference laboratories during outbreaks of FMD. In Sahelian Africa and the connected regions, there was no comprehensive region-based studies done on the role of animal husbandry and trade practises on FMD virus transmission. Moreover, the role and involvement of wildlife in the epidemiology of FMD in these regions are unknown. Presence of the FMD virus serotypes and topotypes needs virological verification and genotype characterisation by collecting epithelium and probang samples after outbreaks are necessary. This will underpin to develop new technologies, including research towards improved tailored vaccines, appropriate vaccine matching methods and diagnostics relating to the conditions of Sahelian Africa and the connected regions.

To date, with the exception of few countries, in the wide areas of Sahelian Africa and connected regions have not yet implemented the global FAO/OIE progressive control pathway strategy (FAO 2011) for the control of FMD. The first step in this progressive control pathway is the identification of risk factors for virus transmission based on a serological survey and epidemiological studies. It is essential to establish an FMD response to detect, control and contain FMD outbreaks in animals as quickly as possible. Cross-border pastoralism/transhumance is an inherited culture of Sahelian Africa and difficult to circumvent. In such type of pastoralist transhumance animal husbandry system, FMD control needs a region-based ecosystem approach through implementing intensive and active surveillance for monitoring the outbreaks. In Sahelian Africa and the connected regions vaccination is the most preferred approach in FMD control strategy.

For a possible achievable FMD control using vaccination in the FMD endemic settings of Sahelian Africa and connected regions, it is imperative to have data and information on the occurrence and distribution of various FMD virus serotypes, topotypes and strains. A decision on the most effective region-based control strategy should focus on an ecosystem and social approach, identification of primary endemic areas, animal husbandry practises, climate and animal movement (Maree et al. 2014). Within Sahelian Africa and the connected regions, the nomadic, semi-nomadic and transhumance traditional livestock management systems and the unregulated livestock movements are potentially the source for the spread of FMD virus. Although outbreaks occur, vaccination is not practised in all parts of the Sahel and connected regions and if present at irregular intervals.

In FMD endemic settings, vaccination remains an option as part of an effective control strategy. Vaccination is widely used to prevent, control and eradicate FMD (Garland 1999; Bergmann et al. 2005). Current vaccines are inactivated whole virus preparations of a particular strain and the immunity they induce will only protect against a limited range of field strains. This range is maximised by selecting vaccine strains that are as immunogenic and cross-reactive as possible. FMD vaccine matching is required in the event of new FMD virus field strain incursion and during routine vaccination programs in FMD endemic region/countries. Selecting vaccine batches of certified potency (OIE standards) with approved strains is essential to establish the relationship between the field isolates and the vaccine strains. It is proposed that for FMD vaccine strain selection purposes, strains need to be selected, which induce a cross-reactive antibody response within the serotype. A vaccine should not only be selected on the basis of r_1-values (relationship coefficient) but also on the height of the antibody response. Vaccine strain selection for emerging foot-and-mouth disease virus (FMDV) outbreaks in endemic countries can be addressed through antigenic and genetic characterisation of recently circulating viruses.

Conclusions

The key challenges for FMD control in the Sahelian Africa and the connected regions is the trade-related and pastoral livestock unregulated movements that coexist at local, national and regional levels. The unregulated and undocumented livestock movements due to the permeability of national borders is the main cause for FMD virus new incursions. In North, West, Central and East Africa, livestock movement is one of the main risk factors involved in the transmission of FMDV (Di Nardo et al. 2011). In Sahelian Africa and the connected regions, it has been shown that FMD virus serotypes/topotypes distribution can be explained by animal husbandry systems and trade routes and it is believed that the direction of trade often limits transmission of FMD virus, for example the Arabian Peninsula and the Gulf States import live animals from both Africa and Asia, but African FMDV strains have not spread further East. This is most likely due to the fact that no animals have moved eastwards from the Arabian Peninsula and the Gulf States (Tekleghiorghis et al. 2014).

Control by vaccination seems the best option, as control of animal movement will be difficult. Use of broadly cross-protective vaccines for FMD is recommended as variant viruses could be introduced from neighbouring regions. Twice annual FMD vaccination using quality and matching vaccine antigens and effective coverage (>80%) with good vaccine delivery system are the main strategic means to control FMD.

An all-inclusive region-based FMDV control strategy along the OIE/FAO progressive control pathway for FMDV control in Africa, when implemented in a well-coordinated manner, would effectively reduce the occurrence and transmission of FMDV. In the long run, these efforts would improve national and regional economies and food security and protect livelihoods.

The future of FMD control in Africa is uncertain due to a number of challenging issues including (1) presence of multiple FMDV serotypes having great genetic and antigenic diversity, which makes the application of vaccine challenging; (2) poor quality vaccines having low stability and lacking matching with field strains; (3) involvement of wildlife (African buffalo) in maintenance of the virus and disease transmission with the SAT serotypes; (4) unregulated cross-border animal movement for grazing, water and trade practises; (5) poor veterinary services and inadequate infrastructure and (6) inadequate data on FMD epidemiology.

Acknowledgements I am grateful to Dr. Jean Paul Gonzalez for his helpful suggestions on the manuscript and preparing the map of Africa. My great thank goes to Dr. Ahmed Lugelo at the Sokoine University of Agriculture for providing the pictures of FMD lesions.

References

Acharya R, Fry E, Stuart D, Fox G, Rowlands D, Brown F. The three-dimensional structure of foot-and-mouth disease virus at 2.9 A resolution. Nature. 1989;337:709–16.

Ahmed HA, Salem SAH, Habashi AR, Arafa AA, Aggour MGA, Salem GH, Gaber AS, Selem O, Abdelkader SH, Knowles NJ, Madi M, Valdazo-González B, Wadsworth J, Hutchings GH, Mioulet V, Hammond JM, King DP. Emergence of foot-and-mouth disease virus SAT 2 in Egypt during 2012. Transbound Emerg Dis. 2012;59:476–81.

Aidaros HA. Regional status and approaches to control and eradication of foot-and-mouth disease in the Middle East and North Africa. Rev Sci Tech Off Int Epiz. 2002;21:451–8.

Alexandersen S, Zhang Z, Donaldson AI. Aspects of the persistence of foot-and-mouth disease virus in animals—the carrier problem. Microbes Infect. 2002;4:1099–110.

Ayebazibwe C, Mwiine FN, Tjornehoj K, Balinda SN, Muwanika VB, Okurut ARA, Belsham GJ, Normann P, Siegismund HR, Alexandersen S. The role of African buffalos (*Syncerus caffer*) in the maintenance of foot-and-mouth disease in Uganda. BMC Vet Res. 2010;6:1–13.

Ayelet G, MAHAPATRA M, Gelaye E, Egziabher BG, Rufeal T, Sahle M, Ferris NP, Wadsworth J, Hutchings GH, Knowles NJ. Genetic characterization of foot-and-mouth disease viruses, Ethiopia, 1981-2007. Emerg Infect Dis. 2009;15:1409–17.

Bachrach HL. Foot-and-mouth disease. Annu Rev Microbiol. 1968;22:201–44.

Bachrach H, Hess W, Callis J. Foot-and-mouth disease virus: its growth and cytopathogenicity in tissue culture. Science. 1955;122:1269–70.

Balinda SN, Sangula AK, Heller R, Muwanika VB, Belsham GJ, Masembe C, Siegismund HR. Diversity and transboundary mobility of serotype O foot-and-mouth disease virus in East Africa: implications for vaccination policies. Infect Genet Evol. 2010;10:1058–65.

Barasa M, Catley A, Machuchu D, Laqua H, Puot E, Tap Kot D, Ikiror D. Foot-and-mouth disease vaccination in South Sudan: benefit–cost analysis and livelihoods impact. Transbound Emerg Dis. 2008;55:339–51.

Barros JJF, Malirat V, Rebello MA, Costa EV, Bergmann IE. Genetic variation of foot-and-mouth disease virus isolates recovered from persistently infected water buffalo (*Bubalus bubalis*). Vet Microbiol. 2007;120:50–62.

Bastos ADS, Haydon DT, Forsberg R, Knowles NJ, Anderson EC, Bengis RG, Nel LH, Thomson GR. Genetic heterogeneity of SAT-1 type foot-and-mouth disease viruses in southern Africa. Arch Virol. 2001;146:1537–51.

Bastos AD, Haydon DT, Sangaré O, Boshoff CI, Edrich JL, Thomson GR. The implications of virus diversity within the SAT 2 serotype for control of foot-and-mouth disease in sub-Saharan Africa. J Gen Virol. 2003a;84(Pt 6):1595–606.

Bastos ADS, Anderson EC, Bengis RG, Keet DF, Winterbach HK, Thomson GR. Molecular epidemiology of SAT3-type foot-and-mouth disease. Virus Genes. 2003b;27:283–90.

Bergmann IE, Malirat V, Falczuk AJ. Evolving perception on the benefits of vaccination as foot-and-mouth disease control policy: contributions of South America. Expert Rev Vaccines. 2005;4:903–13.

Bouguedour R, Ripani A. Review of the foot and mouth disease situation in North Africa and the risk of introducing the disease into Europe. OIE Rev Sci Tech. 2016;35:757–68.

Bronsvoort BM, Radford AD, Tanya VN, Nfon C, Kitching RP, Morgan KL. Molecular epidemiology of foot-and-mouth disease viruses in the Adamawa province of Cameroon. J Clin Microbiol. 2004;42:2186–96.

Brooksby JB. The virus of foot-and-mouth disease. Adv Virus Res. 1958;5:1–37.

Brooksby JB. Portraits of viruses: foot-and-mouth disease virus. Intervirology. 1982;18:1–23.

Brooksby JB, Rogers J. Methods used in typing the virus of foot-and-mouth disease at Pirbright, 1950–55. In: Methods of typing and cultivation of foot-and-mouth disease virus: Project No. 208. Paris: European Productivity Agency of the Organization of European Cooperation (OEEC); 1957.

Carroll AR, Rowlands DJ, Clarke BE. The complete nucleotide sequence of the RNA coding for the primary translation product of foot-and-mouth disease virus. Nucleic Acids Res. 1984;12: 2461–72.

Condy JB, Hedger RS, Hamblin C, Barnett ITR. The duration of the foot-and-mouth disease virus carrier state in African buffalo (I) in the individual animal and (II) in a free-living herd. Comp Immunol Microb Infect Dis. 1985;8:259–65.

Dawe PS, Flanagan FO. Natural transmission of foot-and-mouth disease virus from African buffalo (*Syncerus caffer*) to cattle in a wildlife area of Zimbabwe. Vet Rec. 1994;134:230–2.

Dhikusooka MT, Ayebazibwe C, Namatovu A, Belsham GJ, Siegismund HR, Wekesa SN, Balinda SN, Muwanika VB, Tjarnehaj K. Unrecognized circulation of SAT 1 foot-and-mouth disease virus in cattle herds around Queen Elizabeth National Park in Uganda. BMC Vet Res. 2016;12: 5.

Di Nardo A, Knowles NJ, Paton DJ. Combining livestock trade patterns with phylogenetics to help understand the spread of foot and mouth disease in sub-Saharan Africa, the Middle East and Southeast Asia. Rev Sci Tech Off Int Epiz. 2011;30:63–85.

Domingo E, Mateu MG, Martinez MA, Dopazo J, Moya A, Sobrino F. Genetic variability and antigenic diversity of foot-and-mouth disease virus. In: Kurstak E, Marusyk RG, Murphy SA, Regenmortel MHVV, editors. Applied virology research, virus variation and epidemiology. New York: Plenum; 1990.

Domingo E, Escarmís C, Baranowski E, Ruiz-Jarabo CM, Carrillo E, Núñez JI, Sobrino F. Evolution of foot-and-mouth disease virus. Virus Res. 2003;91:47–63.
Donaldson AI, Sellers RF. Foot-and-mouth disease. In: Aitken ID, editor. Diseases of sheep. Oxford: Blackwell Science; 2000.
Ehizibolo DO, Haegeman A, Vleeschauwer ARD, Umoh JU, Kazeem HM, Okolocha EC, Borm SV, Clercq KD. Foot-and-mouth disease virus serotype SAT1 in cattle, Nigeria. Transbound Emerg Dis. 2017;64:683–90.
FAO. Progressive control pathway for foot-and-mouth disease control (PCP-FMD) (2011)—Principles, stage descriptions and standards. Rome: FAO; 2011.
Fenner FJ, Gibbs PJ, Murphy FA, Rott R, Studdert MJ, White DO, editors. Veterinary virology. New York: Academic; 1993.
Fevre EM, Bronsvoort BMDC, Hamilton KA, Cleaveland SA. Animal movements and the spread of infectious diseases. Trends Microbiol. 2006;14:125–31.
Forss S, Strebel K, Beck E, Schaller H. Nucleotide sequence and genome organization of foot-and-mouth disease virus. Nucleic Acids Res. 1984;12:6587–601.
Fracastorius H. De contagione et contagiosis morbis et curatione, Venecia; 1546.
Gao Y, Sun S-Q, Guo H-C. Biological function of foot-and-mouth disease virus non-structural proteins and non-coding elements. Virol J. 2016;13:107.
Garland AJM. Vital elements for the successful control of foot-and-mouth disease by vaccination. Vaccine. 1999;17:1760–6.
Geering WA. Foot-and-mouth disease in sheep. Aust Vet J. 1967;43:485–9.
Gorna K, Houndjè E, Romey A, Relmy A, Blaise-Boisseau S, Kpodékon M, Saegerman C, Moutou F, Zientara S, Bakkali Kassimi L. First isolation and molecular characterization of foot-and-mouth disease virus in Benin. Vet Microbiol. 2014;171:175–81.
Grubman MJ, Robertson BH, Morgan DO, Moore DM, Dowbenko D. Biochemical map of polypeptides specified by foot-and-mouth disease virus. J Virol. 1984;50:579–86.
Haydon DT, Bastos AD, Knowles NJ, Samuel AR. Evidence for positive selection in foot-and-mouth disease virus capsid genes from field isolates. Genetics. 2001;157:7–15.
Hedger RS. Foot-and-mouth disease in wildlife with particular reference to the African buffalo (*Syncerus caffer*). New York: Plenum; 1976.
Hedger RS. Foot-and-mouth disease. In: Davis JW, Karstad LH, Trainer DO, editors. Infectious diseases of wild mammals. Ames, IA: Iowa State University Press; 1981.
Hedger RS, Forman AJ, Woodford MH. Foot-and-mouth disease in east African buffalo. Bull Epiz Dis Afr. 1973;21:99–101.
House C, House JA. Evaluation of techniques to demonstrate foot-and-mouth disease virus in bovine tongue epithelium: comparison of the sensitivity of cattle, mice, primary cell cultures, cryopreserved cell cultures and established cell lines. Vet Microbiol. 1989;20:99–109.
Jori F, Caron A, Thompson PN, Dwarka R, Foggin C, Garine-Wichatitsky M, Hofmeyr M, Heerden J, Heath L. Characteristics of foot-and-mouth disease viral strains circulating at the wildlife/livestock interface of the great Limpopo Transfrontier conservation area. Transbound Emerg Dis. 2016;63:e58–70.
Juleff N, Windsor M, Reid E, Seago J, Zhang Z, Monaghan P, Morrison IW, Charleston B. Foot-and-mouth disease virus persists in the light zone of germinal centres. PLoS One. 2008;3:e3434.
Kitching RP. A recent history of foot-and-mouth disease virus. J Comp Pathol. 1998;118:89–108.
Kitching RP, Hutber AM, Thrusfield MV. A review of foot-and-mouth disease with special consideration for the clinical and epidemiological factors relevant to predictive modelling of the disease. Vet J. 2005;169:197–209.
Knight-Jones TJD, Rushton J. The economic impacts of foot and mouth disease—what are they, how big are they and where do they occur? Prev Vet Med. 2013;112:161–73.
Knowles NJ, Samuel AR. Molecular epidemiology of foot-and-mouth disease virus. Virus Res. 2003;91:65–80.
Loeffler F, Frosch P. Summarischer bericht uber die ergebnisse der untersuchungen der kommoission zur erforchung der maul-und-klamenseuche. Zentbl Bakteriol Parasitenkd Infektionskr. 1897;22:257–9.

Loeffler F, Frosch P. Report of the commission for research on foot-and-mouth disease. Zentbl Bakteriol Parasitkd Infektionskr. 1898;23:371–91.
Longjam N, Tayo T. Antigenic variation of foot and mouth disease virus—an overview. Vet World. 2011;4:475–9.
Mackay IM, Arden KE, Nitsche A. Real-time PCR in virology. Nucleic Acids Res. 2002;30: 1292–305.
Maree FF, Kasanga CJ, Scott KA, Opperman PA, Chitray M, Sangula AK, Sallu R, Sinkala Y, Wambura PN, King DP, Paton DJ, Rweyemamu MM. Challenges and prospects for the control of foot-and-mouth disease: an African perspective. Vet Med Res Rep. 2014;5:119–38.
Mullis KB, Faloona FA. Specific synthesis of DNA in vitro via a polymerase-catalyzed chain reaction. Methods Enzymol. 1987;155:335–50.
Pacheco JM, Smoliga GR, O'Donnell V, Brito BP, Stenfeldt C, Rodriguez LL, Arzt J. Persistent foot-and-mouth disease virus infection in the nasopharynx of cattle; tissue-specific distribution and local cytokine expression. PLoS One. 2015;10:e0125698.
Parida S, Fleming L, Gibson D, Hamblin PA, Grazioli S, Brocchi E, Paton DJ. Bovine serum panel for evaluating foot-and-mouth disease virus nonstructural protein antibody tests. J Vet Diagn Investig 2007;19:539–44.
Paton DJ, Sumption KJ, Charleston B. Options for control of foot-and-mouth disease: knowledge, capability and policy. Philos Trans R Soc B. 2009;364:2657–67.
Pereira HG. Subtyping of foot-and-mouth disease virus. In: International Symposium on Foot-and-Mouth Disease, Development Biological Standardards. Basel: S. Karger; 1977.
Radostits O, Arundel J, Gay C. Veterinary medicine, a textbook of the diseases of cattle, sheep, pigs, goats and horses. London: W B Saunders Co.; 2000.
Reid SM, Ferris NP, Hutchings GH, Samuel AR, Knowles NJ. Primary diagnosis of foot-and-mouth disease by reverse transcription polymerase chain reaction. J Virol Methods. 2000;89: 167–176.
Reid SM, Ferris NP, Hutchings GH, Zhang Z, Belsham GJ, Alexandersen S. Diagnosis of foot-and-mouth disease by real-time fluorogenic PCR assay. Vet Rec. 2001;149:621.
Reid S, Grierson S, Ferris N, Hutchings G, Alexandersen S. Evaluation of automated RT-PCR to accelerate the laboratory diagnosis of foot-and-mouth disease virus. J Virol Methods. 2003;107: 129–39.
Roeder PL, Knowles NJ. Foot-and-mouth disease virus type C situation: the target for eradication? Report of the session of the Research Group of the Standing Technical Committee of EUFMD. Sicily: Erice; 2008.
Rushton J, Knight-Jones T. Socio-economics of foot and mouth disease. In: FAO/OIE Global Conference on Foot-and-Mouth Disease Control, Bangkok, Thailand. 2012.
Sahle M, Dwarka RM, Venter EH, Vosloo W. Comparison of SAT-1 foot-and-mouth disease virus isolates obtained from East Africa between 1971 and 2000 with viruses from the rest of sub-Saharan Africa. Arch Virol. 2007;152:797–804.
Salt JS. The carrier state in foot and mouth disease—an immunological review. Br Vet J. 1993;149: 207–23.
Samuel AR, Knowles NJ. Foot-and-mouth disease type O viruses exhibit genetically and geographically distinct evolutionary lineages (topotypes). J Gen Virol. 2001;82:609–21.
Sangaré O. Study of the molecular epidemiology of foot-and-mouth disease virus in West Africa, PhD. Pretoria, South Africa: University of Pretoria; 2002.
Sangula AK. Evolutionary genetics of foot-and-mouth disease virus in Kenya. PhD: Makerere University; 2010.
Sellers RF. Growth and titration of the viruses of foot-and-mouth disease and vesicular stomatitis in kidney monolayer tissue cultures. Nature. 1955;176:547–9.
Shaw AE, Reid SM, Ebert K, Hutchings GH, Ferris NP, King DP. Implementation of a one-step real-time RT-PCR protocol for diagnosis of foot-and-mouth disease. J Virol Methods, 2007; 143:81–85.
Snowdon WA. Growth of foot-and-mouth disease virus in monolayer cultures of calf. Thyroid Cells. 1966;210:1079–80.

Sørensen KJ, Madsen KG, Madsen ES, Salt JS, Nqindi J, Mackay DKJ. Differentiation of infection from vaccination in foot-and-mouth disease by the detection of antibodies to the non-structural proteins 3D, 3AB and 3ABC in ELISA using antigens expressed in baculovirus. Arch Virol. 1998;143:1461–76.

Sumption KJ, Pinto J, Lubroth J, Morzaria S, Murray T, Rocque S. Foot-and-mouth disease: situation worldwide and major epidemiological events in 2005–2006. EMPRES Focus Bull. 2007;1

Sutmoller P, Gaggero A. Foot-and mouth diseases carriers. Vet Rec. 1965;77:968–9.

Tekleghiorghis T, Moormann RJM, Weerdmeester K, Dekker A. Foot-and-mouth disease transmission in Africa: implications for control, a review. Transbound Emerg Dis. 2014;63:136–51.

Thomson GR. Foot and mouth disease. In: Coetzer JAW, Thomas GR, Tustin RC, Kriek NPJ, editors. Infectious diseases of livestock with special reference to Southern Africa. Cape Town: Oxford University Press; 1994.

Thomson GR, Vosloo W, Bastos ADS. Foot and mouth disease in wildlife. Virus Res. 2003;91: 145–61.

Vallée H, Carré H. Sur la pluralité du virus aphteux (in French). C R Hebd Acad Sci Paris. 1922;174:1498–500.

van Bekkum J, Frenkel H, Frederiks H, Frenkel S. Observations on the carrier state of cattle exposed to foot-and-mouth disease virus. Bull Off Int Epizoot. 1959;51:917–22.

van Regenmortel MHV, Fauquet CM, Bishop DHL, Carsten EB, Estes MK, Lemon SM, Maniloff J, Mayo MA, Mcgeoch DJ, Pringle CR, Wickner RB. Virus taxonomy: classification and nomenclature of viruses: seventh report of the international committee on taxonomy of viruses. San Diego: Academic; 2000.

Vosloo W, Thomson GR. Natural habitats in which foot-and-mouth disease virus is maintained. In: Sobrino F, Domingo E, editors. Foot and mouth disease: current perspectives. Wymondham: Horizon Bioscience; 2004.

Vosloo W, Bastos AD, Kirkbride E, Esterhuysen JJ, Van Rensburg DJ, Bengis RG, Keet DW, Thomson GR. Persistent infection of African buffalo (*Syncerus caffer*) with SAT-type foot-and-mouth disease viruses: rate of fixation of mutations, antigenic change and interspecies transmission. J Gen Virol. 1996;77:1457–67.

Vosloo W, Bastos ADS, Sangare O, Hargreaves SK, Thomson GR. Review of the status and control of foot and mouth disease in sub-Saharan Africa. Rev Sci Tech. 2002;21:437–49.

Vosloo W, De Klerk LM, Boshoff CI, Botha B, Dwarka RM, Keet D, Haydon DT. Characterisation of a SAT-1 outbreak of foot-and-mouth disease in captive African buffalo (*Syncerus caffer*): clinical symptoms, genetic characterisation and phylogenetic comparison of outbreak isolates. Vet Microbiol. 2007;120:226–40.

Wernery U, Kaaden OR. Foot-and-mouth disease in camelids: a review. Vet J. 2004;168:134–42.

World Organisation for Animal Health (OIE). Foot-and-mouth disease. In: OIE, editor. Manual of diagnostic tests and vaccines for terrestrial animals. 6th ed. Paris: OIE; 2012.

WRLFMD. Molecular epidemiology/genotyping. OIE/FAO FMD Reference Laboratory Network Reports. http://www.wrlfmd.org/fmd_genotyping/2017.htm. Accessed 10 Sept 2017 [Online]. http://www.wrlfmd.org/fmd_genotyping/index.html (2003–2017). Accessed 10 Sept 2017.

Young JR, Suon S, Andrews CJ, Henry LA, Windsor PA. Assessment of financial impact of foot and mouth disease on smallholder cattle farmers in southern Cambodia. Transbound Emerg Dis. 2013;60:166–74.

Chapter 12
Peste des Petits Ruminants

Adama Diallo, Arnaud Bataille, Renaud Lancelot, and Geneviève Libeau

Abstract Peste des petits ruminants (PPR) is a contagious viral disease of domestic and wild small ruminants. Clinically, it is characterized by fever, gastroenteritis, erosive lesions of mucous membranes, and respiratory distress due to severe bronchopneumonia. PPR is a transboundary animal disease (TAD) with mortality rates varying considerably but as high as 60–70%. It is in the list of the group of economically important animal diseases to be notified to the World Organisation for Animal Health (OIE). Described for the first time in 1942 in Côte d'Ivoire, PPR has steadily expanded its geographical distribution throughout Africa, the Middle and Near East, and Asia, from China to Kazakhstan. It has now become the most important sheep and goat infectious disease. It is estimated that productions of nearly two billion of sheep and goats, and many vulnerable wild small ruminants such as Saiga in Mongolia, are threatened by PPR. As sheep and goats are vital for day-to-day livelihoods of small farmers, the fight against PPR should be seen as a program for the reduction of poverty in the world, one of the Millennium Development Goals. Taking lessons from the success of the Global Rinderpest Eradication Programme (GREP), and as PPR shares with rinderpest some favorable technical attributes that have facilitated rinderpest eradication, FAO and the OIE have jointly developed a Strategy for the Global Eradication of PPR by the year 2030.

Keywords PPR · Morbillivirus · SLAM · Sheep · Goat · Wildlife · Lineage · PPR-GCES · GREP · Eradication

A. Diallo (✉)
UMR ASTRE, CIRAD, Montpellier, France

ASTRE, Montpellier University, CIRAD, INRA, Montpellier, France

ISRA-LNERV, Dakar-Hann, Senegal
e-mail: Adama.diallo@cirad.fr

A. Bataille · R. Lancelot · G. Libeau
UMR ASTRE, CIRAD, Montpellier, France

ASTRE, Montpellier University, CIRAD, INRA, Montpellier, France
e-mail: arnaud.bataille@cirad.fr; Renaud.lancelot@cirad.fr; Genevieve.libeau@cirad.fr

M. Kardjadj et al. (eds.), *Transboundary Animal Diseases in Sahelian Africa and Connected Regions*, https://doi.org/10.1007/978-3-030-25385-1_12

Abbreviations

CCPP	Contagious caprine pleuropneumonia
cELISA	Competitive enzyme-linked immunosorbent assay
FAO	Food and Agriculture Organization of the United Nations
GREP	Global Rinderpest Eradication Programme
OIE	World Organisation for Animal Health
PPR-GCES	PPR Global Control and Eradication Strategy
PPRV	Peste des petits ruminants virus
SLAM	Signaling lymphocyte activation molecules
TAD	Transboundary animal disease

History

In 1942, Gargadennec and Lalanne described a new disease of sheep and goats for the first time in mid-1930 in Côte d'Ivoire. It was affecting goats more than sheep with symptoms similar to rinderpest, but cattle in contact of sick small ruminants remained apparently healthy. As this new disease looked like rinderpest, they named "peste des petits ruminants" (PPR) (Gargadennec and Lalanne 1942). At the same time, a similar syndrome was reported in Dahomey (now Benin) by Cathou under the name of "peste des espèces ovine et caprine" (plague of ovine and caprine species) (Mornet et al. 1956a). In both cases, sheep and goats were involved in outbreaks. For about 10–15 years, only those two countries reported outbreaks of that new disease. But in 1955, Mornet et al. (1956b) noted the disease for the first time in Senegal. Later on, a similar syndrome was described in Nigeria with goats being the main affected species. This was called "Kata" (Nigerian local name, pidgin for "catarrh"), goat plague, pseudo-rinderpest, and finally stomatitis-pneumoenteritis complex (Whitney et al. 1967; Isoun and Mann 1972). Further investigation proved PPR and stomatitis-pneumoenteritis complex are the same disease (Hamdy et al. 1976; Rowland et al. 1969, 1971; Rowland and Bourdin 1970). Since then, the original French name "peste des petits ruminants" is used as the scientific name of this disease. However, "stomatitis-pneumoenteritis complex" is the denomination which fits best with the clinical signs because it takes into consideration the bronchopneumonia characteristic of PPR, a sign not found in the case of rinderpest. The causal agent of PPR was seen by Mornet et al. (1956a, b) as a variant of the rinderpest virus. But further cross-neutralization and protection studies proved that those two diseases are caused by two different but closely related viruses belonging to the morbillivirus genus that includes measles and canine distemper viruses (Gibbs et al. 1979). Protein gel and nucleic acid probe hybridization analyses proved definitively that rinderpest and PPR viruses are different (Diallo et al. 1987, 1989a). Sequencing data analyses showed PPR virus (PPRV) is less closely related to rinderpest virus than measles virus (Diallo et al. 1994).

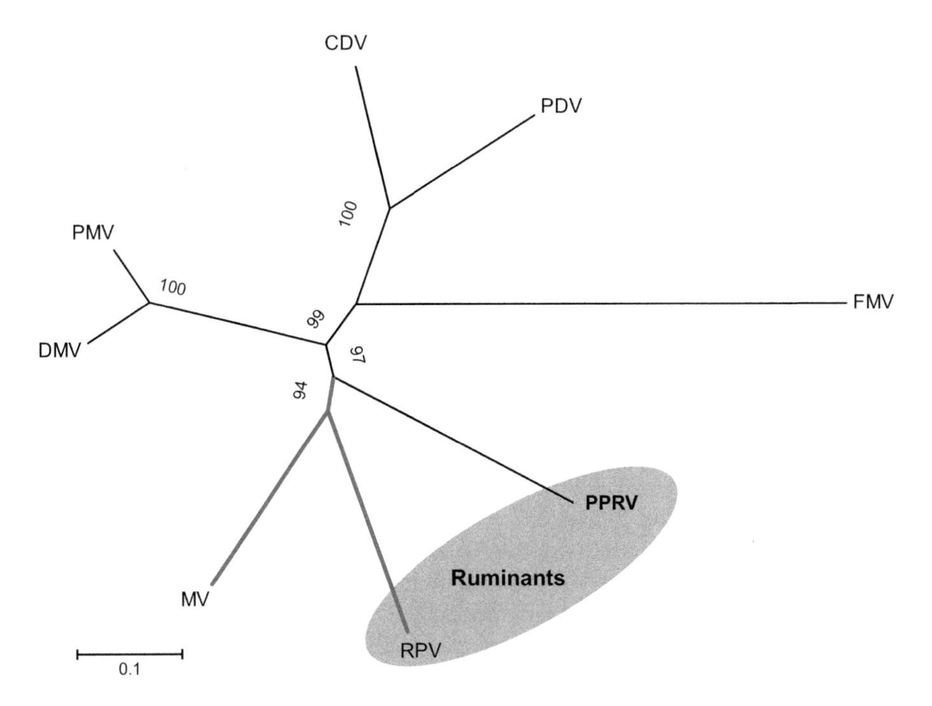

Fig. 12.1 Phylogenetic relationship between morbilliviruses. The tree was constructed using the complete nucleoprotein gene sequence of peste des petits ruminants virus (PPRV, GenBank accession number KY628761), rinderpest virus (RPV, X98291), measles virus (MV, NC_001498), feline morbillivirus (FMV, JQ411015), porpoise morbillivirus (PMV, AY949833), dolphin morbillivirus (DMV, NC_005283), canine distemper virus (CDV, AF014953), and phocine distemper virus (PDV, NC_028249). The analysis was performed on MEGA 6 using the neighbor-joining method with the Tamura-Nei substitution model. The numbers at the nodes are bootstrap values obtained from 1000 replicates. The section of the tree joining RPV and MV is highlighted in dark gray. Morbilliviruses infecting ruminants (PPRV and RPV) are indicated in light gray

The Etiology

The causative agent of peste des petits ruminants, the PPR virus (PPRV), is a member of the morbillivirus genus, family *Paramyxoviridae* (Gibbs et al. 1979). All members of this virus group are highly pathogenic infecting agents: (1) humans (measles virus), (2) dogs (canine distemper virus), (3) cattle and buffaloes (rinderpest virus), (4) marine mammals (dolphin morbillivirus and porpoise distemper virus), and (5) cat (feline morbillivirus) (see Fig. 12.1) (Diallo and Libeau 2014; Libeau et al. 2014; Baron et al. 2016). PPRV is an enveloped, pleomorphic particle the size of which varies between 150 and 700 nm, the mean size being at about 500 nm (Bourdin and Laurent-Vautier 1967; Durojaiye et al. 1985). The genome is a single-stranded ribonucleic acid (RNA) of negative sense. It is composed of 15,948 nucleotides residues and is encapsidated by the nucleoprotein (N) associated with the polymerase complex formed by the phosphoprotein (P) and the large protein (L) which possesses the nucleic acid polymerase activity (Bailey et al. 2005; Baron et al. 2016). The viral envelope is derived from the host cell membrane into which are incorporated the virus matrix protein (M) and the two external virus glycoproteins: the fusion (F) and the hemagglutinin (H) proteins. The M protein binds the nucleocapsid to the envelope membrane and thereby plays an important role during the virion budding process (Baron et al. 2016; Peeples 1991). Co-expression of the PPRV N and M proteins results in the formation of virus-like particles (Liu et al. 2013). The external envelope proteins, H and F, are involved in the viral first step infection of the host cell. Indeed, the infection of the target cell starts by its attachment to that cell through the binding of the H protein to a cell surface protein it uses as a receptor. Then F protein mediates the fusion of the viral envelope with the cell membrane (Diallo and Libeau 2014; Baron et al. 2016). The cell proteins used as receptors by wild-type morbilliviruses are the signaling lymphocyte activation molecule (SLAM) for lymphoid cells and the nectin-4 protein for the epithelial cells (Tatsuo et al. 2000; Baron 2005; Birch et al. 2013; Adombi et al. 2011; Diallo and Libeau 2014; Baron et al. 2016). The interaction of the H protein with the SLAM and nectin-4 protein is responsible for the tropism and host specificity of morbilliviruses (Baron et al. 2016).

In addition to the N, P, L, M, F, and L proteins that are components of the viral particles, morbilliviruses produce in infected cells two other proteins that are called nonstructural proteins as they do not integrate the virus structure. Those proteins, V and C proteins, appear to play multiple functions in morbilliviruses. Indeed, not only they are involved in the replication of the virus but also they modulate the host cell immune response by blocking the IFN production in different ways (Baron et al. 2014a, b, 2016; Kumar et al. 2014; Sanz-Bernardo et al. 2017).

The Epidemiology

Host Range

PPR is primarily a disease of sheep and goats. The majority of reports on PPR in Africa are related to outbreaks in goat populations (Lefèvre and Diallo 1990). Based on that observation it is believed that goat, and in particular the West African dwarf goat breed, is the most sensitive species to PPRV, morbidity and mortality rates being low in sheep in general compared to goats (Lefèvre and Diallo 1990; Obi et al. 1983a, b; Roeder et al. 1994). However, the first PPR cases reported in India were related to sheep (Shaila et al. 1989). But some subsequent reports in India mentioned PPR outbreaks in which only goats were involved; sheep in contact with the affected animals were apparently healthy (Shaila et al. 1996a; Kulkarni et al. 1996). An epidemiological survey was carried out in the Andhra Pradesh state in India between 1994 and 1998. Among 147 PPR outbreaks that were recorded during that survey, 132 were restricted to sheep. In four cases, both sheep and goats were involved (Taylor et al. 2002). Serological surveys also showed some discrepancies. Indeed, some of the results showed higher prevalence in sheep populations than in goats.

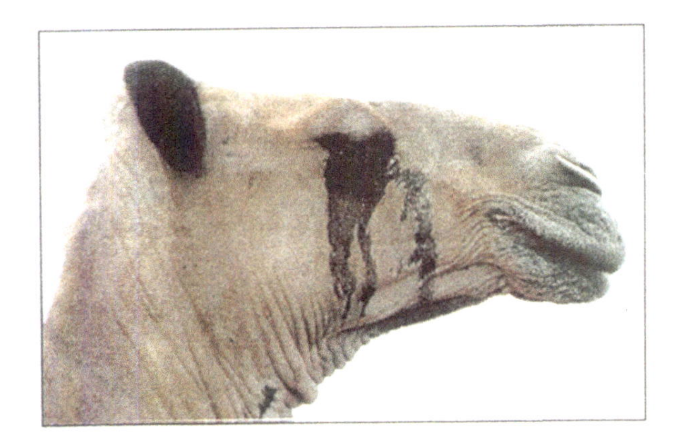

Fig. 12.2 Suspected PPR Ethiopia dromedaries: ocular discharges (from Roger F.)

They are interpreted by the authors as a consequence of the high survival rate of the former animals to PPRV infection compared to the second animal species (Abraham et al. 2005; Ozkul et al. 2002). Other reports provided opposite results which, for the authors, are interpreted as an indication of the high susceptibility of goats to PPRV with high probability in developing antibodies against this virus compared to sheep (Al-Majali et al. 2008; Ayari-Fakhfakh et al. 2011; Delil et al. 2012). In fact, contradictions noted at present between reports related to the susceptibility of sheep and goats to PPRV are not understood yet. There is no indication of the existence of virus variants more adapted to sheep or goats. In addition to animal species and breeds, other factors such as farming practices, animal densities, and trade may influence the rate of infection and its consequences (Abubakar et al. 2009; Diop et al. 2005; Ezeokoli et al. 1986; Hammouchi et al. 2012). Obviously, dynamic demographic and epidemiological models are needed to explain these different patterns.

In addition to sheep and goats, PPRV was considered as the causative agent of an epidemic respiratory disease of dromedaries that occurred in Ethiopia and Sudan in the late 1990s and early 2000s. Respiratory syndrome, ocular discharge (Fig. 12.2), and abortion were observed (Khalafalla et al. 2010; Roger et al. 1998, 2001). In addition, paresis was reported (Fig. 12.3) (Roger et al. 1998, 2001). This clinical sign has never been described for PPR in affected sheep and goats. PPRV antigens and nucleic acids were detected in some pathological samples collected during those outbreaks. In the same period, similar results were found in testing lung samples of camels suffering from pneumonia and slaughtered in Sokoto abattoir in Nigeria (Shamaki, personal communication). In each of these cases, no virus could be isolated from the samples. Also, the disease could not be reproduced in camels after an experimental infection (Wernery 2011). The presence of PPR antibodies in dromedary sera collected has been reported in different countries (Abraham et al. 2005; Ismail et al. 1992; Intisar et al. 2017). But there is no evidence of the involvement of camels in the transmission of PPRV to other animals.

Many serological surveys have shown the presence of PPRV antibodies in cattle sera in PPR endemic areas without observation of any PPR-like sign in those animals. This indicates that PPRV can be transmitted from small ruminants to cattle

Fig. 12.3 Suspected PPR Ethiopia dromedaries: animal with paresis (from Roger F.)

in the field without any disease (Anderson and McKay 1994; Balamurugan et al. 2012; Abubakar et al. 2015). Experimentally, Sen et al. (2010) have obtained subclinical infection of cattle inoculated with PPRV although it has persisted in infected animals for more than a year without any excretion. Morbilliviruses being lymphotropic viruses, virulent PPRV may grow in bovine lymphocytes although the efficiency is less than in sheep and goat cells (Rossiter and Wardley 1985). Therefore, it might be possible that occasionally cattle in poor conditions develop lesions after PPRV infection, clinical signs that would be assimilated to rinderpest. Results obtained by Mornet et al. (1956a) might support this assumption. These authors recorded disease and death in experimentally PPRV-infected calves. PPRV was also isolated from an outbreak of rinderpest-like disease in buffaloes in India in 1995 (Govindarajan et al. 1997). Apart from rinderpest virus which can infect and cause disease with different levels of severity in many Artiodactyls, morbilliviruses are generally host specific: measles virus for humans, canine distemper virus for carnivores, and PPRV for small ruminants. However, sequence data have indicated that morbilliviruses that cause diseases in wild large felids in the 1990s were not feline morbilliviruses but true canine distemper viruses (Harder et al. 1996). Thus, for some reasons which are unknown at present, it appears that morbilliviruses can, from time to time, cross species barriers and cause disease in species believed to be not sensitive. This might happen probably with PPRV in large ruminants. An

Fig. 12.4 Mongolia PPR-affected gazelle (from Kock R.)

explanation for the apparition of disease in large ruminants following PPRV infection would be the involvement of a highly pathogenic virus strain to overcome the innate relative resistance of those animals. If the variation in pathogenicity is well known with RPV, different animal experiments carried out with PPRV so far have shown inconsistent results: a same isolate will prove very pathogenic today in goats for example but will give mild symptoms in the same species in another experiment (Pope et al. 2013). In light of these findings, there is a need for more investigations to determine the eventual role, passive or active, of cattle, buffaloes, and camels in the epidemiology of PPR.

Concerning pig, an experiment that was carried out in Nigeria showed that this species undergoes only subclinical infection and it was classified as a dead-end host (Nawathe and Taylor 1979). However, during recent investigations, Schulz et al. have succeeded to reproduce disease in European breed pigs inoculated with the PPRV lineage IV strain isolated from wild goat in Kurdistan. The sick pigs were able to transmit the virus to in-contact animals (Schulz et al. 2018). This result calls for further investigations by using different pig breeds and different PPRV strains.

In addition to sheep and goats, PPR can cause severe losses in different species of wild ungulate (Figs. 12.4 and 12.5). The first cases of PPR in wildlife were observed on different species of gazelles (*Ovis orientalis laristanica, Oryx gazella*) kept in a zoo in the Middle East (Furley et al. 1987). Since then, they are reports on PPR outbreaks in different wildlife species in the field in Kurdistan, Iran, Pakistan, China, and Mongolia, putting in danger some of them that are already close to extinction such as Saiga antelope in Mongolia (Abubakar et al. 2011; Bao et al. 2011; Hoffmann et al. 2012; Marashi et al. 2017; Shatar et al. 2017; Rahman et al. 2018;

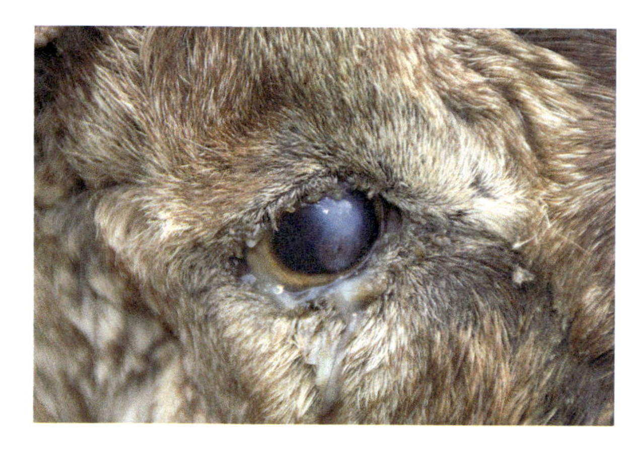

Fig. 12.5 Kurdistan PPR-affected wild goat: ocular discharges (from Hoffmann B.)

Aziz-Ul-Rahman et al. 2018). In sub-Saharan Africa, no clinical case of PPR in wildlife has been reported so far. The presence of the infection in those populations has been noted only by serology results (Couacy-Hymann et al. 2005; Mahapatra et al. 2015). For the moment, there is no evidence of an active role of wildlife in the epidemiology of PPR. Infection of those animals is considered to be a spillover of sheep and goat infection by PPRV (Mahapatra et al. 2015).

Current Geographical Distribution

From 1940, date of its first description (Gargadennec and Lalanne 1942) to the mid-1970s, PPR outbreaks were reported only in some West African countries bordering the Atlantic Ocean: Nigeria, Benin, Ghana, and Côte d'Ivoire (Gargadennec and Lalanne 1942; Mornet et al. 1956a, b; Bourdin et al. 1970, Bourdin 1973; Durtnell 1972; Isoun and Mann 1972; Johnson and Ritchie 1968; Whitney et al. 1967). As of the 1980s, reports on PPR outbreaks in other African countries accumulated rapidly. The first suspicion of PPR outside its original West African stronghold was made by Provost et al. in Chad in 1972 based on serological data and the observation of cases of goats not reacting to the inoculation of the live caprinized rinderpest vaccine (Provost et al. 1972). But it was only in 1995 that the isolation of PPRV was reported for the first time in Chad (Bidjeh et al. 1995). In 1971–1972, Babiker reported on rinderpest-like outbreaks in sheep and goats in Sudan. The virus was first identified as rinderpest virus based on clinical signs and the detection of "rinderpest antigen" in pathological samples by the agar gel immunodiffusion (AGID) test (El Hag Ali 1973). The virus strains that were isolated at that time were reexamined 10 years later and were proved to be PPRV (El Hag Ali and Taylor 1984). Until 1994, Sudan was the only East African country where PPR outbreaks were regularly reported from. But in that year, the disease was identified in Ethiopia (Roeder et al. 1994). Nearly at the same time, Wamwayi et al. (1995) reported the identification of PPR antibodies in the sera of sheep and goats collected in regions at the border of Kenya and Uganda. But it was 10 years later, 2006–2007, that PPR outbreaks were reported in those two countries, followed by Tanzania in 2008 (Swai et al. 2009; Luka et al. 2011). With those new cases, PPR endemic areas in Africa, limited to the region between Sahara and the Equator until 1994, were then extended beyond the Equator southern limit. Afterward the disease was reported in other countries of the region: Gabon, Republic of Congo, Democratic Republic of Congo, Angola, Comoros Islands in 2012, and Burundi in 2017 (Cêtre-Sossah et al. 2016; Libeau et al. 2014; Maganga et al. 2013; Niyokwishimira et al. 2019). In northern Africa, Egypt is the first country where PPR was reported as a rinderpest (Ismail and House 1990). It was nearly 20 years later that Morocco experienced is first PPR outbreak in 2008 (Kwiatek et al. 2011; Muniraju et al. 2013). But PPR antibodies were detected in sera collected in 2006 in Tunisia (Ayari-Fakhfakh et al. 2011). Since then clinical cases with virus identification were reported in Algeria and Tunisia (De Nardi et al. 2012; Kardjadj et al. 2015; Baazizi et al. 2017). In 2012 and 2013, Libya declared to the OIE the existence of PPR infection based on serological data (Dayhum et al. 2018). But no PPR outbreak has been reported in that country yet.

The first evidence of the presence of PPR outside of Africa was reported by Hedger et al. (1980) who identified PPR antibodies in small ruminant sera collected in Oman. This was followed by the isolation of PPRV during an outbreak in gazelles kept in a zoo in 1983 (Furley et al. 1987). In Asia, the disease was recognized for the first time through a publication in 1989. In that year, Shaila et al. (1989), by DNA probe testing, detected PPRV nucleic acid in pathological samples collected from sheep that were affected by a rinderpest-like disease in India. Currently, PPR is endemic in all African regions extending from Morocco to Burundi, the Near and Middle East including Turkey, and in Asia from China to Tajikistan and Kazakhstan. In addition to the European part of Turkey, two other European countries have now experienced PPR outbreaks: Georgia in 2015–2016 and Bulgaria in 2018 (Donduashvili et al. 2018; Almeshay et al. 2017). Today PPR is threatening small ruminant productions in more than 65 countries in Africa, Asia, and the Middle and Near East (see Fig. 12.6). The intensification of international animal trade is one of the keys factors of the dramatic PPR spread that has been noted since the 1990s onward. However, better knowledge on the disease geographical distribution started to build up quickly when PPR-specific diagnostic tools were made available as of the late 1980s. Many rinderpest outbreaks that were reported in small ruminant populations in the past were certainly PPR. And it is not too speculative to state that PPR is not a recent disease but has existed long before 1942. Likely, it has always been overlooked in favor of pasteurellosis due to respiratory signs or of rinderpest because both diseases are clinically similar and have shared most of the same endemic areas. A case of PPR misdiagnosis was the first report of Sudan PPR mentioned above. Another example is the situation in India. If PPR was identified in that country for the first time in 1987 (Shaila et al. 1989), Gopalakrhisna mentioned in 1940 and 1942 already the presence of rinderpest-like disease of sheep and goats in Assam (Taylor et al. 2002). While the pathological material was highly pathogenic for the small ruminants, it was very mild for calves (Taylor et al. 2002). From

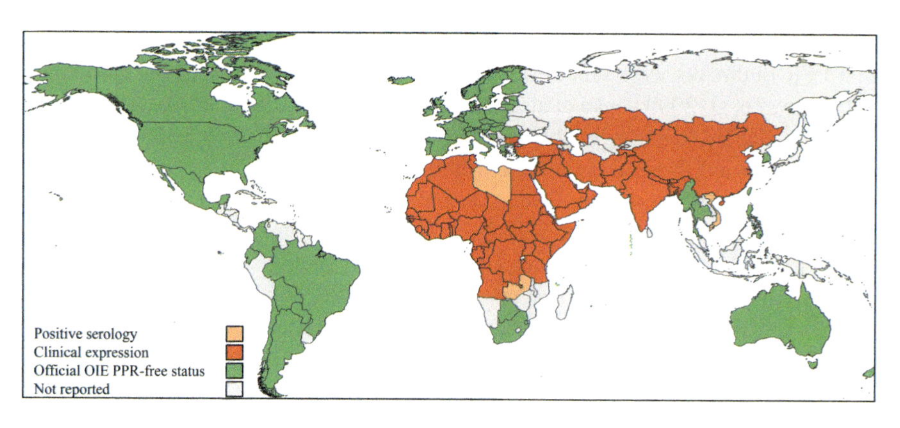

Fig. 12.6 World PPR endemic map (July 2018 update)

that period onward, "rinderpest" was regularly reported in sheep and goats in India, based on results obtained by agar immunodiffusion test with rinderpest hyperimmune serum, or by animal inoculation experiments and disease control by rinderpest vaccine. Considering the fact that PPRV might cause disease, even mild, in cattle and buffaloes in some unknown circumstances (El Hag Ali and Taylor 1984, 1988; Mornet et al. 1956b), some of these reported rinderpest cases in sheep and goats were probably PPR. Indeed, with the success of rinderpest vaccination campaign in cattle and buffaloes, along with the gradual implementation of specific differential diagnostic tests in the country, it appears that PPR is widespread all over India from South to North (Taylor et al. 2002). Now, with the availability of PPR-specific diagnostic tests, all rinderpest-like cases in small ruminants are identified as PPR.

Molecular epidemiology analysis of virus strains identified in the PPR different endemic areas has allowed grouping them into four genotypes based on the partial sequence of the N and F protein genes (see Fig. 12.7) (Kwiatek et al. 2007; Shaila et al. 1996b). N-based nomenclature was slightly different from the F-based one. Indeed, the first one was done according to the apparent expansion of our knowledge in PPR geographical distribution, from West to East Africa, lineage I to III, and Asia, lineage IV. The lineage I of N-based nomenclature corresponds to the lineage II of the F-based nomenclature. But now the N-based nomenclature is the globally adopted PPRV lineage classification (Kumar et al. 2014; Parida et al. 2015). Historically PPRV samples that were collected from Africa until the mid-1990s belonged to lineage I for samples from Burkina Faso to Senegal, lineage II for Nigeria, Benin, and Ghana samples, and then lineage III for East African PPRV isolates. All Asian PPRV were in the lineage IV. This latter lineage is also the dominant lineage in the Middle East as lineage III was detected there only once: the virus responsible of the PPR outbreak in wildlife in a zoo in 1983 (Furley et al. 1987) belongs to that lineage, its origin being probably East Africa. While all the currently identified Asia and the Middle East PPRV strains are of lineage IV, the distribution of lineages is much more complex and dynamic in Africa. Indeed, West African

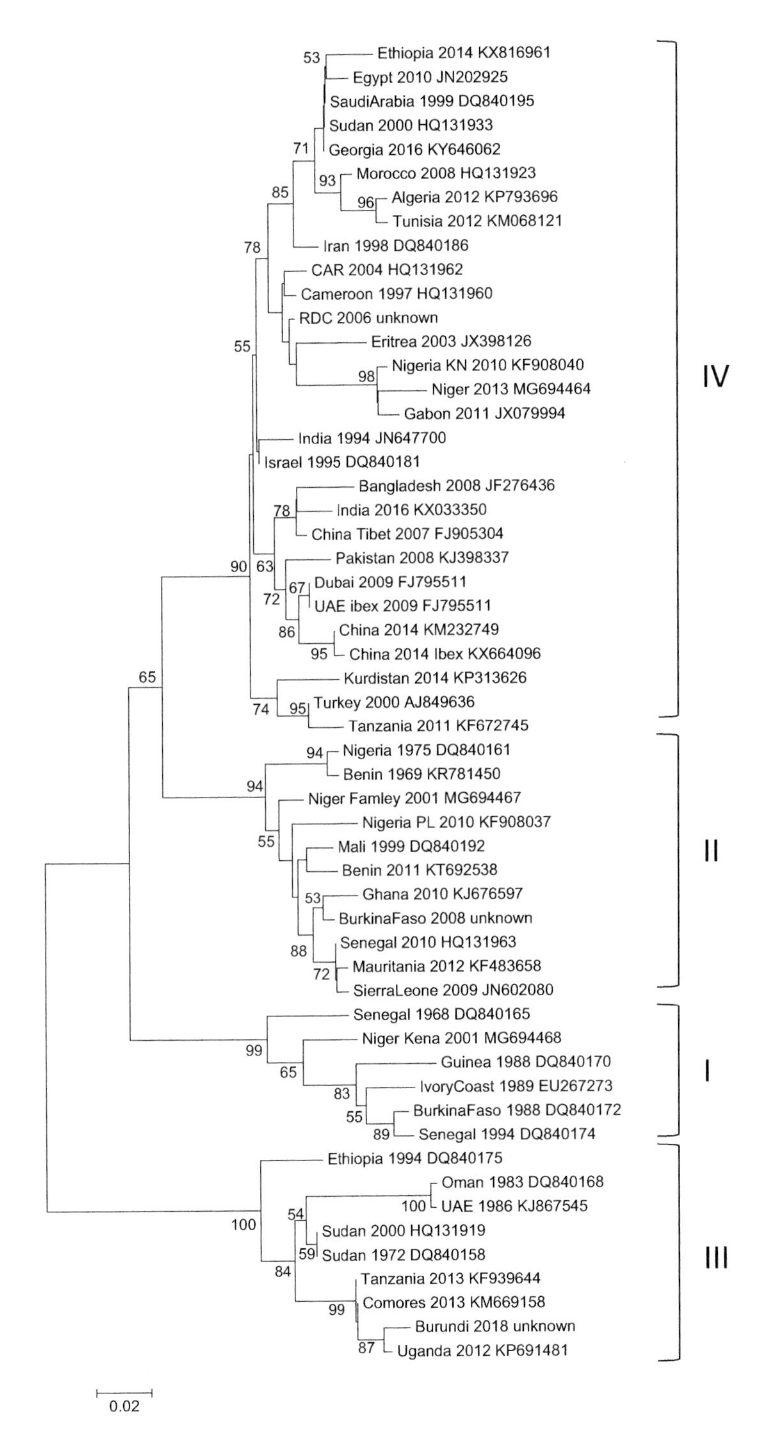

Fig. 12.7 Phylogenetic tree showing the relationships between strains of peste des petits ruminants virus (PPRV). The tree was constructed from an alignment of partial sequences (255 base pairs) of the PPRV nucleoprotein gene. Country of origin, date of collection, and GenBank accession

PPRV strains that were of lineage I are now "pushed away" by those of lineage II (Libeau et al. 2014; Parida et al. 2015; Tounkara et al. 2018). The lineage IV, after its expansion in North Africa, is also replacing slowly the lineage III in Sudan and Ethiopia (Kwiatek et al. 2011; Muniraju et al. 2016). It is now replacing also lineage II in Nigeria (Woma et al. 2016). It has also been identified in Chad, Niger (Tounkara et al. 2018), Gabon, Democratic Republic of Congo, and Angola.

Apparition and Evolution of Outbreaks

At the individual level, PPRV transmission, like for other morbilliviruses, is through the respiratory tract. The source of contamination is infected animals which excrete the virus in feces, as well as nasal and ocular discharges (Abegunde and Adu 1977; Liu et al. 2015). PPRV is a fragile virus and it cannot survive for long outside the host. Its half-life has been estimated to be 3.3 h at 37 °C (Rossiter and Taylor 1994). So in hot climate conditions, PPR is transmitted mainly by close contact between susceptible animals and those excreting the virus. Results of experimental infections with PPRV showed virus excretion can start at least 2 days before the onset of any clinical sign and can last up to 21–26 post infection, at the time when infected animals have recovered (Couacy-Hymann et al. 2009; Liu et al. 2015; Pope et al. 2013; Truong et al. 2014). This means an infected animal can be a source of PPRV for at least 3 weeks after the infection even without any clinical sign.

High-risk areas for contamination are market places and common grazing points where animals of different origins have many contacts during several hours each day (Waret-Szkuta et al. 2011). Therefore, the higher the small ruminant density, the higher the risk of PPRV infection (El Arbi et al. 2014). Moreover, a classical reaction of farmers who are confronted to a PPR outbreak in their flocks is to very quickly get rid of the remaining animals by selling them in the market. Several of those animals might be infected, at the incubation stage and already excreting "silently" the virus. This results in secondary foci a few days after the introduction of such infected animals in their destination flocks.

In sub-Saharan Africa, PPR appears as a seasonal disease, its incidence being recorded in three main periods of the year (Ezeokoli et al. 1986; Obi et al. 1983a, b; Whitney et al. 1967):

- Around the period of the Muslim festival Eïd during which, for religious sacrifice, there is a dramatic increase of sheep trade and therefore an increase of pathogens dissemination by close contacts between animals. The epidemiological

Fig. 12.7 (continued) number are indicated for each PPRV sequence. The analysis was performed on MEGA 6 using the neighbor-joining method with the Tamura-Nei substitution model. The numbers at the nodes are bootstrap values obtained from 1000 replicates. Only bootstrap values >60 are shown

importance of this phenomenon was overlooked because of the lack of quantitative data. The situation starts improving thanks to the implementation of regional projects with coordinated data collection and advanced data analysis. For example, in Mauritania, a retrospective study targeted small ruminants, cattle, and camels during the period from June 2014 to May 2015 (Apolloni et al. 2018). Regarding the international trade of small ruminants, three million heads were exported mostly by foot to Senegal and Mali from March to June. In addition, 750,000 sheep were exported to Senegal mostly by truck during the few weeks before the Eid. These figures can be used in risk analyses for the introduction of infectious diseases, subsequently allowing us to design risk-based surveillance and control measures.

- During the cold dry season with the dusty wind "Harmattan" (January-February) in West Africa. Not only the relatively low temperatures constitute a stress for animals but also they may allow longer survival of the virus, increasing the risk of transmission. The dusty wind will facilitate contamination of the respiratory track by bacteria, the major complication of PPR infection. In addition, this time period coincides with the end of the parturition peak in Sahelian sheep and goats (Lesnoff 1999). Because the (putative) colostral antibodies only persist 2 or 3 months on average (Ata et al. 1989), many highly sensitive young animals are exposed to these bad weather conditions, thus making the floor to PPRV epidemics.
- At the beginning of the rainy season. After a long period of drought, animals are in poor conditions and the start of heavy rains might constitute another stress to weaken the resistance of the animals.

Like rinderpest virus, PPRV provides lifelong immunity in host and no carrier state has been demonstrated yet. Therefore, the virus can maintain itself in an area only upon constitution of a relatively important number of susceptible animals and a frequent supply of virus source. In a given region, the disease flares up in about a 3-year cyclic manner. The rate of turnover in small ruminants' population is about 30%. This means that in 3 years' time, nearly all animals in a flock are new and susceptible to PPRV if they have not been infected or vaccinated previously. This creates conditions favorable for the start of a new outbreak.

In sub-Saharan Africa, higher morbidity and mortality rates are noted in dwarf goat populations compared to the Sahelian goat breeds. This difference was attributed to the high innate resistance of the latter breeds (Rossiter and Taylor 1994). Ezeokoli et al. (1986) also proposed that the husbandry system might be responsible for the contrasted PPR epidemiology between the humid forest and the dry Sahel areas. In the latter regions, animals roam freely most of the time to find forage and water, a situation allowing frequent contacts between animals of various origins. In that case, young animals have many opportunities to experience PPRV infection at an age where they still have maternal antibodies protecting them from clinical PPR. Then they might acquire some active immunity that could be re-enforced by subsequent reinfections. In contrast to this husbandry system in the Sahelian regions, animals have less opportunity to mix in the southern humid regions and the numbers of susceptible hosts build up very quickly because of higher birth rates and higher removal/mortality rates (Ezeokoli et al. 1986; Hammami et al. 2018). In those regions, the high relative humidity may constitute better conditions for longer survival of PPRV in contaminated environment and thereby increases risk of animal infection.

Pathogenesis

As other morbilliviruses, PPRV is both lymphotropic and epitheliotropic, resulting in damage of the lymphoid and epithelial cells (Pope et al. 2013; Truong et al. 2014; Yang et al. 2018). Those cells possess the morbillivirus receptors: the SLAM protein for lymphoid cells and nectin-4 protein for epithelial cells (Tatsuo et al. 2000, 2001; Baron 2005; Adombi et al. 2011, Birch et al. 2013). The natural route of infection of these viruses is the respiratory tract. Following recent investigations on pathogenesis of PPRV infection, it was suggested that the virus, following the infection by the nasal route, is taken up by the immune cells of the respiratory mucosa, alveolar macrophages, and dendritic cells, which migrate into the T-cell- rich areas of local lymphoid organs including the tonsil for amplification before entering into the general circulation and then the epithelial cells, the digestive tract being one of the predominant virus replication sites along with the lymphoid tissues (Pope et al. 2013; Truong et al. 2014). Infections of those cells result in necrotic lesions in both lymphoid and epithelial tissues that can be observed in necropsies. The destruction of the lymphoid cells causes lymphopenia, one of the factors responsible of the profound but transient immunosuppressive effect of morbillivirus infection (Rajak et al. 2005). PPRV infection involves the modulation of the cytokine production and thus the modulation of the host innate immune response (Baron et al. 2014a, b, 2016; Sanz-Bernardo 2017; Yang et al. 2018). A consequence of that effect is the increased susceptibility of the host to opportunistic infections and increased mortality.

Clinical Signs

PPR is known mainly as an acute disease. The incubation period lasts for 5–6 days. It is broken by a sudden dullness of the animal with pyrexia and loss of appetite. Ocular and oral membranes are congested. The ocular and nasal discharges which were serous at the beginning of the disease gradually become purulent. They may stick together parts of the eyelids (ocular discharges) or may block partially the nose (nasal discharges) (Fig. 12.8) and thereby make breathing difficult. It is the time where the bronchopneumonia is well established with moist and productive cough. In the oral cavity, discrete, tiny, grayish necrotic foci develop over the reddish background. The signs are accompanied by a fever that lasts for 4–5 days. When the fever starts to drop, the necrotic spots expand and coalesce to make extensive diphtheritic plaques. These

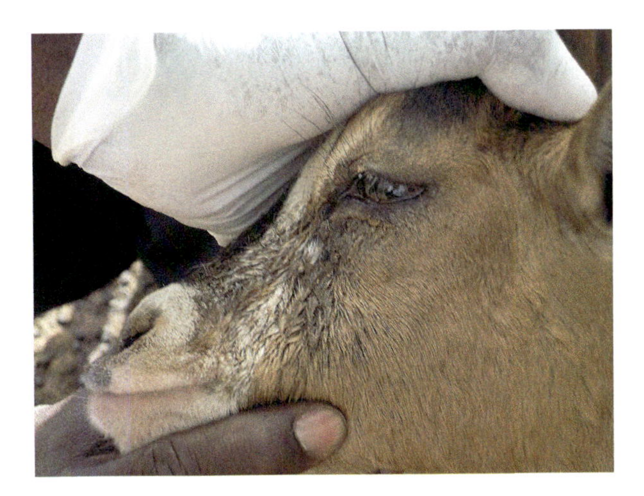

Fig. 12.8 PPR-affected goat: ocular discharges (from Salami H.)

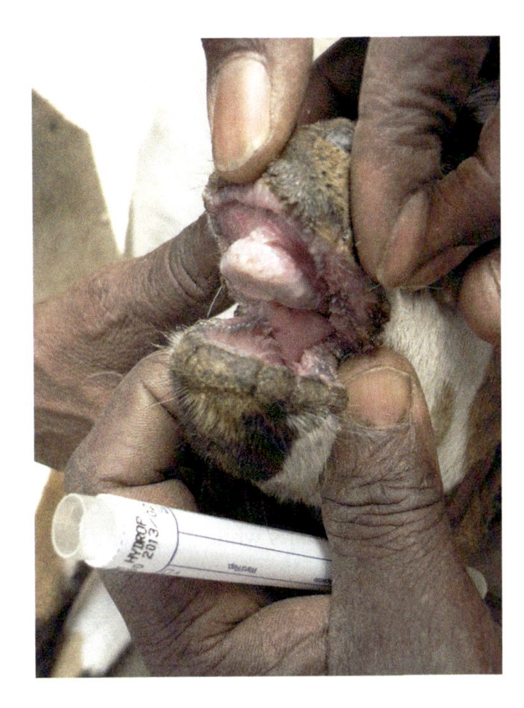

Fig. 12.9 PPR-affected goat: mouth lesions (from Salami H.)

lesions cover the tongue, the dental pad, the hard palate, and the cheeks (Fig. 12.9). Their removal will leave shallow irregular nonhemorrhagic erosive lesions. At that time, the animal is very depressed and is less and less interested by feed. It starts to pass liquid feces (Fig. 12.10), which is sometimes dysenteric. The severity of the diarrhea is correlated in many cases with the outcome of the disease. The necrotic lesions in the mouth give the animal an unpleasant and fetid odor when it breathes. They are also present on the vulvar membrane. Pregnant females abort. In 70–80% of

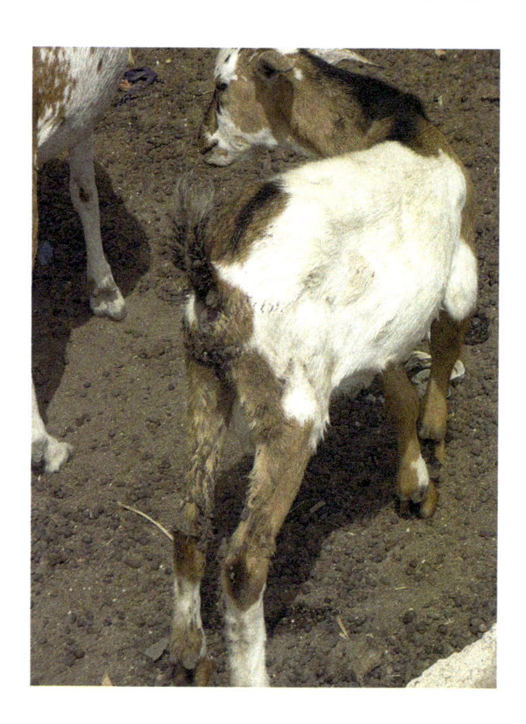

Fig. 12.10 PPR-affected goat: diarrhea (from Salami H.)

cases, the sick animals will die in 10–12 days after the onset of the disease. Those which survive will fully recover in 1-week period.

In young animals in most cases PPRV infection results in the superacute form for which the course of the disease is more rapid than the acute form. It leads 100% mortality rates. After an incubation period of about 3 days after the infection, the disease starts suddenly by high fever, the rectal temperature of the animal being between 40 and 42 °C. The animal is depressed, and it ceases eating. Mucous membranes, in particular those in the mouth and eyes, are very congested. One to two days after the onset of the disease, ocular and nasal discharges become apparent. Profuse diarrhea starts while the fever declines. In 100% of cases, animal will die in 5–6 days after the disease starts.

Another form of PPR infection is the subacute form: the symptoms are less severe than in the acute form. The diarrhea is slight and will last for 2–3 days. Ocular and nasal discharges are less abundant and will make crusts around the mouth and nostril orifices, symptoms similar to those of contagious ecthyma. The prognostic is good in that case because all affected animals always recover.

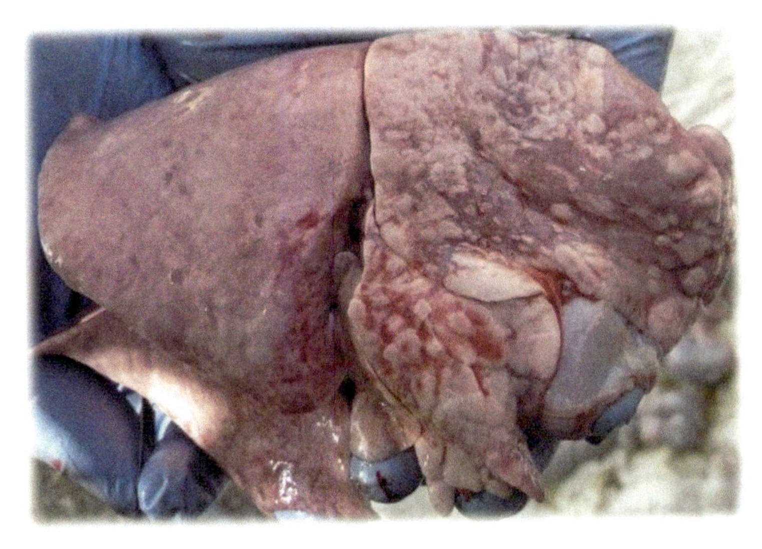

Fig. 12.11 PPR-affected goat: lung with pneumonia lesions (from Salami H.)

Lesions

At necropsy, the main striking features are the white necrotic tissue with erosive lesions in the oral cavity. In many cases, mucopurulent or frothy exudates are found in the trachea. The lung presents diffuse edema and bronchopneumonia lesions (Fig. 12.11), mainly on the anterior lobes (Taylor 1984; Truong et al. 2014). Linear hemorrhagic lesions are seen in the abomasum, the cecum, the colon, and along the folds of the rectum (Aruni et al. 1998; Bundza et al. 1988; Hamdy et al. 1976; Obi et al. 1983a, b). Because of this particular aspect, these lesions in the rectum are called zebra strips. The spleen is congested. The lymph nodes are also congested, edematous, and slightly enlarged (Aruni et al. 1998; Truong et al. 2014). Microscopic observation of hematoxylin-eosin stained cells reveals giant cells with eosinophilic, intracytoplasmic, and intranuclear inclusion bodies in the cells of many organs. Vacuolation and coagulation of cell cytoplasm together with pyknosis and karyorrhexis are present (Isitor et al. 1984). Bronchial epithelium shows squamous cell metaplasia. The alveolar septae are thickened and infiltrated by mononuclear cells. In the alimentary tract, the degenerated epithelium is eroded leaving scanty debris above the stratum germinativum. The lymph nodes are depleted from lymphoid cells. Sometimes they present areas of necrosis.

Diagnosis

There is no single pathognomonic clinical sign of PPR and in many cases not all signs are present in a given sick animal. Therefore, clinical diagnosis is done considering all signs observed in sick animals present in the affected flock. Information of epidemiology of the outbreak will also help in the diagnosis. For a well-informed person, the clinical diagnosis may be easy. Such a diagnosis is provisional until confirmation by a laboratory test, in particular in case of an outbreak in a new area. This confirmation is important because PPR can be confused clinically with many other small ruminant diseases. The differential diagnosis must be made between PPR and diseases with stomatitis lesions, diseases with enteritis symptoms, and diseases with respiratory distress. In the past, there was confusion with rinderpest. As that disease is now eradicated worldwide, the differential diagnosis of PPR is now done with other small ruminant diseases:

1. Diseases with stomatitis lesions such as foot and mouth disease and bluetongue. Upon closer examination of the oral cavity, the absence of foot lesions in the case of PPR will allow ruling out these two similar diseases.
2. Diseases with enteritis signs such as diarrhea caused by parasites (nematodes and coccids in the case of hemorrhagic feces) or by bacteria (*E. coli*, *Salmonella*). Latent infections with those pathogens might turn to overt disease after the animal infection by PPRV, an immunosuppressive agent (Adetosoye and Ojo 1983; Obi et al. 1983a, b);
3. Diseases with respiratory distress. With the respiratory difficulty that is met in all acute cases, PPR can be confused with lung diseases such as capripox (goat pox and sheep pox) or ecthyma but in those cases there are skin lesions. Contagious caprine pleuropneumonia (CCPP) is another respiratory disease to be differentiated from PPR. It is widespread in many sub-Saharan African countries and in the Middle East. However, in this disease caused by a mycoplasma, there are neither mucosal lesions nor diarrhea. The main respiratory disease which has obscured the diagnosis of PPR in many countries for a long time is pasteurellosis. Indeed, *Pasteurella haemolytica* but also *Staphylococcus sp.*, *Proteus sp.*, *Pseudomonas aeruginosa*, *Proteus sp.* have been isolated from nasal discharges or lung samples collected from PPR-affected animals. The presence of those bacteria is a result of secondary infections and they are responsible for bronchopneumonia.

For the confirmation of the PPR diagnosis, the samples to be submitted to the laboratory are:

- From live animals, ocular, nasal, and oral swabs. Samples must be collected from at least 3–4 animals.
- From dead animals (if possible animal euthanized at the fever period): lymph nodes, pieces of lung, gut mucosa, and spleen should be collected.

The collected samples must be dispatched to the laboratory in cold conditions. They will be submitted to antigen or nucleic acid detection and eventually virus isolation.

Blood samples must be collected from many animals in the suspected flocks for PPR antibody detection.

Today, the most used, rapid, and specific test for the detection of PPR antigen is the immunocapture ELISA (Libeau et al. 1994; Diallo et al. 1995). It is very sensitive, easy to perform, and rapid because it can be completed in 2 h time.

Hemagglutination test has been applied to PPRV identification as this virus, contrary to rinderpest virus, has the hemagglutination capability (Wosu 1991). This is a cheap and easy to use test that may be useful for laboratories of low settings.

The nucleic acid-based diagnostic technique currently used for PPR diagnosis is the polymerase chain reaction (PCR): classical and quantitative reverse transcription PCR (QRT-PCR). It is the most sensitive and specific test for PPRV identification (Forsyth and Barrett 1995; Couacy-Hymann et al. 2002; Balamurugan et al. 2006, 2012; George et al. 2006; Bao et al. 2008; Kwiatek et al. 2010; Batten et al. 2011; Baron et al. 2016). LAMP-PCR was also developed with the objective to be used as field test in the future (Li et al. 2010).

Immuno-chromatography tests are currently available for field use (Baron et al. 2014a, b; IDvet, personal communication).

It is also possible to directly analyze the pathological tissue and swabs by immunofluorescence or immunochemical test after their fixation in formalin solution (Brown et al. 1991; Bundza et al. 1988; Sumption et al. 1998).

The gold standard test for PPR diagnosis is the virus isolation. Swab material in suspension in PBS or 10% tissue suspension is inoculated onto primary lamb kidney/lung cell or Vero cell. The last one is a cell line easy to maintain in culture and is common in virology laboratories. It is the most used cell for PPRV isolation in vitro. But the chance of isolating wild-type PPRV in Vero cells from pathological samples is low. Because of that inconvenience and following the discovery of the natural morbillivirus receptors, the SLAM and nectin-4, new cell lines expressing these proteins have been developed following genetic engineering. With those modified cells expressing either the goat/sheep SLAM (Adombi et al. 2011), the dog SLAM (Baazizi et al. 2017), or human nectin-4 (Fakri et al. 2016), isolation of PPRV from pathological has been dramatically improved.

Finally, the presence of PPRV infection can be made by serology. For animal trade purpose, the OIE prescribed test for PPR antibody detection is the virus neutralization test (VNT) (Rossiter et al. 1985). It is described in the OIE *Terrestrial Manual*. In the past, and because of the cross reaction existing between rinderpest and PPR viruses, the VNT was performed simultaneously against both viruses. The higher titer with a serum will indicate the homologous virus (Taylor 1979). But now rinderpest being successfully eradicated worldwide, the cross-neutralization test with RPV is not needed anymore. Moreover, handling of RPV is very restricted and is limited to few OIE/FAO authorized high security laboratories. Although VNT is the OIE prescribed test, it is not suitable for testing a high number of sera at the same time. Thus, serum testing for the detection of PPR antibodies is performed by use of competitive ELISA tests that are allowed by the OIE as alternative test for international trade. They are based on the use of monoclonal antibodies anti-PPRV N or H protein (Anderson and McKay 1994; Libeau et al. 1995; Saliki et al. 1993; Bodjo et al. 2018).

Economic Importance

Sheep and goats, and in particular goats, are important for the livelihoods of small farmers in the developing world. Therefore, PPR, a contagious disease which is threatening the productivity of those animals, appears as a major constraint to food security in all areas where it occurs. In a report of an international survey conducted in 2001 in Africa and Asia to identify priorities in animal health research for potential benefits to the poor in developing countries, PPR is rated in the list of the top ten priority animal diseases to be considered for poverty alleviation in sub-Saharan Africa and South Asia (Perry et al. 2002). This study has clearly highlighted the economic importance of PPR and its impact on the livelihood of the poor in many developing countries. Already in 1976, Hamdy et al. (1976) evaluated at USD1.5 million the annual loss induced by PPR in Nigeria. In 1993, Stem (1993) published results of his study on PPR consequence in Niger. He concluded that investment of USD2 million in PPR vaccination would generate USD24 million as return. Now there is an important body of publications on the economic impact of PPR (Abubakar and Munir 2014; Opasina and Putt 1985; Singh et al. 2004, 2014; Awa et al. 2000; Jones et al. 2016). The most important document on PPR economic impact is certainly the document elaborated by FAO and OIE for the Global Control and Eradication of PPR (Anonymous 2015). In that document, experts have estimated that PPR is causing an annual loss of 1.2–1.7 billion of USD in endemic countries worldwide. This value is certainly underestimated today as PPR has continued to expand since the publication of the document.

Treatment/Control

Treatment

Treatment of PPR-affected animals by the administration of anti-PPR serum or by antibiotics in association with anti-diarrhea medicines has been mentioned by some authors (Ihemelandu et al. 1985). Such reports might not have a practical interest as it is too expensive in regard to the price of individual sheep and goats. Synthetic short interfering RNAs (siRNAs) have been tested on PPRV-infected cells in vitro as an approach of PPR treatment (Servan de Almeida et al. 2007; Keita et al. 2008). However, despite the good performance of this technology to shut down PPRV replication by 10,000-fold in vitro, its use in vivo seems to be complicated and not possible for the moment.

Sanitary Measures

All sheep and goats of the affected stock should be under quarantine until at least 1 month after the last clinical case (Rossiter and Taylor 1994). The animal movements must be strictly controlled in the area of the infection. Unfortunately, all these sanitary measures are difficult to maintain in countries where PPR is endemic. Therefore, the only effective way to control PPR is the vaccination.

Control: Vaccination

Current Vaccines Since PPRV and RPV are closely related viruses, the attenuated rinderpest vaccine was used in the past as heterologous vaccine to protect against PPR (Bourdin et al. 1970; Taylor 1979). However, following the success of the global rinderpest eradication, the use of that vaccine is strictly forbidden. There are live attenuated PPR vaccines that, like rinderpest vaccine, provide also lifelong immunity in sheep and goats (Liu et al. 2014; Diallo and Singh, manuscript submitted for publication). Indeed, despite the profound immunosuppression morbilliviruses may induce, this effect is transient and the recovery from the disease is usually followed by the establishment of a strong, specific, and long-term protective immune response of the host (Servet-Delprat et al. 2003; Rajak et al. 2005; Cosby et al. 2006). Morbillivirus-attenuated vaccines seem to have less immune suppression capacity compared to wild type but have conserved the strong protective characteristic (Cosby et al. 2006).

Currently, there are at least six PPR live attenuated vaccines (Diallo and Singh, manuscript submitted for publication): PPRV Nigeria 75/1 (Nigeria), PPRV Sungri 96 (India), PPRV Arasur 87 (India), PPRV Coimbatore (India), PPRV Titu (Bangladesh), and PPRV 45G37/35-K (Kazakhstan). Among all these attenuated PPRV vaccines, the PPRV Nigeria 75/1, lineage II, and the PPRV Sungri 96, lineage IV, are currently the most used vaccines. They are the strains for which most of the information on PPR vaccine is available (Diallo et al. 1989b; Diallo 2004; Singh et al. 2009; Singh 2012; Singh and Bandyopadhyay 2015). They provide protection in inoculated animals for at least 3 years (Zahur et al. 2015). Both vaccines have been extensively tested and used in the field, experiences which have proved their efficiency in protecting sheep and goats against infections against all PPRV wild types currently known, whatever the lineage they belong to.

Once produced, the current PPR live attenuated vaccine should be maintained in a cold chain from manufacturer premises until delivery to animals in the field as it is a thermolabile product. Unfortunately, most of countries where PPR is endemic are in regions of hot climate and have poor infrastructure with inconstant electric power supply. This issue has been addressed by many investigators by ever improving the freeze-drying procedure in using cryoprotectants or diluents (Worrwall et al. 2000; Sarkar et al. 2003; Sen et al. 2014; Riyesh et al. 2011; Silva et al. 2011, 2014). With those improvements, it was possible to keep the vaccine at 45 °C for at least 14 days with minimal loss of potency (Worrwall et al. 2000).

Recombinant-Based PPR Vaccines The current PPR live attenuated vaccines do not allow differentiating infected animals from those that have been vaccinated. Thus, the sero-epidemiosurveillance of the disease cannot be carried out in endemic areas where a vaccination program has been or is being implemented. A way to combine vaccination and disease sero-surveillance activities for the better management of the disease would be the use of a vaccine which enables differentiation between infected and vaccinated animals (DIVA vaccine). With the advent of recombinant DNA technology, different approaches are being followed to develop effective PPR marker vaccines to enable such differentiation and which would allow countries to implement both vaccination and disease surveillance programs at the same time (Diallo et al. 2007).

A way to develop PPR marker vaccine is the use of a vector to express PPRV immunizing proteins, the F and H proteins. This type of vaccine does not contain the PPRV N protein. Therefore, animals that are inoculated with it do not produce antibodies anti-PPRV N and would be differentiated with PPRV-infected animals by a PPRV N-based serological test. For that purpose, the live attenuated capripox virus vaccine has been used as a vector and the obtained recombinant is a potential dual vaccine to protect animals against two important small ruminant diseases: PPR and capripox. (Diallo et al. 2002, 2007; Berhe et al. 2003; Chen et al. 2010; Caufour et al. 2014). Adenovirus and plant have also been used as vectors for the expression of PPRV H protein to develop PPR recombinant vaccines (Prasad et al. 2004; Khandelwal et al. 2011; Qin et al. 2012; Wang et al. 2013; Rojas et al. 2014; Herbert et al. 2014).

Another way that has been used to develop PPR DIVA vaccine is to engineer an attenuated PPRV vaccine in such a way to introduce the marker into its genome. Such a vaccine in which the hemagglutinin protein has been modified is now available (Muniraju et al. 2015). However, that vaccine as all other recombinant PPR vaccines indicated above are not available for use yet. Indeed, most of them are under evaluation.

Control and Eradication Strategy The control and eventually its eradication are based on mass vaccination of small ruminant populations in endemic countries. It is estimated that a 70% post vaccination immunity rate PIR is needed in a given epidemiological unit to prevent PPR virus spread (OIE and FAO 2015). However, implementing mass vaccination is difficult and costly in smallholder farming systems with scattered livestock and limited facilities. Regarding this, controlling PPR is a special challenge in Sahelian Africa. Hammami et al. (2018) used a modeling approach to assess the effect of several variables on PIR in two contrasted smallholder farming systems: (1) goats reared in subhumid areas (ca. 900 mm of annual rainfall) with high reproduction, mortality, and offtake rates and (2) sheep reared in semiarid areas (ca. 250 mm of annual rainfall), with lower reproduction, offtake, and mortality rates. The PVIR dynamic was simulated over a 4-year vaccination program thus corresponding to a recommended duration for the control stage. Because of the huge difference in population dynamics, the annual vaccination campaign was made of a single round of vaccination in semiarid environment, and two rounds in subhumid area. The ability of different vaccination scenarios to reach the 70% PVIR was assessed throughout the program. The tested scenarios differed in (1) their overall schedule, (2) their delivery month, and (3) their vaccination coverage.

In sheep reared in semiarid areas, the vaccination month did affect the PVIR decay (best month for vaccination generally in September) though it did not in goats in subhumid regions. In both cases, the study highlighted (1) the importance of targeting the whole eligible population at least during the first 2 years of the vaccination program and (2) the importance of reaching a vaccination coverage as high as 80% of this population. This study confirmed the relevance of the vaccination schedules recommended by international organizations.

In Sahelian Africa and in the corresponding coastal areas of West Africa, there is strong evidence that PPRV transmission is intense everywhere from arid to humid agro-ecosystems. Therefore, in this area, there is no other option than mass vaccination, campaign coordinated at the regional level (Lancelot et al. 2014). However in other locations, the lack of knowledge about the transmission potential of PPRV may compromise eradication efforts. By fitting a metapopulation model simulating PPRV spread to the results of a nationwide serological survey in Ethiopia, Fournié et al. (2018) estimated the level of viral transmission in an endemic setting and the vaccination coverage required for elimination. Results suggest that the pastoral production system as a whole acts as a viral reservoir, from which PPRV spills over into the sedentary production system, where viral persistence is uncertain. Estimated levels of PPRV transmission indicate that viral spread could be prevented if the proportion of immune small ruminants is kept permanently above 37% in at least 71% of pastoral village populations. However, due to the high turnover of these populations, maintaining the fraction of immune animals above this threshold would require high vaccine coverage within villages and vaccination campaigns to be conducted annually. Adapting vaccination strategies to the specific characteristics of the local epidemiological context and small ruminant population dynamics would result in optimized allocation of limited resources and increase the likelihood of PPR eradication.

For that purpose, a general framework was defined by OIE and FAO (Anonymous 2015). It is made of four stages implemented at a national level with a regional and global coordination: (1) assessment of the epidemiological situation, (2) control (vaccination), (3) eradication, and (4) post-eradication.

1. In Stage 1, the epidemiological situation of PPR is thoroughly assessed: spatial distribution and its drivers (animal densities, livestock trade network, etc.), molecular epidemiology, as well as the study of the sociotechnical networks to be accounted for at the control stage. The epidemiological investigations and

surveillance are an official public veterinary services (VS) responsibility and are public goods.

2. In Stage 2, the control activities, particularly vaccination, are implemented or supervised by the VS in the targeted geographical areas or production systems. This is done via a public–private partnership, where possible, and in line with the national control plan. At this stage, the goal is to stop the transmission of PPRV using mass vaccination of the target small ruminant populations during a time period ranging from 2 to 4 years in general. In parallel, a nation-wide PPR surveillance program is implemented to:

    (a). Assess the post-vaccination immunity rate in the targeted small ruminant population.
    (b). Assess the reduction of PPR clinical incidence in this population.
    (c). Detect any PPR foci outside of the targeted population (in cases where PPR-free areas had been defined).

3. At the eradication stage, stage 3, area-wide mass vaccination is stopped and reinforced risk-based surveillance is implemented. If a PPR focus is detected, the restriction of sheep and goat movements applies, and ring vaccination is immediately triggered. Indeed, any health events that could be related to PPRV need to be promptly detected and reported and appropriate measures immediately put in place to control them. The country must develop and have the capacity to implement the contingency plan that forms part of the eradication strategy. If a new risk of introducing PPRV in the area or production system arises, the results of the surveillance system and epidemiological analysis must identify and qualify the risks and appropriate measures should be rapidly implemented to mitigate the risk of introduction. The recommended duration of stage 3 is 3 years on average (from 2 to 5 years).
4. Entry into Stage 4 means that a country is ready to start implementing a full set of activities that should lead to it being recognized as officially free from PPR. Eradication and prevention measures are based on early detection and reporting of any new outbreak occurrence, emergency response, and contingency planning. Vaccination is prohibited. If emergency vaccination needs to be implemented, the country or the vaccinated zone ("zone" as defined in the OIE Terrestrial Code) is downgraded to Stage 3.

Conclusion: The FAO/OIE PPR Global Control and Eradication Strategy

After the declaration of the successful global rinderpest eradication in 2011, many experts called for the launch of a similar program for the global eradication of PPR (Albina et al. 2013; Anderson et al. 2011; Baron et al. 2011; Thomson and Penrith 2017). In 2015, about 65 countries, with more than 1.5 billion of small ruminants, were recorded as PPR endemic countries (Anonymous 2015; Mariner et al. 2016). When considering countries at risk, it can be stated that about two billion of sheep, goats, and wild small ruminants are at risk of PPR. Considering the high importance of sheep and goats for the poor in the developing world, and the dramatic expansion of PPR endemic zones in the past 10 years, FAO and OIE jointly convened an international conference on PPR in April 2015 in Abidjan, Côte d'Ivoire. At that conference, a Strategy for the Global Control and Eradication of PPR was adopted by 2030 (Anonymous 2015). This strategy was developed by taking into consideration lessons learnt from the success of the Global Rinderpest Eradication (GREP). Technical features which have allowed that success are shared with PPR: contamination by close contact, no vector in the epidemiological cycle, no carrier state after infection, and strong and lifelong immunity following recovery from infection or after vaccination. Despite these common technical characteristics favorable to the control and eradication of an infectious disease, eradication of a small ruminant disease will need addressing many challenges:

1. The individual marketing value of a goat is far less than that of cattle and this fact is not attractive for funding. However, the social value of goat, its importance in meeting the needs of the poor, is higher than that of cattle.
2. The high number of animals, between 1.5 and 2 billion, to be targeted in the PPR eradication program; this is more than 3 times the number of cattle of GREP.
3. The fact that small ruminant populations are far scattered than those of cattle.
4. The high turnover rate of small ruminant flocks, 30–35%, compared to cattle herds, about 10%, which needs an intensification of the vaccination activities, twice a year in some cases as planned in the PPR-GCEP. It is estimated that in view of eradication of the circulation of PPRV, the herd immunity should be at least 70% for at least 3 years (Anonymous 2015; Fournié et al. 2018; Hammami et al. 2018).

Because of the three last challenges indicated above, the PPR eradication program will be far more expensive than GREP, a cost that may make that program not attractive for fund donors. The availability of required funds for GREP that was ensured through strong political commitments of countries, international organizations, and donors has been a key element in the success of GREP. Such commitments are needed for PPR-GCES to save a vital asset of the poor, small ruminant production, and thereby to achieve one of the Millennium Development Goals (MDG), the reduction of the poverty in the world.

References

Abegunde AA, Adu F. Excretion of the virus of peste des petits ruminants by goats. Bull Anim Health Prod Afr. 1977;25:307–11.

Abraham G, Sintayehu A, Libeau G, et al. Antibody seroprevalences against peste des petits ruminants (PPR) virus in camels, cattle, goats and sheep in Ethiopia. Prev Vet Med. 2005;70:51–7.

Abubakar M, Jamal SM, Arshed MJ, et al. Peste des petits ruminants virus (PPRV) infection; its association with species, seasonal variations and geography. Trop Anim Health Prod. 2009;41:1197–202. https://doi.org/10.1007/s11250-008-9300-9.

Abubakar M, Rajput ZI, Arshed MJ, Sarwar G, Ali Q. Evidence of peste des petits ruminants virus (PPRV) infection in Sindh Ibex (Capra aegagrus blythi) in Pakistan as confirmed by detection of antigen and antibody. Trop Anim Health Prod. 2011;43:745–7.

Abubakar M, Munir M. Peste des Petits ruminants virus: an emerging threat to goat farming in Pakistan. Transbound Emerg Dis. 2014;61(Suppl 1):7–10. https://doi.org/10.1111/tbed.12192.

Abubakar M, Rajput ZI, Arshed MJ, et al. Evidence of peste des petits ruminants virus (PPRV) infection in Sindh Ibex (Capra aegagrus blythi) in Pakistan as confirmed by detection of antigen and antibody. Trop Anim Health Prod. 2015;43:745–7.

Adetosoye AI, Ojo MO. Characteristic of *Escherichia coli* isolated from goats suffering from peste des petits ruminants. Trop Anim Health Prod Afr. 1983;16:119–22.

Adombi CM, Mamadou L, Charles EL, Shamaki D, Koffi YM, Traoré A, Silber R, Couacy-Hymann E, Bodjo SC, Djaman JA, Luckins AG, Diallo A. Monkey CV1 cell line expressing the sheep–goat SLAM protein: a highly sensitive cell line for the isolation of peste des petits ruminants virus from pathological specimens. J Virol Methods. 2011;173:30.

Albina E, Kwiatek O, Minet C, et al. Peste des petits ruminants, the next eradicated animal disease? Vet Microbiol. 2013;165:38.

Al-Majali AM, Hussain NO, Amarin NM, et al. Seroprevalence of, and risk factors for, peste des petits ruminants in sheep and goats in Northern Jordan. Prev Vet Med. 2008;85(1–2):1–8. https://doi.org/10.1016/j.prevetmed.2008.01.002.

Almeshay MD, Gusbi A, Eldaghayes I, Mansouri R, Bengoumi M, Dayhum AS. An epidemological study on Peste des petits ruminants in Tripoli Region, Lybia. Vet Ital. 2017;53:235–42. https://doi.org/10.12834/VetIt.964.5025.3.

Anderson J, McKay JA. The detection of antibodies against peste des petits ruminants virus in cattle, sheep and goats and the possible implications to rinderpest control programmes. Epidemol Infect. 1994;112:225–31.

Anderson J, Baron M, Cameron A, et al. Rinderpest eradicated; what next? Vet Rec. 2011;169:10.

Anonymous. PPR control and global strategy. 2015. http://www.fao.org/3/a-i4460e.pdf

Apolloni A, Nicolas G, Coste C, et al. Towards the description of livestock mobility in Sahelian Africa: some results from a survey in Mauritania. PLoS One. 2018;13:e0191565.

Aruni AW, Lalitha PS, Mohan AC, et al. Histopathological study of a natural outbreak of peste des petits ruminants in goats of Tamilnadu. Small Rum Res. 1998;28:233–40.

Ata FA, Al Sumry HS, King GJ, et al. Duration of maternal immunity to peste des petits ruminants. Vet Rec. 1989;124:590–1.

Awa DN, Njoya A, Tama ACN. Economics of prophylaxis against Peste des Petits ruminants and gastrointestinal helminthosis in small ruminants in North Cameroon. Trop Anim Health Prod. 2000;32:391–403.

Ayari-Fakhfakh E, Ghram A, Bouattour A, et al. First serological investigation of peste-des-petits-ruminants and Rift Valley fever in Tunisia. Vet J. 2011;187:402.

Aziz-Ul-Rahman, Wensman JJ, Abubakar M, et al. Peste des petits ruminants in wild ungulates. Trop Anim Health Prod. 2018;50(8):1815–9. https://doi.org/10.1007/s11250-018-1623-6.

Baazizi R, Mahapatra M, Clarke BD, et al. Peste des petits ruminants (PPR): a neglected tropical disease in Maghreb region of North Africa and its threat to Europe. PLoS One. 2017;12(4): e0175461. https://doi.org/10.1371/journal.pone.0175461.

Bailey D, Banyard A, Dash P, et al. Full genome sequence of peste des petits ruminants virus, a member of the Morbillivirus genus. Virus Res. 2005;110:119–24.

Balamurugan V, Sen A, Saravanan P, et al. One-step multiplex RT-PCR assay for the detection of peste des petits ruminants virus in clinical samples. Vet Res Commun. 2006;30:655.

Balamurugan V, Sen A, Venkatesan G, et al. A rapid and sensitive one step-SYBR green based semi quantitative real time RT-PCR for the detection of peste des petits ruminants virus in the clinical samples. Virol Sin. 2012;27:1.

Bao J, Li L, Wang Z, Barrett T, et al. Development of one-step real-time RT-PCR assay for detection and quantitation of peste des petits ruminants virus. J Virol Methods. 2008;148:232.
Bao J, Wang Z, Li L, et al. Detection and genetic characterization of peste des petits ruminants virus in free-living bharals (*Pseudois nayaur*) in Tibet, China. Res Vet Sci. 2011;90:238–40.
Baron MD. Wild-type rinderpest virus uses SLAM (CD150) as its receptor. J Gen Virol. 2005;86:1753.
Baron MD, Parida S, Oura CA. Peste des petits ruminants: a suitable candidate for eradication? Vet Rec. 2011;169:16.
Baron J, Bin-Tarif A, Herbert R, et al. Early changes in cytokine expression in peste des petits ruminants disease. Vet Res. 2014a;45:22.
Baron J, Fishbourne E, Couacy-Hyman E, et al. Development and testing of a field diagnostic assay for peste des petits ruminants virus. Transbound Emerg Dis. 2014b;61:390.
Baron MD, Diallo A, Lancelot R, et al. Peste des petits ruminants virus. Adv Virus Res. 2016;95:1–42. https://doi.org/10.1016/bs.aivir.2016.02.001.
Batten CA, Banyard AC, King DP, et al. A real time RT-PCR assay for the specific detection of peste des petits ruminants virus. J Virol Methods. 2011;171:401.
Berhe G, Minet C, Le Goff C, et al. Development of a dual recombinant vaccine to protect small ruminants against peste des petits ruminants and capripox infections. J Virol. 2003;77:1571–7.
Bidjeh K1, Bornarel P, Imadine M, et al. First- time isolation of the peste des petits ruminants (PPR) virus in Chad and experimental induction of the disease. Rev Elev Med Vet Pays Trop. 1995;48 (4):295–300.
Birch J, Juleff N, Heaton MP, et al. Characterization of ovine Nectin-4, a novel peste des petits ruminants virus (PPRV) receptor. J Virol. 2013;87(8):4756.
Bodjo SC, Baziki JD, Nwankpa N, et al. Development and validation of an epitope-blocking ELISA using an anti-haemagglutinin monoclonal antibody for specific detection of antibodies in sheep and goat sera directed against peste des petits ruminants virus. Arch Virol. 2018;163 (7):1745–56.
Bourdin P. La peste des petits ruminants (PPR) et sa prophylaxie au Sénégal et en Afrique de l'ouest. Rev Elev Med Vet Pays Trop. 1973;26:71a–4a.
Bourdin P, Laurent-Vautier A. Note sur la structure du virus de la peste des petits ruminants. Rev Elev Méd Vét Pays Trop. 1967;20:383–6.
Bourdin P, Rioche M, Laurent A. Emploi d'un vaccin anti-bovipestique produit sur cultures cellulaires dans la prophylaxie de la peste des petits ruminants au Dahomey: notes préliminaires. Rev El Méd Vét Pays Trop. 1970;23:295–300.
Brown CC, Mariner JC, Olander HJ. An immunohistochemical of the pneumonia caused by peste des petits ruminants virus. Vet Pathol. 1991;28:166–70.
Bundza A, Afshar A, Dukes TW, et al. Experimental peste des petits ruminant (goat plague) in goats and sheep. Can J Vet Res. 1988;52:46–52.
Caufour P, Rufael T, Lamien CE, et al. Protective efficacy of a single immunization with capripoxvirus-vectored recombinant peste des petits ruminants vaccines in presence of pre-existing immunity. Vaccine. 2014;32:3772.
Cêtre-Sossah C, Kwiatek O, Faharoudine A, et al. Impact and epidemiological investigations into the incursion and spread of peste des petits ruminants in the Comoros Archipelago: an increased threat to surrounding islands. Transbound Emerg Dis. 2016;63:452–9. https://doi.org/10.1111/tbed.12296.
Chen W, Hu S, Qu L, Hu Q, Zhang Q, Zhi H, Huang K, Bu Z. A goat poxvirus-vectored peste-des-petits-ruminants vaccine induces long-lasting neutralization antibody to high levels in goats and sheep. Vaccine. 2010;28:4742–50. https://doi.org/10.1016/j.vaccine.2010.04.102.
Cosby SL, Chieko K, Yamanouchi K. Immunology of rinderpest—an immunosuppression but a lifelong vaccine protection. In: Barrett T, Pastoret PP, Taylor WP, editors. Rinderpest and peste des petits ruminants virus: plague of large and small ruminants; Biology of animal infections. London: Academic Press; 2006. p. 197–221.

Couacy-Hymann E, Roger F, Hurard C, et al. Rapid and sensitive detection of peste des petits ruminants virus by a polymerase chain reaction assay. J Virol Meth. 2002;100:17–25.

Couacy-Hymann E, Bodjo C, Danho T, et al. Surveillance of wildlife as a tool for monitoring rinderpest and peste des petits ruminants in West Africa. Rev Sci Tech. 2005 Dec;24(3):869–77.

Couacy-Hymann E, Bodjo SC, Koffi MY, et al. The early detection of peste-des-petits-ruminants (PPR) virus antigens and nucleic acid from experimentally infected goats using RT-PCR and immunocapture ELISA techniques. Res Vet Sci. 2009;87(2):332–5.

Dayhum A, Sharif M, Eldaghayes I, Kammon A, Calistri P, Danzetta ML, Di Sabatino D, Petrini A, Ferrari G, Grazioli S, Pezzoni G, Brocchi E. Sero-prevalence and epidemiology of peste des petits ruminants in Libya. Transbound Emerg Dis. 2018;65(1):e48–54. https://doi.org/10.1111/tbed.12670.

De Nardi M, Lamin Saleh SM, Batten C, et al. First evidence of peste des petits ruminants (PPR) virus circulation in Algeria (Sahrawi territories): outbreak investigation and virus lineage identification. Transbound Emerg Dis. 2012;59:214.

Delil F, Asfaw Y, Gebreegziabher B. Prevalence of antibodies to peste des petits ruminants virus before and during outbreaks of the disease in Awash Fentale district, Afar, Ethiopia. Trop Anim Health Prod. 2012;44(7):1329–30. https://doi.org/10.1007/s11250-012-0110-8.

Diallo A. Control of Peste des Petits ruminants: vaccination for the control of Peste des Petits ruminants. Dev Biol (Basel). 2004;119:93–8.

Diallo A, Libeau G. Peste des petits ruminants. In: Liu D, editor. Manual of security sensitive microbes and toxins. Boca Raton: CRC Press; 2014. p. 703–14.

Diallo A, Barrett T, Lefèvre PC, et al. Comparison of proteins induced in cells infected with rinderpest and peste des petits ruminants viruses. J Gen Virol. 1987;68:2033–8.

Diallo A, Barrett T, Barbron M, et al. Differentiation of rinderpest and peste des petits ruminants viruses using specific cDNA clones. J Virol Meth. 1989a;23:127–36.

Diallo A, Taylor WP, Lefèvre PC, et al. Atténuation d'une souche de virus de la peste des petits ruminants: candidat pour un vaccin homologue vivant. Rev Elev Méd Vét Pays Trop. 1989b;42:311–9.

Diallo A, Barrett T, Barbron M, et al. Cloning of the nucleocapsid protein gene of the peste-des-petits-ruminants virus: relationship to other morbilliviruses. J Gen Virol. 1994;75:233–7.

Diallo A, Libeau G, Couacy-Hymann E, et al. Recent developments in the diagnosis of rinderpest and peste des petits ruminants. Vet Microbiol. 1995;44:307–17.

Diallo A, Minet C, Berhe G, et al. Goat immune response to capripox vaccine expressing the hemagglutinin protein of peste des petits ruminants. Ann N Y Acad Sci. 2002;969:88–91.

Diallo A, Minet C, Le Goff C, et al. The threat of peste des petits ruminants: progress in vaccine development for disease control. Vaccine. 2007;25:5591–7.

Diop M, Sarr J, Libeau G. Evaluation of novel diagnostic tools for peste des petits ruminants virus in naturally infected goat herds. Epidemiol Infect. 2005;133(4):711–7.

Donduashvili M, Goginashvili K, Toklikishvili N, et al. Identification of peste des petits ruminants virus, Georgia, 2016. Emerg Infect Dis. 2018;24(8):1576–8. https://doi.org/10.3201/eid2408.170334.

Durojaiye OA, Taylor WP, Smale C. The ultastructure of peste des petits ruminants virus. Zentralbl Veterinärmed. 1985;32:460–5.

Durtnell RE. A diseases of Sokoto goats resembling "peste des petits ruminants". Trop Anim Health Prod. 1972;4(3):162–4.

El Arbi AS, El Mamy AB, Salami H, et al. Peste des petits ruminants, Mauritania. Emerg Infect Dis. 2014;20:334–6.

El Hag Ali B. A natural outbreak of rinderpest involving sheep, goats and cattle in Sudan. Bull Epizoot Dis Afr. 1973;12:421.

El Hag Ali B, Taylor WP. Isolation of peste des petits ruminants virus from the Sudan. Res Vet Sci. 1984;36:1–4.

El Hag Ali B, Taylor WP. An investigation on rinderpest virus transmission and maintenance by sheep, goats and cattle. Bull Anim Hlth Prod Afr. 1988;36:290–4.

Ezeokoli CD, Umoh JU, Chineme CN, et al. Clinical and epidemiological features of peste des petits ruminants in Sokoto red goats. Rev Elev Méd Vét Pays Trop. 1986;39:269–73.
Fakri F, Elarkam A, Daouam S, et al. VeroNectin-4 is a highly sensitive cell line that can be used for the isolation and titration of peste des petits ruminants virus. J Virol Methods. 2016;228:135–9. https://doi.org/10.1016/j.jviromet.2015.11.017.
Forsyth MA, Barrett T. Evaluation of polymerase chain reaction for the detection and characterization of rinderpest and peste des petits ruminants viruses for epidemiological studies. Virus Res. 1995;39:151–63.
Fournié G, Waret-Szkuta A, Camacho A, et al. A dynamic model of transmission and elimination of peste des petits ruminants in Ethiopia. Proc Natl Acad Sci U S A. 2018;115:8454–9.
Furley CW, Taylor WP, Obi TU. An outbreak of peste des petits ruminants in a zoological collection. Vet Rec. 1987;121:443–7.
Gargadennec L, Lalanne A. La peste des petits ruminants. Bull Serv Zoot Epiz AOF. 1942;5:16–21.
George A, Dhar P, Sreenivasa BP, Singh RP, Bandyopadhyay SK. The M and N genes-based simplex and multiplex PCRs are better than the F or H gene based simplex PCR for Peste-des-petits-ruminants virus. Acta Virol. 2006;50:217–22.
Gibbs EPJ, Taylor WP, Lawman MPJ, et al. Classification of peste des petits ruminants virus as the fourth member of the Genus *Morbillivirus*. Intervirology. 1979;11:268–74.
Govindarajan R, Koteeswaran A, Venugopalan AT, et al. Isolation of peste des petits ruminants virus (PPRV) from an outbreak in Indian Buffalo (*Bubalus bubalis*). Vet Rec. 1997;141:573–4.
Hamdy FM, Dardiri AH, Nduaka O, Breese SS Jr, Ihemelandu EC. Etiology of the stomatitis pneumoenteritis complex in Nigerian dwarf goats. Can J Comp Med. 1976;40:276–84.
Hammami P, Lancelot R, Domenech J, et al. Ex-ante assessment of different vaccination-based control schedules against the peste des petits ruminants virus in sub-Saharan Africa. PLOS One. 2018;13:1–20.
Hammouchi M, Loutfi C, Sebbar G, Touil N, Chaffai N, Batten C, Harif B, Oura C, El Harrak M. Experimental infection of alpine goats with a Moroccan strain of peste des petits ruminants virus (PPRV). Vet Microbiol. 2012;160:240–4. https://doi.org/10.1016/j.vetmic.2012.04.043.
Harder TC, Kenter M, Vos H, et al. Canine distemper virus from diseased large felids: biological properties and phylogenetic relationships. J Gen Virol. 1996;77:397–405.
Hedger RS, Barnett IT, Gray DF. Some virus diseases of domestic animals in the Sultanate of Oman. Trop Anim Health Prod. 1980;12:107–14.
Herbert R, Baron J, Batten C, et al. Recombinant adenovirus expressing the haemagglutinin of peste des petits ruminants virus (PPRV) protects goats against challenge with pathogenic virus; a DIVA vaccine for PPR. Vet Res. 2014;45:24.
Hoffmann B, Wiesner H, Maltzan J, et al. Fatalities in wild goats in Kurdistan associated with peste des petits ruminants virus. Transbound Emerg Dis. 2012;59(2):173–6.
Ihemelandu EC, Nduaka O, Ojukwu EM. Hyperimmune serum in the control of peste des petits ruminants. Trop Anim Hlth Prod. 1985;17:83–8.
Intisar KS, Ali YH, Haj MA, Sahar MA, Shaza MM, Baraa AM, Ishag OM, Nouri YM, Taha KM, Nada EM, Ahmed AM, Khalafalla AI, Libeau G, Diallo A. Peste des petits ruminants infection in domestic ruminants in Sudan. Trop Anim Health Prod. 2017;49:747–54. https://doi.org/10.1007/s11250-017-1254-3.
Isitor GN, Ezeokoli CD, Chineme CM. A histopathological and ulstrasuctural study of lesions of peste des petits ruminants. Trop Vet. 1984;2:151–8.
Ismail IM, House J. Evidence of identification of peste des petits ruminants from goats in Egypt. Arch Exp Vet. 1990;44:471.
Ismail TM, Hassan HB, Youssef NMA, et al. Studies on prevalence of rinderpest and peste des petits ruminants antibodies in camel sera in Egypt. Vet Med J. 1992;40:49–53.
Isoun TT, Mann ED. A stomatitis and pneumoenteritis complex of sheep in Nigeria. Bull Epizoot Dis Afr. 1972;20:167–74.
Johnson RH, Ritchie JSD. A virus associated with pseudo-rinderpest in Nigerian dwarf goats. Bull Epizoot Dis Afr. 1968;16:411–7.

Jones BA, Rich KM, Mariner JC, Anderson J, Jeggo M, Thevasagayam S, Cai Y, Peters AR, Roeder P. The economic impact of eradicating Peste des Petits ruminants: a benefit-cost analysis. PLoS One. 2016;11:e0149982. https://doi.org/10.1371/journal.pone.0149982.

Kardjadj M, Ben-Mahdi MH, Luka PD. First serological and molecular evidence of PPRV occurrence in Ghardaia district, center of Algeria. Trop Anim Health Prod. 2015;47:1279.

Keita D, Servan de Almeida R, Libeau G, Albina E. Identification and mapping of a region on the mRNA of morbillivirus nucleoprotein susceptible to RNA interference. Antivir Res. 2008;80 (2):158–67.

Khalafalla AI, Saeed IK, Ali YH, et al. An outbreak of peste des petits ruminants (PPR) in camels in the Sudan. Acta Trop. 2010;116:161–5.

Khandelwal A, Renukaradhya GJ, Rajasekhar M, Sita GL, Shaila MS. Immune responses to hemagglutinin-neuraminidase protein of peste des petits ruminants virus expressed in transgenic peanut plants in sheep. Vet Immunol Immunopathol. 2011;140(3–4):291–6. https://doi.org/10.1016/j.vetimm.2010.12.007.

Kulkarni DD, Bhikane AU, Shaila MS, et al. Peste des petits ruminant in goats in India. Vet Rec. 1996;138:187–8.

Kumar KS, Babu A, Sundarapandian G, et al. Molecular characterisation of lineage IV peste des petits ruminants virus using multi gene sequence data. Vet Microbiol. 2014;174:39–49.

Kwiatek O, Minet C, Grillet C, et al. Peste des petits ruminants (PPR) outbreak in Tajikistan. J Comp Pathol. 2007;136:111–9.

Kwiatek O, Keita D, Gil P, et al. Quantitative one-step real-time RT-PCR for the fast detection of the four genotypes of PPRV. J Virol Methods. 2010;165:168–77.

Kwiatek O, Ali YH, Saeed IK, et al. Asian lineage of peste des petits ruminants virus in Africa. Emerg Infect Dis. 2011;17:1223–31.

Lancelot R, Bouyer F, Peyre M, et al. Vaccine standards and pilot approach to PPR control in Africa (VSPA). Component 3: epidemiological assessment of PPR vaccination strategies in Burkina Faso and Ghana. Technical report; 2014. Montpellier: Cirad/OIE. 278p.

Lefèvre PC, Diallo A. La peste des petits ruminants. Rev Sci Tech Off Int Epiz. 1990;9:935–50.

Lesnoff M. Dynamics of a sheep population in a Sahelian area (Ndiagne district in Senegal): a periodic matrix model. Agric Syst. 1999;61:207–21.

Li L, Bao J, Wu X, et al. Rapid detection of peste des petits ruminants virus by a reverse transcription loop-mediated isothermal amplification assay. J Virol Methods. 2010;170:37.

Libeau G, Diallo A, Colas F, et al. Rapid differential diagnosis of rinderpest and peste des petits ruminants using an immunocapture Elisa. Vet Rec. 1994;134:300–4.

Libeau G, Prehaud C, Lancelot R, et al. Development of a competitive ELISA for peste des petits ruminants virus antibody detection using a recombinant N protein. Res Vet Sci. 1995;58:50–5.

Libeau G, Diallo A, Parida S. Evolutionary genetics underlying the spread of peste des petits ruminants virus. Anim Front. 2014;4:14–20.

Liu W, Wu X, Wang Z, Bao J, Li L, Zhao Y, Li J. Virus excretion and antibody dynamics in goats inoculated with a field isolate of peste des petits ruminants virus. Transbound Emerg Dis. 2013;60(Suppl 2):63–8. https://doi.org/10.1111/tbed.12136.

Liu F, Wu X, Liu W, Li L, Wang Z. Current perspectives on conventional and novel vaccines against peste des petits ruminants. Vet Res Commun. 2014;38:307–22.

Liu F, Wu X, Zou Y, Li L, Wang Z. Peste des petits ruminants virus-like particles induce both complete virus-specific antibodies and virus neutralizing antibodies in mice. J Virol Methods. 2015;213:45–9. https://doi.org/10.1016/j.jviromet.2014.11.018.

Luka PD, Erume J, Mwiine FN, Ayebazibwe C, Shamaki D. Molecular characterization and phylogenetic study of peste des petits ruminants viruses from north central states of Nigeria. BMC Vet Res. 2011;7:32. https://doi.org/10.1186/1746-6148-7-32.

Maganga GD, Verrier D, Zerbinati RM, et al. Molecular typing of PPRV strains detected during an outbreak in sheep and goats in south-eastern Gabon in 2011. Virol J. 2013;10:82. https://doi.org/10.1186/1743-422X-10-82.

Mahapatra M, Sayalel K, Muniraju M, et al. Spillover of peste des petits ruminants virus from domestic to wild ruminants in the Serengeti ecosystem, Tanzania. Emerg Infect Dis. 2015 Dec;21(12):2230–4. https://doi.org/10.3201/eid2112.150223.

Marashi M, Masoudi S, Moghadam MK, et al. Peste des petits ruminants virus in vulnerable wild small ruminants, Iran, 2014–2016. Emerg Infect Dis. 2017;23(4):704–6. https://doi.org/10.3201/eid2304.161218.

Mariner JC, Jones BA, Rich KM, Thevasagayam S, Anderson J, Jeggo M, Cai Y, Peters AR, Roeder PL. The opportunity to eradicate Peste des Petits ruminants. J Immunol. 2016;196:3499–506. https://doi.org/10.4049/jimmunol.1502625.

Mornet P, Orue J, Gilbert Y, et al. La peste des petits ruminants en Afrique Occidentale Française. Ses rapports avec la peste bovine. Rev Elev Méd Vét Pays Trop. 1956a;9:313–42.

Mornet P, Orue J, Gilbert Y. Unicité et plasticité du virus bovipestique. A propos d'un virus naturel adapté sur petits ruminants. CR Acad Sci. 1956b;242:2886–9.

Muniraju M, El Harrak M, Bao J, et al. Complete genome sequence of a peste des petits ruminants virus recovered from an alpine goat during an outbreak in Morocco in 2008. Genome Announc. 2013;1(3):e00096–13. https://doi.org/10.1128/genomeA.00096-13.

Muniraju M, Mahapatra M, Buczkowski H, et al. Rescue of a vaccine strain of peste des petits ruminants virus: in vivo evaluation and comparison with standard vaccine. Vaccine. 2015;33:465–71.

Muniraju M, Mahapatra M, Ayelet G, et al. Emergence of lineage IV peste des petits ruminants virus in Ethiopia: complete genome sequence of an Ethiopian isolate 2010. Transbound Emerg Dis. 2016;63:435–42.

Nawathe DR, Taylor WP. Experimental infection of domestic pigs with the virus of peste des petits ruminants. Trop Anim Health Prod. 1979;11:120–2.

Niyokwishimira A, de D Baziki J, Dundon WG, Nwankpa N, Njoroge C, Boussini H, Wamwayi H, Jaw B, Cattoli G, Nkundwanayo C, Ntakirutimana D, Balikowa D, Nyabongo L, Zhang Z, Bodjo SC. Detection and molecular characterization of peste des petits ruminants virus from outbreaks in Burundi, December 2017-January 2018. Transbound Emerg Dis. 2019;66:2067–73. https://doi.org/10.1111/tbed.13255.

Obi TU, Ojo MO, Durojaiye OA, et al. Peste des petits ruminants (PPR) in goats in Nigeria: clinical, microbiological and pathological features. Zentralbl Veterinärmed. 1983a;30:751–61.

Obi TU, Ojo MO, Taylor WP, et al. Studies on the epidemiology of peste des petits ruminants in Southern Nigeria. Trop Vet. 1983b;1:209–17.

OIE and FAO. PPR global control and eradication strategy (PPR-GCES). 2015. http://www.fao.org/3/a-i4460e.pdf.

Ozkul A, Akca Y, Alkan F, Barrett T, Karaoglu T, Dagalp SB, Anderson J, Yesilbag K, Cokcaliskan C, Gencay A. Prevalence, distribution, and host range of Peste des petits ruminants virus, Turkey. Emerg Infect Dis. 2002;8:708–12.

Opasina BA, Putt SN. Outbreaks of peste des petits ruminants in village goat flocks in Nigeria. Trop Anim Health Prod. 1985 Nov;17(4):219–24.

Parida S, Muniraju M, Mahapatra M, et al. Peste des petits ruminants. Vet Microbiol. 2015;181:90–106.

Peeples ME. Paramyxovirus M protein. Pulling it all together and taking it on the road. In: Kingsbury DW, editor. The Paramyxoviruses. New York: Plenum Press; 1991. p. 427–56.

Perry BD, Randolph TF, McDermott JJ, et al. Investing in animal health research to alleviate poverty; 2002. Nairobi: ILRI (International Livestock Research Institute). 148p.

Pope RA, Parida S, Bailey D, et al. Early events following experimental infection with peste-des-petits ruminants virus suggest immune cell targeting. PLoS One. 2013;8:e55830.

Prasad V, Satyavathi VV, Sanjaya KMV, Khandelwal A, Shaila SM, Lakshmi Sita G. Expression of biologically active hemagglitinin-neuraminidase protein of Peste des petits ruminants virus in transgenic pigeon pea (Cajanus cajan (L) Mill sp.). Plant Sci. 2004;166:199–205.

Provost A, Maurice Y, Bourdin P. La peste des petits ruminants existe-elle en Afrique centrale ? 40th general conference of the committtee of the OIE. Report, 202; 1972. 9p.

Rajak KK, Sreenivasa BP, Hosamani M, et al. Experimental studies on immunosuppressive effects of peste des petits ruminants (PPR) virus in goats. Comp Immunol Microbiol Infect Dis. 2005;28:287–96.

Riyesh T, Balamurugan V, Sen A, et al. Evaluation of efficacy of stabilizers on the thermostability of live attenuated thermo-adapted peste des petits ruminants vaccines. Virol Sin. 2011;26:324–37.

Rahman MZ, Haider N, Gurley ES, Ahmed S, Osmani MG, Hossain MB, Islam A, Khan SA, Hossain ME, Epstein JH, Zeidner N, Rahman M. Epidemiology and genetic characterization of Peste des petits ruminants virus in Bangladesh. Vet Med Sci. 2018; https://doi.org/10.1002/vms3.98.

Roeder PL, Abraham G, Kenfe G, et al. Peste des petits ruminants in Ethiopian goats. Trop Anim Health Prod. 1994;26:69–73.

Roger F, Yigezu LM, Hurard C, et al. Investigations on a new pathology of camels in Ethiopia. 3rd Annual Meeting for Animal Production under Arid Conditions. Al-Ain: Camel Production and Future Pesrpectives; 1998. p. 2–3.

Roger F, Gebre Yesus M, Libeau G, et al. Detection of antibodies of rinderpest and peste des petits ruminants viruses (*Paramyxoviridae*, *Morbillivirus*) during a new epizootic disease in Ethiopian camels (*Camelus dromedarius*). Rev Méd Vét. 2001;152:265–8.

Rojas JM, Moreno H, Valcárcel F, et al. Vaccination with recombinant adenoviruses expressing the peste des petits ruminants virus F or H proteins overcomes viral immunosuppression and induces protective immunity against PPRV challenge in sheep. PLoS One. 2014;9(7):e101226.

Rossiter PB, Taylor WP. Peste des petits ruminants. In: Coetzer JAW, Thomson GR, Tustin RC, editors. Infectious diseases of liviestock, vol. 2. Cape Town: Oxford University Press; 1994. p. 758–65.

Rossiter PB, Wardley RC. The differential growth of virulent and avirulent strains of rinderpest virus in bovine lymphocytes and macrophages. J Gen Virol. 1985;66:969–75.

Rossiter PB, Jessett DM, Taylor WP. Microneutralisation systems for use with different strains of peste des petits ruminants and rinderpest virus. Trop Anim Health Prod. 1985;17:75–81.

Rowland AC, Bourdin P. The histological relationship between peste des petits ruminants and kata in West Africa. Rev Elev Méd Vét Pays Trop. 1970;21:301–7.

Rowland AC, Scott GR, Hill HD. The pathology of an erosive stomatitis and enteritis in Wet African dwarf goats. J Pathol. 1969;98:83–7.

Rowland AC, Scott GR, Ramachandran S, et al. A comparative study of peste des petits ruminants and kata in West African dwarf goats. Trop Anim Health Prod. 1971;3:241–7.

Saliki JT, Libeau G, House JA, et al. Monoclonal antibody-based blocking enzyme-linked immunosorbent assay for specific detection and titration of peste des petits ruminants virus antibody in caprine and ovine sera. J Clin Microbiol. 1993;31:1075–82.

Sanz-Bernardo B, Goodbourn S, Baron MD. Control of the induction of type I interferon by peste des petits ruminants virus. PLoS One. 2017;12(5):e0177300. https://doi.org/10.1371/journal.pone.0177300.

Sarkar J, Sreenivasa BP, Singh RP, et al. Comparative efficacy of various chemical stabilizers on the thermostability of a live-attenuated peste des petits ruminants (PPR) vaccine. Vaccine. 2003;21:4728–35.

Schulz C, Fast C, Schlottau K, Hoffmann B, Beer M. Neglected hosts of small ruminant Morbillivirus. Emerg Infect Dis. 2018;24:2334–7. https://doi.org/10.3201/eid2412.180507.

Sen A, Saravanan P, Balamurugan V, Rajak KK, Sudhakar SB, Bhanuprakash V, Parida S, Singh RK. Vaccines against peste des petits ruminants virus. Expert Rev Vaccines. 2010;9:785–96.

Sen A, Saravanan P, Balamurugan V, Bhanuprakash V, Venkatesan G, Sarkar J, Rajak KK, Ahuja A, Yadav V, Sudhakar SB, Parida S, Singh RK. Detection of subclinical peste des petits ruminants virus infection in experimental cattle. Virusdisease. 2014;25:408–11. https://doi.org/10.1007/s13337-014-0213-0.

Servan de Almeida R, Keita D, Libeau G, et al. Control of ruminant morbillivirus replication by small interfering RNA. J Gen Virol. 2007;88(Pt 8):2307–11.

Servet-Delprat C, Vidalain PO, Valentin H, Rabourdin-Combe C. Measles virus and dendritic cell functions: how specific response cohabits with immunosuppression. Curr Top Microbiol Immunol. 2003;276:103–23.
Shaila MS, Purushothaman V, Bhavasar D, et al. Peste des petits ruminants of sheep in India. Vet Rec. 1989;125:602.
Shaila MS, Peter AB, Varalakshmi P, et al. Peste des petits ruminants in Tamilnadu goats. Ind Vet J. 1996a;73:587–8.
Shaila MS, Shamaki D, Forsyth MA, et al. Geographic distribution and epidemiology of peste des petits ruminants viruses. Virus Res. 1996b;43:149–53.
Shatar M, Khanui B, Purevtseren D, Khishgee B, Loitsch A, Unger H, Settypalli TBK, Cattoli G, Damdinjav B, Dundon WG. First genetic characterization of peste des petits ruminants virus from Mongolia. Arch Virol. 2017;162:3157–60. https://doi.org/10.1007/s00705-017-3456-4.
Silva AC, Carrondo MJ, Alves PM. Strategies for improved stability of peste des petits ruminants vaccine. Vaccine. 2011;29:4983–91.
Silva AC, Yami M, Libeau G, et al. Testing a new formulation for Peste des Petits Ruminants vaccine in Ethiopia. Vaccine. 2014;32:2878–81.
Singh RP. Strategic control of Peste des Petits ruminants. In: Garg SR, editor. Veterinary and livestock sector: a blueprint for capacity building. Delhi: Satish Serial Publishing House; 2012. p. 327–45.
Singh RP, Bandyopadhyay SK. Peste des petits ruminants vaccine and vaccination in India: sharing experience with disease endemic countries. Virusdisease. 2015;26:215–24. https://doi.org/10.1007/s13337-015-0281-9.
Singh RP, Sreenivasa BP, Dhar P, Shah LC, Bandyopadhyay SK. Development of monoclonal antibody based competitive-ELISA for detection and titration of antibodies to peste des petits ruminants virus. Vet Microbiol. 2004;98:3–15.
Singh RK, Balamurugan V, Bhanuprakash V, Sen A, Saravanan P, Pal Yadav M. Possible control and eradication of peste des petits ruminants from India: Technical aspects. Vet Ital. 2009;45:449–62.
Singh B, Bardhan D, Verma MR, Prasad S, Sinha DK. Estimation of economic losses due to Peste de Petits ruminants in small ruminants in India. Veterinary World. 2014;7(4):194–9.
Stem C. An economic analysis of the prevention of peste des petits ruminants in Nigerien goats. Prev Vet Med. 1993;16(2):141–50.
Sumption KJ, Aradom G, Libeau G, et al. Detection of peste des petits ruminants antigen in conjunctival smears of goats by indirect immunofluorescence. Vet Rec. 1998;142:421–4.
Swai ES, Kapaga A, Kivaria F, et al. Prevalence and distribution of peste des petits ruminants virus antibodies in various districts of Tanzania. Vet Res Commun. 2009;33:927–36. https://doi.org/10.1007/s11259-009-9311-7.
Tatsuo H, Ono N, Tanaka K, Yanagi Y. SLAM (CDw150) is a cellular receptor for measles virus. Nature. 2000;406:893–7.
Tatsuo H, Ono N, Yanagi Y. Morbilliviruses use signaling lymphocyte activation molecules (CD150) as cellular receptors. J Virol. 2001;75(13):5842–50.
Taylor WP. Serological studies with the virus of peste des petits ruminants in Nigeria. Res Vet Sci. 1979;26:236–42.
Taylor WP. The distribution and epidemiology of peste des petits ruminants. Prev Vet Med. 1984;2:157–66.
Taylor WP, Diallo A, Gopalakrishna S, et al. Peste des petits ruminants has been widely present in southern India since, if not before, the late 1980s. Prev Vet Med. 2002;52:305–12.
Thomson GR, Penrith ML. Eradication of transboundary animal diseases: can the rinderpest success story be repeated? Transbound Emerg Dis. 2017;64(2):459–75. https://doi.org/10.1111/tbed.12385.
Tounkara K, Bataille A, Adombi CM, et al. First genetic characterization of peste des petits ruminants from Niger: on the advancing front of the Asian virus lineage. Transbound Emerg Dis. 2018;65(5):1145–51. https://doi.org/10.1111/tbed.12901.

Truong T, Boshra H, Embury-Hyatt C, et al. Peste des petits ruminants virus tissue tropism and pathogenesis in sheep and goats following experimental infection. PLoS One. 2014;9(1): e87145. https://doi.org/10.1371/journal.pone.0087145.

Qin J, Huang H, Ruan Y, Hou X, Yang S, Wang C, Huang G, Wang T, Feng N, Gao Y, Xia X. A novel recombinant Peste des petits ruminants-canine adenovirus vaccine elicits long-lasting neutralizing antibody response against PPR in goats. PLoS One. 2012;7:e37170. https://doi.org/10.1371/journal.pone.0037170.

Wamwayi HM, Rossiter PB, Kariuki DP, et al. Peste des petits ruminants antibodies in east Africa. Vet Rec. 1995;136(8):199–200.

Wang Y, Liu G, Chen Z, et al. Recombinant adenovirus expressing F and H fusion proteins of peste des petits ruminants virus induces both humoral and cell-mediated immune responses in goats. Vet Immunol Immunopathol. 2013;154:242.

Waret-Szkuta A, Ortiz-Pelaez A, Pfeiffer DU, Roger F, Guitian FJ. Herd contact structure based on shared use of water points and grazing points in the Highlands of Ethiopia. Epidemiol Infect. 2011;139:875–85. https://doi.org/10.1017/S0950268810001718.

Wernery U. Peste des petits ruminants (PPR) in camelids with own investigation. J Camel Pract Res. 2011;18:219.

Whitney JC, Scott GR, Hill DH. Preliminary observations on a stomatitis and enteritis of goats in Southern Nigeria. Bull Epizoot Dis Afr. 1967;15:331–41.

Woma TY, Adombi CM, Yu D, et al. Co-circulation of peste-des-petits-ruminants virus Asian lineage IV with lineage II in Nigeria. Transbound Emerg Dis. 2016;63(3):235–42. https://doi.org/10.1111/tbed.12387.

Worrwall EE, Litamoi JK, Seck BM, et al. Xerovac: an ultra rapid method for the dehydration and preservation of live attenuated rinderpest and peste des petits ruminants vaccines. Vaccine. 2000;19:834–9.

Wosu LO. Haemagglutination test for diagnosis of peste des petits ruminants disease in goats with samples from live animals. Small Rum Res. 1991;5:169–71.

Yang B, Qi X, Chen Z, et al. Binding and entry of peste des petits ruminants virus into caprine endometrial epithelial cells profoundly affect early cellular gene expression. Vet Res. 2018;49:8. https://doi.org/10.1186/s13567-018-0504-3.

Zahur AB, Irshad H, Ullah A, et al. Peste des petits ruminants vaccine (Nigerian strain 75/1) confers protection for at least three years in sheep and goats. J Biosci Med Sci. 2015;360:23.

Chapter 13
Lumpy Skin Disease and Vectors of LSDV

Esayas Gelaye and Charles Euloge Lamien

Abstract Lumpy skin disease (LSD) is a viral disease of cattle caused by lumpy skin disease virus (LSDV). LSDV shares high degree of sequence homology with goatpox virus (GTPV) and sheeppox virus (SPPV), the two other members of the genus *Capripoxvirus* of the family *Poxviridae*. Genetically LSDV is a double-stranded DNA genome of approximately 151 kbp. LSD is an economically important and notifiable animal disease by the World Organisation for Animal Health (OIE). Clinically LSD is characterized by fever and the appearance of nodules on the skin and mucous membranes. In severe and chronic cases, nodular skin lesions cover the entire body and become deep scab and eroded. Transmission of the disease occurs predominantly by insects possibly through mechanical vectors, contaminated feed and water, infected saliva, and rarely natural contact. LSD is endemic in many African countries and mostly coexists with sheeppox and goatpox. Recently, LSD has been rapidly spreading to the Middle East, Turkey, and Russia, the Balkan and European Union countries. Diagnosis is mainly based on observation of clinical signs and the detection of virus genome using conventional and real-time PCR methods. In Africa, prevention and control of LSD relies on vaccination using live attenuated vaccines derived from Kenyan or South African LSDV strains. Vaccine that can allow the differentiation of infected from vaccinated animals (DIVA) and high-throughput serological method for the detection of specific antibody need to be developed. Alternatively, a 12 nucleotide deletion that exists only on the G-protein-coupled chemokine receptor gene of LSDV field isolates can be used to differentiate wild-type from vaccine strains by sequencing and real-time PCR.

E. Gelaye (✉)
Research and Development Directorate, National Veterinary Institute (NVI), Debre zeit, Ethiopia
e-mail: esayas.gelaye@nvi.gov.et

C. E. Lamien
Animal Production and Health Laboratory, FAO/IAEA Agriculture and Biotechnology Laboratory, International Atomic Energy Agency, Vienna, Austria
e-mail: c.lamien@iaea.org

M. Kardjadj et al. (eds.), *Transboundary Animal Diseases in Sahelian Africa and Connected Regions*, https://doi.org/10.1007/978-3-030-25385-1_13

Keywords Capripoxvirus · Diagnosis · Genotyping · Lumpy skin disease · Poxviridae · Vector

History

Lumpy skin disease (LSD) is caused by lumpy skin disease virus (LSDV), a member of the genus *Capripoxvirus* (CaPV) of the *Poxviridae* family (Tulman et al. 2001; Andrew et al. 2012; OIE 2016). LSDV is a double-stranded DNA genome of approximately 151 kilo base pair (kpb) with a central coding region of 156 putative genes bounded by identical 2.4 kbp inverted terminal repeats (Tulman et al. 2001). LSDV shares the genus *Capripoxvirus* with sheeppox virus (SPPV) and goatpox virus (GTPV), which are closely related, but phylogenetically distinct (Tulman et al. 2001; Lamien et al. 2011a; Gelaye et al. 2015). There is only one serological type of LSDV, and LSD, SPP, and GTP viruses cross-react serologically and are difficult to differentiate using serological methods (Diallo and Viljoen 2007). LSDV genome is very stable; thus, very little genetic variability occurs.

The origin of LSDV has remained a mystery ever since it was first identified in a geographical region free of SPP and GTP viruses. Lumpy skin disease was first diagnosed in northern Rhodesia/Zambia in 1929 (MacDonald 1931) and then spread to Botswana in 1943 (Von Backstrom 1945), South Africa in 1944 (Thomas and Maré 1945), southern Rhodesia/Zimbabwe in 1945 (Houston 1945), Kenya in 1957 (MacOwan 1959), Sudan in 1971 (Ali and Obeid 1977), Chad and Niger in 1973 (Nawathe et al. 1978), Nigeria in 1974 (Nawathe et al. 1978), and Ethiopia in 1983 (Mebratu et al. 1984). Lumpy skin disease spread steadily to almost all sub-Saharan countries by the end of the 1970s and remained only in this region till 1987 (Diallo and Viljoen 2007). From 1988, the diseases spread to Egypt (Ali et al. 1990).

Initially it was difficult to know the etiological agent of LSD. LSD was first described by MacDonald in 1931 as "pseudo-urticaria" (MacDonald 1931). In 1945, Thomas and Maré were the first to demonstrate the transmission of the infectious agent by subcutaneous inoculation of cattle with a suspension of skin nodules and thus concluded on the infectious nature of the disease (Thomas and Maré 1945). Furthermore they have demonstrated that the infectious agent was not transmitted to sheep (Thomas and Maré 1945) and provided initial data on the histological studies on the skin lesions produced by LSD (Thomas and Maré 1945). During the same period, Van Den Ende and colleagues successfully isolated LSD virus in embryonating chicken eggs and showed that the virus could be neutralized by sera from LSD-convalescent bovines (Van Den Ende et al. 1948).

The efforts for the isolation of LSDV using cell culture were frustrating due to the presence of contaminating viruses such as orphan viruses (bovine herpes virus-4) and Allerton virus (bovine herpes virus-2 causing of pseudo-lumpy skin disease). Alexander and collaborators were the first to isolate the Neethling type of LSD virus using tissue culture in South Africa in 1957 (Alexander et al. 1957).

Geographic Distribution and Economic Impact

Lumpy skin disease is currently endemic and widespread in almost all African countries except for a few northern African countries: Libya, Tunisia, Algeria, and Morocco (Tuppurainen et al. 2017). After being confined for almost 60 years in the sub-Saharan African region and Egypt, the disease has spread to the Middle East countries such as Israel, the Palestinian Autonomous Territories, Jordan, Lebanon, Kuwait, Iraq, Iran, Saudi Arabia, Bahrain, Oman, Yemen, and the United Arab Emirates (Nawathe et al. 1982; Yeruham et al. 1995; Stram et al. 2008; Tuppurainen and Oura 2012; Tuppurainen et al. 2014, 2017). From the year 2013, LSD spread to Turkey, and then Azerbaijan in 2014, Armenia and Kazakhstan in 2015, and the southern Russian federation (Dagestan, Chechnya, Krasnodar Kray and Kalmykia) and Georgia in 2016. LSDV infection also advanced to northern Cyprus and Greece in 2015, Bulgaria, Macedonia, Serbia, Montenegro, Albania, and Kosovo in 2016 (Agianniotaki et al. 2017a). Currently, Central Asia and Western and Eastern Europe are at high risk of the infection. The geographical distribution of LSD differs from that of SPP and GTP which are endemic in many African countries, excluding southern Africa, Asia, the Middle East, and Turkey with sporadic outbreaks in Greece and some eastern European countries (Tantawi et al. 1979; Oguzoglu et al. 2006; Verma et al. 2011; Yan et al. 2012; Zhou et al. 2012).

With the exception of few northern and southern African countries, all three capripox diseases coexist in Africa, creating an urgent need for a precise identification of the circulating virus genotypes during outbreaks. This can help to design and implement better prevention and control measures to limit the spread of disease. It is difficult to explain why SPP and GTP viruses have not spread south of the Equator as did LSD (Babiuk et al. 2008). It is possible that the ability of LSDV to provide heterologous cross-protection against SPP and GTP is one possible factor limiting the spread of these diseases southward, although this assumption is not consistent with the coexistence of LSD, SPP, and GTP in many equatorial African countries.

LSDV infections are notifiable to the World Organisation for Animal Health (OIE) and as such can negatively impact the national economy of the affected country due to international trade restrictions on exportation of live cattle and animal products to the global market. Occurrence of LSD outbreak has a serious socioeconomic impact. Even though the morbidity and mortality rates of LSD are usually low, it is an economically important disease of cattle in Africa because of the prolonged loss of productivity of dairy and beef cattle, decrease in body weight, decrease in milk production, mastitis, and severe orchitis, which may result in temporary infertility and sometimes permanent sterility (Weiss 1968; Babiuk et al. 2008). Pregnant cows may abort and infertility of cows can last for several months (Weiss 1968). In severely affected animals damage to hides is permanent, resulting in rejected or poor quality hides, eventually the leather industry is greatly damaged (Green 1959; Babiuk et al. 2008). In countries where LSD is exotic, the economic costs because of disease eradication and trade restrictions would be substantial and comparable to foot and mouth disease outbreaks (Garner and Lack 1995; De Clercq and Goris 2004). Since LSD is a transboundary animal disease with significant impediments to trade of livestock and livestock products, it affects the economy and well-being of farmers in developing countries and would have a substantial economic impact on disease-free/developed countries if the disease is introduced (Babiuk et al. 2008).

A report estimated a total economic loss of around USD667,785.60 due to LSD outbreaks between June and December 2011 in feedlot farms in central Ethiopia (Alemayehu et al. 2013). Additionally, Ayelet et al. (2013) also estimated a milk price loss of USD44.70 per cow during an average of 20 days when an outbreak occurs on a dairy farm. Although mortality rate caused by LSD is usually low, the disease is of major economic importance owing to costs incurred for veterinary service, cow death, abortion, and still birth of calves. Dairy cattle are mostly severely affected and experiencing a 50% drop in milk production and occurrence of secondary mastitis originating from the development of lesions on the teats. The financial cost, calculated as the sum of the average production losses due to morbidity and mortality, arising from milk loss, beef loss, traction power loss, and treatment and vaccination measures cost at the herd level, in LSDV-infected herds is estimated to be USD6.43 (5.12–8) per head for local zebu and USD58 (42–73) per head for HF/crossbred cattle (Gari et al. 2011).

Epidemiology

Susceptible Hosts

Lumpy skin disease virus is host specific, causing natural infection mainly in cattle and Asian water buffalo (*Bubalus bubalis*), although the morbidity rate is significantly lower in buffalo (1.6%) than in cattle (30.8%) (El-Nahas et al. 2011). Although mixed herds of cattle, sheep, and goats are common in most of the LSD endemic areas, to date no epidemiological evidence on the role of small ruminants as a reservoir for LSD has been reported (Tuppurainen et al. 2017). Few studies have been conducted so far to see the role of wildlife as a reservoir for transmission of LSD virus. Experimentally, it was possible to reproduce the clinical signs of LSD by infecting impala (*Aepyceros melampus*) and giraffe (*Giraffa camelopardalis*) (Tuppurainen et al. 2017). Natural infection by LSDV has also been reported in an Arabian oryx (*Oryx leucoryx*) in Saudi Arabia, springbok (*Antidorcas marsupialis*) in Namibia, Oryx gazelle in South Africa, and Asian water buffalo in Egypt (Tuppurainen et al. 2017). LSD virus genome was also characterized in skin lesion samples collected from South African Springbok antelope (Lamien et al. 2011a). Capripoxvirus antibody was also detected from serum samples collected from African buffalo, greater kudu, water buck, reed buck, impala, springbok, and giraffe (Barnard 1997). The susceptibility of wild ruminants or their possible role in the transmission and epidemiology of LSDV infection is not known. However, Hedger and Hamblin (1983) reported that wildlife do not play significant role in the epidemiology of the disease. Small ruminants have been suspected to be susceptible to LSDV, but so far nothing has been confirmed. Lumpy skin disease virus does not affect humans.

Transmission

The existence of a specific reservoir for the virus is not known, nor is how and where the virus survives between epidemics. LSD usually occurs at regular intervals in endemic areas or it may cause epidemics, which spread fairly rapidly throughout a region or country (Davies 1991). Infected animals shed the virus in oral, nasal, and ocular secretions and transmission occurs through aerosols and direct contact (Bowden et al. 2008; MacLachlan and Dubovi 2011; OIE 2016). The skin is the primary portal of entry for most poxviruses; however, respiratory and mucosal routes have been associated with poxvirus infections such as orthopoxviruses. Due to the stability of the poxvirus, the virus may persist in the environment for prolonged periods of time, leading to infection of naïve animals. The primary cause of infection is usually associated with the introduction of infected animal into or in close proximity to a herd of susceptible animals. Susceptibility of the host depends on immune status, age, and breed. Thus, the spread of LSDV into new areas is predominantly associated with the increase of illegal animal movement through trade (Domenech et al. 2006) as well as inadequate or breakdown of veterinary services (Rweyemamu et al. 2000). Climatic conditions, such as heavy rainfall, humid and warm weather, and drought periods, affect the insect populations and will support or suppress the spread of the disease. Outbreaks are usually seasonal but may occur any time because in many affected countries no season is completely vector free (Tuppurainen et al. 2017).

In cattle affected with the disease, the most common source of the virus are the skin nodules, the crusts of skin lesions, blood, saliva, nasal, and ocular discharge, semen, and milk (Weiss 1968; Lamien et al. 2011a; Gelaye et al. 2015). The virus was not found in urines and feces. Infectious LSDV remains well-protected inside crusts, particularly when these drop off from the skin lesions; although no experimental data are available, it is likely that the farm environment remains contaminated for a long time without thorough cleaning and disinfection (Tuppurainen et al. 2017). In experimentally infected animals, infectious LSDV has been detected in saliva and nasal discharged for up to 20 days post-infection (Lamien et al. 2011a; Tuppurainen et al. 2017); even though it needs more studies to investigate how long the infectious LSD virus is excreted. The virus persists in the semen of infected bulls so that natural mating or artificial insemination may be a source of infection for females. The virus may be transmitted to suckling calves through milk, or from skin lesion in the teats.

LSD varied from mild with few secondary skin nodules to generalized infection of varying severity, characterized by high morbidity with low mortality; thus, LSD is highly variable disease. There is a knowledge gap on the effect of climatic and environmental changes to the insect and tick populations and spread of LSDV, and the role of vectors for LSDV transmission and epidemiology should be further investigated in both controlled environment and the field conditions. How much LSD virus load, minimum infectious dose, is needed to initiate infection into susceptible hosts should be addressed. The role of birds as carriers of insects and ticks is not studied.

Field and experimental evidence has proven that LSD is not highly contagious. The morbidity rates in natural outbreaks vary from 3 to 85%. In experimentally produced infections, only 40–50% of inoculated animals showed clinical signs of the disease (Weiss 1968). The mortality rate is usually low, less than 10% (Thomas and Maré 1945; Tuppurainen et al. 2017). The annual cumulative incidence of LSD infection in Holstein-Friesian (HF)/crossbred and local zebu cattle were 33.9% and 13.4%, respectively, and significantly different ($p < 0.5$) (Gari et al. 2011). Variation in the morbidity and mortality rates could be due to the involvement of strains of different pathogenicity, efficiency of the transmission of the disease by the vector, and the route of virus infection (Carn and Kitching 1995a). Annual mortality was also significantly higher in HF/crossbred cattle 7.43% than in local zebu cattle 1.25% (Gari et al. 2011). In general, high milk-producing European cattle breeds are highly susceptible compared to indigenous African and Asian animals. No carrier state was detected.

Vectors of LSDV

Transmission of LSDV between animals by contagion is extremely inefficient and that parenteral inoculation of virus is required to establish infection, and the high proportion of animals with generalized disease following intravenous inoculation implies that naturally occurring cases of generalized LSD may follow spread by intravenously feeding arthropods (Carn and Kitching 1995a).

The transmission of LSDV is believed to occur mainly by blood-feeding arthropod vectors and large biting Diptera (Haig 1957; Weiss 1968; Kitching and Mellor 1986; Chihota et al. 2001; Tuppurainen et al. 2017). Although no scientific evidence still exists for the multiplication of the LSD virus in blood-feeding arthropod vectors, experimental data and field reports showed that no or limited transmission of LSD is observed in the absence of arthropods (Nawathe et al. 1978; Carn and Kitching 1995b; Yeruham et al. 1995). In addition, a mathematical model explained that flying and blood-sucking insects play a role in the transmission of the virus (Magori-Cohen et al. 2012). Due to high viral loads in the skin lesions, mechanical transmission (no development/propagation of pathogen in the vector) may occur by insect vector like *Stomoxys calcitrans* for some poxvirus species (Kitching and Mellor 1986; MacLachlan and Dubovi 2011). Flies are thought to transmit mechanically LSDV from infected animal to susceptible one by depositing the virus intradermally (Carn and Kitching 1995a). Nevertheless, experimentally it has been demonstrated that intradermal inoculation of LSD virus predominately produces only localized lesion at the site of inoculation, whereas intravenous inoculation results in generalized lesions and more severe disease (Carn and Kitching 1995a). In contrast to flies, blood-feeding arthropods, such as mosquitoes and sand flies which feed intravenously, are thought to be associated with outbreaks of LSD characterized by generalized lesions.

An experimental study has shown that *Stomoxys calcitrans* (stable flies) does not play a significant role in the transmission of LSDV (Chihota et al. 2003). In contrast, it was suggested that LSD has spread from Egypt to Israel in 1989 through *Stomoxys calcitrans* carried by wind or inside vehicles of cattle merchants (Yeruham et al. 1995). Additionally, LSD virus was isolated from *Stomoxys calcitrans* and *Biomyia fasciata* caught after they have been fed on infected cattle (Weiss 1968). Chihota et al. (2001) reported that experimentally *Aedes aegypti* female mosquitoes were able to transmit, mechanically, LSDV from infected to susceptible cattle. Such mechanical transmission of LSDV likely occurs by any vector feeding frequently and changing host between feedings. Chihota et al. (2003) detected LSDV genome in mosquitoes (*Anopheles stephensis* and *Culex quinquefasciatus*) and biting midge (*Culicoides nubeculosus*) fed on LSD-positive animals but did not observe any LSDV transmission by these insects.

The transstadial and transovarial transmission of LSDV by African blue tick *Rhipicephalus (Boophilus) decoloratus* and mechanical or transstadial transmission by brown tick *Rhipicephalus appendiculatus* and bont tick *Amblyomma hebraeum* were recently reported based on molecular evidence (Tuppurainen et al. 2011). These three ticks are sub-Saharan ticks. The closely related species in the Middle East region are *Rhipicephalus (Boophilus) annulatus, R. praetextatus, A. variegatum*, and *Hyalomma excavatum.* No research has been conducted on European tick species on their possible role in LSDV transmission. The ticks acquire the virus through direct feeding of blood on an infected animal, co-feeding from infected tick during co-feeding of many ticks at a time on one host, or vertical transmission through transovarian (from adult to eggs) or transstadial (from larvae to nymphs).

Biological transmission and overwintering were recorded when the laboratory-bred *R. appendiculatus* and *A. hebraeum* nymphs and adults and *R. decoloratus* larvae feed on the experimentally infected cattle with LSDV (Tuppurainen et al. 2013). Feeding on viremic cattle without showing skin lesions was sufficient for the successful transmission of the virus by *R. appendiculatus* males. No evidence was obtained on the replication of LSDV in vitro in tick cell lines. LSDV DNA was detected from ticks (*Rhipicephalus*, *Amblyomma*, and *Hyalomma* species) collected from naturally infected Holstein-Friesian or Sanga cattle in Egypt and South Africa (Tuppurainen et al. 2015).

No data are available explaining for how long and/or how much the virus persists in the mouth parts as mechanical transmission of artificially fed arthropod and on the leftover blood. Female ticks may get infected either orally through infected blood or during copulation where it is not necessarily biological transmission. Intra- or extracellular survival of the virus in tick tissues is likely to be more important than the actual replication of the virus in the tick vectors (Tuppurainen et al. 2015).

Although other blood-sucking arthropods, *Simuliidae* (black flies), *Haematobia irritans* (horn fly), Sand flies, and non-biting *Muscidae* (*M. autumnalis* and *M. domestica*) are suspected, there is no practical evidence of their potential role as vectors in the transmission of LSDV. Experimental data suggest that there is no LSDV transmission by lacrimal sac of a non-biting Muscidae flies (Carn and Kitching 1995a).

Clinical Signs and Lesions

The incubation period of LSD under field condition is between 2 and 4 weeks, while the experimentally induced disease is between 4 and 14 days (Carn and Kitching 1995b). Experimental study showed that viremia is observed in cattle from 6 to 15 days post-infection by real-time PCR, and shedding of the virus is detected in nasal, oral, and conjunctival swabs from 12 to 18 days post-infection by real-time PCR and the virus was isolated from skin nodules and nasal mucosa (Tuppurainen et al. 2005).

The major clinical signs include lacrymation and nasal discharge—usually observed first, subscapular and prefemoral lymph nodes become enlarged and are easily palpable, high fever (>40.5 °C) may persist approximately for 1 week, sharp drop in milk yield and mastitis are common complications during secondary bacterial infections, appearance of characteristic nodular skin lesions of 10–50 mm in diameter all over the body, sometimes painful ulcerative lesions develop in the cornea of one or both eyes, leading to blindness in worst cases, subclinical infections are common in the field, and pox lesions can be found throughout the entire digestive and respiratory tracts and on the surface of almost any internal organs. The skin nodules could persist for many months (MacLachlan and Dubovi 2011; OIE 2016; Tuppurainen et al. 2017).

Poxviruses have tropism of epithelial cells. The term "pox" relates to the formation of pocks in the skin, with subsequent scarring and formation of pockmarks, a classic finding among smallpox survivors (Smith et al. 2010). LSD virus persists for at least 30 days in the skin lesions of infected animals. It remains viable for 18 days in the skin lesions and superficial epidermal scrapings of a hide, which are air-dried and kept at room temperature. Virus infection in immunologically naïve animals leads to concurrent fever and skin papules, followed by rhinitis, conjunctivitis, and hypersalivation (MacLachlan and Dubovi 2011; OIE 2016). Although pox lesions can be widespread, the more common presentation is a few nodules on body parts with limited hair. Pox lesions also develop in the lungs and gastrointestinal tract. High viral loads occur in the skin, and viremia is probably cell associated. Observed lesions at necropsy include tracheal congestion and patchy discoloration of the lungs. The spleen and lymph nodes are enlarged, with multifocal to coalescing areas of necrosis (Bowden et al. 2008). The essential histological lesion is necrosis, with depletion of lymphocytes in the paracortical regions and absence of germinal centers in the spleen and lymph nodes. Lumpy skin disease is most commonly recognized by nodular skin lesions covering the entire body as shown in Figs. 13.1 and 13.2.

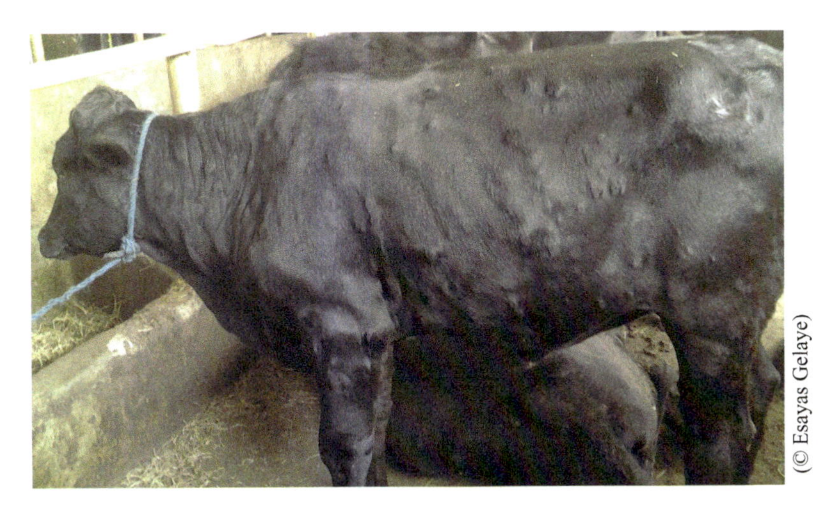

Fig. 13.1 Characteristic acute lesions of LSD. Note the appearance of multiple active skin nodules covering the entire body in severely affected Holstein-Friesian cross-bred dairy heifer in Ethiopia

Fig. 13.2 Severe and chronic case of LSD. Note the skin lesion showing mixture of characteristic nodular, deep scar, and crusted lesions existing on the entire body in zebu cattle in fattening farm in Ethiopia

Diagnosis

Clinically

A presumptive diagnosis of the disease can be undertaken clinically by observing the active skin nodular lesions that occur mostly on the neck and legs regions. Sometimes in severe and chronic cases the nodules develop deep scar and crusted lesions covering the entire body of the animal. Additionally, recording the disease history through animal owner's interview such as affected animal species, number of animals showing clinical signs, number of dead animals, presence of flies and ticks, season/month of disease occurrence, and vaccination status could be used as primary source of information to reach an LSD diagnosis.

Sample Collection and Transport

The preferred clinical samples for further laboratory confirmation of LSDV are active nodular skin lesions and scabs, nasal, oral, and ocular swabs, saliva, and EDTA blood from clinically diseased or suspected animals (Bowden et al. 2008; Lamien et al. 2011a; Gelaye et al. 2015). Skin lesions are the best samples for the diagnosis of LSDV, whereas blood is not ideal since viremia in blood is short-lived. In general, poxviruses are mainly cell associated, and therefore there may be fewer virus particles in the blood. Representative skin lesion or swab samples from suspected or clinically diseased animals should be collected aseptically and transferred into sterile labeled containers. Serum samples can be collected from diseased or suspected animals for CaPV antibody detection or from vaccinated animals to evaluate sero-conversion.

Diagnosis of LSD should be conducted urgently and representative suspected samples should be transported to the diagnostic laboratory through maintaining the cold chain. The samples must be transported following the regulations set for the transport of dangerous goods and should be packed using triple packaging (primary, secondary, and outer containers). Suspected LSDV-infected field samples are classified as Infectious Substances Class B (Division 6.2) and the International Air Transport Association (IATA) packing instruction P650 using the UN-approved packaging material. The packaging must be followed as Biological Substance (UN3373, Category B): hazard for animal health, not for humans. All required shipping documents should be enclosed with the sample package during transport. It is strictly forbidden to transport infectious substances as carry-on baggage, checked baggage, or in person. Prior to shipping the samples, the destination/consignee laboratory must be informed in advance about the shipment. Import permit must be obtained from the diagnostic laboratory or authorized organization from destination country and should be included with the sample transfer document.

Virus Isolation

Infectious LSD virus can be isolated and multiplied using a variety of sensitive mammalian cell cultures: primary cell culture (skin, kidney, and testis cells) of bovine or ovine origin or cell line culture (Vero or ESH-L or OA3.Ts cells) (OIE 2016). LSDV can also be cultivated on the chorioallantoic membrane of embryonated chicken egg (Kalra and Sharma 1981; Binepal et al. 2001; Babiuk et al. 2007; OIE 2016). Tissue samples processing and virus isolation could be conducted in disease endemic countries under the Biosafety level-II laboratory facilities, whereas in disease-free countries isolation of LSDV should be conducted using Biosafety level-III facilities. The virus induces the formation of distinct plaques with cytopathic effects (CPEs) characterized by elongated cells. However, primary cells have several disadvantages including the need to constantly establish new cultures, cell lot variation, and contamination with extraneous agents (Babiuk et al. 2008). Characteristic poxvirus-induced CPEs can be observed under the inverted microscope in infected cells within a week, although sometimes the procedure needs several blind passages (Diallo and Viljoen 2007; Lamien et al. 2011a; OIE 2016; Gelaye et al. 2015). LSDV isolation can be confirmed by immunostaining using anti-LSDV serum (Gulbahar et al. 2006; Babiuk et al. 2007). The virus can be recovered from intact skin nodules kept at −80 °C for 10 years and from infected tissue culture fluid kept at +4 °C for 6 months. In tissue culture fluid stored in dry ice, virus remains viable for at least 10 years (Weiss 1968).

Histopathology and Electron Microscopy

Histology, immunohistochemistry, and electron microscopy of skin nodule could also be used for the diagnosis of LSDV infection, although not commonly used by the diagnostician. Hematoxylin and Eosin staining of skin nodule from acute stage can show the presence of eosinophilic intracytoplasmic inclusion bodies in epidermal cells with mild hydropic degeneration (Babiuk et al. 2008). Immunohistochemistry counterstained with Mayer's hematoxylin of skin nodule of subacute-chronic stage can show a positive reaction in epidermal cells and inside macrophages infiltrating the dermis (Babiuk et al. 2008).

Antibody Detection

Immunity against LSD is predominantly cell mediated but humoral antibodies are also detectable. Currently available serological assays may not be sensitive to detect mild and chronic infections. Seroconversion after vaccination and serum samples from disease suspected cattle can be tested using virus neutralization test (OIE 2016) and indirect fluorescent antibody test (Gari et al. 2008). Even though virus neutralization is considered a gold standard test by OIE in detecting anti-LSDV antibodies, it is slow and labor intensive and requires live LSDV, access to which is often not permitted in disease-free countries (Babiuk et al. 2008; OIE 2016). A recombinant capripoxvirus that express green fluorescent protein has been evaluated for use in a virus neutralization assay and has decreased the length of time required for detection of virus neutralization activity from 6 to 2 days (Wallace et al. 2007). Western blotting assay can be used since the method is specific and sensitive; however, it is difficult to perform and interpret (Chand et al. 1994). Agar gel immunodiffusion test for LSD diagnosis is not a recommended test because of the cross-reaction with antibodies to parapoxviruses (OIE 2016). To overcome this diagnostic limitation, several antibody detection ELISAs were developed using capripoxvirus recombinant proteins (Carn et al. 1995; Heine et al. 1999; Bowden et al. 2009; Tian et al. 2010). Assays based on the recombinant mature virion envelope protein P32 expressed in *E.coli* have been developed for the detection of CaPV-specific antibodies in cattle (Carn et al. 1995) and sheep (Heine et al. 1999). However, difficulties with the expression and the stability of the recombinant antigens have compromised these assays. Babiuk et al. (2009) later developed an indirect ELISA using inactivated, sucrose gradient-purified SPPV as coating antigen, for detection of antibodies to SPPV, GTPV, and LSDV. Nevertheless, although suited to screening sera from all three host species, the viral antigen is difficult and expensive to produce in large quantities and not applicable in disease-free countries. Bowden et al. (2009) also developed an indirect ELISA for the detection of antibodies based on selected CaPV antigenic recombinant virion core proteins. This assay performs well on sheep and goat sera collected from experimentally infected with virulent virus isolates, but was unable to detect antibodies from vaccinated sheep and goat sera. An indirect ELISA for the diagnosis of sheeppox and goatpox was later developed by Tian et al. (2010) based on two synthetic peptides corresponding to the major antigen P32 of CaPVs, but this assay performed well only on sera from immunized sheep. No commercial validated ELISA kit is available on the market for the detection of LSDV antibody even though many development activities are ongoing. With the recent expansion of capripox disease in new geographical areas, there is an urgent need for a high-throughput serological assay to facilitate the serological surveillance of the diseases in countries under threat. The availability of such an assay will also facilitate animal screening during live animal export and post-vaccination monitoring during vaccination campaigns.

Genetic Detection

Two main categories of conventional PCR and real-time PCR methods have been developed for the detection of LSDV and other capripoxviruses: those allowing the generic detection of capripoxviruses and those that will allow the specific identification of LSDV. For the generic detection of capripoxviruses, a gel-based PCR was developed by Ireland and Binepal (1998). Another gel-based PCR method targeting the 30 kDa RNA polymerase subunit (RPO30) gene of capripoxviruses (Lamien et al. 2011a) is available for the detection of LSDV. Although this method could differentiate LSDV from SPPV, it is unable to differentiate LSDV from GTPV. Several real-time PCR methods (Balinsky et al. 2008; Bowden et al. 2008; Stubbs et al. 2012; Haegeman et al. 2013) are available for the generic detection of capripoxviruses, including LSDV; however, none of the abovementioned methods can differentiate LSD virus from SPPV and GTPV. Two LAMP PCR methods (Das et al. 2012; Murray et al. 2013) and a field-ready nucleic acid extraction and real-time PCR platform (Amson et al. 2015) are also available for the generic detection of capripoxviruses. For the specific detection of LSDV, two approaches were used in various reports. A first approach, using gel-based PCR, consists of detecting only LSDV and not detecting other Capripoxviruses such as SPPV and GTPV (Stram et al. 2008). A second approach consists of simultaneously detecting capripoxviruses and differentiating LSDV from GTPV and SPPV. For instance, a dual hybridization probe assay targeting the G-protein-coupled chemokine receptor (GPCR) gene (Lamien et al. 2011b) and the snapback assay targeting the RPO30 gene (Gelaye et al. 2013) are currently available. In addition to the real-time PCR-based methods, the sequencing of the RPO30 and GPCR genes of LSDV genome could also be used for the phylogenetic classification of LSDV isolates and differentiation from SPPV and GTPV (Le Goff et al. 2009; Lamien et al. 2011a; Gelaye et al. 2015). Similarly, a sequencing-based method using the P32 gene for capripoxvirus differentiation was developed by Hosamani et al. (2004), and another based on the LSDV002 gene was reported by Stram et al. (2008).

Due to the frequent report of adverse effects following vaccination against LSD with live attenuated LSDV vaccines (Yeruham et al. 1994; Brenner et al. 2009; Gelaye et al. 2015; Abutarbush et al. 2016; Agianniotaki et al. 2017a), several molecular approaches were developed to differentiate LSDV field isolates from LSDV vaccine strains.

For instance, a 12 nucleotide deletion present in the GPCR gene of LSDV field isolates only (Le Goff et al. 2009; El-Tholoth and El-Kenawy 2016) is used to differentiate LSDV field isolates from live attenuated LSDV vaccines through sequencing and multiple alignment of the GPCR gene (Gelaye et al. 2015; Agianniotaki et al. 2017a), or using a duplex real-time PCR with hydrolysis probes (Agianniotaki et al. 2017b). Methods, using gel-based PCR in combination with restriction enzymatic digestion of the PCR products, have also been reported (Menasherow et al. 2014; Agianniotaki et al. 2017a). Recently, real-time PCR methods, targeting the LSDV126 gene encoding for the EEV glycoprotein, were developed to differentiate LSDV wild-type and live attenuated using high-resolution melting curve analysis (Menasherow et al. 2016; Katsoulos et al. 2017) and a probe-based assay (Vidanović et al. 2016).

Differential Diseases

Skin diseases that can confuse with LSD and are considered as differential diagnosis of LSD include the skin lesions of pseudo-lumpy skin disease (caused by bovine herpes virus type-2), insect bites, Demodex infection, onchocerciosis, besnoitiosis, and dermatophilosis (Barnard et al. 1994). Diseases causing mucosal lesions that can be confused with LSD are rinderpest, bovine viral diarrhea, mucosal disease, and bovine malignant catarrhal fever (Barnard et al. 1994). The early stage of infection and mild cases of LSD may be difficult to differentiate clinically from cowpox (Orthopoxvirus), pseudocowpox (Parapoxvirus), bovine papular stomatitis (Parapoxvirus), pseudo-lumpy skin disease (Bovine herpesvirus-2), dermatophilosis, and insect bites. Recently, a real-time PCR method has been developed and can simultaneously detect and differentiate LSDV from other poxviruses affecting cattle such as cowpox, bovine papular stomatitis, and pseudocowpox viruses (Gelaye et al. 2017).

Prevention and Control

Only live attenuated vaccines are currently available against LSDV. The two live attenuated LSDV vaccine strains for the production of LSD vaccine by the different vaccine manufacturers are the Kenyan isolate—KS1 O180/O240 and the South African isolate—Neethling. The South African LSDV Neethling field isolate, prototype virus of LSDV from cattle, was attenuated by 60 serial passages in lamb kidney cells, followed by 20 serial passages in the chorioallantoic membranes of embryonated chicken eggs (Weiss 1968; Van Rooyen et al. 1969). Many veterinary vaccine production laboratories in Africa are using the Neethling vaccine strain for the production of LSD vaccine; currently, this vaccine is widely used by many European and Middle East countries where the disease is newly reported. The Kenyan O180/O240 vaccine strain, also named as KS1, was isolated from a sheep, but is genetically an LSDV. It was attenuated by 18 serial passages in bovine fetal muscle cell and followed by four serial passages in lamb testis cells (Kitching et al. 1986). The KS1 vaccine has been used for many years for the prevention and control of capripoxvirus infections in sheep, goats, and cattle (Gelaye et al. 2015). However, the KS1 vaccine was reported to cause mild skin reactions in the European cattle breeds and in naïve breeds in newly affected area of the Middle East. Vaccinations of the bulls with the live attenuated Neethling strain prevented shedding of LSDV in the semen in animals challenged with LSDV after vaccination, and vaccinated animals did not shed vaccine virus in the semen (Osuagwuh et al. 2007).

In general, poxvirus infections are self-limiting and treatment is primarily supportive care. Antibacterial and antifungal therapy is warranted when secondary infection is present (Smith et al. 2010). LSD is a notifiable disease as recommended by OIE; therefore, any suspicion of disease must be disclosed to the appropriate veterinary authorities. Antigenically, poxviruses are very complex, including both specific and cross-reacting antibodies, hence the possibility of vaccinating against one disease with another virus within the same genus.

During an LSD outbreak, movements of cattle should be strictly regulated, but in practice effective control is often difficult and appropriate legal powers should be in place to allow veterinary authorities to act as soon as any illegal transport of cattle is detected (Tuppurainen et al. 2017). Insect control on cattle and limiting vector breeding in surrounding environment may reduce mechanical transmission of the virus. New animals should be examined and declared free of clinical signs prior to movement and on arrival and should be kept in quarantine/separated from the herd for at least 28 days (Tuppurainen et al. 2017).

Endemic Countries

In LSD endemic areas, vaccination against LSD is the only efficient and cost-effective method to control the spread of disease and annual vaccination is recommended in affected countries. Clinically diseased cattle should be separated from apparently healthy ones and be treated for secondary complication until recovered. Harmonized vaccination campaigns across regions provide best control measures and should be carried out before large-scale movements of cattle prior to seasonal grazing and watering.

Enzootic Countries

In enzootic areas, annual vaccination of susceptible animals with a live vaccine could help to control the disease together with the strict regulation of animal movement and implementation of sanitary measures. Movement of cattle across the border or within the country should be controlled and if, the movement is authorized, it is essential to be accompanied with health certificate from the local certified veterinarian. In the case of an LSD outbreak in non-endemic areas the slaughter of infected and in-contact animals, ring vaccination in a radius of 25–50 km, movement restriction of animals, and destruction of contaminated hides should usually be sufficient to eradicate the disease (Carn 1993). Where nomadic and seasonal farming is practiced in enzootic countries, cattle should be vaccinated at least 28 days before going on the move (Tuppurainen et al. 2017).

Newly Affected Countries

Prophylactic vaccination of the entire cattle population, at least 80% coverage, is the best option to prevent the introduction of infection into a new area. Destruction of diseased or in-contact animals is recommended to prevent the spread of the virus and provide compensation to the animal owners. Thorough cleaning and disinfection with appropriate products of the affected farm premises and surrounding contaminated environment is highly recommended to kill the virus released from the diseased animals. Newly purchased cattle should be quarantined at least for 3 weeks before introducing into the farm. In the event of LSD entering a country, farm biosecurity should be seriously implemented. Treatment of cattle regularly for flies and ticks is recommended to prevent vector role in the transmission of the virus.

Disease-Free Countries

LSD prevention in countries where the virus is absent relies on the restriction of the importation of live animals and hides and skin from infected areas, but if introduced accidentally then stamping-out policy can be applied. Vaccination is not recommended. Cattle should be treated regularly with insect repellents to minimize the risk of vector transmission of the disease. Surveillance programs based on active and passive clinical surveillance and laboratory testing of blood samples, nasal and ocular swabs, and skin biopsies collected from suspected cases should be implemented regularly whenever there is suspicion of introduction of LSD in a free-area/country. Regular virus surveillance programs should be implemented in disease-free country bordering with countries where LSD is reported.

In general, in LSD-affected areas it is recommended to avoid using communal grazing and watering points to prevent further spread of infection to the susceptible cattle population. Cattle from endemic countries should be vaccinated against LSD and be kept in quarantine station for at least 30 days prior to shipment to disease-free countries. Control of vectors should be highly considered during importation and exportation of live animals between disease endemic and free countries to prevent the spread of the virus. Vaccine failure or breakdown has been observed when cattle were immunized against LSD with live attenuated vaccines associated with the appearance of small-sized skin nodules between 8 and 18 days post-vaccination (Ben-Gera et al. 2015; Abutarbush et al. 2016; Katsoulos et al. 2017). Skin lesions caused by live attenuated virus are usually local reaction at the vaccination site, and the lesions are superficial, small, and different from those caused by the fully virulent field strain; they disappear within 2–3 weeks without converting into necrotic scabs or ulcers (Tuppurainen et al. 2017). Development of full protection from the vaccine takes approximately 3 weeks and during this time cattle may still get infected by the field virus and may show clinical signs despite being vaccinated. Some animals may also be incubating the virus when vaccinated, and in such cases clinical signs are detected less than 10 days after vaccination (Tuppurainen et al. 2017). Calves born to immunized cows will have passive immunity that persists for about 6 months (Weiss 1968).

Conclusion

LSD is a transboundary animal disease which has been steadily expanding into new geographical area during this last decade. The movement of people and goods in the Middle East countries due to civil unrest and war has probably been one of the major factors of LSD expansion in this region. Nevertheless, the increased number of outbreaks within newly affected areas in the Middle East and Europe also suggests the influence of other factors such as illegal movement of animal from infected areas to disease-free areas and the presence of compatible vectors for LSD transmission. This expansion also highlights the difficulties in the newly affected areas to implement the adequate control measures. Active surveillance, rapid detection and prompt culling of animals from the infected herds, and the restriction of animal movement from infected to non-infected areas are effective measures for LSD control, although it is difficult to implement efficiently these control measures in many African countries because of economic and social reasons.

In this situation, where LSD is now endemic in a region, vaccination remains the only means to reduce the incidence of the disease. Only live attenuated vaccines against LSD are available in the market in Africa. Homologous LSD vaccines are more effective than sheeppox virus or goatpox virus strain vaccines. Nevertheless, the safety and efficacy of the currently available LSD vaccines should be improved. The development of a safe, effective, and non-replicating "differentiating infected from vaccinated animals" (DIVA) vaccines for differentiating between infected and vaccinated animals is greatly recommended. Recombinant vaccines using the LSDV vaccine strain as a vector expressing the immunogenic protein/gene from another pathogen affecting cattle (like RVFV, FMDV, and others) can also provide protective immunity targeting two or more pathogens and thus reduce the cost for the control of the targeted diseases.

In general, prevention and control of LSD in endemic sub-Saharan countries should be based on (1) regular annual vaccination of the herds and newly introduced cattle, (2) isolation and movement restriction of diseased animals, (3) prompt diagnosis of clinically suspected cattle followed by the immediate report to the nearby veterinary service in case of suspected cases or outbreak, and (4) provision of palatable feed to diseased animals and supportive treatment to avoid secondary microbial complications. In countries, where the disease was not yet reported, cattle should be discarded safely whenever they show clinical signs of the disease. Duration of humoral response after natural infection and vaccination and maternal immunity for protection of young calves should be further studied.

References

Abutarbush SM, Hananeh WM, Ramadan W, Al Sheyab OM, Alnajjar AR, Al Zoubi IG, Knowles NJ, Bachanek-Bankowska K, Tuppurainen ES. Adverse reactions to field vaccination against lumpy skin disease in Jordan. Transbound Emerg Dis. 2016;63(2):213–9.

Agianniotaki EI, Chaintoutis SC, Haegeman A, Tasioudi KE, De Leeuw I, Katsoulos PD, Sachpatzidis A, De Clercq K, Alexandropoulos T, Polizopoulou ZS, Chondrokouki ED, Dovas CI. Development and validation of a TaqMan probe-based real-time PCR method for the differentiation of wild type lumpy skin disease virus from vaccine virus strains. J Virol Methods. 2017a;249:48–57.

Agianniotaki EI, Tasioudi KE, Chaintoutis SC, Iliadou P, Mangana-Vougiouka O, Kirtzalidou A, Alexandropoulos T, Sachpatzidis A, Plevraki E, Dovas CI, Chondrokouki E. Lumpy skin disease outbreaks in Greece during 2015–16, implementation of emergency immunization and genetic differentiation between field isolates and vaccine virus strains. Vet Microbiol. 2017b;201:78–84.

Alemayehu G, Zewde G, Admassu B. Risk assessments of lumpy skin diseases in Borena bull market chain and its implication for livelihoods and international trade. Trop Anim Health Prod. 2013;45:1153–9.
Alexander RA, Plowright W, Haig DA. Cytopathogenic agents associated with lumpy skin disease of cattle. Bull Epiz Dis Afr. 1957;5:489–92.
Ali BH, Obeid HM. Investigation of the first outbreaks of lumpy skin disease in the Sudan. Br Vet J. 1977;133(2):184–9.
Ali AA, Esmat M, Attia H, Selim A, Abdel-Hamid YM. Clinical and pathological studies of lumpy skin disease in Egypt. Vet Rec. 1990;127:549–50.
Amson B, Fowler VL, Tuppurainen ESM, Howson ELA, Madi M, Sallu R, Kasanga CJ, Pearson C, Wood J, Martin P, Mioulet V, King DP. Detection of Capripox virus DNA using a field-ready nucleic acid extraction and real-time PCR platform. Transbound Emerg Dis. 2015;64(3):994–7.
Andrew MQK, Michael JA, Eric BC, Elliot JL. Family *Poxviridae*. In: Virus taxonomy classification and nomenclature of viruses: ninth report of the international committee on taxonomy of viruses. London: Elsevier; 2012. p. 14–309.
Ayelet G, Abate Y, Sisay T, Nigussie H, Gelaye E, Jenberie S, Asmare K. Lumpy skin disease: preliminary vaccine efficacy assessment and overview on outbreak impact in dairy cattle at Debre Zeit, central Ethiopia. Antivir Res. 2013;98:261–5.
Babiuk S, Parkyn G, Copps J, Larence JE, Sabara MI, Bowden TR, Boyle DB, Kitching RP. Evaluation of an ovine testis cell line (OA3.Ts) for propagation of Capripox virus isolates and development of an immunostaining technique for viral plaque visualization. J Vet Diagn Investig. 2007;19:486–91.
Babiuk S, Bowden TR, Boyle DB, Wallace DB, Kitching RP. Capripoxviruses: an emerging worldwide threat to sheep, goats and cattle. Transbound Emerg Dis. 2008;55:263–72.
Babiuk S, Bowden TR, Parkyn G, Dalman B, Hoa DM, Long NT, Vu PP, Bieu DX, Copps J, Boyle DB. Yemen and Vietnam capripoxviruses demonstrate a distinct host preference for goats compared with sheep. J Gen Virol. 2009;90:105–14.
Balinsky CA, Delhon G, Smoliga G, Prarat M, French RA, Geary SJ, Rock DL, Rodriguez LL. Rapid preclinical detection of sheeppox virus by real-time PCR assay. J Clin Microbiol. 2008;46(2):438–42.
Barnard BJH. Antibodies against some viruses of domestic animals in South African wild animals. Onderstepoort J Vet Res. 1997;64:95–110.
Barnard B, Munz E, Dumbell K, Prozesky L. Lumpy skin disease. In: Coetzer JAW, Thomson GR, Tustin RC, editors. Infectious diseases of livestock with special reference to Southern Africa. Cape Town: Oxford University Press; 1994. p. 604–12.
Ben-Gera J, Klement E, Khinich E, Stram Y, Shpigel NY. Comparison of the efficacy of Neethling lumpy skin disease virus and x10RM65 sheep-pox live attenuated vaccines for the prevention of lumpy skin disease. The results of a randomized controlled field study. Vaccine. 2015;33:3317–23.
Binepal YS, Ongadi FA, Chepkwony JC. Alternative cell lines for the propagation of lumpy skin disease virus. Onderstepoort J Vet Res. 2001;68(2):151–3.
Bowden TR, Babiuk SL, Parkyn GR, Copps JS, Boyle DB. Capripoxvirus tissue tropism and shedding: a quantitative study in experimentally infected sheep and goats. Virology. 2008;371 (2):380–93.
Bowden TR, Coupar BE, Babiuk SL, White JR, Boyd V, Duch CJ, Shiell BJ, Ueda N, Parkyn GR, Copps JS, Boyle DB. Detection of antibodies specific for sheeppox and goatpox viruses using recombinant capripoxvirus antigens in an indirect enzyme-linked immunosorbent assay. J Virol Methods. 2009;161(1):19–29.
Brenner J, Bellaiche M, Gross E, Elad D, Oved Z, Haimovitz M, Wasserman A, Friedgut O, Stram Y, Bumbarov V, Yadin H. Appearance of skin lesions in cattle populations vaccinated against lumpy skin disease: statutory challenge. Vaccine. 2009;27:1500–3.
Carn VM. Control of capripoxvirus infections. Vaccine. 1993;11:1275–9.

Carn VM, Kitching RP. An investigation of possible routes of transmission of lumpy skin disease virus (Neethling). Epidemiol Infect. 1995a;114:219–26.
Carn VM, Kitching RP. The clinical response of cattle experimentally infected with lumpy skin disease (Neethling) virus. Arch Virol. 1995b;140:503–13.
Carn VM, Kitching RP, Hammond JM, Chand P, Aderson J, Black DN. Use of a recombinant antigen in an indirect ELISA for detecting bovine antibody to capripoxvirus. J Virol Methods. 1995;53:273.
Chand P, Kitching RP, Black DN. Western blot analysis of virus-specific antibody responses for capripox and contagious pustular dermatitis viral infections in sheep. Epidemiol Infect. 1994;113(2):377–85.
Chihota CM, Rennie LF, Kitching RP, Mellor PS. Mechanical transmission of lumpy skin disease virus by *Aedes aegypti* (Diptera: Culicidae). Epidemiol Infect. 2001;126:317–21.
Chihota CM, Rennie LF, Kitching RP, Mellor PS. Attempted mechanical transmission of lumpy skin disease virus by biting insects. Med Vet Entomol. 2003;17(3):294–300.
Das A, Babiuk S, Mclntosh MT. Development of a loop-mediated isothermal amplification assay for rapid detection of Capripox viruses. J Clin Microbiol. 2012;50(5):1613–20.
Davies FG. Lumpy skin disease of cattle: a growing problem in Africa and the Near East. World Anim Rev. 1991;68(3):37–42.
De Clercq K, Goris N. Extending the foot-and-mouth disease module to the control of other diseases. Dev Biol. 2004;119:333–40.
Diallo A, Viljoen GJ. Genus *Capripoxvirus*. In: Mercer A, Schmidt A, Weber O, editors. Poxviruses. Basel: Birkhauser; 2007. p. 167–81.
Domenech J, Lubroth J, Eddi C, Martin V, Roger F. Regional and international approaches on prevention and control of animal transboundary and emerging diseases. Ann N Y Acad Sci. 2006;1081:90–107.
El-Nahas EM, El-Habbaa AS, El-Bagoury GF, Radwan MEI. Isolation and identification of lumpy skin disease virus from naturally infected buffaloes at Kaluobia, Egypt. Glob Vet. 2011;7:234–7.
El-Tholoth M, El-Kenawy AA. G-protein-coupled chemokine receptor gene in lumpy skin disease virus isolates from cattle and water buffalo (*Bubalus bubalis*) in Egypt. Transbound Emerg Dis. 2016;63(6):288–95.
Gari G, Biteau-Coroller F, Le Goff C, Caufour P, Roger F. Evaluation of indirect fluorescent antibody test (IFAT) for the diagnosis and screening of lumpy skin disease using Bayesian method. Vet Microbiol. 2008;129(3–4):269–80.
Gari G, Bonnet P, Roger F, Waret-Szkuta A. Epidemiological aspects and financial impact of lumpy skin disease in Ethiopia. Prev Vet Med. 2011;102(4):274–83.
Garner MG, Lack MB. Modelling the potential impact of exotic diseases on regional Australia. Aust Vet J. 1995;72(3):81–7.
Gelaye E, Lamien CE, Silber R, Tuppurainen ES, Grabherr R, Diallo A. Development of a cost-effective method for capripoxvirus genotyping using snapback primer and dsDNA intercalating dye. PLoS One. 2013;8(10):e75971.
Gelaye E, Belay A, Ayelet G, Jenberie S, Yami M, Loitsch A, Tuppurainen E, Grabherr R, Diallo A, Lamien CE. Capripox disease in Ethiopia: genetic differences between field isolates and vaccine strain, and implications for vaccination failure. Antivir Res. 2015;119:28–35.
Gelaye E, Mach L, Kolodziejek J, Grabherr R, Loitsch A, Achenbach JE, Nowotny N, Diallo A, Lamien CE. A novel HRM assay for the simultaneous detection and differentiation of eight poxviruses of medical and veterinary importance. Sci Rep. 2017;7:42892.
Green HF. Lumpy skin disease: its effect on hides and leather and a comparison on this respect with some other skin diseases. Bull Epiz Dis Afr. 1959;7:63.
Gulbahar MY, Davis WC, Yuksel H, Cabalar M. Immunohistochemical evaluation of inflammatory infiltrate in the skin and lung of lambs naturally infected with sheeppox virus. Vet Pathol. 2006;43(1):67–75.

Haegeman A, Zro K, Vandenbussche F, Demeestere L, Campe W, Van Ennaji MM, De Clercq K. Development and validation of three Capripoxvirus real-time PCRs for parallel testing. J Virol Methods. 2013;193(2):446–51.

Haig DA. Lumpy skin disease. Bull Epiz Dis Afr. 1957;5:421–30.

Hedger RS, Hamblin C. Neutralizing antibodies to lumpy skin disease virus in African wildlife. Comp Immunol Microbiol Infect Dis. 1983;6(3):209–13.

Heine HG, Stevens MP, Foord AJ, Boyle DB. A capripoxvirus detection PCR and antibody ELISA based on the major antigen P32, the homolog of the vaccinia virus H3L gene. J Immunol Methods. 1999;227(1–2):187–96.

Hosamani M, Mondal B, Tembhurne PA, Bandyopadhyay SK, Singh RK, Rasool TJ. Differentiation of sheep pox and goat poxviruses by sequence analysis and PCR-RFLP of P32 gene. Virus Genes. 2004;29(1):73–80.

Houston PD. Report of Chief Veterinary Surgeon, Southern Rhodesia; 1945.

Ireland DC, Binepal YS. Improved detection of capripoxvirus in biopsy samples by PCR. J Virol Methods. 1998;74(1):1–7.

Kalra SK, Sharma VK. Adaptation of Jaipur strain of sheeppox virus in primary lamb testicular cell culture. Indian J Exp Biol. 1981;19(2):165–9.

Katsoulos PD, Chaintoutis SC, Dovas CI, Polizopoulou ZS, Brellou GD, Agianniotaki EI, Tasioudi KE, Chondrokouki E, Papadopoulos O, Karatzias H, Boscos C. Investigation on the incidence of adverse reactions, viraemia and haematological changes following field immunization of cattle using a live attenuated vaccine against lumpy skin disease. Transbound Emerg Dis. 2017;65(1):174–85. https://doi.org/10.1111/tbed.12646.

Kitching RP, Mellor PS. Insect transmission of capripoxvirus. Res Vet Sci. 1986;40(2):255–8.

Kitching RP, McGrane JJ, Taylor WP. Capripox in the Yemen Arab Republic and the sultanate of Oman. Trop Anim Health Prod. 1986;18:115–22.

Lamien CE, Le Goff C, Silber R, Wallace DB, Gulyaz V, Tuppurainen E, Madani H, Caufour P, Adam T, El Harrak M, Luckins AG, Albina E, Diallo A. Use of the Capripoxvirus homologue of Vaccinia virus 30 kDa RNA polymerase subunit (RPO30) gene as a novel diagnostic and genotyping target: development of a classical PCR method to differentiate Goat poxvirus from Sheep poxvirus. Vet Microbiol. 2011a;149:30–9.

Lamien CE, Lelenta M, Goger W, Silber R, Tuppurainen E, Matijevic M, Luckins AG, Diallo A. Real time PCR method for simultaneous detection, quantitation and differentiation of capripoxviruses. J Virol Methods. 2011b;171:134–40.

Le Goff C, Lamien CE, Fakhfakh E, Chadeyras A, Aba-Adulugba E, Libeau G, Tuppurainen E, Wallace DB, Adam T, Silber R, Gulyaz V, Madani H, Caufour P, Hammami S, Diallo A, Albina E. Capripoxvirus G-protein-coupled chemokine receptor: a host-range gene suitable for virus animal origin discrimination. J Gen Virol. 2009;90:1967–77.

Macdonald RAS. Pseudourticaria of cattle. In: Northern Rhodesia Department of Animal Health Annual Report, 1930; 1931. p. 20–1.

MacLachlan NJ, Dubovi EJ. Fenner's veterinary virology. 4th ed. London: Elsevier; 2011. p. 153–65.

MacOwan RDS. Observation on the epizootiology of lumpy skin disease during the first year of its occurrence in Kenya. Bull Epiz Dis Afr. 1959;7:7–20.

Magori-Cohen R, Louzoun Y, Herziger Y, Oron E, Arazi A, Tuppurainen E, Shpigel NY, Klement E. Mathematical modelling and evaluation of the different routes of transmission of lumpy skin disease virus. Vet Res. 2012;43(1):1–13.

Mebratu GY, Kassa B, Fikre Y, Berhanu B. Observation on the outbreak of lumpy skin disease in Ethiopia. Rev Elev Med Vet Pays Trop. 1984;37:395–9.

Menasherow S, Rubinstein-Giuni M, Kovtunenko A, Eyngor Y, Fridgut O, Rotenberg D, Khinich E, Stram Y. Development of an assay to differentiate between virulent and vaccine strains of lumpy skin disease virus (LSDV). J Virol Methods. 2014;199:95–101.

Menasherow S, Erster O, Rubinstein-Giuni M, Kovtunenko A, Eyngor E, Gelman B, Khinich E, Stram Y. A high-resolution melting (HRM) assay for the differentiation between Israeli field and Neethling vaccine lumpy skin disease viruses. J Virol Methods. 2016;232:12–5.
Murray L, Edwards L, Tuppurainen ESM, Bachanek-Bankowska K, Oura CAL, Mioulet V, King DP. Detection of Capripox virus DNA using a novel loop-mediated isothermal amplification assay. BMC Vet Res. 2013;9:90–7.
Nawathe DR, Gibbs EPJ, Asagba MO, Lawman MJP. Lumpy skin disease in Nigeria. Trop Anim Health Prod. 1978;10:49–54.
Nawathe DR, Asagba MO, Abegunde A, Ajayi SA, Durkwa L. Some observations on the occurrence of lumpy skin disease in Nigeria. Zentralbl Veterinarmed B. 1982;29:31–6.
Oguzoglu TC, Alkan F, Ozkul A, Vural SA, Gungor AB, Burgu I. A sheeppox virus outbreak in Central Turkey in 2003: isolation and identification of capripoxvirus ovis. Vet Res Commun. 2006;30:965–71.
Osuagwuh UI, Bagla V, Venter EH, Annandale CH, Irons PC. Absence of lumpy skin disease virus in semen of vaccinated bulls following vaccination and subsequent experimental infection. Vaccine. 2007;25:2238–43.
Rweyemamu M, Paskin R, Benkirane A, Martin V, Roeder P, Wojciechowski K. Emerging diseases of Africa and the Middle East. Ann N Y Acad Sci. 2000;916:61–70.
Smith GL, Beard P, Skinner MA. Poxviruses. In: Mahy BWJ, Van Regenmortel MHV, editors. Desk encyclopedia of human and medical virology. London: Elsevier; 2010. p. 239–46.
Stram Y, Kuznetzova L, Friedgut O, Gelman B, Yadin H, Rubinstein-Guini M. The use of lumpy skin disease virus genome termini for detection and phylogenetic analysis. J Virol Methods. 2008;151:225–9.
Stubbs S, Oura CAL, Henstock M, Bowden TR, King DP, Tuppurainen ESM. Validation of a high-throughput real-time polymerase chain reaction assay for the detection of capripoxviral DNA. J Virol Methods. 2012;179(2):419–22.
Tantawi HH, Shony MO, Hassan FK. Isolation and identification of the Sersenk strain of goat pox virus in Iraq. Trop Anim Health Prod. 1979;11:208–10.
The World Organisation for Animal Health (OIE). Manual of diagnostic tests and vaccines for terrestrial animals. Paris: The World Organisation for Animal Health (OIE); 2016.
Thomas AD, Maré CVE. Knopvelsiekte. Am J Vet Med Assoc. 1945;16:36–43.
Tian H, Chen Y, Wu J, Shang Y, Liu X. Serodiagnosis of sheeppox and goatpox using an indirect ELISA based on synthetic peptide targeting for the major antigen P32. Virol J. 2010;7:245.
Tulman ER, Afonso CL, Lu Z, Zsak L, Kutish GF, Rock DL. Genome of lumpy skin disease virus. J Virol. 2001;75(15):7122–30.
Tuppurainen ES, Oura CA. Review: lumpy skin disease: an emerging threat to Europe, the Middle East and Asia. Transbound Emerg Dis. 2012;59:40–8.
Tuppurainen ESM, Venter EH, Coetzer JAW. The detection of lumpy skin disease virus in samples of experimentally infected cattle using different diagnostic techniques. Onderstepoort J Vet Res. 2005;72(2):153–64.
Tuppurainen ES, Stoltsz WH, Troskie M, Wallace DB, Oura CA, Mellor PS, Coetzer JA, Venter EH. A potential role for ixodid (hard) tick vectors in the transmission of lumpy skin disease virus in cattle. Transbound Emerg Dis. 2011;58(2):93–104.
Tuppurainen ES, Lubinga JC, Stoltsz WH, Troskie M, Carpenter ST, Coetzer JA, Venter EH, Oura CA. Mechanical transmission of lumpy skin disease virus by *Rhipicephalus appendiculatus* male ticks. Epidemiol Infect. 2013;141(2):425–30.
Tuppurainen ES, Pearson CR, Bachanek-Bankowska K, Knowles NJ, Amareen S, Frost L, Henstock MR, Lamien CE, Diallo A, Mertens PP. Characterization of sheep pox virus vaccine for cattle against lumpy skin disease virus. Antivir Res. 2014;109:1–6.
Tuppurainen ES, Venter EH, Shisler JL, Gari G, Mekonnen GA, Juleff N, Lyons NA, De Clercq K, Upton C, Bowden TR, Babiuk S, Babiuk LA. Review: Capripoxvirus diseases: current status and opportunities for control. Transbound Emerg Dis. 2015;64(3):729–45.

Tuppurainen E, Alexandrov T, Beltrán-Alcrudo D. Lumpy skin disease field manual—a manual for veterinarians. FAO Animal Production and Health Manual No. 20. Rome: Food and Agriculture Organization of the United Nations (FAO); 2017.
Van Den Ende M, Don AP, Kipps A, Alexander R. Isolation in chicken embryos of a filtrable agent possibly related etiologically to Lumpy skin disease of cattle. Nature. 1948;161:526.
Van Rooyen PJ, Munz EK, Weiss KE. The optimal conditions for the multiplication of the Neethling-type lumpy skin disease virus in embryonated eggs. Ondersepoort J Vet Res. 1969;36(2):165–74.
Verma S, Verma LK, Gupta VK, Katoch VC, Dogra V, Pal B, Sharma M. Emerging Capripoxvirus disease outbreaks in Himachal Pradesh, a northern state of India. Transbound Emerg Dis. 2011;58:79–85.
Vidanović D, Šekler M, Petrović T, Debeljak Z, Vasković N, Matović K, Hoffmann B. Real-time PCR assays for the specific detection of field Balkan strains of lumpy skin disease virus. Acta Vet-Beogr. 2016;66(4):444–54.
Von Backstrom U. Ngamiland cattle disease: Preliminary report on a new disease, the etiological agent being probably of an infectious nature. J South Afr Vet Med Assoc. 1945;16:29–35.
Wallace DB, Weyer J, Nel LH, Viljoen GJ. Improved method for the generation and selection of homogeneous lumpy skin disease virus (SA-Neethling) recombinants. J Virol Methods. 2007;146(1–2):52–60.
Weiss KE. Lumpy skin disease virus. Virol Monogr. 1968;3:111–31.
Yan XM, Chu YF, Wu GH, Zhao ZX, Li J, Zhu HX, Zhang Q. An outbreak of sheep pox associated with goat poxvirus in Gansu province of China. Vet Microbiol. 2012;156:425–8.
Yeruham I, Perl S, Nyska A, Abraham A, Davidson M, Haymovitch M, Zamir O, Grinstein H. Adverse reactions in cattle to a capripox vaccine. Vet Rec. 1994;135:330–2.
Yeruham I, Nir O, Braverman Y, Davidson M, Grinstein H, Haymovitch M, Zamir O. Spread of lumpy skin disease in Israeli dairy herds. Vet Rec. 1995;137:91–3.
Zhou T, Jia H, Chen G, He X, Fang Y, Wang X, Guan Q, Zeng S, Cui Q, Jing Z. Phylogenetic analysis of Chinese sheeppox and goatpox virus isolates. Virol J. 2012;9:25–32.

Chapter 14
Sheep and Goat Pox

Esayas Gelaye and Charles Euloge Lamien

Abstract Sheep pox (SPP) and goat pox (GTP) are viral diseases of sheep and goats caused by sheep pox virus (SPPV) and goat pox virus (GTPV), respectively. SPPV and GTPV belong to the genus *Capripoxvirus* of the family *Poxviridae*, together with lumpy skin disease virus (LSDV) of cattle. They have double-stranded DNA genomes of approximately 134–147 kbp. SPPV and GTPV are closely related to LSDV though they possess specific nucleotide differences suggesting distinct phylogeny. SPP and GTP are notifiable diseases to the World Organisation for Animal Health (OIE). They are highly contagious diseases: the viruses spread through direct contact with lesions or contaminated objects, feed, and wool. SPP and GTP are endemic in Africa (except southern Africa), central Asia, the Indian subcontinent, the Middle East, Turkey, Greece, and some eastern European countries. Clinically, the presence of nodular skin lesions, mostly around the mouth and perineum regions, is a typical sign of the disease. Capripoxviruses classification and their nomenclature have been mainly based on the affected host species, creating a challenge for isolates naming. For a more accurate naming, it is better to use molecular methods as support to identify and classify capripoxvirus isolates. Conventional and real-time PCR methods are available that could help with the simultaneous detection and genotyping of the viruses. SPPV and GTPV as well as LSDV cross-react serologically, making it difficult to differentiate them using serological methods. To prevent and control SPP and GTP, illegal animal movement restrictions and vaccination campaigns with adequate vaccines and sufficient vaccination coverage are two very effective measures. The development of a high-throughput serological assay (ELISA) with better sensitivity and specificity and the development of a safe and effective vaccine, which can support the differentiation of infected from vaccinated animals (DIVA), are highly required.

E. Gelaye (✉)
Research and Development Directorate, National Veterinary Institute (NVI), Debre zeit, Ethiopia
e-mail: esayas.gelaye@nvi.gov.et

C. E. Lamien
Animal Production and Health Laboratory, FAO/IAEA Agriculture and Biotechnology Laboratory, International Atomic Energy Agency, Vienna, Austria
e-mail: c.lamien@iaea.org

M. Kardjadj et al. (eds.), *Transboundary Animal Diseases in Sahelian Africa and Connected Regions*, https://doi.org/10.1007/978-3-030-25385-1_14

Keywords Capripoxvirus · Diagnosis · Genotyping · Poxviridae · Sheep and goat pox

History

Sheep pox virus (SPPV) and goat pox virus (GTPV) are members of the genus *Capripoxvirus* (CaPV), subfamily *Chordopoxvirinae*, of the *Poxviridae* family (Tulman et al. 2002; Andrew et al. 2012; OIE 2016). They have large, complex, double-stranded DNA genomes of approximately 134–147 kbp size with 147 putative genes which encode proteins. The genome has a conserved central region bounded by two identical inverted terminal repeats (ITR) at the ends (Tulman et al. 2002). They share a high degree of sequence homology, with 96% identity between SPPV and GTPV (Tulman et al. 2002). The viruses primarily affect sheep and goats causing sheep pox (SPP) and goat pox (GTP), respectively, which collectively constitute the most severe poxvirus infections of small ruminants. SPP and GTP are reportable animal diseases to OIE due to their potential for significant economic impact on small ruminant production industry (OIE 2016).

Sheep pox virus (SPPV) and goat pox virus (GTPV) share the genus *Capripoxvirus* with lumpy skin disease virus (LSDV) which is closely related, though they possess specific nucleotide differences suggesting that they are phylogenetically distinct (Tulman et al. 2002; Lamien et al. 2011a; Gelaye et al. 2015). SPPV and GTPV as well as LSDV cross-react serologically, making it difficult to differentiate them using serological methods (Diallo and Viljoen 2007). Capripoxviruses classification and their nomenclature have been mainly based on the affected host species, creating a challenge for isolates naming. Indeed if some GTPV and SPPV strains produce disease in only either sheep or goats, there are also cases of strains that can cause disease in both animal species.

SPP started in Central Asia and then spread to African countries (Hutyra et al. 1946), while GTP was first reported in Norway in 1879 by Hansen (Rafyi and Ramyar 1959). SPP appeared soon after human smallpox, and its history dates back to second century AD (Hutyra et al. 1946). African, Asian, and European Countries reported SPP through the mid-twentieth century. Both SPP and GTP have been eradicated from many developed countries, yet they are still present, and creating serious health and economic problem in Africa, north of the equator, through Asia, and occasionally spreading from Turkey into Greece (Murray et al. 1973; Kitching 2003; OIE 2016).

Geographic Distribution and Economic Impact

SPP and GTP are endemic in Africa (except for southern Africa), central Asia, the Indian subcontinent, the Middle East, and Turkey, causing high morbidity and mortality in susceptible sheep and goats (Kitching 2003; Bhanuprakash et al.

2005; Zro et al. 2014; OIE 2016). From time to time, Greece and some Eastern European countries reported SPP incursions, with reported outbreak cases from August 2013 until April 2014 (Oguzoglu et al. 2006; Verma et al. 2011; Yan et al. 2012; Zhou et al. 2012; EFSA 2014). Last SPP outbreak occurred in Greece mainland in February 2015; however, it reoccurred in Lesvos Island between December 2016 and December 2017 (European Commission 2017).

SPP and GTP may cause significant damage to wool and hide quality and decrease in mutton and milk production (Babiuk et al. 2008). The mortality rates of SPP and GTP range from 5 to 10% in local goat and sheep breeds in endemic areas. However, imported exotic breeds may display higher rates (OIE 2016). The morbidity rate of SPP and GTP can reach as high as 100% particularly in young lambs, kids, yearlings, and immunologically naïve sheep and goats in natural outbreaks. Factors like hosts (age, sex, breed, nutritional and immunological status), agent (strain, pathogenicity, virulence), harsh environment, poor management, feed scarcity, and inadequate veterinary services have a direct influence on the epidemiology of the diseases. The presence of SPP and GTP in a country limits the export of live animals and animal products to the global trade and also causes potential economic losses due to costs associated with disease control and eradication (Aparna et al. 2016; OIE 2016).

SPPV and GTPV are registered as animal bioterrorist agents by the United States Department of Agriculture, since they (1) produce high morbidity and mortality with sharp production losses, (2) expand quickly to reach wide areas within a few days or weeks, (3) create severe socioeconomic consequences due to the death of the afflicted animals, and (4) could considerably restrict the international trade of animals and animal products (Babiuk et al. 2008; Aparna et al. 2016). SPP and GTP are notifiable animal diseases to OIE (2016).

Epidemiology

Susceptible Hosts

SPP and GTP affect sheep and goats of all ages, both sexes and all breeds, yet are more common and severe in young and old animals (Aparna et al. 2016). Sheep and goats are the natural hosts for SPPV and GTPV, respectively (OIE 2016). Though SPPV and GTPV generally display a host preference/specific for either sheep or goats, some strains can infect and equally cause disease in both species or affect heterologous hosts (Bhanuprakash et al. 2006, 2010; Babiuk et al. 2009a; Lamien et al. 2011a; Gelaye et al. 2015). Unlike LSDV (Lamien et al. 2011a), there is no documented report on the existence of SPPV and GTPV in wild ruminants (Tuppurainen et al. 2015). Additionally, both viruses are considered nonhazardous to human health (OIE 2016).

Transmission

SPP and GTP are highly contagious diseases. The causal agents are transmitted from animals to animals through direct contact with lesions or contaminated objects, feed, and wool. Environmental contamination leads to the virus introduction into the skin wounds. Excreted viruses are detectable in the nasal secretions, milk, feces, and possibly urine from infected animals. The common practice of herding sheep and goats together in one barn at night in endemic countries provides adequate exposure for the circulation of the virus and its maintenance in an area. During an outbreak, the virus is probably transmitted among animals by the inhalation of virus-contaminated droplets. SPPV and GTPV can persist up to 3 months on the wool or hair after the onset of clinical signs and possibly for a prolonged period in the skin nodules and scabs (Bowden et al. 2008). SPPV and GTPV infections do not lead to a carrier stage in infected animals (Bhanuprakash et al. 2006, 2011). Wading of sheep and goat skin through bushes and thorny plants, like Acacia, to nibble leaves, damages the skin, facilitating disease transmission from infected to susceptible animals. Additionally, wounded areas are easily accessible to biting flies, which suck blood helping in transmitting the virus quickly.

Owing to the intentness of the skin, and the significant viral load in the lesions, vectors can spread the viruses indirectly through mechanical transmission. Experimental studies showed that *Stomoxys calcitrans* could transmit, mechanically, SPPV and GTPV (Kitching and Mellor 1986; Mellor et al. 1987). Pre-infected flies spread the virus to susceptible goats, and the virus remains alive for up to 4 days in some flies. The inherent resistance of the virus, the significant virus load in skin nodules of sick animals, and the involvement of vectors able to keep the virus alive for prolonged periods are the essential factors favoring mechanical transmission (Bhanuprakash et al. 2006). Similarly, Nigerian and Oman isolates of SPPV were successfully transmitted between sheep by *S. calcitrans*. In contrast, biting (*Mallophaga* species) and suckling lice (*Damalimia* species), sheep head flies (*Hydrotaea irritans*), and midges (*Culicoides nubeculosus*) fail to transmit the virus (Kitching and Mellor 1986).

The occurrence of global climate change could impact the further extent of these diseases into naive geographic regions due to the spread of insects (Aparna et al. 2016). The appearance of SPP and GTP in disease-free areas is predominantly associated with the illegal animal movement through trade, from infected to previously free regions (Domenech et al. 2006), as well as the lack of adequate or breakdown of veterinary services and regulatory policies (Rweyemamu et al. 2000).

Clinical Signs and Lesions

In natural circumstances, SPP and GTP have an incubation period of 8–14 days, following contacts between infected and susceptible animals. The infections can exhibit mild to severe clinical signs, depending on the immune status of the host and strain of the virus involved (Davies and Otema 1981). Both SPPV and GTPV have tropism for skin, lung, and discrete sites within the mucosal surfaces of oro-nasal tissues and the gastrointestinal tract, and to a lesser extent, the lymphoid tissue (Bowden et al. 2008). Hence, the tropism of both viruses for the skin as well as minor involvement of liver and spleen suggests that the pathogenesis of capripox disease closely resembles smallpox and monkeypox diseases (Fenner 1988; Zaucha et al. 2001; Jahrling et al. 2004; Babiuk et al. 2008).

Some breeds of sheep perish during the acute infection without showing any skin lesion. In other breeds, the disease begins with an initial rise in rectal temperature to above 40 °C, followed by the development of macules—small circumscribed areas of hyperamia—within 2–5 days. Those macules, which are mainly detectable on unpigmented skin, will evolve into papules, hard swellings of between 0.5 and 1 cm in diameter, covering the body or restricted to the groin, axilla, and perineum.

During SPPV and GTPV infections, cell-associated viremia develops concurrently with the development of macules and papules in the skin of susceptible animals. The viremia persists until the host develops adequate antibodies against the virus (Kitching and Taylor 1985).

Rhinitis, conjunctivitis, and excessive salivation also occur throughout the infection. Pox lesions can widely spread, affecting over 50% of the skin surface. More commonly in enzootic areas, the lesions are restricted to a few nodules under the tail. Internal organs such as the lung and the stomach also develop characteristic pox-like lesions. Infected sheep and goats show fever, ocular and purulent nasal discharge, and cutaneous papules and nodules in areas of the skin with less hair, such as the head and the perineum as shown in Fig. 14.1.

The clinical signs and postmortem lesions vary considerably among breeds and depend on the strain of capripoxvirus (OIE 2016). Indigenous breeds are less susceptible and frequently present few lesions which could be confused with insect bites or contagious pustular dermatitis. However, naive lambs, animals kept isolated or brought into endemic areas, due to the stress of moving over long distances, are more susceptible, often showing generalized lesions and fatalities.

Clinically, it is challenging to distinguish SPPV infection from that of GTPV. Nevertheless, most strains of SPPV and GTPV display host preferences and produce

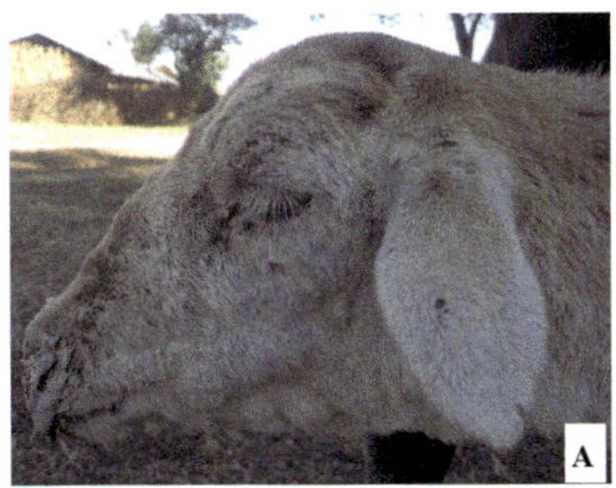

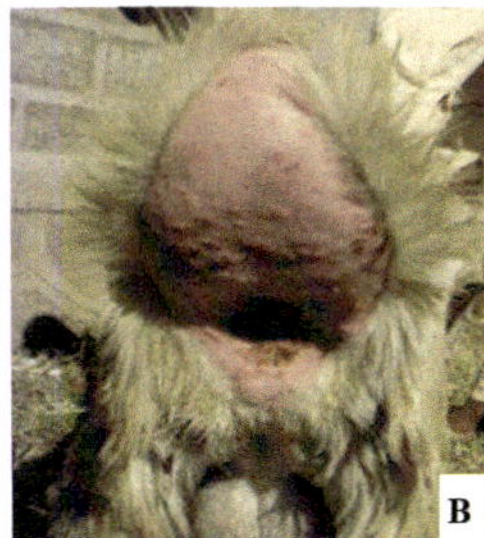

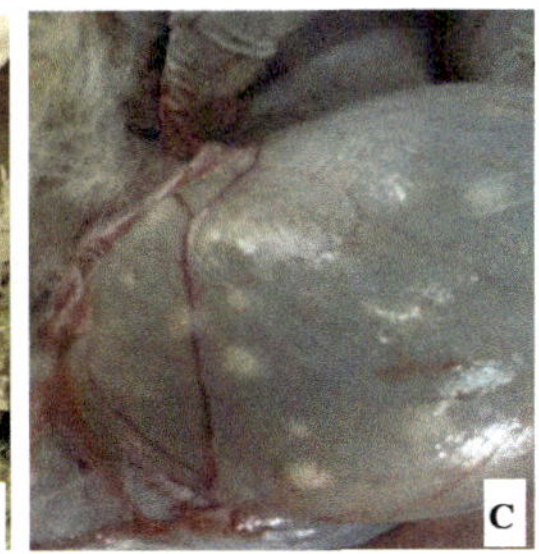

Fig. 14.1 Characteristic lesions of sheeppox observed during outbreak investigation. (**a**) Cutaneous nodular lesions around the face and dewlap area of the ewe with nasal secretion. (**b**) Severely affected male sheep with cutaneous popular and nodular lesions covering the perineum area. (**c**) Multifocal nodular lesion on the rumen mucosa of a sheep. *Courtesy © Esayas Gelaye*

Table 14.1 Evidence of cross infections by capripoxviruses

Strain name	Origin	Species of origin	Genotyping
GTPV Saudi Arabia/93	Saudi Arabia	Goat	SPPV[a]
SPPV OMAN/84	Oman	Sheep	GTPV[a]
SPPV KS-1	Kenya	Sheep	LSDV[a]
LSDV RSA 06 Springbok	South Africa	Springbok	LSDV[a]
LSDV RSA/00 OP126402	South Africa	Springbok	LSDV[a]
GTPV Nigeria goat vaccine	Nigeria	Goat	SPPV[a]
O58/2011	Kenya	Sheep	GTPV[un]
O59/2011	Kenya	Sheep	GTPV[un]
Akaki/2008	Ethiopia	Sheep	GTPV[b]
Metekel/2010	Ethiopia	Sheep	GTPV[b]
Chagni O06/2012	Ethiopia	Sheep	GTPV[b]

The genotyping of outbreak isolates revealed GTPVs collected from sheep in Ethiopia, Kenya, and Oman and SPPVs collected from goat in Saudi Arabia and Nigeria
[a]Lamien et al. (2011a)
[b]Gelaye et al. (2015)
un unpublished

more severe disease in the homologous host (Kitching et al. 1986; Bhanuprakash et al. 2006). Some studies also revealed natural infections, by either SPPV or GTPV, with similar severity and clinical signs in sheep and goats (Lamien et al. 2011a; Gelaye et al. 2015) as shown in Table 14.1.

Diagnosis

Clinical Observation

Clinically, the presence of nodular skin lesions mostly observed around the mouth and perineum regions is helpful to diagnose SPP and GTP (OIE 2016). In endemic areas and chronic cases, the nodular lesions develop scars, and sometimes, in severe situations, the nodules cover the entire body of the animal. Besides, owner's interview and disease history records such as affected host species, morbidity and mortality rates, month/season of disease occurrence, and vaccination history could be used as primary source of information to reach a diagnosis.

Sample Collection and Transport

Active nodular skin or postmortem lesions from skin papules, lung lesions, and lymph node are good samples for virus isolation and antigen detection (OIE 2016). It is advisable to collect samples within the first week following the occurrence of clinical signs and lesions, before the rise of antibodies (Rao and Bandyopadhyay 2000; OIE 2016). Nasal, oral, and ocular swabs as well as saliva, from clinically diseased or suspected animals, are also good clinical samples for virus isolation and antigen detection (Bowden et al. 2008; Lamien et al. 2011a; Gelaye et al. 2015). Buffy coat obtained from blood collected in an anticoagulant medium through the viremia or within 4 days can be used for virus isolation (Bhanuprakash et al. 2006). However, poxviruses are mainly cell associated; consequently, there may be fewer virus particles in the blood. Representative skin lesions or swab samples from suspected or clinically diseased animals should be collected aseptically, using sterile and labeled containers. Serum samples can be collected from diseased or suspected animals for CaPV antibody detection or from vaccinated animals to evaluate the seroconversion.

It is essential to conduct urgently SPP and GTP diagnosis on representative suspected samples transported to the diagnostic laboratory through maintaining the cold chain. Samples are packed using triple packaging (primary, secondary, and outer containers) and carried following the regulations set for the transport of dangerous goods. Samples from SPP and GTP suspected outbreaks are classified as Infectious Substances Class B (Division 6.2) and must follow the International Air Transport Association (IATA) packing instruction P650, using the UN-approved packaging material. The packaging must bear the labels UN3373, Biological Substance, Category B: hazard for animals, not for human health. It is strictly forbidden to carry infectious substances as carry-on baggage, checked baggage, or in person.

Virus Isolation

Primary lamb kidney and testis cells are commonly used for the isolation and multiplication of infectious SPPV and GTPV (Plowright and Ferris 1958; Kalra and Sharma 1981; Bowden et al. 2008). However, the use of primary cells for virus isolation presents several disadvantages such as the need to continuously establish new cultures, cell lot variation, and contamination with extraneous agents (Babiuk et al. 2008). SPPV and GTPV can also be multiplied using established cell lines such as ESH-L cells (Lamien et al. 2011b; OIE 2016), Vero cells (Singh and Rai 1991; Prakash et al. 1994; Gelaye et al. 2015), MDBK cells (Pandey et al. 1985; Joshi et al. 1995), and OA3Ts cell lines (Babiuk et al. 2007). SGPV can also be propagated on the chorioallantoic membrane of embryonated chicken egg (Kalra and Sharma 1981; Babiuk et al. 2007; OIE 2016). In SPPV and GTPV endemic countries, a biosafety level-II grade laboratory is sufficient for handling clinical samples and for virus isolation, whereas in disease-free countries, virus isolation should be conducted inside biosafety level-III facilities. Infectious viruses induce the formation of distinct plaques with cytopathic effects (CPEs) characterized by elongated cells, ballooning, high refractility, rounding, intracytoplasmic inclusion bodies, plaque formation, and detachment from the tissue culture flask (Soman and Singh 1980). Characteristic poxvirus-induced CPEs can be observed, in infected cells within 7 days, using an inverted microscope, although sometimes the procedure may need several blind passages (Diallo and Viljoen 2007; Lamien et al. 2011a; Gelaye et al. 2015; OIE 2016). Isolation of SPPV and GTPV can be further confirmed by immunostaining using anti-SPPV and GTPV serum (Gulbahar et al. 2006; Babiuk et al. 2007). Histopathology, immunohistochemistry, and electron microscopy examination of skin nodules are additional options for the SPP and GTP diagnosis (Gulbahar et al. 2006; Bowden et al. 2008).

Antibody Detection

Serological assays can only identify SPPV and GTPV as Capripoxviruses, without discriminating the two viruses from each other. Immunity against SPPV and GTPV is predominantly cell mediated, though humoral antibodies are also detectable. Virus neutralization test can be used to examine serum samples for antibodies in disease-suspected sheep and goats (OIE 2016) and seroconversion following vaccination. Even though virus neutralization is referred to as a gold standard in the OIE Manual for detecting anti-SGP antibodies, it is slow, labor intensive, and not sensitive and requires handling of live virus, which is often not permitted in disease-free countries (Babiuk et al. 2008; OIE 2016). A recombinant capripoxvirus, expressing the green fluorescent protein, has been evaluated for virus neutralization assay (Wallace et al. 2007). The results showed a decline in the time required for the detection of virus neutralization activity from 6 to 2 days (Wallace et al. 2007). Western blotting assays are specific and sensitive enough for virus detection; however, they are expensive and difficult to perform and interpret (Chand et al. 1994). Agar gel immunodiffusion tests are less specific, due to cross-reactivity with orf virus antibodies; consequently, they are not recommended for SGP diagnosis (OIE 2016). In the pursuit for high-throughput and specific serological test, researchers developed several ELISAs using capripoxvirus recombinant proteins for antibody detection (Carn et al. 1995; Heine et al. 1999; Bowden et al. 2009; Tian et al. 2010). For instance, an indirect ELISA, based on the recombinant mature virion envelop protein P32 expressed in *E.coli* and yeast (Bhanot et al. 2009), has been reportedly used for the detection of CaPV-specific antibodies sheep (Heine et al. 1999). However, difficulties for the expression and the instability of the recombinant antigens have compromised these assays.

Babiuk et al. (2009b) developed an indirect ELISA for the detection of antibodies to SPPV, GTPV, and LSDV using sucrose gradient-purified inactivated SPPV as coating antigen. This ELISA is suited for screening sera from all three host species; however, the viral antigen is difficult and expensive to produce in large quantities. Moreover, such an approach is not applicable in disease-free countries. Bowden et al. (2009) also developed an indirect ELISA for the detection of antibodies based on selected capripoxvirus antigenic recombinant virion core proteins. This assay performed well on sera collected from sheep and goat that were infected experimentally with virulent virus isolates; however, the test was unable to detect antibodies in sera from vaccinated sheep and goat. An indirect ELISA for the diagnosis of SPP and GTP using two synthetic peptides corresponding to the major antigen P32 of capripoxvirus was also reported (Tian et al. 2010); however, this assay performed well only on sera from immunized sheep. Currently, in 2019, there is only one ELISA kit commercially available for the detection of antibodies against SPPV, GTPV, and LSDV (ID Screen® Capripox Double Antigen Multi-species, IDVet, France), even though many development activities are ongoing elsewhere. With the recent spread of capripox diseases into new geographical areas, there is an urgent need for a high-throughput serological test to facilitate the serological surveillance of capripox in countries under threat. The availability of such a test will also facilitate animal screening during live animal export and post-vaccination monitoring during vaccination campaigns.

Nucleic Acid Detection

Various molecular techniques (conventional and real-time PCR) are available for specific and sensitive detection and differentiation of capripoxviruses such as SPPV and GTPV (Verma et al. 2011; Venkatesan et al. 2014). A gel-based PCR, for the generic detection of capripoxviruses, was described by Ireland and Binepal (1998). Similarly, a highly sensitive multiplex conventional PCR method is available for the detection and differentiation of SPPV, GTPV, and orf virus in clinical samples. This assay targets the DNA binding phosphoprotein (I3L) coding gene of capripoxviruses and DNA polymerase (E9L) gene of orf virus (Venkatesan et al. 2014). Another gel-based PCR method targets the 30 kDa RNA polymerase subunit (RPO30) gene (Lamien et al. 2011a) to detect capripoxvirus and differentiate SPPV from GTPV. However, this method is unable to differentiate GTPV from LSDV. Several real-time PCR methods (Balinsky et al. 2008; Bowden et al. 2008; Stubbs et al. 2012; Haegeman et al. 2013) are available for the generic detection of capripoxviruses; however, they do not intend to differentiate SPPV from GTPV and LSDV. Two LAMP PCR methods (Das et al. 2012; Murray et al. 2013) and a field-ready nucleic acid extraction and real-time PCR platform (Amson et al. 2015) are also reportedly used for the generic detection of capripoxviruses. A dual hybridization probe assay, targeting the G-protein-coupled chemokine receptor (GPCR) gene (Lamien et al. 2011b), and a snapback assay targeting the RPO30 gene (Gelaye et al. 2013) offer the possibility to simultaneously detect the three capripoxviruses and differentiate SPPV from GTPV and LSDV. In addition to the above real-time PCR-based methods, the RPO30 and GPCR genes could be sequenced for the phylogenetic classification of the capripoxvirus isolates (Le Goff et al. 2009; Lamien et al. 2011a; Gelaye et al. 2015). Similarly, a sequencing-based method reported by Hosamani et al. (2004) targets the P32 gene for capripoxvirus differentiation. Both conventional and real-time PCR are useful molecular tools for active clinical surveillance of capripoxviruses in an endemic situation, or newly affected countries and regions.

Differential Diagnosis

Diseases that can be confused with SPP and GTP are bluetongue (caused by bluetongue virus), peste des petits ruminants (caused by peste des petits ruminants virus) for respiratory symptoms, contagious ecthyma (caused by orf virus causing proliferative pox lesions on the muzzle and eyes), insect bites, and mange infestation (e.g., psoroptic mange/sheep scab) (Rao and Bandyopadhyay 2000; Bhanuprakash et al. 2006; OIE 2016). They cause similar kind of skin lesions, in affected hosts requiring a differential diagnosis from SPP and GTP. Hence, laboratory confirmation using conventional methods including antigen or antibody-based tests and molecular diagnostic techniques is necessary to confirm the cause of the diseases or outbreaks. Recently, a real-time PCR method able to simultaneously detect and differentiate SPPV and GTPV from other poxviruses affecting sheep and goats such as orf virus has been developed (Gelaye et al. 2017).

Prevention and Control

The immunity developed against poxvirus infection is predominantly cell mediated; thus, the immune status of animals does not correlate with neutralizing antibody titers in serum (Carn 1993). Previous exposure to SPPV and GTPV results in substantial and long-lasting protective immunity against subsequent reinfection with the virus. Live attenuated and inactivated strains of SPPV or GTPV are the most common vaccines in disease-endemic countries. There is only a partial cross-protection when sheep and goat are vaccinated with GTPV vaccine against SPPV and or vice versa (Kitching et al. 1987; Hosamani et al. 2004; Bhanuprakash et al. 2012). In disease-endemic countries, vaccination of small ruminants using a vaccine containing a virus homologous to the circulating isolates is an economical and sustainable means of disease prevention and control (Bhanuprakash et al. 2012; Hosamani et al. 2008). Consequently, better protection against locally prevalent strain for either SPPV or GTPV is achieved using homologous vaccines (Rao and Bandyopadhyay 2000; Bhanuprakash et al. 2005). Generally, inactivated vaccines do not provide adequate and long-lasting protective immunity; however, an inactivated SPPV vaccine would provide a safe and valuable tool to protect sheep and goat against SPPV and GTPV infections, particularly in the case of a first incursion of the virus in the previously disease-free area, or for preventive vaccination in region threatened by SPP or GTP (Boumart et al. 2016). For instance, inactivated SPPV vaccine produced using the Roumanian Fanar (RF) strain showed potential to replace live attenuated vaccine for the prevention and control of SPP in disease-endemic or disease-free countries (Boumart et al. 2016).

LSDV-derived vaccines are also widely used for the prevention and control of SPP and GTP. A single vaccine, through intradermal or subcutaneous route using the OIE recommended dose ($10^{2.5}$ $TCID_{50}$), using the Kenyan sheep and goat pox virus

(O180/KS-1 or O240) strain, protected both sheep and goats against the virulent strains of SPPV and GTPV. This vaccine, used in many countries in the Middle East and Africa, presented proper safety and protection result, though the vaccinal strain is, in fact, an LSDV (Gelaye et al. 2015; OIE 2016). Several recombinant vectored vaccines have been developed based on capripoxvirus. A recombinant capripoxvirus vaccine harboring the F or H genes of PPR virus (rCPV-PPR) provided adequate protection against both capripoxvirus and PPR virus (Diallo et al. 2002; Berhe et al. 2003; Caufour et al. 2014). Currently, a vaccine allowing the differentiation of infected from vaccinated animals (DIVA) is not commercially available against SPP and GTP.

In general, SPPV and GTPV infections are self-limiting; treatment with antibacterial and antifungal against secondary bacterial or fungal infections and supportive care help to improve the health status of infected animals (Smith et al. 2010). SPP and GTP are notifiable diseases as recommended by OIE; therefore, it is mandatory to notify suspicion of infection to the appropriate veterinary authorities and strictly regulate the movements of small ruminants. To implement an effective prevention and control strategy, proper veterinary services with moderately equipped resources, adequate infrastructure and logistic support, appropriate disease surveillance, and diagnostic activities are essential. It is also critical, achieving good vaccinations coverage using effective vaccines to build the herd immunity (at least 80% coverage), controlling the illegal animal movements and animal products. Political stability and economic development are also essential factors for the implementation of effective prevention and successful control strategy. Countries that do not report the occurrence of SPP and GTP diseases should strictly implement testing of animal and animal products before importing from disease-affected countries.

Conclusion

SPP and GTP are transboundary and OIE notifiable small ruminant diseases of low- and middle-income countries. SPP and GTP incidences have steadily increased in new geographical areas of South East Asia and Europe. The primary sources of virus spread into disease-free countries are the trade of infected animals and animal products such as wool and hides and the movement of people and goods due to civil war and unrest. The socioeconomic impact of SPP and GTP in the agricultural development and the livelihood of the small ruminant holders should be studied in endemic countries to support for the design of prevention and control strategies, to allocate resources, and to draw policy maker's attention. The implementation of well-organized vaccination campaigns, based on effective vaccines and achieving sufficient coverage, can help to reduce the burden of SPPV and GTPV infections in disease-endemic regions. Monitoring by active surveillance and the genotyping of outbreak viruses can guide for the selection of the most appropriate vaccines. The development of a high-throughput serological test, with enhanced sensitivity and specificity, and the availability of a safe and effective vaccine allowing for the differentiation of infected from vaccinated animals (DIVA) are highly required.

References

Amson B, Fowler VL, Tuppurainen ESM, Howson ELA, Madi M, Sallu R, Kasanga CJ, Pearson C, Wood J, Martin P, Mioulet V, King DP. Detection of Capripox virus DNA using a field-ready nucleic acid extraction and real-time PCR platform. Transbound Emerg Dis. 2015;64(3):994–7.

Andrew MQK, Michael JA, Eric BC, Elliot JL. Family *Poxviridae*. In: Virus taxonomy classification and nomenclature of viruses, ninth report of the International Committee on Taxonomy of Viruses. London: Elsevier; 2012. p. 14–309.

Aparna M, Gnanavel V, Amit K. Capripoxviruses of small ruminants: current updates and future perspectives. Asian J Anim Vet Adv. 2016;11(12):757–70.

Babiuk S, Parkyn G, Copps J, Larence JE, Sabara MI, Bowden TR, Boyle DB, Kitching RP. Evaluation of an ovine testis cell line (OA3.Ts) for propagation of Capripox virus isolates and development of an immunostaining technique for viral plaque visualization. J Vet Diagn Investig. 2007;19:486–91.

Babiuk S, Bowden TR, Boyle DB, Wallace DB, Kitching RP. Capripoxviruses: an emerging worldwide threat to sheep, goats and cattle. Transbound Emerg Dis. 2008;55:263–72.

Babiuk S, Bowden TR, Parkyn G, Dalman B, Hoa DM, Long NT, Vu PP, Bieu do X, Copps J, Boyle DB. Yemen and Vietnam capripoxviruses demonstrate a distinct host preference for goats compared with sheep. J Gen Virol. 2009a;90:105–14.

Babiuk S, Wallace DB, Smith SJ, Bowden TR, Dalman B, Parkyn G, Copps J, Boyle DB. Detection of antibodies against capripoxviruses using an inactivated sheeppox virus ELISA. Transbound Emerg Dis. 2009b;56(4):132–41.

Balinsky CA, Delhon G, Smoliga G, Prarat M, French RA, Geary SJ, Rock DL, Rodriguez LL. Rapid preclinical detection of sheeppox virus by real-time PCR assay. J Clin Microbiol. 2008;46(2):438–42.

Berhe G, Minet C, Le Goff C, Barrett T, Ngangnou A, Grillet C, Libeau G, Fleming M, Black DN, Diallo A. Development of a dual recombinant vaccine to protect small ruminants against peste des petits ruminants virus and capripoxvirus infections. J Virol. 2003;77:1571–7.

Bhanot V, Balamurugan V, Bhanuprakash V, Venkatesan G, Sen A, Yadav V, Yogisharadhya R, Singh RK. Expression of P32 protein of goatpox virus in *Pichia pastoris* and its potential use as diagnostic antigen in ELISA. J Virol Methods. 2009;162:251–7.

Bhanuprakash V, Moorthy AR, Krishnappa G, Srinivasa Gowda RN, Indrani BK. An epidemiological study of sheep pox infection in Karnataka State, India. Rev Sci Tech. 2005;24 (3):909–20.

Bhanuprakash V, Indrani BK, Hosamani M, Singh RK. The current status of sheep pox disease. Comp Immunol Microbiol Infect Dis. 2006;29(1):27–60.

Bhanuprakash V, Venkatesan G, Balamurugan V, Hosamani M, Yogisharadhya R, Chauhan RS, Pande A, Mondal B, Singh RK. Pox outbreaks in sheep and goats at Makhdoom (Uttar Pradesh), India: evidence of sheeppox virus infection in goats. Transbound Emerg Dis. 2010;57 (5):375–82.

Bhanuprakash V, Hosamani M, Singh RK. Prospects of control and eradication of capripox from the Indian subcontinent: a perspective. Antivir Res. 2011;91(3):225–32.

Bhanuprakash V, Hosamani M, Venkatesan G, Balamurugan V, Yogisharadhya R, Singh RK. Animal poxvirus vaccines: a comprehensive review. Expert Rev Vaccines. 2012;11:1355–74.

Boumart Z, Daouam S, Belkourati I, Rafi L, Tuppurainen E, Tadlaoui KO, El Harrak M. Comparative innocuity and efficacy of live and inactivated sheeppox vaccines. BMC Vet Res. 2016;12:133–8.

Bowden TR, Babiuk SL, Parkyn GR, Copps JS, Boyle DB. Capripoxvirus tissue tropism and shedding: a quantitative study in experimentally infected sheep and goats. Virology. 2008;371 (2):380–93.

Bowden TR, Coupar BE, Babiuk SL, White JR, Boyd V, Duch CJ, Shiell BJ, Ueda N, Parkyn GR, Copps JS, Boyle DB. Detection of antibodies specific for sheeppox and goatpox viruses using recombinant capripoxvirus antigens in an indirect enzyme-linked immunosorbent assay. J Virol Methods. 2009;161(1):19–29.

Carn VM. Control of capripoxvirus infections. Vaccine. 1993;11:1275–9.

Carn VM, Kitching RP, Hammond JM, Chand P, Anderson J, Black DN. Use of a recombinant antigen in an indirect ELISA for detecting bovine antibody to capripoxvirus. J Virol Methods. 1995;53(2–3):273.

Caufour P, Rufael T, Lamien CE, Lancelot R, Kidane M, Awel D, Sertse T, Kwiatek O, Libeau G, Sahle M, Diallo A, Albina E. Protective efficacy of a single immunization with capripoxvirus-vectored recombinant peste des petits ruminants vaccines in presence of pre-existing immunity. Vaccine. 2014;32:3772–9.

Chand P, Kitching RP, Black DN. Western blot analysis of virus-specific antibody responses for capripox and contagious pustular dermatitis viral infections in sheep. Epidemiol Infect. 1994;113(2):377–85.

Das A, Babiuk S, Mclntosh MT. Development of a loop-mediated isothermal amplification assay for rapid detection of capripox viruses. J Clin Microbiol. 2012;50(5):1613–20.

Davies FG, Otema C. Relationships of capripox viruses found in Kenya with two Middle Eastern strains and some orthopox viruses. Res Vet Sci. 1981;31(2):253–5.

Diallo A, Viljoen GJ. Genus *Capripoxvirus*. In: Mercer A, Schmidt A, Weber O, editors. Poxviruses. Basel: Birkhauser; 2007. p. 167–81.

Diallo A, Minet C, Berhe G, Le Goff C, Black DN, Fleming M, Barrett T, Grillet C, Libeau G. Goat immune response to capripox vaccine expressing the hemmagglutinin protein of Peste des petits ruminants. Ann N Y Acad Sci. 2002;969:88–91.

Domenech J, Lubroth J, Eddi C, Martin V, Roger F. Regional and international approaches on prevention and control of animal transboundary and emerging diseases. Ann N Y Acad Sci. 2006;1081:90–107.

EFSA (European Food Safety Authority). Scientific opinion on Sheep and goat pox. EFSA Panel on Animal Health and Welfare. EFSA J. 2014;12(11):3885.

European Commission. Lumpy skin disease and Sheep pox. Animal Health Advisory Committee. 18 Dec 2017, Brussels. 2017. Retrieved 3 Jan 2019, from https://ec.europa.eu/food/sites/food/files/animals/docs/comm_ahac_20171218_pres05.pdf

Fenner F. The pathogenesis, pathology and immunology of smallpox and vaccinia. In: Fenner F, Henderson DA, Arita I, Jezek Z, Ladnyi ID, editors. Smallpox and its eradication. Geneva: World Health Organization; 1988. p. 121–68.

Gelaye E, Lamien CE, Silber R, Tuppurainen ES, Grabherr R, Diallo A. Development of a cost-effective method for capripoxvirus genotyping using snapback primer and dsDNA intercalating Dye. PLoS One. 2013;8(10):e75971.

Gelaye E, Belay A, Ayelet G, Jenberie S, Yami M, Loitsch A, Tuppurainen E, Grabherr R, Diallo A, Lamien CE. Capripox disease in Ethiopia: genetic differences between field isolates and vaccine strain, and implications for vaccination failure. Antivir Res. 2015;119:28–35.

Gelaye E, Mach L, Kolodziejek J, Grabherr R, Loitsch A, Achenbach JE, Nowotny N, Diallo A, Lamien CE. A novel HRM assay for the simultaneous detection and differentiation of eight poxviruses of medical and veterinary importance. Sci Rep. 2017;7:42892.

Gulbahar MY, Davis WC, Yuksel H, Cabalar M. Immunohistochemical evaluation of inflammatory infiltrate in the skin and lung of lambs naturally infected with sheeppox virus. Vet Pathol. 2006;43(1):67–75.

Haegeman A, Zro K, Vandenbussche F, Demeestere L, Campe W, Van Ennaji MM, De Clercq K. Development and validation of three capripoxvirus real-time PCRs for parallel testing. J Virol Methods. 2013;193(2):446–51.

Heine HG, Stevens MP, Foord AJ, Boyle DB. A capripoxvirus detection PCR and antibody ELISA based on the major antigen P32, the homolog of the vaccinia virus H3L gene. J Immunol Methods. 1999;227(1–2):187–96.
Hosamani M, Mondal B, Tembhurne PA, Bandyopadhyay SK, Singh RK, Rasool TJ. Differentiation of sheep pox and goat poxviruses by sequence analysis and PCR-RFLP of P32 gene. Virus Genes. 2004;29(1):73–80.
Hosamani M, Bhanuprakash V, Kallesh DJ, Balamurugan V, Pande A, Singh RK. Cell culture adapted sheeppox virus as a challenge virus for potency testing of sheeppox vaccine. Indian J Exp Biol. 2008;46:685–9.
Hutyra F, Marek J, Manninger R. Special pathology and therapeutics of the diseases of domestic animals. 5th ed. London: Bailliere, Tindall and Cox; 1946. p. 353–66.
Ireland DC, Binepal YS. Improved detection of capripoxvirus in biopsy samples by PCR. J Virol Methods. 1998;74(1):1–7.
Jahrling PB, Hensley LE, Martinez MJ, Le Duc JW, Rubins KH, Relman DA, Huggins JW. Exploring the potential of variola virus infection of cynomolgus macaques as a model for human smallpox. Proc Natl Acad Sci USA. 2004;101:15196–200.
Joshi RK, Garg SK, Chandra R, Sharma VD. Growth and cytopathogenicity of goat pox virus in MDBK cell line. Indian J Virol. 1995;11:31–3.
Kalra SK, Sharma VK. Adaptation of Jaipur strain of sheeppox virus in primary lamb testicular cell culture. Indian J Exp Biol. 1981;19(2):165–9.
Kitching RP. Vaccines for lumpy skin disease, sheep pox and goat pox. Dev Biol. 2003;114:161–7.
Kitching RP, Mellor PS. Insect transmission of capripoxvirus. Res Vet Sci. 1986;40(2):255–8.
Kitching RP, Taylor WP. Clinical and antigenic relationship between isolates of sheep and goat pox viruses. Trop Anim Health Prod. 1985;17:64–74.
Kitching RP, McGrane JJ, Taylor WP. Capripox in the Yemen Arab Republic and the Sultanate of Oman. Trop Anim Health Prod. 1986;18:115–22.
Kitching RP, Hammond JM, Taylor WP. A single vaccine for the control of capripox infection in sheep and goats. Res Vet Sci. 1987;42:53–60.
Lamien CE, Le Goff C, Silber R, Wallace DB, Gulyaz V, Tuppurainen E, Madani H, Caufour P, Adam T, El Harrak M, Luckins AG, Albina E, Diallo A. Use of the capripoxvirus homologue of Vaccinia virus 30 kDa RNA polymerase subunit (RPO30) gene as a novel diagnostic and genotyping target: development of a classical PCR method to differentiate goat poxvirus from Sheep poxvirus. Vet Microbiol. 2011a;149:30–9.
Lamien CE, Lelenta M, Goger W, Silber R, Tuppurainen E, Matijevic M, Luckins AG, Diallo A. Real time PCR method for simultaneous detection, quantitation and differentiation of capripoxviruses. J Virol Methods. 2011b;171:134–40.
Le Goff C, Lamien CE, Fakhfakh E, Chadeyras A, Aba-Adulugba E, Libeau G, Tuppurainen E, Wallace DB, Adam T, Silber R, Gulyaz V, Madani H, Caufour P, Hammami S, Diallo A, Albina E. Capripoxvirus G-protein-coupled chemokine receptor: a host-range gene suitable for virus animal origin discrimination. J Gen Virol. 2009;90:1967–77.
Mellor PS, Kitching RP, Wilkinson PJ. Mechanical transmission of capripox virus and African swine fever virus by *Stomoxys calcitrans*. Res Vet Sci. 1987;43(1):109–12.
Murray M, Martin WB, Koylu A. Experimental sheeppox. A histological and ultrastructural study. Res Vet Sci. 1973;15:201–8.
Murray L, Edwards L, Tuppurainen ESM, Bachanek-Bankowska K, Oura CAL, Mioulet V, King DP. Detection of capripox virus DNA using a novel loop-mediated isothermal amplification assay. BMC Vet Res. 2013;9:90–7.
Oguzoglu TC, Alkan F, Ozkul A, Vural SA, Gungor AB, Burgu I. A sheeppox virus outbreak in Central Turkey in 2003: isolation and identification of capripoxvirus ovis. Vet Res Commun. 2006;30:965–71.
Pandey KD, Rai A, Goel AC, Mishra SC, Gupta BK. Adaptation and growth cycle of sheep pox virus in MDBK cell line. Indian J Virol. 1985;1:133–8.
Plowright W, Ferris RD. The growth and cytopathogenicity of sheep-pox virus in tissue cultures. Br J Exp Pathol. 1958;39:424–35.

Prakash V, Chandra R, Rao VDP, Garg SK, Ved Prakash RC. Cultivation of goat pox virus in established cell line. Indian J Virol. 1994;10:60–3.

Rafyi A, Ramyar H. Goat pox in Iran; serial passage in goats and the developing egg, and relationship with sheep pox. J Comp Pathol. 1959;69(2):141–7.

Rao TVS, Bandyopadhyay SK. A comprehensive review of goat pox and sheep pox and their diagnosis. Anim Health Res Rev. 2000;1:127–36.

Rweyemamu M, Paskin R, Benkirane A, Martin V, Roeder P, Wojciechowski K. Emerging diseases of Africa and the Middle East. Ann N Y Acad Sci. 2000;916:61–70.

Singh B, Rai A. Adaptation and growth of sheep pox virus in Vero cell culture. Indian Vet Med J. 1991;15:245–50.

Smith GL, Beard P, Skinner MA. Poxviruses. In: Mahy BWJ, Van Regenmortel MHV, editors. Desk encyclopedia of human and medical virology. London: Elsevier; 2010. p. 239–46.

Soman JP, Singh IP. Plaque formation by sheep pox virus adapted to lamb kidney cell culture. Indian J Exp Biol. 1980;18:313–4.

Stubbs S, Oura CAL, Henstock M, Bowden TR, King DP, Tuppurainen ESM. Validation of a high-throughput real-time polymerase chain reaction assay for the detection of capripoxviral DNA. J Virol Methods. 2012;179(2):419–22.

The World Organisation for Animal Health (OIE). Manual of diagnostic tests and vaccines for terrestrial animals. Paris: The World Organisation for Animal Health (OIE); 2016.

Tian H, Chen Y, Wu J, Shang Y, Liu X. Serodiagnosis of sheeppox and goatpox using an indirect ELISA based on synthetic peptide targeting for the major antigen P32. Virol J. 2010;7:245–8.

Tulman ER, Afonso CL, Lu Z, Zsak L, Sur JH, Sandybaev NT, Kerembekova UZ, Zaitsev VL, Kutish GF, Rock DL. The genomes of sheeppox and goatpox viruses. J Virol. 2002;76:6054–61.

Tuppurainen ES, Venter EH, Shisler JL, Gari G, Mekonnen GA, Juleff N, Lyons NA, De Clercq K, Upton C, Bowden TR, Babiuk S, Babiuk LA. Review: Capripoxvirus diseases: current status and opportunities for control. Transbound Emerg Dis. 2015;64(3):729–45.

Venkatesan G, Balamurugan V, Bhanuprakash V. TaqMan based real-time duplex PCR for simultaneous detection and quantitation of capripox and orf virus genomes in clinical samples. J Virol Methods. 2014;201:44–50.

Verma S, Verma LK, Gupta VK, Katoch VC, Dogra V, Pal B, Sharma M. Emerging capripoxvirus disease outbreaks in Himachal Pradesh, a northern state of India. Transbound Emerg Dis. 2011;58:79–85.

Wallace DB, Weyer J, Nel LH, Viljoen GJ. Improved method for the generation and selection of homogeneous lumpy skin disease virus (SA-Neethling) recombinants. J Virol Methods. 2007;146(1–2):52–60.

Yan XM, Chu YF, Wu GH, Zhao ZX, Li J, Zhu HX, Zhang Q. An outbreak of sheep pox associated with goat poxvirus in Gansu province of China. Vet Microbiol. 2012;156:425–8.

Zaucha GM, Jahrling PB, Geisbert TW, Swearengen JR, Hensley L. The pathology of experimental aerosolized monkeypox virus infection in cynomolgus monkeys (*Macaca fascicularis*). Lab Investig. 2001;81(12):1581–600.

Zhou T, Jia H, Chen G, He X, Fang Y, Wang X, Guan Q, Zeng S, Cui Q, Jing Z. Phylogenetic analysis of Chinese sheeppox and goatpox virus isolates. Virol J. 2012;9:25–32.

Zro K, Zakham F, Melloul M, El Fahime E, Ennaji MM. A sheeppox outbreak in Morocco: isolation and identification of virus responsible for the new clinical form of disease. BMC Vet Res. 2014;10:31–8.

Chapter 15
Bluetongue Disease

Stefano Cappai, Mario Forzan, Federica Loi, Sandro Rolesu, Soufien Sghaier, Antonio Petrini, Giovanni Savini, and Alessio Lorusso

Abstract Bluetongue (BT) is a noncontagious OIE-listed disease of domestic and wild ruminants caused by a virus (Bluetongue virus—BTV) of the *Orbivirus* genus within the family *Reoviridae* and transmitted by biting midges of the genus *Culicoides*. BT is a considerable socioeconomic concern and of major importance for the international trade of animals and animal products. In the past, BT endemic areas were considered those between latitudes 40 °N and 35 °S; however, BT has spread far beyond this traditional range. BTV has multiple serotypes and these serotypes exist in a complex network of serological cross-relationships, varying from partial to no protection between heterologous strains. This chapter summarizes several aspects of BT and BTV with particular emphasis for BTV epidemiology in Sahelian Africa.

Keywords Bluetongue · Bluetongue virus · Epidemiology · Sahel · Vectors

Economic Impact, Etiology, and History

Bluetongue (BT) is a vector-borne disease of domestic and wild ruminants listed as a notifiable disease by the World Organisation for Animal Health (OIE). BT is a disease of considerable socioeconomic concern and of major importance in the

S. Cappai · F. Loi · S. Rolesu
Istituto Zooprofilattico Sperimentale della Sardegna, Cagliari, Italy

M. Forzan
Department of Veterinary Science, University of Pisa, Pisa, Italy

S. Sghaier
Laboratoire de virologie, Institut de la Recherche Vétérinaire de Tunisie (IRVT), Université de Tunis El Manar, Tunis, Tunisia

A. Petrini · G. Savini · A. Lorusso (✉)
OIE Reference Laboratory for Bluetongue, Istituto Zooprofilattico Sperimentale dell'Abruzzo e del Molise (IZSAM), Teramo, Italy
e-mail: a.lorusso@izs.it

M. Kardjadj et al. (eds.), *Transboundary Animal Diseases in Sahelian Africa and Connected Regions*, https://doi.org/10.1007/978-3-030-25385-1_15

international trade of animals and animal products. Losses due to any livestock disease may be classified as losses in production (direct losses), expenditure, and lost revenue (indirect losses) (Rushton 2009). The former may be visible, such as reduced milk yield or increased mortality, weight loss, reduced fertility rate, abortion, reduced meat production, efficiency, and death (Sperlova and Zendulkova 2011). Indirect losses include costs of vaccines or lost revenue, such as through trade restrictions limiting access to higher value markets. Multiple outbreaks of BTV from 1998 in North Africa and in southern Europe caused the death of hundreds of thousand animals causing enormous economic losses.

With regard to BTV-8, the epidemic in northern Europe has probably caused greater economic damage than any previous single-serotype BT outbreak (Wilson and Mellor 2009). A study commissioned by the Scottish Government assessed the impact on production as direct costs estimated an amount of £30 million per year (Scottish Government 2008). The deterministic economic model used to evaluate the cost of the BTV-8 epidemic in the Netherlands estimated an overall cost as 32.4 million euros and 164–175 million euros in 2006 and 2007, respectively (Velthuis et al. 2010). In 2012, an economic evaluation of the surveillance and control program applied in 2008–2009 in Switzerland estimated an amount of the disease costs as 12.2 million euros and 3.6 million euros, respectively (Häsler et al. 2012).

BT is caused by Bluetongue virus (BTV), the prototype species of the genus *Orbivirus* within the family *Reoviridae*, which includes 30 genera divided into two subfamilies infecting a wide variety of plants, vertebrates, and invertebrates, including crustaceans, fish, insects, reptiles, and mammals (Mertens et al. 2005; Attoui et al. 2009; ICTV 2018, https://talk.ictvonline.org/taxonomy/). The genus *Orbivirus* contains 22 recognized viral species. As the most widespread and of economic importance, BTV has been studied more extensively than the other orbiviruses. BTV is a non-enveloped virus, 90 nm in diameter, structurally organized into three concentric shells enclosing the viral genome consisting of ten linear dsRNA segments (Seg-1 to Seg-10) encoding for 7 structural (VP1 to VP7) and 5 nonstructural proteins (NS1, NS2, NS3/NS3A, NS4, and S10-ORF2) (Belhouchet et al. 2011; Ratinier et al. 2011; Stewart et al. 2015). The outer capsid is composed by two major proteins, namely VP2 and VP5 (coded by Seg-2 and Seg-5, respectively), which are essential in host receptor binding and virus entry. VP2 is responsible for the generation of neutralizing antibodies and determines, mainly, serotype specificity. Up to 2008, only 24 classical serotypes of BTV were officially recognized (Maan et al. 2008). These 24 serotypes exist in a complex network of serological cross relationships, varying from partial to no protection between heterologous strains. Any of these classical BTV serotypes has potential to cause BT. Except for serotype diversity, the localized circulation of the virus in different ecosystems throughout the world has also led to the evolution of distinct geographical variants or topotypes. BTVs are, indeed, broadly divided into western (w) and eastern (e) topotypes, based on phylogenetic analysis of nucleotide sequences from the majority of the genome segments. Viruses from the western topotypes circulate in Africa, Europe, the Caribbean, and the Americas, whereas those from eastern topotypes are endemic in Asia, Indonesia, and Australia (Gould and Hyatt 1994; Carpi et al. 2010). In the last few years, novel and generally asymptomatic BTV serotypes have been discovered by researchers in the field. These include so far BTV-25 (TOV strain) from Switzerland, BTV-26 from Kuwait, BTV-27 (variants 01, 02 and 03) from Corsica (France), BTV-XJ1407 from China, a BTV strain isolated from a sheep pox vaccine (SP vaccine derived BTV) from the Middle East, BTV-X ITL2015, BTV-Z ITA2017 from Italy, and BTV-Y TUN2017 from Tunisia (Hofmann et al. 2008a; Maan et al. 2011; Zientara et al. 2014; Schulz et al. 2016; Sun et al. 2016; Bumbarov et al. 2016; Savini et al. 2017; Marcacci et al. 2018; Lorusso et al. 2018). A putative novel BTV serotype has also been described from an Alpaca in South Africa (Belbis et al. 2017).

BT was historically considered as an African disease and it was first described in the late eighteenth century, following the importation of European Merinos sheep in Cape Colony in South Africa. Initially, the illness was thought to be caused by a parasite and was initially referred as fever, malarial catarrhal fever of sheep, or epizootic malignant catarrhal fever of sheep (Hutcheon 1902). In 1902, Spreull provided the first scientific detailed description of infected sheep, including details of typical lesions, such as the presence of a dark-colored (blue) tongue. As a consequence of this observation, the scientist introduced the name of "bluetongue," which is the English translation of the name for the disease of "Blaauwtong," which was named by South African farmers who noticed tongue cyanosis in seriously diseased animals (Spreull 1902; MacLachlan et al. 2009). In 1905, by performing experimental infections in goats and cattle, Spreull discovered that the disease was transmissible to other ruminants but without production of clear clinical signs. The viral cause of the disease was determined in the same year by Sir Arnold Theiler who demonstrated that the etiological agent of Bluetongue was filterable. The scientist, by serial passages of the virus on susceptible sheep, formulated the first vaccine for BTV that was used in South Africa for several years (Theiler 1906). The fact that BTV might be transmitted by vectors was initially supposed by Spreull (1902) and Hutcheon (1902) which observed the seasonal incidence of the disease and the protection of sheep to the infection stabling them during summer nights. The role of insects in disease transmission was first confirmed by Du Toit in 1944, who indicated the hematophagous *Culicoides* midges as the biological competent vectors of BTV. Subsequently, Foster in 1963 definitively demonstrated the transmission of BTV by *Culicoides*. Information regarding the morphology of BTV was obtained by using the first electron microscopy images of the virus suggesting a Reovirus-like structure (Owen and Munz 1966; Studdert et al. 1966). Following the first report of BTV purification from infected cell lines (Vervoerd 1969), extensive and more detailed studies on the structure of BTV were performed in the 1970s providing essential information regarding the identification of viral structural proteins, their organization in two concentric capsids, and their possible role in virus infection (Vervoerd 1970). The development of molecular biology techniques, as well as the use of cell culture systems as alternative to inoculation of susceptible animals, greatly improved the knowledge on BTV etiology. The nonstructural proteins NS1 and NS2 were identified in 1979 (Huismans 1979), NS3 in 1981 (Gorman et al. 1981). Only recently a fourth and a fifth viral protein have been characterized

(Belhouchet et al. 2011; Ratinier et al. 2011; Stewart et al. 2015). Since the beginning of this century, several discoveries have been made to elucidate BTV biology, replication and evolution. These discoveries had a tremendous impact on the development of diagnostic tools and control strategies and improved the understanding of the complex epidemiology and pathogenesis of the virus.

Distribution of BTV in Sahel Region, Africa

In the 1940s, it was thought that BT was restricted to southern Africa and that BT outbreaks in other regions of the world reflected the emergence of the virus from the African continent (Gibbs and Greiner 1994). This assumption was soon challenged by the recognition of the disease in geographically widespread regions outside Africa. The first recognized outbreak of BT outside Africa occurred in Cyprus in 1943 (Gambles 1949). Since 1998, BTV has moved considerably northward reaching the 50th parallel in some parts of the world such as regions of Asia, North America, and North Europe where the virus was never reported before (Clavijo et al. 2000; Lundervold et al. 2003; Tabachnick 2004; Jafari-Shoorijeh et al. 2010; MacLachlan 2010). BTV circulation has been described in several North Africa countries including Morocco, Algeria, Tunisia, Libya, and Egypt. Remarkably, BTV emergence in Europe is related to the wind-driven dissemination of infected midges from northern African countries (Calistri et al. 2004; Lorusso et al. 2013, 2014, 2017, 2018; Sghaier et al. 2017; Mahmoud et al. 2018; Cappai et al. 2019). BTV has also been described from Sahelian Africa countries, such as Senegal, Sudan, Mauritania, Cameroon, Nigeria, and Gambia. Generally in African countries BTV outbreaks occur during the wet seasons and, in these regions, BTV can be considered endemic (Sellers 1984; Bakhoum et al. 2013; Fall et al. 2015a, b). In Sudan, BT was first reported in 1953, when samples from the Blue Nile province were confirmed by the Veterinary Research Laboratory at Onderstepoort, South Africa, to contain BTV (Anon 1953). Several studies have lately been performed; they confirmed that different BTV serotypes were endemic in Sudan in domestic and wild animals (Eisa et al. 1979, 1980; Mellor et al. 1984; Abu Elzein and Tag Eldin 1985). BT seroprevalence in Sudan was 61% during the 1980s (Abu Elzein 1983) and 50–60% in recent years (Saeed 2017; Adam et al. 2014). In Nigeria, no BT outbreaks were reported until 1966, when the first BT epidemic started (Bida and Eid 1974). BTV serotypes 6, 7, 10, 12, and 16 were identified in Nigeria between 1966 and 1974 by virus isolation (Moore and Kemp 1974). During the same period, BTV outbreaks were recorded in Egypt sustained by serotypes 1, 4, 10, 12, and 16 (Hafez and Ozawa 1973; Ismail et al. 1987). BTV was also diagnosed by the Animal Research Institute of Mankon in Cameroon from samples collected from five sheep succumbed after clinical disease between June and October 1982. BTV 1, 4, 5, 12, 14, and 16 were also isolated in the same period (Ekue et al. 1985). The first official BTV outbreak in Senegal was notified in 1982 and a serological survey revealed specific BTV-10 antibodies in goats nearby the border with Guinea.

The overall seroprevalence in Senegal ranged, at that time, between 35 and 48% (Lefevre and Taylor 1983). The first overview upon BTV epidemiology in Africa and the Mediterranean region has been published by Sellers in 1984, which described the BTV serotypes that were circulating in African countries and the several cases which could have remained unnoticed as clinical signs were mild. A survey performed in Gambia in 1998 revealed an overall prevalence of BTV antibodies of 62–66% in goats and 55–58% in sheep (Goossens et al. 1998). Specific antibodies to BTV-26 or to a BTV-26-like virus have been recently described in camelids and cattle from Mauritania (Lorusso et al. 2016).

Vectors

Disease spread is affected by different factors that interplay and regulate the presence of insect vectors (De Liberato et al. 2003). The availability of susceptible hosts, along with climatic conditions, may play a decisive role in spatial and temporal distribution of BTV (Purse et al. 2004a, b, 2005, 2008). Cattle play an important role in the epidemiology of BT, since they generally are infected (viremic) without showing clinical symptoms. However, BTV-8 in Northern Europe in 2006–2008 recapitulated this assumption. The infectious period for cattle was considered to be equal to 60 days post-infection, or more (Sellers and Taylor 1980). Also wild ruminants can play an important role to maintain the infection. This role is proved in Africa and in North America, where animals such as the American deer (*Odocoileus virginianus*), the American antelogoat (*American antelogoat*), and the desert bighorn sheep (*Ovis canadensis*) can also show clinical symptoms (Ruiz-Fons et al. 2008).

The climatic conditions and the presence of geographical barriers may influence the vector and host presence. Furthermore, BT prevalence is seasonal and strongly correlated with the abundance of insect vectors and other environmental conditions involved in their life cycle (Conte et al. 2007). The increasing of the temperature, rainfall and climate changes, and wind patterns variability are also predicted to have effect on vector abundance as discussed in many studies (Wittmann and Baylis 2000; Sellers and Mellor 1993; Rawlings et al. 1998; Walton 2004; Staubach et al. 2007; De Koeijer and Elbers 2007; Purse et al. 2008). Distribution of BTV throughout the world is strictly linked to the spatial and temporal distribution of species of *Culicoides* (Diptera: Ceratopogonidae) biting midges, which are the biological vector of the virus (Tatem et al. 2003; Purse et al. 2005; Mellor et al. 2000; Carpenter et al. 2015; Foxi et al. 2016). Adult midges can only become infected by ingestion of a blood meal from an infected mammalian host and are only capable of transmitting the virus to a new host when they have a subsequent blood meal. Although more than 1400 *Culicoides* species have been identified worldwide (Mellor et al. 2000; Borkent 2005), currently nearly 30 of them have been reported as vectors of BTV (Meiswinkel et al. 2004). Adult *Culicoides* fly during the night (from sunset to sunrise) and females bite animals feeding on their blood. However, studies conducted in 2006 during the BTV-8 epidemic in central Europe indicated that a certain level of vector activity can be detected also during daylight (Meiswinkel et al. 2008). The insects become infected biting viremic animals and they remain infected along all their life. Humid areas, little water pods drastically favor the adult *Culicoides* reproduction (Conte et al. 2007), and it is also supposed that adults individuals remain around the place where they were born for all their life. The effect of temperature on the abundance of *C. imicola,* the main vector of BT in the Mediterranean basin, has been discussed in many studies (Sellers and Mellor 1993; De Koeijer and Elbers 2007; Purse et al. 2008). In particular, Wittmann in 2001 showed that high mean monthly temperatures (>12.5°) in winter favor the survival of *Culicoides* larvae and temperatures between 25 and 30 °C constitute their ideal habitat. In each geographic region, different *Culicoides* species may be involved in BTV transmission, and specific links between the *Culicoides* species distribution and BTV serotypes prevalence have been demonstrated (Tabachnick 2004). In the USA, the main vectors are *C. sonorensis* and *C. insignis*. In northern and eastern Australia, the main BTV vector is *C. brevitarsis*, whereas in Africa, southern Europe, and the Middle East is *C. imicola*. In northern Europe where *C. imicola* is absent, the transmission of BTV-8 was therefore assigned to Palearctic species of *Culicoides* midges including *C. obsoletus*, *C. pulicaris*, *C. dewulfi*, and *C. chipterus* (Conte et al. 2003; Purse et al. 2004b).

Although biting midges are ubiquitous (Mellor et al. 2000), they are most frequently present in warm, damp, and muddy areas. Temperature conditions are essential not only for the survival of the vector but they can also have a role in the replication cycle of the virus (Sellers et al. 1979). Another essential factor promoting vector density is represented by the level of humidity (Cappai et al. 2018). Precipitation may promote the presence of moisture of the soil generating microhabitat that can enhance vector life cycle and avoid desiccation (Purse et al. 2004a, b, 2005). In addition, the wind also affects vector survival (Mellor et al. 2000). Cold winters should halt vector-borne diseases by limiting the presence of competent vectors and, as a result, the transmission of the involved virus. In Europe, BTV outbreaks are usually reported during summer-autumn when the vector is available. The mechanism of overwintering (the way the virus survives in a given area between vectors seasons) is not entirely clear. Competent insect vectors have a complex life cycle: after the blood meal, the eggs mature inside the female in a variable time, depending on the temperature and *Culicoides* species. For example, if the temperature is 27 °C the eggs maturation time for *C. imicola* is about 2 days, but could be 3 or 4 days if the temperature is low than 22 °C. For each blood meal, a group of eggs is deposited. Insects must survive about five days in order to lay their eggs (EFSA 2007). The vital circle includes four larval stages, the pupa stage, and the adult stage. The transition to the pupa stage usually occurs in a period ranging from 10 to 30 days, depending on the *Culicoides* species, the environmental temperature, and the amount of nutrients present in the environment (EFSA 2007). The pupae remain in this stage from 2 days up to 4 weeks; furthermore, they do not need nourishment and in most species they perform very limited movements (Meiswinkel et al. 1994).

The average life of adults is about 3–6 weeks, but they can survive for up to 9 weeks (EFSA 2007). In order to have viral transmission from the host to the insect, it is necessary that the blood meal occurs during the viremic period of the host, which corresponds with the feverish period of the animal: up to 11 days post-infection in sheep and 49 days post-infection in cattle (Bonneau et al. 2002). Recently, other studies have shown that the viremia resulting from experimental infections can last up to 45 days in sheep, more than 31 days for goats, and for more than 78 days in cattle (EFSA 2007). Once the BTV is ingested, it passes into the posterior part of the arthropod esophagus, bypassing all intestinal diverticula. In the first two days post-infection the viral titer in *Culicoides* decreases due to inactivation and fecal excretion which are superior to viral replication (eclipse phase). After 7–9 days post-infection the viral title reaches a plateau and this concentration remains so for the whole life of the insect. Transmission to the host is possible from 10 to 14 days post-infection (Wittmann et al. 2002; Mecham and Nunamaker 1994; Venter et al. 1991). The virion dose transmitted through the puncture itself can infect a receptive host by extrusion and exocytosis, resulting in cell damage (Schwartz-Cornil et al. 2008).

Symptoms and lesions

Following the bite, the virus establishes primary infection in fibroblast and mononuclear phagocytes, dendritic cells, lymphocytes, and endothelial cells (MacLachlan et al. 2009). Virus is transported to the local lymph nodes where replication starts (Drew et al. 2010a) and then spreads to blood circulation inducing a primary viremia reaching secondary organs, such as spleen and lungs, and then diffuse to oral mucosa and the hooves (Barratt-Boyes and MacLachlan 1994; Sanchez-Cordon et al. 2010; Melzi et al. 2016). The virus replicates in vascular endothelial cells, macrophages, and lymphocytes (MacLachlan et al. 1990, 2009; MacLachlan 2004; Barratt-Boyes and MacLachlan 1994; Drew et al. 2010b). The severity of the clinical signs of BT depends on serotype involved and by immunological status of the affected animals. Interaction of specific BTV viral proteins with the host triggers the activation of humoral and cellular immune-response which in turn determines host damage (Huismans et al. 2004; Caporale et al. 2014; Coetzee et al. 2014; Janowicz et al. 2015).

BT in sheep is manifested as hemorrhagic disease; acute symptoms appear after an incubation period of 3–6 days, which include fever that may last several days, serous and then mucopurulent nasal discharge, excess salivation, lymph node swelling, erosions and ulcers of the oral and nasal cavity, edema of the ears and neck, respiratory distress, hyperemia, and coronitis. Pathological lesions such as petechiae, ecchymosis, or hemorrhages in the pulmonary artery, cardiac lesions including pericardial effusion and myocardial necrosis, particularly the left papillary muscle, coronary bands around the hooves, vascular congestion, erosion and ulceration of the mucosa of the upper gastrointestinal tract, pulmonary edema, pleural and/or pericardial effusion, and endothelial hypertrophy with associated perivascular hemorrhage and perivascular edema are observed in severe acute BT in sheep (Spreull 1905; Moulton 1961; Erasmus 1975; Verwoerd and Erasmus 2004).

A swollen cyanotic tongue, from which originated the name Bluetongue, although characteristic is rarely detected. Mortality rate can be as high as 70%, and it is usually reached when a naïve population of sheep is introduced into an endemic area, when novel serotypes/strains emerge, or when stressful environmental factors are present. Hyperemia at the coronary bands of the feed and ulceration of the oral cavity may be extremely painful, determining reluctance to move and to eat resulting in weakness and prostration. Other common symptoms include "wool break" and pronounced torticollis. Usually sheep can die from respiratory distress and bacterial complications. Acute infection could lead to death within 14 days.

BTV infection of sheep during initial stages of gestation with field, vaccine or low passaged field strains may result in abortion, stillbirth, birth of viraemic animals, fetus cerebral and skeletal malformations, or even fetus death (Gard et al. 1987; MacLachlan et al. 2009; Rasmussen et al. 2013; Savini et al. 2014; Spedicato et al. 2019).

Infection in cattle and goats is usually asymptomatic or subclinical with prolonged viremia (Tweedle and Mellor 2002; MacLachlan et al. 2009). However, outbreaks of BTV-8 in European cattle have induced clinical symptoms like those described in sheep, including decreased milk production and reproductive disorders (abortion, stillbirth, and congenital abnormalities).

Diagnosis of BTV: a rapid overview

Prompt diagnosis of BTV infection is essential for the activation of specific control and restriction measures as established by the OIE terrestrial manual (OIE 2018). In recent years, along with standard and traditional techniques, sophisticated, fast, and sensitive methods have been used for the diagnosis of BTV. Diagnosis of BTV usually involves detection and identification of specific antigens, antibodies, or RNA in diagnostic samples taken from potentially infected animals using virus isolation, serological and molecular assays able to identify and characterize the involved serotype/strain. Furthermore, recognition of the clinical signs of BT can provide an early indication of infection and forms a basis for the passive surveillance. However, none of clinical signs are pathognomonic and its severity and range may be influenced by several factors such as species, age, virus strains, and immune status of the host.

Currently, the OIE Manual of standard for diagnostic tests and vaccines (OIE 2018) cites the competitive ELISA as a prescribed test for the detection of BTV group-specific antibodies. Then, serum neutralization test (SNT) is regularly used to detect and quantify neutralizing antibodies, that are specific for each BTV serotype, in serum samples (Jeggo et al. 1986).

Molecular assay methods are widely used to identify viral RNA of BTV from biological specimens (whole blood, spleen, lymph nodes, midges) or from cell-culture isolates by targeting specific viral segments (Wade-Evans et al. 1990; Katz et al. 1993; Jimenez-Clavero et al. 2006; Orru et al. 2006; Anthony et al. 2007; Shaw et al. 2007; Wilson and Mellor 2009; Hofmann et al. 2008b). A pan RT real-time PCR targeting Seg-10 of all known BTV serotypes is available and commonly used in reference laboratories (Hofmann et al. 2008b). Genotyping is generally performed after genogroup identification with several typing assays which are normally performed according to the epidemiological scenario of a given region. These assays, either PCR-based or microarrays, target the Seg-2 of the viral genome (Curini et al. 2019). Sequencing-based techniques of viral genes are nowadays so improved that the complete nucleotide sequence of a single gene can be achieved in few hours with a very low error rate. Sequencing of the entire Seg-2 allows the identification of BTV serotypes and the characterization of the topotype (Johnson et al. 2000; Zientara et al. 2006; Mertens et al. 2007; Maan et al. 2012; Lorusso et al. 2017). Due to the reassortment capability of BTV, whole genome sequencing has become essential to identify reassortant strains. The understanding of this mechanism is important to predict possible generation of emergent strains in area where two or more serotypes/strains normally circulate. In this perspective, next-generation sequencing technique is becoming an essential tool to obtain, in a fast manner, the complete genome constellation of a BTV isolate or directly from nucleic acids purified from biological specimens (Savini et al. 2017; Marcacci et al. 2018; Cappai et al. 2019).

Application to Prevention and Control/Adopted Surveillance and Control Strategies

The application of control measures for a vector-borne disease such as BT can be difficult to adopt and are greatly influenced by the constraints of relevant nation and international legislation and agreements. Diagnostic methods are essential to collect fundamental information about origins, distribution, and prevalence in BT monitoring and surveillance, to identify routes of incursion and movements of viruses. These data are on the basis of risk assessment and for the implementation of appropriate control strategies.

In 2008, the European Food Safety Authority (EFSA) published the scientific opinion of the Panel on Animal Health and Welfare (Question No EFSA-Q-2007-201), which provides strategic guidelines for the urgent strengthening of insect vector control measures, as a key approach to preventing BT disease and responding to epidemics. Since the virus is primarily transmitted by infected *Culicoides*, it is necessary, where possible, to either limit the exposition of susceptible animals during maximum vector activity or apply strictly farm management measures (i.e., store animals during the night, avoid water stagnation and night pasture, and use insecticides and repellents) aimed to reduce the number of insects in the farming area (Braverman and Chizov-Ginzburg 1997; Tweedle and Mellor 2002). Entomological surveillance throughout vector sampling can be used to collect essential data about the presence, abundance, proportions, and seasonal variations in the numbers of adult *Culicoides* or to identify new potential vector species.

The application of restriction of animal movement and trading and the introduction of surveillance strategies in restriction zones are necessary for the monitoring and the control of disease spreading from unaffected to affected areas (Caporale and Giovannini 2010; MacLachlan and Mayo 2013). Movement restriction could be rather impossible to achieve or even to control, in areas such as sub-Saharan Africa where camelids, which have been proven to be competent host for the virus, are used by nomadic tribes as transport vehicles during long-distance travels to reach either markets of farming areas.

Since clinical signs of BT may be unrecognized or go unnoticed, particularly in cattle and wild ruminants, serological surveillance is currently in use more than clinical surveillance. BTV circulation can be monitored using "sentinel animals," selected and tested seronegative, monitored regularly to confirm the absence of virus circulation. The rapid nature of currently available assays and their potential for high-throughput automation are particularly valuable for large-scale surveillance by national and international reference laboratories. Vaccination is the main control measure for BTV. Modified live vaccines (MLV), developed in endemic areas in South Africa, have been used in several parts of the world for seasonal vaccination campaigns and have been also used during past European outbreaks (Savini et al. 2004; Roy et al. 2009; McVey and MacLachlan 2015). However, due to safety issues such as reassortment with field strains, reversion to virulence, transmission to insect vectors and by them to unvaccinated animals, side effects in inoculated animals such as development of clinical signs, reduction in milk production, semen secretion of the virus, and abortion, MLV were withdrawn and replaced by inactivated vaccines (Dungu et al. 2004; Monaco et al. 2004; Veronesi et al. 2010; Savini et al. 2007, 2008; 2014). Inactivated vaccine can be considered safer and are currently used as monovalent, bivalent, and tetravalent formula (Savini et al. 2008; Reddy et al. 2010). Their efficiency and the presence of only few side effects have induced the pharmaceutical companies to produce copious quantities of inactivated vaccines overcoming major downsides such as large cost of production.

To date, several approaches have been used for the development of new generation vaccines able to offer major protection, low-cost production, and almost none safety issues when compared to the traditional live attenuated or inactivated vaccines. Modern vaccines are designed to accomplish what is called DIVA strategy (differentiating infected from vaccinated animals) which is normally achieved by designing a serological test detecting a viral protein that is not present in the vaccine. DIVA strategy is mainly based on the production of recombinant vaccines.

The achievement of reverse genetics system for BTV is an important landmark for the research and development of new generation vaccines, which are able to offer good protection and safety and can be used in a DIVA strategy (Celma et al. 2013). Reverse genetics allow the generation of replication competent virus starting from plasmids (Boyce et al. 2008).

This approach gives the possibility to "play" with the viral segments to either understand their role in virus replication and to generate safe replication competent vaccines. By reverse genetics it is possible to engineer disabled infectious single-cycle (DISC) and disabled infectious single-animal (DISA) vaccines.

DISA vaccines do not produce viremia in sheep and there is no risk of transmission by insect vectors (Feenstra et al. 2015). Although DISC and DISA vaccines seem to be very promising, cost production for the generation of large quantities of synthetic viruses is still an issue (Feenstra and van Rijn 2017). Overall, it is important to perform correct surveillance in order to predict as much as possible outbreaks of novel BTV strains. This can be achieved only by strong collaboration between farmers, veterinarians, and researchers.

References

Abu Elzein EME. Precipitating antibodies against bluetongue and foot and mouth disease viruses in cattle between the two Niles in Khartoum Province, Sudan. Rev Sci Tech Off Int Epiz. 1983;2:1059–66.

Abu Elzein EME, Tag Eldin MH. The first outbreaks of sheep bluetongue in Khartoum Province, Sudan. Rev Sci Tech Off Int Epiz. 1985;4:509–15.

Adam IA, Abdalla MS, Mohamed MEH, Aradaib IE. Prevalence of bluetongue virus infection and associated risk factors among cattle in North Kordufan State, Western Sudan. BMC Vet Res. 2014;10:94–101.

Anon. Records of the Animal Production Department. Sudan: Government; 1953.

Anthony S, Jones H, Darpel KE, Elliott H, Maan S, Samuel A, Mellor PS, Mertens PP. A duplex RT-PCR assay for detection of genome segment 7 (VP7gene) from 24 BTV serotypes. J Virol Methods. 2007;141:188–97.

Attoui H, Maan SS, Anthony SJ, Mertens PPC. Bluetongue virus, other orbiviruses and other reoviruses: their relationships and taxonomy. 1st ed. London: Elsevier/Academic Press; 2009. p. 23–552.

Bakhoum M, Fall M, Fall A, Bellis G, Gottlieb Y, Labuschagne K. First record of Culicoides oxystoma Kieffer and diversity of species within the Schultzei group of Culicoides Latreille (Diptera: Ceratopogonidae) biting midges in Senegal. PLoS One. 2013;8:1–8.

Barratt-Boyes SM, MacLachlan NJ. Dynamics of viral spread in bluetongue virus infected calves. Vet Microbiol. 1994;40:361–71.

Belbis G, Zientara S, Brèard E, Sailleau C, Caignard G, Vitour D, Attoui H. Bluetongue virus: from BTV-1 to BTV-27. Adv Virus Res. 2017;99:161–97.

Belhouchet M, Mohd Jaafar F, Firth AE, Grimes JM, Mertens PPC, Attoui H. Detection of a fourth orbivirus non-structural protein. PLoS One. 2011;6(10):e25697. https://doi.org/10.1371/journal.pone.0025697.

Bida SA, Eid FIA. Bluetongue of sheep in Nigeria. J Nig Vet Med Assoc. 1974;3:12–6.

Bonneau KR, Demaula CD, Mullens BA, MacLachlan NJ. Duration of viraemia infectious to *Culicoides sonorensis* in Bluetongue virus-infected cattle and sheep. Vet Microbiol. 2002;88(2):115–25.

Borkent A. The biting midges, the Ceratopogonidae (Diptera). In: Marquardt WH, editor. Biology of disease vectors. Burlington, MA: Elsevier Academic Press; 2005. p. 113–26.

Boyce M, Celma CCP, Roy P. Development of reverse genetics systems for Bluetongue virus: recovery of infectious virus from synthetic RNA transcripts. J Virol. 2008;82:8339–48.

Braverman Y, Chizov-Ginzburg A. Repellency of synthetic and plant-derived preparations for *Culicoides imicola*. Med Vet Entomol. 1997;11(4):355–60.

Bumbarov V, Golender N, Erster O, Khinich Y. Detection and isolation of Bluetongue virus from commercial vaccine batches. Vaccine. 2016;34:3317–23.

Calistri P, Giovannini A, Conte A, Nannini D, Santucci U, Patta C, Rolesu S, Caporale V. Bluetongue in Italy: Part I. Vet Ital. 2004;40(3):243–51.

Caporale V, Giovannini A. Bluetongue control strategy, including recourse to vaccine: a critical review. Rev Sci Tech. 2010;29(3):573–91.

Caporale M, Di Gialleonorado L, Janowicz A, Wilkie G, Shaw A, Savini G, Van Rijn PA, Mertens P, Di Ventura M, Palmarini M. Virus and host factors affecting the clinical outcome of Bluetongue virus infection. J Virol. 2014;88:10399–411.

Cappai S, Loi F, Coccollone A, Contu M, Capece P, Fiori M, Canu S, Foxi C, Rolesu S. Retrospective analysis of Bluetongue farm risk profile definition, based on biology, farm management practices and climatic data. Prev Vet Med. 2018;155:75–85.

Cappai S, Rolesu S, Loi F, Liciardi M, Leone A, Marcacci M, Teodori L, Mangone I, Sghaier S, Portanti O, Savini G, Lorusso A. Western Bluetongue virus serotype 3 in Sardinia, diagnosis and characterization. Transbound Emerg Dis. 2019;66(3):1426–31.

Carpenter S, Veronesi E, Mullens B, Venter G. Vector competence of Culicoides for arboviruses: three major periods of research, their influence on current studies and future directions. Rev Sci Tech Off Int Epiz. 2015;34(1):97–112.

Carpi G, Holmes EC, Kitchen A. The evolutionary dynamics of Bluetongue virus. J Mol Evol. 2010;70:583–92. https://doi.org/10.1007/s00239-010-9354-y.

Celma CCP, Boyce M, van Rijn PA, Eschbaumer M, Wernike K, Hoffmann B, Beer M, Haegeman A, De Clercq K, Roy P. Rapid generation of replication-deficient monovalent and multivalent vaccines for Bluetongue virus: protection against virulent virus challenge in cattle and sheep. J Virol. 2013;87:9856–64.

Clavijo A, Heckert RA, Dulac GC, Afshar A. Isolation and identification of Bluetongue virus. J Virol Methods. 2000;87(1–2):13–23.

Coetzee P, Stokstad M, Venter EH, Myrmel M, Van Vuuren M. Bluetongue: a historical and epidemiological perspective with the emphasis on South Africa. Virol J. 2014;9:198.

Conte A, Giovannini A, Savini G, Goffredo M, Calistri P, Meiswinkel R. The effect of climate on the presence of *Culicoides imicola* in Italy. J Veterinary Med Ser B. 2003;50:139–47.

Conte A, Goffredo M, Ippoliti C, Meiswinkel R. Influence of biotic and abiotic factors on the distribution and abundance of *Culicoides imicola* and the Obsoletus Complex in Italy. Vet Parasitol. 2007;150(4):333–44.

Curini V, Marcacci M, Tonelli A, Di Teodoro G, Di Domenico M, D'Alterio N, Portanti O, Ancora M, Savini G, Panfili M, Cammà C, Lorusso A. Molecular typing of Bluetongue virus using the nCounter analysis system platform. J Virol Methods. 2019;269:64–9.

De Koeijer AA, Elbers ARW. Modelling of vector-borne diseases and transmission of Bluetongue virus in North-West Europe, Chapter 6. In: Takken W, Knols BGJ, editors. Ecology and control of vector-borne disease, vol. 1. Wageningen: Wageningen Academic; 2007. p. 99–112.

De Liberato C, Purse BV, Goffredo M, Scholl F, Scaramozzino P. Geographical and seasonal distribution of the Bluetongue virus vector, *Culicoides imicola*, in central Italy. Med Vet Entomol. 2003;17:388–94.

Drew CP, Gardner IA, Mayo CE, Matsuo E, Roy P, MacLachlan NJ. Bluetongue virus infection alters the impedance of monolayers of bovine endothelial cells as a result of cell death. Vet Immunol Immunopathol. 2010a;138:108–15.

Drew CP, Heller MC, Mayo C, Watson JL, MacLachlan NJ. Bluetongue virus infection activates bovine monocyte-derived macrophages and pulmonary artery endothelial cells. Vet Immunol Immunopathol. 2010b;136:292–6.

Du Toit RM. The transmission of Bluetongue and horse sickness by Culicoides. Onderstepoort J Vet Sci Anim Ind. 1944;19:7–16.

Dungu B, Gerdes T, Smit T. The use of vaccination in the control of Bluetongue in southern Africa. Vet Ital. 2004;40:616–22.
Eisa M, Karrar AE, Elrahim A. Incidence of Bluetongue virus precipitating antibodies in sera of some domestic animals in the Sudan. J Hyg. 1979;83(3):539–45.
Eisa M, Osman OM, Karrar AE, Abdel Rahim AH. An outbreak of Bluetongue in sheep in the Sudan. Vet Rec. 1980;106(23):481–2.
Ekue FN, Nfi AN, Tsangue P, et al. Bluetongue in exotic sheep in Cameroon. Trop Anim Health Prod. 1985;17:187.
Erasmus BJ. The control of Bluetongue in an enzootic situation. Aust Vet J. 1975;51:209–10.
European Food Safety Authority (EFSA). Scientific Opinion of the Scientific Panel on Animal Health and Welfare on request from the European Commission on Bluetongue Vectors and Vaccines. EFSA J. 2007;479:1–29.
Fall M, Diarra M, Fall AG, Balenghien T, Seck MT, Bouyer J, Garros C, Gimonneau G, Allène X, Mall I, Delécolle JC, Rakotoarivony I, Bakhoum MT, Dusom AM, Ndao M, Konaté L, Faye O, Baldet T. Culicoides (Diptera: Ceratopogonidae) midges, the vectors of African horse sickness virus—a host/vector contact study in the Niayes area of Senegal. Parasit Vectors. 2015a;8:39.
Fall M, Fall AG, Seck MT, Bouyer J, Diarra M, Lancelot R, Gimonneau G, Garros C, Bakhoum MT, Faye O, Baldet T, Balenghien T. Host preferences and circadian rhythm of Culicoides (Diptera: Ceratopogonidae), vectors of African horse sickness and Bluetongue viruses in Senegal. Acta Trop. 2015b;149:239–45.
Feenstra F, van Rijn PA. Current and next-generation Bluetongue vaccines: requirements, strategies, and prospects for different field situations. Crit Rev Microbiol. 2017;43:142–55.
Feenstra F, Eenstraab JSP, van Rijnac PA. Application of Bluetongue Disabled Infectious Single Animal (DISA) vaccine for different serotypes by VP2 exchange or incorporation of chimeric VP2. Vaccine. 2015;33:812–8.
Foxi C, Delrio G, Falchi G, Marche MG, Satta G, Ruiu L. Role of different Culicoides vectors (Diptera: Ceratopogonidae) in bluetongue virus transmission and overwintering in Sardinia (Italy). Parasit Vectors. 2016;9(1):440.
Gambles RM. Bluetongue of sheep in Cyprus. J Comp Pathol. 1949;59:176–90.
Gard GP, Shorthose JE, Weir RP, Erasmus BJ. The isolation of a Bluetongue serotype new to Australia. Aust Vet J. 1987;64:87–8.
Gibbs EP, Greiner EC. The epidemiology of Bluetongue. Comp Immunol Microbiol Infect Dis. 1994;17:207–20.
Goossens B, Osaer S, Kora S, Chandler KJ, Petrie L, Thevasagayam JA, Woolhouse T, Anderson J. Abattoir survey of sheep and goats in the Gambia. Vet Rec. 1998;142(11):277–81.
Gorman BM, Taylor J, Walker PJ, Davidson WL, Brown F. Comparisons of Bluetongue type 20 with certain viruses of the Bluetongue and Eubenangee serological groups of orbiviruses. J Gen Virol. 1981;57:251–61.
Gould AR, Hyatt AD. The Orbivirus genus. Diversity, structure, replication and phylogenetic relationships. Comp Immunol Microbiol Infect Dis. 1994;17:163–88. https://doi.org/10.1016/0147-9571(94)90041-8.
Hafez SM, Ozawa Y. Serological survey of Bluetongue in Egypt. Bull Epiz Dis Afr. 1973;21 (3):297–304.
Häsler B, Howe KS, Di Labio E, Schwermer H, Stärk KDC. Economic evaluation of the surveillance and intervention programme for Bluetongue virus serotype 8 in Switzerland. Prev Vete Med. 2012;103:93–111.
Hofmann MA, Renzullo S, Mader M, Chaignat V, Worwa G, Thuer B. Genetic characterization of Toggenburg Orbivirus, a new Bluetongue virus, from goats, Switzerland. Emerg Infect Dis. 2008a;14:1855–61.
Hofmann M, Griot C, Chaignat V, Perler L, Thur B. Bluetongue disease reaches Switzerland. Schweiz Arch Tierheilkd. 2008b;150:49–56.
Huismans H. Protein synthesis in Bluetongue virus infected cells. Virology. 1979;92:385–96.

Huismans H, Van Staden V, Fick WC, Van Niekerk M, Meiring TL, Texeira L. A comparison of different Orbivirus proteins that could affect virulence and pathogenesis. In : Proceedings of the third OIE Bluetongue international symposium, Oct 26–29; 2004.

Hutcheon D. Malarial catarrhal fever of sheep. Vet Res. 1902;14:629–33.

ICTV. Virus taxonomy: 2014 release; 2018. Retrieved 15 Jun 2015.

Ismail JM, Martin J, Nazmi A. Bluetongue neutralization test with different virus under variable condition. Agr Res Rev Egypt. 1987;65(5):867–72.

Jafari-Shoorijeh S, Ramin AG, MacLachlan NJ, Osburn BI, Tamadon A, Behzadi MA, Mahdavi M, Araskhani A, Samani D, Rezajou N, Amin-Pour A. High seroprevalence of Bluetongue virus infection in sheep flocks in West Azerbaijan, Iran. Comp Immunol Microbiol Infect Dis. 2010;33:243–7.

Janowicz A, Caporale M, Shaw A, Gulletta S, DiGialleonardo L, Ratinier M, Palmarini M. Multiple genome segments determine virulence of Bluetongue virus serotype 8. J Virol. 2015;89:5238–49.

Jeggo MH, Wardley C, Brownlie J, Corteyn AH. Serial inoculation of sheep with two Bluetongue virus types. Res Vet Sci. 1986;40:386–92.

Jiménez-Clavero MA, Agüero M, San Miguel E, Mayoral T, López MC, Ruano MJ, Romero E, Monaco F, Polci A, Savini G, Gomez-Tejedor C. High throughput detection of Bluetongue virus by a new real-time fluorogenic reverse transcription-polymerase chain reaction: application on clinical samples from current Mediterranean outbreaks. J Vet Diagn Investig. 2006;18:7–17.

Johnson DJ, Wilson WC, Paul PS. Validation of a reverse tran-scriptase multiplex PCR test for the serotype determination of U.S. isolates of Bluetongue virus. Vet Microbiol. 2000;76:105–15.

Katz JB, Alstad AD, Gustafson GA, Moser KM. Sensitive identification of Bluetongue virus serogroup by a colorimetric dual oligonucleotide sorbent assay of amplified viral nucleic acid. J Clin Microbiol. 1993;31:3028–30.

Lefevre PC, Taylor WP. Situation épidemiologique de la fièvre catarrhale du mouton (Blue tongue) au Sénégal. Rev Elev Méd Vét Pays Trop. 1983;36(3):241–5.

Lorusso A, Sghaier S, Carvelli A, Di Gennaro A, Leone A, Marini V, Pelini S, Marcacci M, Rocchigiani AM, Puggioni G, Savini G. Bluetongue virus serotypes 1 and 4 in Sardinia during autumn 2012: new incursions or re-infection with old strains? Infect Genet Evol. 2013;19:81–7. https://doi.org/10.1016/j.meegid.2013.06.028.

Lorusso A, Sghaier S, Ancora M, Marcacci M, Di Gennaro A, Portanti O, Savini G. Molecular epidemiology of Bluetongue virus serotype 1 circulating in Italy and its connection with northern Africa. Infect Genet Evol. 2014;28:44–9.

Lorusso A, Baba D, Spedicato M, Teodori L, Bonfini B, Marcacci M, Di Provvido A, Isselmou K, Marini V, Carmine I, Scacchia M, Di Sabatino D, Petrini A, Bezeid BA, Savini G. Bluetongue virus surveillance in the Islamic Republic of Mauritania: is serotype 26 circulating among cattle and dromedaries? Infect Genet Evol. 2016;40:109–12. https://doi.org/10.1016/j.meegid.2016.02.036.

Lorusso A, Guercio A, Purpari G, Cammà C, Calistri P, D'Alterio N, Hammami S, Sghaier S, Savini G. Bluetongue virus serotype 3 in Western Sicily, November 2017. Vet Ital. 2017;53(4):273–5.

Lorusso A, Sghaier S, Di Domenico M, Barbria ME, Zaccaria G, Megdich A, Portanti O, Seliman IB, Spedicato M, Pizzurro F, Carmine I, Teodori L, Mahjoub M, Mangone I, Leone A, Hammami S, Marcacci M, Savini G. Analysis of Bluetongue serotype 3 spread in Tunisia and discovery of a novel strain related to the Bluetongue virus isolated from a commercial sheep pox vaccine. Infect Genet Evol. 2018;59:63–71. https://doi.org/10.1016/j.meegid.2018.01.025.

Lundervold M, Milner-Gulland EJ, O'Callaghan CJ, Hamblin C. First evidence of Bluetongue virus in Kazakhstan. Vet Microbiol. 2003;92:281–7.

Maan S, Maan NS, Ross-Smith N, Batten CA, Shaw AE, Anthony SJ, Samuel AR, Darpel KE, Veronesi E, Oura CA, Singh KP, Nomikou K, Potgieter AC, Attoui H, van Rooij E, van Rijn P, De Clercq E, Vandenbussche F, Zientara S, Bréard E, Sailleau C, Beer M, Hoffman B, Mellor PS, Mertens PP. Sequence analysis of Bluetongue virus serotype 8 from the Netherlands 2006 and comparison to other European strains. Virology. 2008;377:308–18.

Maan S, Maan NS, Nomikou K, Veronesi E, Bachanek-Bankowska K, Belaganahalli MN, Attoui H, Mertens PP. Complete genome characterisation of a novel 26th Bluetongue virus serotype from Kuwait. PLoS One. 2011;6:e26147.
Maan NS, Maan S, Belaganahalli MN, Ostlund EN, Johnson DJ, Nomikou K, Mertens PP. Identification and differentiation of the twenty six Bluetongue virus serotypes by RT-PCR amplification of the serotype-specific genome segment 2. PLoS One. 2012;7:e32601.
MacLachlan NJ. Bluetongue: pathogenesis and duration of viraemia. Vet Ital. 2004;40(4):462–7.
MacLachlan NJ. Global implications of the recent emergence of Bluetongue virus in Europe. Vet Clin North Am Food Anim Pract. 2010;26:163–71.
MacLachlan NJ, Mayo CE. Potential strategies for control of Bluetongue, a globally emerging, Culicoides-transmitted viral disease of ruminant livestock and wildlife. Antivir Res. 2013;99 (2):79–90.
MacLachlan NJ, Jagels G, Rossitto PV, Moore PF, Heidner HW. The pathogenesis of experimental Bluetongue virus infection of calves. Vet Pathol. 1990;27:223–9.
MacLachlan NJ, Drew CP, Darpel KE, Worwa G. The pathology and pathogenesis of Bluetongue. J Comp Path. 2009;141:1–16.
Mahmoud AS, Savini G, Spedicato M, Monaco F, Carmine I, Lorusso A, Francesco T, Mazzei M, Forzan M, Eldaghayes I, Dayhum A. Exploiting serological data to understand the epidemiology of Bluetongue virus serotypes circulating in Libya. Vet Med Sci. 2018;5(1):79–86. https://doi.org/10.1002/vms3.136.
Marcacci M, Sant S, Mangone I, Goria M, Dondo A, Zoppi S, van Gennip RGP, Radaelli MC, Cammà C, van Rijn PA, Savini G, Lorusso A. One after the other: a novel Bluetongue virus strain related to Toggenburg virus detected in the Piedmont region (North-western Italy), extends the panel of novel atypical BTV strains. Transbound Emerg Dis. 2018;65(2):370–4. https://doi.org/10.1111/tbed.12822.
McVey DS, MacLachlan NJ. Vaccines for prevention of Bluetongue and epizootic hemorrhagic disease in livestock—a North American perspective. Vector Borne Zoonotic Dis. 2015;15:385–96.
Mecham JO, Nunamaker RA. Complex interactions between vectors and pathogens: Culicoides variipennis sonorensis (Diptera: Ceratopogonidae) infection rates with Bluetongue viruses. J Med Entomol. 1994;31(6):903–7.
Meiswinkel R, Nevill EM, Venter GJ. Vectors: Culicoides spp. In: Coetzer JAW, Tustin RC, editors. Infectious diseases of livestock with special reference to Southern Africa, vol. 1. Cape Town: Oxford University Press; 1994. p. 69–89.
Meiswinkel R, Gomulski LM, Delecolle JC, Goffredo M, Gasperi G. The taxonomy of Culicoides vector complexes—unfinished business. Vet Ital. 2004;40:151–9.
Meiswinkel R, Baldet T, de Deken R, Takken W, Delecolle JC, Mellor PS. The 2006 outbreak of Bluetongue in northern Europe—the entomological perspective. Prev Vet Med. 2008;87:55–63.
Mellor PS, Osborne R, Jennings DM. Isolation of Bluetongue and related viruses from Culicoides spp. in the Sudan. J Hyg. 1984;93(3):621–8.
Mellor PS, Boorman J, Baylis M. Culicoides biting midges: their role as Arbovirus vectors. Annu Rev Entomol. 2000;45:307–40.
Melzi E, Caporale M, Rocchi M, Martín V, Gamino V, Di Provvido A, Marruchella G, Entrican G, Sevilla N, Palmarini M. Follicular dendritic cell disruption as a novel mechanism of virus-induced immunosuppression. PNAS. 2016;113(41):E6238–47.
Mertens PPC, Maan S, Samuel A, Attoui H. Orbiviruses, reoviridae. In: Fauquet CM, Mayo MA, Maniloff J, Desselberger U, Ball LA, editors. Virus taxonomy eighth report of the International Committee on Taxonomy of Viruses. London: Elsevier/Academic Press; 2005. p. 466–83.
Mertens PPC, Maan NS, Prasad G, Samuel AR, Shaw AE, Potgieter AC, Anthony SJ, Maan S. The design of primers and use of RT-PCR assays for typing European BTV isolates: differentiation of field and vaccine strains. J Gen Virol. 2007;88:2811–23.
Monaco F, De Luca N, Spina P, Morelli D, Liberatore I, Citarella R, Savini G, MacLachlan NJ, Pearson JE. Virological and serological responses in cattle following field vaccination with

bivalent modified-live vaccine against Bluetongue virus serotype 2 and 9. Vet Ital. 2004;40:657–60.

Moore DL, Kemp GE. Bluetongue and related viruses in Ibadan, Nigeria. Serologic studies of domesticated and wild animals. Am J Vet Res. 1974;35:1115–20.

Moulton JE. Pathology of Bluetongue of sheep. J Am Vet Med Assoc. 1961;138:493–8.

OIE. Chapter 8.3. Infection with Bluetongue virus. OIE Terrestrial Animal Health Code, 27th ed. 2018. Isbn:978-92-95108-59-2.

Orrù G, Ferrando ML, Meloni M, Liciardi M, Savini G, De Santis P. Rapid detection and quantitation of Bluetongue virus (BTV) using a Molecular Beacon fluorescent probe assay. J Virol Methods. 2006;137:34–42.

Owen NC, Munz EK. Observations on a strain of Bluetongue virus by electron microscopy. Onderstepoort J Vet Res. 1966;33:9–14.

Purse BV, Baylis M, Tatem AJ, et al. Predicting the risk of Bluetongue through time: climate models of temporal patterns of outbreaks in Israel. Rev Sci Tech Off Int Epiz. 2004a;23:761–75.

Purse BV, Tatem AJ, Caracappa S, Rogers DJ, Mellor PS, Baylis M, Torina A. Modelling the distributions of Culicoides Bluetongue virus vectors in Sicily in relation to satellite derived climate variables. Med Vet Entomol. 2004b;18:90–101.

Purse BV, Mellor PS, Rogers DJ, Samuel AR, Mertens PPC, Baulis M. Climate change and the recent emergence of Bluetongue in Europe. Nat Rev Microbiol. 2005;3:171–81.

Purse BV, Brown HE, Harrup L, Mertens PPC, Rogers DJ. Invasion of Bluetongue and other arbovirus infections into Europe: the role of biological and climatic processes. Rev Sci Tech. 2008;27:427–42.

Rasmussen LD, Savini G, Lorusso A, Bellacicco A, Palmarini M, Caporale M, Rasmussen T, Belsham GJ, Bøtner A. Transplacental transmission of field and rescued strains of BTV-2 and BTV-8 in experimentally infected sheep. Vet Res. 2013;44(1):75.

Ratinier M, Caporale M, Golder M, Franzoni G, Allan K, Nunes SF, Armezzani A, Bayoumy A, Rixon F, Shaw A, Palmarini M. Identification and characterization of a novel non-structural protein of Bluetongue virus. PLoS Pathog. 2011;7(12):e1002477. https://doi.org/10.1371/journal.ppat.1002477.

Rawlings P, Capela R, Pro MJ, et al. The relationship between climate and the distribution of *Culicoides imicola* in Iberia. Arch Virol. 1998;14:93–102.

Reddy YK, Manohar BM, Pandey AB, Reddy YM, Prasad G, Chauhan RS. Development and evaluation of inactivated pentavalent adjuvanted vaccine for Bluetongue. Indian Vet J. 2010;87:434–6.

Roy P, Boyce M, Noad R. Prospects for improved Bluetongue vaccines. Nat Rev Microbiol. 2009;7:120–8.

Ruiz-Fons F, Reyes-García AR, Alcaide V, Gortázar C. Spatial and temporal evolution of Bluetongue virus in wild ruminants, Spain. Emerg Infect Dis. 2008;14:951–3.

Rushton J. The economics of animal health and production. Oxfordshire: CAB International; 2009. p. 193–7.

Saeed SI. A survey of Bluetongue virus antibodies and associate risk factors among camels in Khartoum state, Sudan. J Camel Res Prod; 2017.

Sánchez-Cordón PJ, Rodríguez-Sánchez B, Risalde MA, Molina V, Pedrera M, Sánchez-Vizcaíno JM, Gómez-Villamandos JC. Immunohistochemical detection of Bluetongue virus in fixed tissue. J Comp Pathol. 2010;143:20–8.

Savini G, Monaco F, Citarella R, Calzetta G, Panichi G, Ruiu A, Caporale V. Monovalent modified vaccine against Bluetongue virus serotype 2: immunity studies in cows. Vet Ital. 2004;40:664–7.

Savini G, Ronchi GF, Leone A, Ciarelli A, Migliaccio P, Franchi P, Mercante MT, Pini A. An inactivated vaccine for the control of Bluetongue virus serotype 16 infection in sheep in Italy. Vet Microbiol. 2007;124:140–6.

Savini G, MacLachlan NJ, Sanchez-Vizcaino JM, Zientara S. Vaccines against Bluetongue in Europe. Comp Immunol Microbiol Infect Dis. 2008;31:101–20.

Savini G, Lorusso A, Paladini C, Migliaccio P, Di Gennaro A, Di Provvido A, Scacchia M, Monaco F. Bluetongue Serotype 2 and 9 Modified Live Vaccine Viruses as Causative Agents of Abortion in Livestock: A Retrospective Analysis in Italy. Transbound Emerg Dis. 2014;61 (1):69–74.

Savini G, Puggioni G, Meloni G, Marcacci M, Di Domenico M, Rocchiggiani AM, Spedicato M, Oggiano A, Manunta D, Teodori L, Leone A, Portanti O, Cito F, Conte A, Orsini M, Cammà C, Calistri P, Giovannini A, Lorusso A. Novel putative Bluetongue virus in healthy goats from Sardinia, Italy. Infect Genet Evol. 2017;51:108–17.

Schulz C, Breard E, Sailleau C, Jenckel M, Viarouge C, Vitour D, Palmarini M, Gallois M, Höper D, Hoffmann B, Beer M, Zientara S. Bluetongue virus serotype 27: detection and characterization of two novel variants in Corsica, France. J Gen Virol. 2016;97(9):2073–83.

Schwartz-Cornil I, Mertens PP, Contreras V, Hemati B, Pascale F, Bréard E, Mellor PS, MacLachlan NJ, Zientara S. Bluetongue virus: virology, pathogenesis and immunity. Vet Res. 2008;39(5):46.

Scottish Government. Assessing the economic impact of different Bluetongue Virus (BTV) incursion scenarios in Scotland. Technical report Commission Number CR/2007/56; 2008. http://www.gov.scot/Resource/Doc/241420/0067122.pdf. Accessed Mar 2015.

Sellers RF. Bluetongue in Africa, the Mediterranean region and near east—disease, virus and vectors. Prev Vet Med. 1984;2(1–4):371–8.

Sellers RF, Mellor PS. Temperature and the persistence of viruses in Culicoides spp. during adverse conditions. Rev Sci Tech. 1993;12:733–55.

Sellers RF, Taylor WP. Epidemiology of Bluetongue and the import and export of livestock, semen and embryos. Bull Off Int Epiz. 1980;92:587–92.

Sellers RF, Pedgley DE, Tucker MR. Possible windborne spread of Bluetongue to Portugal. J Hyg. 1979;81:181–96.

Sghaier S, Lorusso A, Portanti O, Marcacci M, Orsini M, Barbria ME, Mahmoud AS, Hammami S, Petrini A, Savini G. A novel Bluetongue virus serotype 3 strain in Tunisia, November 2016. Transbound Emerg Dis. 2017;64(3):709–15. https://doi.org/10.1111/tbed.12640.

Shaw AE, Monaghan P, Alpar HO, Anthony S, Darpel KE, Batten CA, Guercio A, Alimena G, Vitale M, Bankowska K, Carpenter S, Jones H, Oura CAL, King DP, Elliott H, Mellor PS, Mertens PPC. Development and initial evaluation of a real-time RT-PCR assay to detect Bluetongue virus genome segment 1. J Virol Methods. 2007;145:115–26.

Spedicato M, Carmine I, Teodori L, Leone A, Casaccia C, Di Gennaro A, Di Francesco G, Marruchella G, Portanti O, Marini V, Pisciella M, Lorusso A, Savini G. Transplacental transmission of the Italian Bluetongue virus serotype 2 in sheep. Vet Ital. 2019;55(2):131–41.

Sperlova A, Zendulkova D. Bluetongue: a review. Vet Med. 2011;56(9):430–52.

Spreull J. Report from veterinary surgeon Spreull on the result of his experiments with the malarial catarrhal fever of sheep. Agric J Cape Good Hope. 1902;20:469–77.

Spreull J. Malarial catarrhal fever (Bluetongue) of sheep in South Africa. J Comp Pathol Ther. 1905;18:321–37.

Staubach C, Conte A, Meiswinkel R, Gethmann J, Unger F, Frohlich A, Gloster J, Purse B. Epidemiological analysis of the 2006 Bluetongue virus serotype 8 epidemic in North-Western Europe: role of environmental factors—small scale environmental analysis. EFSA report; 2007.

Stewart M, Hardy A, Barry G, Pinto RM, Caporale M, Melzi E, Palmarini M. Characterization of a second open reading frame in genome segment 10 of Bluetongue virus. J Gen Virol. 2015;96 (11):3280–93.

Studdert MJ, Pangborn J, Addison RB. Bluetongue virus structure. Virology. 1966;29:509–11.

Sun EC, Huang LP, Xu QY, Wang HX, Xue XM, Lu P, Li WJ, Liu W, Bu ZG, Wu DL. Emergence of a novel Bluetongue virus serotype, China 2014. Transboun Emerg Dis. 2016;63(6):585–9.

Tabachnick WJ. Culicoides and the global epidemiology of Bluetongue virus infection. Vet Ital. 2004;40(3):145–50.

Tatem AJ, Baylis M, Mellor PS, Purse BV, Capela R, Pena I, Rogers DJ. Prediction of Bluetongue vector distribution in Europe and North Africa using satellite imagery. Vet Microbiol. 2003;97:13–29.

Theiler A. Bluetongue in sheep. Ann Rept Dir Agric Transvaal. 1906;1904–1905:110–21.

Tweedle N, Mellor PS. Technical review Bluetongue: the virus, hosts and vectors. Version 1.5. Report to the Department of Health, Social Services and Public Safety UK (DEFRA); 2002. 25 p. http://archive.defra.gov.

Velthuis AGJ, Velthuis AG, Saatkamp HW, Mourits MC, De Koeijer AA, Elbers AR. Financial consequences of the Dutch Bluetongue serotype 8 epidemics of 2006 and 2007. Prev Vet Med. 2010;93:294–304.

Venter GJ, Hill E, Pajor IT, Nevill EM. The use of a membrane feeding technique to determine the infection rate of *Culicoides imicola* (Diptera: Ceratopogonidae) for two Bluetongue virus serotypes in South Africa. Onderstepoort J Vet Res. 1991;63(4):315–25.

Veronesi E, Darpel KE, Hamblin K, Carpenter S, Takamatsu HH, Anthony SJ, Elliott H, Mertens P, Mellor P. Viraemia and clinical disease in Dorset Poll sheep following vaccination with live attenuated Bluetongue virus vaccines serotypes 16 and 4. Vaccine. 2010;28(5):1397–403.

Vervoerd DW. Purification and characterization of Bluetongue virus. Virology. 1969;38:203–12.

Vervoerd DW. Diplornaviruses: a newly recognized group of double-stranded RNA viruses. Progr Med Virol. 1970;12:192–210.

Verwoerd DW, Erasmus BJ. Bluetongue. In: Coetzer JAW, Tustin RC, editors. Infectious diseases of livestock. 2nd Edition. Cape Town: Oxford University Press Southern Africa; 2004. p. 1201–20.

Wade-Evans AM, Mertens PPC, Bostock CJ. Development of the polymerase chain reaction for the detection of Bluetongue virus in tissue samples. J Virol Methods. 1990;30:15–24.

Walton TE. The history of Bluetongue and a current global overview. Vet Ital. 2004;40(3):31–8.

Wilson AJ, Mellor PS. Bluetongue in Europe: past, present and future. Philos Trans R Soc Lond B Biol Sci. 2009;364(1530):2669–81. https://doi.org/10.1098/rstb.2009.0091.

Wittmann EJ, Baylis M. Review. Climate change: effects on Culicoides-transmitted viruses and implications for the UK. Vet J. 2000;160:107–17.

Wittmann EJ, Mellor PS, Baylis M. Using climate data to map the potential distribution of *Culicoides imicola* (Diptera: Ceratopogonidae) in Europe. Rev Sci Tech. 2001;20:731–40.

Wittmann EJ, Mellor PS, Baylis M. Climate change: effects on Culicoides-transmitted viruses and implications for the UK. Vet J. 2002;16:147–56.

Zientara S, Breard E, Sailleau C. Bluetongue: characterization of virus types by reverse transcription-polymerase chain reaction. Dev Biol (Basel). 2006;126:187–96.

Zientara S, Sailleau C, Viarouge C, Höper D, Beer M, Jenckel M, Hoffmann B, Romey A, Bakkali-Kassimi L, Fablet A, Vitour D, Bréard E. Novel Bluetongue virus in goats, Corsica, France. Emerg Infect Dis. 2014;20:2123–5.

Chapter 16
African Swine Fever in Sub-Saharan African Countries

Emmanuel Couacy-Hymann

Abstract African swine fever (ASF) is a dreadful hemorrhagic disease of domestic pig and European wild boars that causes up to 100% mortality in a naive population with a wide range of clinical symptoms and lesions depending upon the virulence of the virus strain involved and host factors. It is due to a unique double-stranded DNA virus, ASF virus (ASFV), an arbovirus harbored by soft ticks of the *Ornithodoros* spp. as vector and maintained in a sylvatic cycle between the soft ticks and the natural hosts, warthog, and bush pigs. However, there is also a domestic cycle for the persistence of the virus involving pig to pig transmission mainly observed in West and Central Africa where soft ticks do not exist and the disease is endemic. From Africa the disease had spread to West Europe, South America, and the Caribbean. The disease had been eradicated from these countries, except Sardinia in Italy. Nowadays the disease is reported in several countries of Eastern Europe including the Caucasus region and the Federation of Russia.

The control of ASF requires some prerequisites such as a laboratory able to diagnose quickly the disease and the veterinary services with adequate capacity to react. Furthermore, the control requires to prevent contact between domestic pigs and any sources of the virus and soft ticks where it exists. In addition, the effective cooperation of all stakeholders in a control and eradication plan is highly needed. Since free-ranging farm system with low biosecurity is common in Africa, approximately 80% of the domestic population, and that contributes to the maintenance of the virus, upgrading the farming system including an improved biosecurity level will effectively help to control the disease.

Keywords African swine fever · Virus · Arbovirus · Africa · DNA · Arthropod · Pig

---

E. Couacy-Hymann (✉)
LANADA/Central Laboratory for Animal Diseases, Bingerville, Ivory Coast

M. Kardjadj et al. (eds.), *Transboundary Animal Diseases in Sahelian Africa and Connected Regions*, https://doi.org/10.1007/978-3-030-25385-1_16

Introduction

African swine fever (ASF) is a highly contagious hemorrhagic disease of pigs that produces a wide range, from acute to chronic, of clinical signs and lesions and causes up to 100% mortality with a dramatic economic consequence (Plowright et al. 1994; Costard et al. 2012). However, in an endemic area the character may change with a reduced mortality rate (Penrith and Vosloo 2009). During chronic infections, most of the pigs die after 1–3 months, still being contagious during this period. ASF virus (ASFV) transmission occurs through direct contact with an infected animal, ingestion of contaminated feed stuffs, and more generally contact with infected fomites, or bites of infected soft ticks of *Ornithodoros* genus (Penrith et al. 2004). The disease is caused by a unique, double-stranded DNA virus (Dixon et al. 1994; Yanez et al. 1995), which is endemic in most sub-Saharan African countries.

ASF must be differentiated from other diseases in particular classical swine fever (CSF) which shows similar clinical signs and lesions and can also generate huge socioeconomic losses (Vandeputte and Chappuis 1999; Edwards et al. 2000). CSF is caused by a RNA virus (CSFV) belonging to the *Pestivirus* genus within the *Flaviviridae* family (Wengler 1991; Heinz et al. 2000). It has never been reported in sub-Saharan African countries except Madagascar. In contrast, Ivory Coast became infected by ASFV during the large 1996 epizootic and invaded the whole West African countries. ASF remains a real burden on the development of pig farms and pig industry in the whole sub-Saharan African countries where the disease is reported yearly and causes huge economical losses. Having taken the importance of the disease, the AU-IBAR, in collaboration with FAO and ILRI, has prepared a global strategic plan for the control and eradication of ASF from Africa based on a regional approach (FAO-AU/IBAR-ILRI 2017).

Etiology of ASF

African swine fever virus (ASFV) is the pathogenic agent of African swine fever. ASFV is a large, enveloped, double-stranded DNA virus, having an icosahedral symmetry (Vinuela 1985) with a linear genome containing at least 150 genes (Wittek and Moss 1980; Baroudy et al. 1982; Brookes et al. 1998). It shares some similarities with the *Iridoviridae* based on the cytoplasmic location of its genome and its morphology (Carrascosa et al. 1984) and with the *Poxviridae* in relation to the genome structure and replication strategy (Wittek and Moss 1980; Baroudy et al. 1982). Now it is the unique member of the *Asfarviridae* family (coming from African swine fever and related viruses) within the *Asfivirus* genus (Penrith et al. 2004; Dixon et al. 2005; Takamatsu et al. 2011).

The replication of ASFV DNA starts, at early times, in the nucleus close to the nucleus membrane of the infected cells. At a later stage the viral DNA is found exclusively in the cytoplasm of those infected cells (Rojo et al. 1999). ASFV is transmitted by arthropods being, in consequence, the sole known DNA virus as arbovirus (arthropod-borne virus) (Plowright et al. 1969; Kleiboeker and Scoles 2001). In addition, ASFV is one of a limited number of DNA viruses where the polyprotein processing occurs (Kuznar et al. 1980; Salas et al. 1981, 1986; Penrith et al. 2004). The main target cells for replication are monocyte and macrophage lineages, but several other cells types can be infected, especially in the later stages of the disease (Blome et al. 2012).

The entry of the virus is receptor mediated using three proteins, p12 and p54 for its attachment and p30 for its internalization (endocytosis) into the cell (Penrith et al. 2004; Oura 2013). DNA replication takes place in the cytoplasm in the infected cell and the peak is observed approximately 8 h post-infection (Blome et al. 2012). The virus encodes for enzymes required for replication, transcription, and translation of its genome using an early and late process phase. This process is taking place in the cytoplasm close to the nucleus (perinuclear regions) of the infected cell, also called “virus factories,” rich in fibrillar, membranous organelles, golgi apparatus, ribosomes, and mitochondria (Brookes et al. 1998). The progeny virions are migrating to the plasma membrane along microtubules where they are released from the cell by budding out or are propelled away along actin projections to infect new cells, approximately 10 h post-infection. During the virus’s lifecycle, most or all of the host cell’s organelles are modified, adapted, or in some cases destroyed (Brookes et al. 1998).

Genome of ASFV contains 151 genes encoding for 113 functional and structural proteins which include proteins of the multigene families and structural, DNA replication enzymes, nucleic acid repair and processing proteins, and immune response modulation and apoptosis proteins (Yanez et al. 1995). The length of ASFV genome ranges between 170 and 190 kbp depending on the virus strain with terminal inverted repeats and hairpin loops (Sogo et al. 1984; Gonzalez et al. 1986). The genome is divided into three parts: a central conserved core region, around 125 kbp in length which is flanked by two regions of variable length. The left flanking region is left variable region (LVR) with 38–47 kbp length and the right variable region (RVR) with 13–16 kbp length. These results come from study on several ASFV strains. It is also shown that the largest degree of length variation is observed in the LRV. The sequencing demonstrates that the genome is composed of 38.95% of G+C and that A+T is predominant with 61.05% (Rodriguez and Vinuela 1995). In further molecular studies of ASFV strains, several structural proteins have been identified such as p12, p14, p30, p54, p150, VP 72/73, with the capsid protein VP72/VP73 as the most important which represents 32% of the total protein mass of the virion. This VP 72/73 protein contains the neutralization site and is antigenically stable and thus well appropriate to be targeted for diagnosis purposes and suitable for molecular epidemiology study (Hess 1981; Wardley 1983). It has been also found that p30 and p54 induce neutralizing antibodies while antigenic variation has been observed on p12, p14, and p150 (Penrith et al. 2004). ASFV remains one of the most complex animal viruses.

Based on the sequencing of the variable region in the 3′ end C-terminal of the B646L gene encoding the major capsid protein VP72/73, 22 (I–XXII), genotypes have been identified (Bastos et al. 2003; Boshoff et al. 2007; Lubisi et al. 2007;

Michaud et al. 2013). The 22 ASFV the *B646L* (p72) genotypes can also be distinguished using microarray (Leblanc et al. 2012). Recently, based on the same molecular differentiation, a new genotype has been described with the Ethiopian isolate, being the XXIII genotype of ASFV (Achenbach et al. 2017) followed by the Mozambique isolate giving the XXIV genotype (Quembo et al. 2017). All these XXIV genotypes have been circulating in eastern and southern Africa while genotype I of ASFV is confined to Europe, South America, the Caribbean, and West Africa (Costard et al. 2013). However, ASFV genotype IX has been described in western Africa (Gallardo et al. 2011) and ASFV genotype II has spread to Madagascar, the Caucasus region, and Russian Federation (Costard et al. 2013). In East and Southern Africa, some ASFV genotypes are country specific, while others have transboundary distributions (Costard et al. 2009, Misinzo et al. 2014).

Evolution and Epidemiology (History and Epidemiology)

History

The virus appears to have evolved around 1700 AD. This date is corroborated by the historical record. It is reported that pigs were initially domesticated in North Africa and Eurasia. They were introduced into southern Africa from Europe by Portuguese, approximately 300–500 years ago, and to the Far East by Chinese, approximately 600 years ago (Levathes 1994; Gilford-Gonzalez and Hanotte 2011). At the end of the nineteenth century, the extensive pig industry in the native region of ASFV (Kenya) started after the massive death of cattle due to rinderpest outbreaks. Pigs were imported on a massive scale for breeding by colonizers from Seychelles in 1904 and from England in 1906. Pig farming was free range at this time. The first outbreak of ASF was reported in 1907. However, it is possible that this disease had caused losses in central and eastern Africa before the first description in Kenya (Montgomery 1921).

The virus was thought to be derivated from a virus of soft tick of the genus *Ornithodoros* that infects wild swine, including giant forest hogs (*Hylochoerus meinertzhageni*), warthogs (*Phacochoerus africanus*), and bushpigs (*Potamochoerus porcus*) through a sylvatic cycle involving these suids as natural hosts and the vector in eastern and southern Africa (De Tray 1957; Sanchez-Botija 1963; Penrith et al. 2004; Burrage 2013). In these wild hosts, infection is generally asymptomatic (Anderson et al. 1998). This sylvatic cycle is the base of the maintenance and transmission of the virus and supports the hypothesis that ASFV circulated in Africa for millennia (Penrith et al. 2004).

Epidemiological Cycles and Stability of ASFV

Africa's population is growing very fast and protein of animal origin is highly in demand and pig is well adapted to provide the needed protein in the diet of these populations. The development of pig farms in sub-Saharan African countries grows up to respond to this demand using also high selected breeds from outside the continent which are very susceptible to ASFV. The disease is endemic or sporadic epidemics are reported in some countries in the period before 1990 such as Casamance region in Senegal, Guinea-Bissau, and Cape Verde in West Africa and the same picture was observed in Central, Eastern, and Southern Africa. However, the epidemiological situation throughout Africa is different based on the presence or absence of the natural hosts, suids and the vector, soft ticks, and pig farming systems. In consequence Penrith et al. (2004) described three different cycles as follows:

- In a sylvatic association between wild suids, particularly warthogs and the *Ornithodoros* sp. ticks that are associated with warthog in the savanna zones of Africa (observed in southern Africa). There ASFV passes between ticks living in warthog burrows and juvenile warthogs (Burrage 2013).
- A cycle involving domestic pigs and *Ornithodoros* sp. ticks that live in pig houses.
- Maintenance of the virus in domestic pig populations independent of the presence of wild suids or ticks such as in West Africa, involving mainly free-range pigs.

Furthermore, the sylvatic cycle can continue in the absence of transmission or acquisition feeding with trans-stadial, venereal, and trans-ovarial transmission of the virus in the tick population allowing the virus to persist even in the absence of viremic hosts. Infected ticks play an important role in the long-term maintenance of the disease, surviving for months in burrows and up to several years after feeding on an infected host (Plowright et al. 1970a, b, 1974; Burrage 2013; Beltrán-Alcrudo et al. 2017). In contrast, it has not been observed that infected warthog can transmit the virus to their offspring in utero or via the colostrums.

Another characteristic which contributes to the maintenance and transmission of the virus is its resistance in the external conditions. ASFV is stable and still infectious in a wide range of temperature, pH, and to freeze/thaw cycle. Examples include (Penrith et al. 2004):

- Virus remains infectious in serum at room temperature for 18 months and blood at 4 °C for at least 6 years, at 37 °C for up to a month, and at 55 °C for 30 min it is inactivated at 60 °C for 30 min.
- Stable from pH 4 to 10 even up to pH 13.
- Putrefaction does not destroy the virus.
- Stable in feces for up to 11 days at room temperature and in decomposed serum for up to 15 weeks.
- Stable in pig meat and processed pig meat products for up to 4–6 months but can be inactivated by heating at 70 °C for 30 min.

Samples in the laboratory can be stored at −70 °C (at least) to safeguard the infectivity of the virus indefinitely. Storage at −20 °C is not suitable and therefore not recommended. Samples are completely inactive at 60 °C for 30 min. The virus is also sensitive to lipid solvent, and detergents such as hypochlorite are recommended. Formalin at 0.5%, Β-propiolactone, acetylethyleneimine, and glycidaldehyde inactivate quickly the virus in 1 h at 37 °C.

Geographical Distribution

Stability of the virus in pig meat and pig meat products is the major factor of ASFV diffusion across long distances. The main source of dissemination of the virus out of Africa remains pig meat and its products. Indeed, the first spread of the disease outside Africa was Portugal in 1957 due to airplane waste used to feed pigs near Lisbon airport. A second outbreak occurred in 1960 and the disease became endemic in the Iberian peninsula from that date up to the mid-1990s and involving *Ornithodoros erraticus* as vector (Sanchez-Botija 1963). Portugal reached the global eradication of the disease in 1995 but reported outbreak in 1999. Outbreaks of ASF were then reported in other European countries during the twentieth century: France (1964, 1967, 1977), Italy (1967, 1980, with Sardinia in 1978), Malta (1978), Belgium (1985), and the Netherlands (1986). The disease was successfully eradicated from these countries with the exception of Sardinia (Italy) where it has remained endemic since its introduction in 1978 where wild boars (*Sus scrofa ferus*) highly susceptible are involved (Penrith 2009; Oura 2013).

In the 1970s, ASF was reported in the Caribbean and South America: Cuba (1971, 1980), south Brazil (1978–1979), and Haiti (1979). The disease was eradicated from these countries where it caused huge economical losses (Oura 2013).

In 1996, an epidemic wave of ASF started from Ivory Coast and spread to Benin, Cape Verde, Togo, and Nigeria in 1997, Senegal and Ghana in 1999 (El-Hicheri et al. 1998; Odemuyiwa et al. 2000; Babalobi et al. 2007), and Burkina Faso in 2003 (OIE 2004). However, it is known that the disease is endemic in southern Senegal since 1979. Madagascar reported ASF for the first time in 1997/1998 already victim of Classical swine fever (CSF) (Rousset et al. 2001; Ravaomanana et al. 2010), Mauritius in 2007, and Ethiopia in 2012 (Couacy-Hymann 2014; Achenbach et al. 2017; FAO-AU/BAR-ILRI 2017). In Ivory Coast, ASFV killed 135,000 pigs (29% of the pig population), mostly from the commercial farming system. The global cost of the epizootic was estimated at US$18 millions in Ivory Coast (El-Hicherie et al. 1998) and US$6 millions in Benin (FAO 1997). In Togo, 62 ASF outbreaks were reported between 1997 and 1999 leading to the destruction of 17,000 pigs (direct losses and stamping out) (Edoukou 2000; FAO 2001). In Ghana, 200,000 pigs were slaughtered in 1999. Since this wave of ASF outbreak in West Africa, the disease is still reported on a regular basis. However, the eradication was obtained in Ivory Coast in 1998 and after 16 years of freedom from the disease, ASF reemerged in the south-western region (San Pedro) in 2014, harboring the second seaport of the country. This sporadic case was rapidly eradicated (Couacy-Hymann et al. 2016; Kouakou et al. 2017). Unfortunately, the disease reemerged in the north of the country, Ferkéssédougou, in June 2017 and spread to Korhogo and Ouangolodougou regions. ASF outbreak was confirmed in August by the Virology Laboratory—Bingerville (Couacy-Hymann, personal communication). First reported to the *World Organisation for Animal Health* (OIE) in late February 2016, the outbreak in Mali began on January 1, 2016. A village herd of more than 4000 pigs in Ségou was affected, with 178 animals showing signs of the disease. Forty-four of them died and a further

35 were destroyed. The source of the infection is uncertain, but it is thought to have been illegal movements of pigs and/or swill feeding. The outbreak was in a region that borders Burkina Faso, where the disease was reported in 2014.

In the period 2003–2016, 267 outbreaks were reported throughout sub-Saharan African countries according to AU-IBAR's Animal Health Information system (Tables 16.1 and 16.2) (AU-IBAR 2017). However, this remains an underestimated reporting of the situation on the ground.

ASF escaped from Africa and was reported in 2007 in the Republic of Georgia. Thereafter it spread to Armenia, Azerbaijan, Iran, Russian, and Belarus. The source of the virus involved in this epidemic feature was characterized as a genotype II, closely related to isolates previously described from Mozambique, Madagascar, and Zambia (Bastos et al. 2004) due probably to ship waste available to free-range pigs around the port of Poti on the Georgian black Sea coast. Wild boars were also concerned adding complication to the epidemiological situation in that region. Furthermore, ASF was confirmed in the following countries: Lithuania (2014), Poland (2014), Latvia (2014, 2017), and Estonia (2015). ASF in these eastern European countries is not yet under control and put at high risk pig population in Eastern Europe and even in Western Europe (Penrith 2009; Rahimi et al. 2010; Dietze et al. 2012; Le Potier and Macé 2013; OIE 2014; Gallardo et al. 2014; Oura 2013; FAO-AU/IBAR-ILRI 2017). It continues to spread and is no more only an African disease and concern. The geographical distribution of ASF from 2005 to 2017 is shown in Fig. 16.1.

Table 16.1 ASF outbreaks in sub-Saharan African countries in the period 2003–2016

Year	No infected countries	No outbreaks	No cases	No deaths	No slaughtered
2000	NA	NA	NA	NA	NA
2001	NA	NA	NA	NA	NA
2002	NA	NA	NA	NA	NA
2003	15	250	28,553	19,639	NA
2004	21	670	118,281	74,667	NA
2005	17	228	19,511	13,717	NA
2006	13	NA	NA	NA	NA
2007	11	101	101,823	100,180	1340
2008	18	207	191,137	96,108	NA
2009	19	130	10,240	7530	NA
2010	21	145	27,529	19,156	7898
2011	22	471	144,950	135,712	15,825
2012	20	361	89,117	72,392	12,375
2013	14	376	40,562	33,892	6867
2014	14	183	31,413	17,012	5465
2015	16	286	23,228	12,375	2079
2016	14	264	38,522	23,404	3426
2017	NA	NA	NA	NA	NA

Table 16.2 ASF: Number of affected countries in sub-Saharan African region in the period 2003–2016

Countries	No outbreaks	No cases	No deaths	No slaughtered
Angola	10	1092	1087	220
Benin	236	35162	20454	136
Burkina Faso	215	24937	16541	
Burundi	30	25705	5118	4
Cameroon	216	51120	43508	881
Chad	8	6435	2737	70
Congo	5	65	53	
Central African Republic	78	42132	18474	
Democratic Republic of Congo	677	42931	350251	111
Ethiopia	7	28	19	
Gambia	5	198	198	
Ghana	87	9872	7300	733
Guinea-Bissau	10	604	557	
Ivory Coast	2	6599	4549	2050
Kenya	24	230	160	
Lesotho	1			
Liberia	3	22	5	3
Madagascar	76	3365	2843	91
Malawi	220	53782	45372	1169
Mali	6	209	126	
Mauritius	3	961	931	930
Mozambique	96	7417	6106	1150
Namibia	24	1169	1161	
Niger	6	79	64	
Nigeria	22	3213	1212	87
Uganda	179	24888	7389	150
Rwanda	202	4040	3121	1903
Senegal	15	919	856	
South Africa	47	316	190	
Swaziland	1	1	1	
Tanzanie	62	15246	4969	1258
Togo	210	11653	7641	1057
Zambia	56	4316	2818	

It has been observed, during the course of an ASF outbreak outside Africa such as in Europe, a difference in the manifestation of the disease which consists of an acute and fatal phase at first followed by the diminution of the severity to attain subclinical cases at the end of the outbreak. In Africa, it is not usual to observe natural occurrence of chronic or mild disease in natural ASF outbreak (Penrith et al. 2004).

Chronic infection can allow us to recover the virus from tissues of infected pigs up to 6 months and those animals are able to transmit the virus up to 1–2 months to in-contact pigs. However, carrier state has not been effectively described (Penrith et al. 2004).

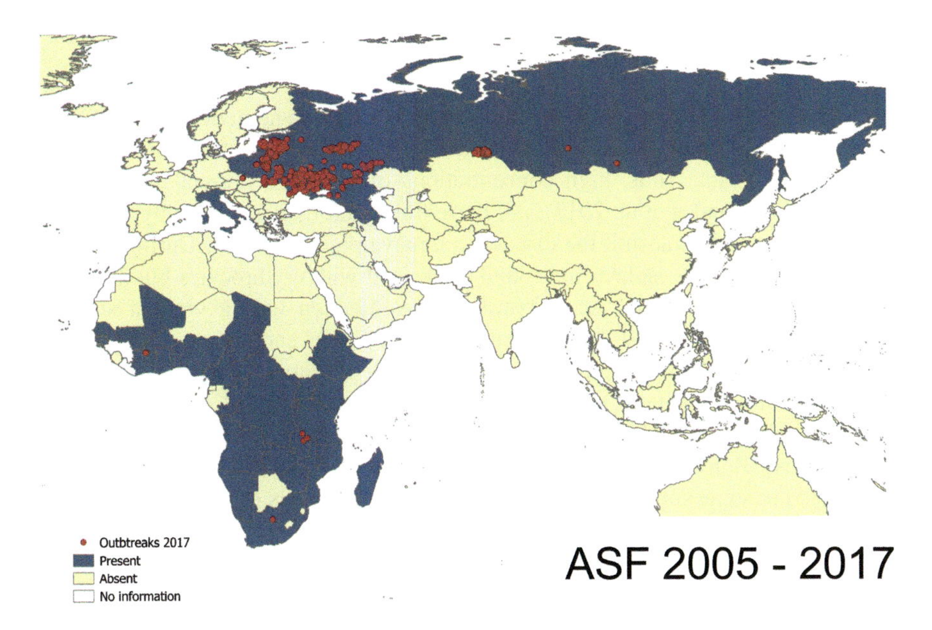

Fig. 16.1 Geographical distribution of ASF in the period 2005–2017. (Source: WOAH/OIE-WAHIS)

Transmission

The main factor of maintenance of the virus in Africa is due to the sylvatic cycle involving the natural hosts, especially warthogs and the soft tick *Ornithodoros sp.* living in the warthog burrows, in the savanna eastern-southern regions and pig to pig transmission in West and part of Central Africa where no vector has been described. ASFV in *Ornithodoros* sp (*O. porcinus porcinus*) ticks is characterized by a low infectious dose, lifelong infection, efficient transmission to both pigs and ticks, and low mortality until after the first oviposition (Kleiboeker and Scoles 2001). In consequence, soft ticks, *Ornithodoros sp,* are probably the most important natural vector of ASFV. The virus in warthogs is characterized by an unapparent infection with transient, low viremic titers (Oura et al. 1998). The spread of the virus from wildlife to domestic pig is probably through infected ticks feeding on pigs (ticks, brought back to villages through the carcasses of warthogs killed for food) or by ingestion of warthog tissues (which seems to be rare).

Virulent virus strains cause peracute/acute disease in domestic pigs and all body fluids (excretions and secretions) and tissues contain large amounts of infectious virus from the onset of clinical disease until death. The virus is present in nasal, oral pharyngeal, conjunctival, genital, and urinary fluids and fecal material. Pig to pig transmission is based on direct contact using the oronasal route or by ingestion of waste food containing unprocessed pig meat or pig meat products (Penrith et al. 2004; Blome et al. 2013).

Pathogenesis of ASF

The virus replicates primarily in the upper respiratory tract and invades the tonsil and lymph nodes draining the head and neck; then the infection is generalized rapidly using bloodstream. Thus, high concentrations of virus are present in all tissues (Gómez-Villamandos et al. 2013). All these fluids and tissues are highly infectious and can be used to transmit the disease to a naïve pig population. During viremia, hemadsorbing ASFV isolates are found associated with erythrocytes but also lymphocytes and neutrophils. Acute phase of the infection with a virulent strain is characterized by profound depletion of lymphoid tissues and apoptosis of lymphocyte subsets. It is now accepted that the massive destruction of macrophage plays a major role in pathogenesis, especially in the impaired hemostasis due to the release of active substances including cytokines, complement factors, and arachidonic acid metabolites (Blome et al. 2013; Gómez-Villamandos et al. 2013). It is established that there are some similarities with viral hemorrhagic fevers of human and animals based on the following:

(a) Primary replication in cells of the monocyte/macrophage lineage
(b) Occurrence of cytokine-mediated lesions including apoptosis of uninfected lymphocytes
(c) Activation of endothelial cells and the coagulation system
(d) Impairment of innate immune functions

However, many questions remain with no clear response such as the role of host factors, ASFV strains with different virulence, mechanisms, and implications of chronic disease courses, and the role of ASF-specific immunity for pathogenesis. Response to these major points might be the key element for the development of a first efficacious ASF vaccine (Blome et al. 2013).

Clinical Signs and Lesions

The disease affects all breeds of domestic pigs and European wild boars. In addition, all age groups are equally susceptible. Peracute, acute, subacute, and chronic forms occur with mortality rates ranging from 0 to 100%, depending on the virulence of the virus strain, swine breed affected, route of exposure, infectious dose, and endemicity status in the area (Beltrán-Alcrudo et al. 2017). The incubation period in natural situation has been observed to vary from 4 to 19 days post-infection.

In Africa, peracute and acute forms of ASF are most common. In those forms, the incubation period can be less than 4 day post-infection. The infected pig may develop a high fever, up to 42 °C, and show only noticeable symptoms such as gradual loss of appetite, depression, and lying down. Sudden death can occur within this time in peracute form (Mebus 1988).

In acute form, the disease is characterized by a short incubation period of 3–7 days, followed by high fever, up to 42 °C, and death in 5–10 days. Observed symptoms are usually loss of appetite, depression, inactivity, lying down with hyperemia of the skins of the ears, abdomen, and legs, respiratory distress, vomiting, bleeding from the nose and rectum, and sometimes constipation or diarrhea. Abortion is sometimes the first event seen in an outbreak (Fig. 16.2).

The severity and distribution of the lesions vary according to virulence of the virus. Hemorrhages occur predominantly in the lymph nodes, kidneys as petechiae,

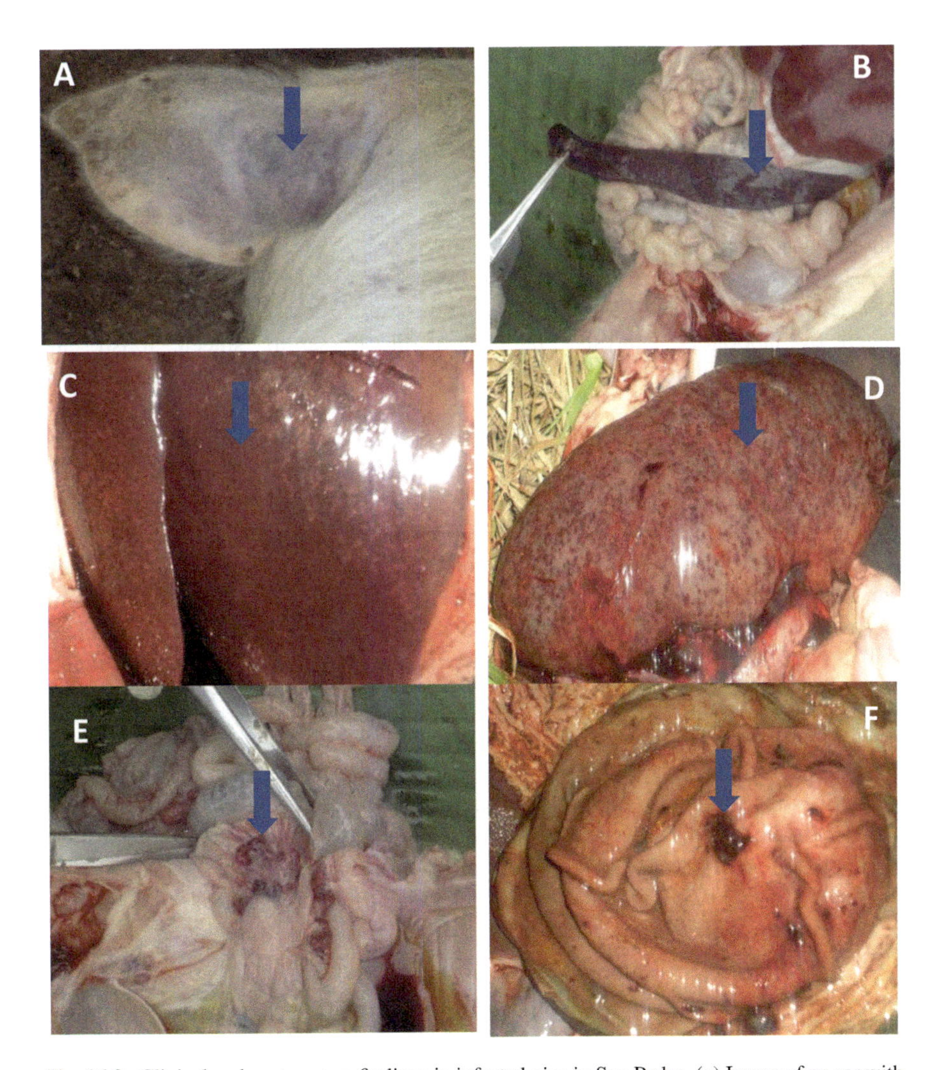

Fig. 16.2 Clinical and postmortem findings in infected pigs in San Pedro. (**a**) Image of an ear with cutaneous congestion common in ill pigs. Postmortem lesions include hypertrophied spleen (**b**), hemorrhagic liver (**c**), petechiation on the kidney (**d**), hemorrhagic mesenteric lymph node (**e**), and in the stomach (**f**) were observed in pigs with ASF. (Source: Couacy-Hymann et al. 2019)

and heart. Hemorrhages in other organs are variable in incidence and distribution such as enlarged and friable spleen, colored blood-stained fluid in pleural, pericardial, and peritoneal cavities, or edema and congestion of the lungs (Blome et al. 2013; Howey et al. 2013; Oura 2013; FAO-AU/IBAR-ILRI 2017).

Some viruses of low virulence have been isolated in Europe and produce nonspecific clinical signs and lesions.

Chronic disease, rarely observed in outbreaks, is characterized by emaciation, swollen joints, and mild respiratory disorder (Oura 2013).

Diagnosis

As for other diseases the diagnosis of ASF should be based on these criteria: epidemiology context, clinical signs and lesions, high mortality regarding pigs of all age groups, and rapid diffusion of the disease.

The clinical symptoms or postmortem lesions of ASF are very similar to those of classical swine fever (CSF, hog cholera) and the two diseases have to be differentiated by laboratory analysis. Samples of blood, serum, spleen, tonsil, and gastrohepatic lymph nodes from suspected cases should be submitted to the laboratory for confirmation.

Several tests are available to confirm the disease:

Hemadsorption cytopathogenesis test: virus isolation by inoculation of a primary culture of monocytes or macrophages from peripheral blood, bone marrow, or lung washing with suspected blood or tissue suspension. The adsorption of erythrocytes to infected cells usually becomes visible in 1–5 days. The virus itself does not have the capacity of hemagglutination. Sometimes it is necessary to subculture to demonstrate weak hemadsorption. There are nonhemadsorbing viral isolates. These isolates produce only a cytopathic effect in macrophage cells. Confirmation of ASF cases can be performed by the detection of:

- Antibody with ELISA and immunoblotting (mostly used to confirm results obtained from ELISA) or indirect immunofluorescence techniques: in acute ASF outbreaks, most pigs die (within 24–48 h) before producing antibodies. The detection of antibodies in that case is of little value since it can be false negative. Serological tests are important in endemic situation.
- Antigen using direct immunofluorescence or ELISA techniques.
- Genome of ASFV using molecular methods: PCR technique, both conventional or real-time PCR.

The detection of ASFV genome serves to confirm ASF but also to trace the origin of that outbreak when it occurs:

Restriction fragment length polymorphism (RFLP) method, based on the specific enzyme restriction profiles, can be used for diagnosis purpose but it serves mainly to determine the origin of outbreaks. Results from previous studies indicate that ASFV strains from Europe and South America have a common origin (Wesley and Tuthill 1984; Blasco et al. 1989; Penrith et al. 2004).

PCR techniques: Conventional PCR or real-time PCR are useful tools for ASF diagnosis purpose. This can be followed by the application of nucleotide sequencing method on the PCR products for molecular epidemiology. All samples collected from sick or dead animals can be used to perform the test. Mainly primers are designed from VP72 gene (Wilkinson 2000; Aguero et al. 2003; Bastos et al. 2003). There are primers designed for only diagnosis purpose and those used for both diagnosis and sequencing. For example, the set of primers, ASF-1 (5-ATGGATACCGAGGGAATAGC-3) and ASF-2 (5-CTTACCGATGAAAATGATAC-3), for the partial amplification of VP72 to confirm the presence of ASF viral DNA (Wilkinson 2000) while primers, P72-U (5-GGCACAAGTTCGGACATGT-3) and P72-D (5-GTACTGTAACGCAGCAC AG-3), for the amplification of the C-terminal region of the VP72 gene, is used for molecular epidemiology (Bastos et al. 2003). In addition, VP72 genotyping when followed by the 9RL Central Variable Region (CVR) characterization is an effective approach for determining relatedness of outbreak and possible sources of infection (Bastos et al. 2003; Penrith et al. 2004; Nix et al. 2006; Achenbach et al. 2017). The partial VP72 gene sequencing confirms the genetic homogeneity of viruses from Europe, South America, the Caribbean, and West Africa, in the group named ESAC-WA genotype (Achenbach et al. 2017). However, high heterogeneity is observed in central, eastern, and southern Africa where the sylvatic cycle is well established and plays an important epidemiological role in the variability of ASFV strains (Plowright et al. 1969; Bastos et al. 2003). With these molecular methods, 23 genotypes (Achenbach et al. 2017) and recently 24 genotypes of ASFV have been found up to now with the Mozambique isolate. This figure also shows three different lineages of ASFV (lineages 1, 2, 3) irrespective of the genotypes (Quembo et al. 2017) (Fig. 16.3).

Several real-time techniques have been also developed for ASF confirmation (McMenamy et al. 2010; Tignon et al. 2011; Fernadez-Pinero et al. 2013; Haines et al. 2013).

Differential Diagnosis

Main Pig Diseases to Differentiate from ASF

Classical Swine Fever (CSF)

Same clinical symptoms and lesions. Differential diagnosis requests a laboratory confirmation.

It is judicious to carry out a differential diagnosis of ASF/CSF on any suspected samples mainly when the disease emerges for the first time in a new area.

Erysipelas

Is a bacterial disease caused by *Erysipelothrix rhusiopathiae*. It affects pigs of all age groups. Usually acute and subacute forms are reported. In young pigs, acute form is

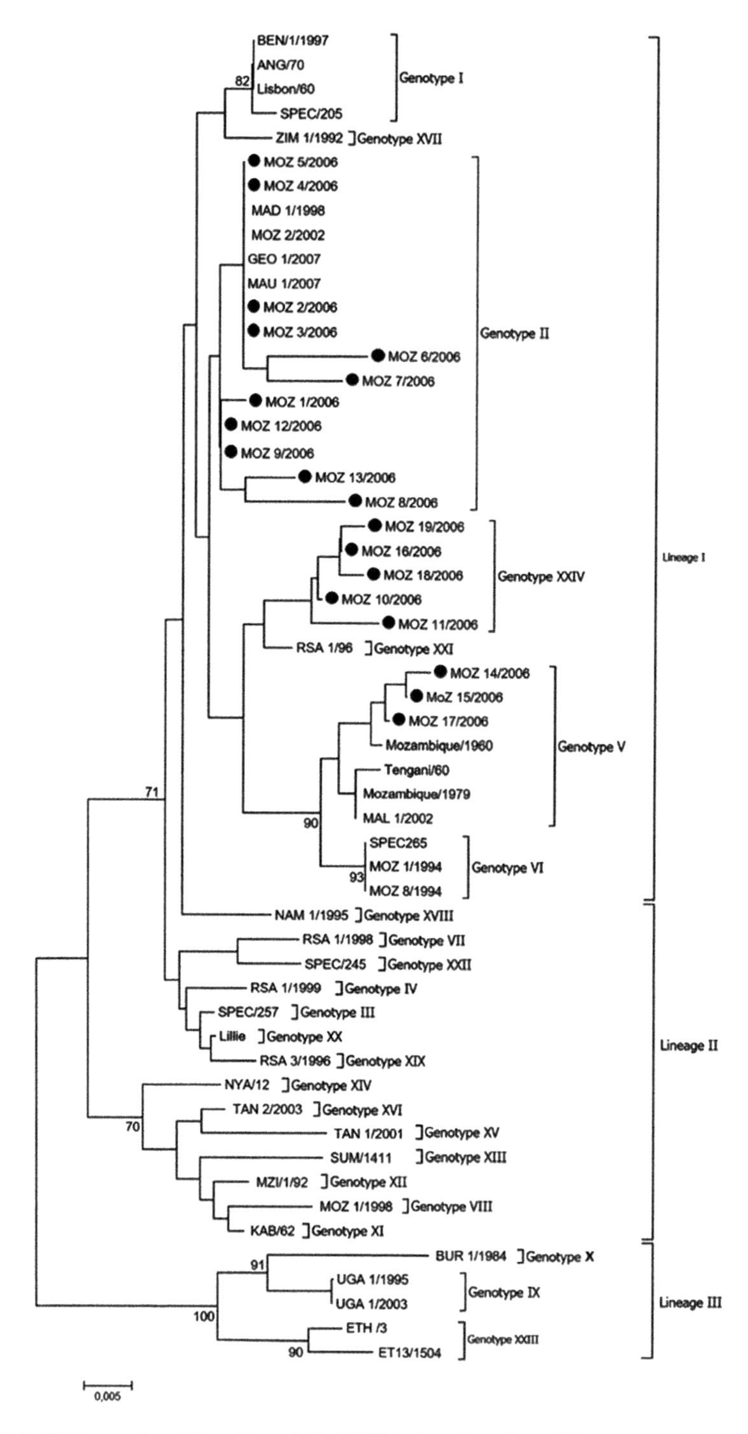

Fig. 16.3 Phylogenetic relationships of 55 ASFV isolates based on p72 gene sequences. (Source: Quembo et al. 2017)

characterized by sudden death followed by mortality but less than in ASF. Affected pigs may show very characteristic diamond-shaped skin lesions associated with necrotizing vasculitis (inflamed blood vessels). Bacterial isolation confirms the diagnosis and pigs respond well to treatment with penicillin (Beltrán-Alcrudo et al. 2017).

Salmonellosis

Younger pigs are usually affected. Animals treated in time may respond to antimicrobial therapy. Laboratory confirmation of the diagnosis is by bacterial culture. Features in common with ASF include fever, loss of appetite, respiratory or gastrointestinal disorders, and a congested, fevered carcass at slaughter. Animals may die 3–4 days post-infection. Pigs dying from septicemic salmonellosis show cyanosis of the ears, feet, tail, and abdomen. Necropsy findings may include petechial hemorrhages in the kidneys and on the heart's surface, enlarged spleen (but with normal color), swelling of mesenteric lymph nodes, enlargement of the liver, and congestion of the lungs (Beltrán-Alcrudo et al. 2017).

Control Measures

It is estimated that around 33 sub-Saharan countries have reported ASF. These reports highlight the importance of the disease and the need to build a concerted and harmonized approach to mitigate the impact and stop further spread in the continent and beyond its borders. Only three African countries eradicated ASF after a single introduction: São Tomé e Principe, Mauritius, and Ivory Coast eradicated the 1996 introduction but experienced a reintroduction of ASF in 2014. This case was rapidly eradicated again (Couacy-Hymann et al. 2016; FAO-AU/IBAR-ILRI 2017). ASF is not a zoonotic disease and affects only pig species. However, it impacts people's livelihoods, food security, and nutrition and so contributes to deeply increase the poverty situation of the populations.

There is no vaccine up to now and no treatment available against ASF. In consequence, all means of prevention remain the only efficient control measure to be implemented. First, it is with regard to prevent contact between domestic pig and all sources of virus (live infected pigs, infected pig meat, pig meat products, arthropod vector, wild natural hosts, contaminated vehicles, infected premises, fomites, swill, contaminated wastes, etc.). Efficient eradication programs rely on a rapid diagnosis and rapid decision to slaughter and disposal of all infected and in-contact animals and premises. However, stamping out should not be the only measure to control ASF outbreak because so dramatic. In addition, application of stamping out without prior preparation and fine awareness of farmers can give the exact opposite result contributing to spread the disease by moving infected pigs away. The control should also involve the movement control of animals and treatment of waste food. Thereafter, a serological survey of all pig farms within a specific control area including sentinel animals should be carried out with a negative result to ensure that there is no circulation of the virus. In Africa, it has been observed that a successful eradication program depends on the adequate compensation of affected farmers who have to cooperate effectively with that program.

The pig farming system such as free-ranging system, with low level of biosecurity and uncontrolled animal movements, existing in most African countries, approximately 80% of the whole pig population of the continent, is a key factor for ASFV maintenance and circulation. The persistence of such practices will keep ASF endemic in these countries. Conversely, upgrading the pig farming production system to higher biosecurity levels and through good husbandry practices and appropriate support policies tailored by the veterinary services could improve controlling ASF in these ASF endemic countries (Couacy-Hymann 2012).

Regional Prevention and Control Strategy

The strategy for preventing and controlling will rely on the epidemiology of the disease, the pig farming systems, the presence or absence of natural wild pigs hosts, and the presence or absence of soft ticks as vector. In most sub-Saharan African countries, more than 80% domestic pigs come from free-ranging production system with low biosecurity level in place. The strategy should be in accordance with the local situations, elaborated by the national veterinary authorities through an ASF contingency plan and implemented the veterinary services with appropriate personnel and financial resources. In addition, the diagnostic laboratory should have experienced personnel and well equipped to carry out rapidly recommended tests to confirm cases of ASF.

Based on this figure, it is possible to divide the sub-Saharan African countries into roughly two zones:

Zone 1: West Africa and first part of Central Africa (including Cameroon, Chad, and Central African Republic)

As a baseline:

- Traditional scavenging pig production system
- No soft ticks of the *Ornithodoros moubata* complex as vector and no wild pigs as natural virus host, playing a significant epidemiological role
- Absence of a sylvatic cycle.

In that situation, the virus can be maintained among the pig populations and the infection is through direct contact between the source of virus and naive animals. This source can be infected animals, pig products, contaminated wastes, premises, fomites, and other facilities such as vehicle. The prevention measures are mainly to focus on the efficient interruption of this way of transmission of the infection by an adequate sensitization of farmers and other stakeholders, control of pig movements (which is not an easy task), and control of personnel's movements between farms. The main concern of the strategy is therefore on preventing outbreaks in domestic pigs by improved organization of the pig sector and identifying and mitigating the risks throughout the pig value chains (FAO-AU/IBAR-ILRI 2017).

Zone 2: second part of Central Africa, Eastern, and Southern Africa

As a baseline:

- Traditional scavenging pig production system
- Presence of soft ticks of the *Ornithodoros moubata* complex as vector and presence of wild pigs as natural virus host, playing a significant epidemiological role
- Presence of a sylvatic cycle

Here, the source of infection includes the vector and the natural host in addition to those already listed above. Following the same strategy, direct contact with the source of virus has to be avoided to prevent outbreaks in domestic pigs. Even in this zone 2 and where the production system is adequately organized in recommended premises with a good biosecurity level, the source of infection remains infected pigs, pig products, contaminated premises or fomites, and uncontrolled movements, like in zone 1 rather than from the sylvatic source. In contrast, scavenging farming production pigs are highly subjected to be in contact and contaminated with the sylvatic source of infection and secondary being the source of infection of other proper penned pigs. In that situation it is important to prevent contact between free-ranging pigs with these advanced pig farms which already implement biosecurity measures regarding ASF and other diseases.

Control of ASF

In both zones 1 and 2, stamping out is the main control measure to be implemented but not alone. It should be applied according to a prepared plan including the sensitization of concerned farmers, personnel of the veterinary services, good communication addressing appropriate messages, zoning of the infected area where all activities concerning pig will be at a standstill, and proper disposal of carcasses. Local cultures and behaviors should be taken into account in the preparation of the stamping out operation.

The compensation of farmers is an important component of any control strategy plan to attain the planned objective. A correct compensation contributes to mitigate the spread of the disease and to bring farmers and animal owners to participate effectively in the control activities.

Conclusion

ASF is a dreadful and highly contagious disease specific to domestic pigs and wild European boars. It remains, nowadays, a real threat to pig production and pig industry worldwide demonstrated by the recent invasion of Europe since 2007.The present rapid spread of the disease across long distances is mainly facilitated by the transportation of live infected pigs, pig products, or contaminated wastes such as swill used to feed these animals without any heat treatment. These means of ASF diffusion are the same in local or regional conditions along with the contact with wild pigs that are the natural host of the virus.

Since there is no vaccine and no treatment against ASF, preventive measures against the introduction of the disease into a free area and control measures where the disease has been reported to avoid its spreading remain the only tools to tackle this disease. The implementation of these measures needs an efficient and organized veterinary services with well-trained personnel along with an animal disease diagnosis laboratory able to confirm promptly any suspicious cases of ASF for a quick response. The prevention and control measures require also the full involvement of all stakeholders of the value chain. Upgrading the pig farming system would improve biosecurity level and will contribute to prevent the occurrence of ASF since the endemicity of the disease is maintained in free-ranging farming system, widely observed in Africa. The control of the disease when it occurs requires mainly the application of the stamping out along with other measures calling for an active collaboration of farmers such as effective compensation and restocking procedure.

References

Achenbach JE, Gallardo C, Nieto-Pelegr E, Rivera-Arroyo B, Degefa-Negi T, Arias M, Jenberie S, Mulisa DD, Gizaw D, Gelaye E, Chibssa TR, Belaye B, Loitsch A, Forsa M, Yami M, Diallo A, Soler A, Lamien CE, Sanchez-Vizca JM. Identification of a new genotype of African swine fever virus in domestic pigs from Ethiopia. Transbound Emerg Dis. 2017;64:1393–404. https://doi.org/10.1111/tbed.12511.

Agüero M, Fernández J, Romero L, Sánchez Mascaraque C, Arias M, Sánchez-Vizcaíno JM. Highly sensitive PCR assay for routine diagnosis of African swine fever virus in clinical samples. J Clin Microbiol. 2003;41(9):4431–4. https://doi.org/10.1128/JCM.41.9.4431-4434. 2003.

Anderson EC, Hutchings GH, Mukarati N, Wilkinson PJ. African swine fever virus infection of the bush pig (*Potamochoerus porcus*) and its significance in the epidemiology of the disease. Vet Microbiol. 1998;62:1–15.

Babalobi OO, Olugasa BO, Oluwayelu DO, Ijagbone IF, Ayoade GO, Agbede SA. Analysis and evaluation of mortality losses of the 2001 African swine fever outbreak, Ibadan, Nigeria. Trop Anim Health Prod. 2007;39:533–42.

Baroudy BM, Venkatesam S, Moss B. Incompletely base-paired flip-flop terminal loops link the two strands of the Vaccinia virus genome into one uninterrupted polynucleotide chain. Cell. 1982;28:315–24.

Bastos ADS, Penrith ML, Crucière C, Edrich JL, Hutchings G, Roger F, Couacy-Hymann E, Thompson GR. Genotyping field strains of African swine fever virus by partial p72 gene characterization. Arch Virol. 2003;148:693–706.

Bastos AD, Penrith ML, Macome F, Pinto F, Thomson GR. Co-circulation of two genetically distinct viruses in an outbreak of African swine fever in Mozambique: no evidence for individual co-infection. Vet Microbiol. 2004;103:169–82. https://doi.org/10.1016/j.vetmic. 2004.09.003.

Beltrán-Alcrudo D, Arias M, Gallardo C, Kramer SA, Penrith M-L, Kamata A. African swine fever-detection and diagnosis. A manual for veterinarians. FAO, N° 19; 2017. 93p.
Blasco R, Aguero M, Almendral IM, Vinuela E. Variable and constant regions in African swine fever virus DNA. Virology. 1989;168:330–8.
Blome S, Gabriel C, Diezte K, Breithaupt A, Beer M. High virulenceof African swine fever virus caucasus isolate in European wild boars of allages. Emerg Infect Dis. 2012;18(4):708.
Blome S, Gabriel C, Beer M. Pathogenesis of African swine fever in domestic pigs and European wild boar. Virus Res. 2013;173:122–30. https://doi.org/10.1016/j.viruses.2012.10.026.
Boshoff CI, Bastos ADS, Gerber LJ, Vosloo W. Genetic characterization of African swine fever viruses from outbreaks in southern Africa (1973–1999). Vet Microbiol. 2007;121:45–55.
Brookes SM, Hyatt AD, Wise T, Parkhouse RME. Intracellular virus DNA distribution and the acquisition of the nucleoprotein core during African swine fever virus particle assembly: ultrastructural *in Situ* hybridisation and DNase-gold labelling. Virology. 1998;249:175–8.
Burrage TG. African swine fever virus infection in Ornithodoros ticks. Virus Res. 2013;173:131–9.
Carrascosa JL, Carazo JM, Carrascosa AL, GarcõÂa N, Santisteban A, Vinuela E. General morphology and capsid fine structure of African swine fever virus. Virology. 1984;132:160–72.
Costard S, Wieland B, de Glanville W, Jori F, Rowlands R, Vosloo W, Roger F, Pfeiffer DU, Dixon KL. African swine fever: how can global spread be prevented? Philos Trans R Soc Lond B Biol Sci. 2009;364:2683–96. https://doi.org/10.1098/rstb.2009.0098.
Costard S, Wieland B, de Glanville W, Jori F, Rowlands R, Vosloo W, Roger F, Pfeiffer DU, Dixon LK. African swine fever: how can global spread be prevented? Philos Trans R Soc Lond B Biol Sci. 2012;364:2683–96.
Costard S, Mur L, Lubroth J, Sanhez-Vizcaino JM, Pfeiffer DU. Epidemiology of African swine fever virus. Virus Res. 2013;173(1):191–7.
Couacy-Hymann E. African swine fever prevention and control strategy in West Africa. Technical workshop on sub-regional strategy for the prevention and control of African swine fever. 4–6 Sept, Accra; 2012.
Couacy-Hymann E. FAO mission report on FAO IDENTIFY project OSRO/INT/902/USA B02 "Support for strengthening animal health laboratory capacities in hot spot regions to combat zoonotic diseases that pose a significant public health threat". Addis-Ababa, 20–27 Jul; 2014.
Couacy-Hymann E, Kouakou V, Godji P. Second meeting of IAEA Coordinated Research Project (CRP)—Early and rapid diagnosis and control of TADs—African swine fever, 20–24 Jun, Vienna; 2016.
Couacy-Hymann E, Kouakou KV, Achenbach JE, Kouadio L, Koffi YM, Godji HP, Kouassi EA, Oulaï J, Pell-Minhiaud HJ, Lamien CE. Re-emergence of genotype I of African swine fever virus in Ivory Coast. Transbound Emerg Dis. 2019;66:882–96. https://doi.org/10.1111/tbed.13098.
De Tray DE. African swine fever in warthog (*Phacochoerus aethiopicus*). J Am Vet Med Assoc. 1957;130:537–40.
Dietze, K., Beltran-Alcrudo, D., Khomenko S, Seck B, Pinto J, Diallo A, Lamien C, Lubroth J, Martin V. African swine fever (ASF) recent developments—timely updates. Focus on No. 6. Rome: FAO; 2012. Available at http://www.fao.org/docrep/016/ap372e/ap372e.pdf. Accessed 30 Sept 2012.
Dixon LK, Twigg SRF, Baylis SA, Vydelingum S, Bristow C, Hammond JM, Smith GL. Nucleotide sequence of a 55 kbp region from the right end of the genome of a pathogenic African swine fever virus isolate (Malawi LIL20/1). J Gen Virol. 1994;75:1655–84.
Dixon LK, Escribano JM, Martins C, Rock DL, Salas ML, Wilkinson PJ. The asfarviridae. In: Fauquet CM, Mayo MA, Maniloff J, Desselberger U, Ball LA, editors. Virus taxonomy. Eighth report of the international committee on taxonomy of viruses. London: Elsevier Academic Press; 2005. p. 135–43.
Edoukou DG. Porciculture et épizootie de peste porcine africaine au Togo. Note d'information. Mars 2000. 12 p.

Edwards S, Fukusho A, Lefèvre PC, Lipowski A, Pejsak Z, Roehe P, Westergaard J. Classical swine fever: the global situation. Vet Microbiol. 2000;73:103–19.
El-Hicheri K, Gómez-Tejedor C, Penrith M-L, Davies G, Douati A, Edoukou GD, Wojciechowski K. L'épizootie de peste porcine africaine de 1996 en Côte d'Ivoire. Rev Sci Tech Off Int Epiz. 1998;17:660–73.
FAO. 1997. FAO: Outbreak of ASF in Benin. Epres, September 1997. www.fao.org/News/1997/970905-e.hym
FAO. Rapport de l'atelier régional sur la peste porcine africaine en Afrique de l'Ouest. Coordination du PACE au Togo. Lomé, Togo, 29–31 Oct; 2001. 21 p.
FAO, AU/IBAR, ILRI. Regional strategy for the control of African swine fever in Africa; 2017. p. 46. http://www.fao.org/africa
Fernandez-Pinero J, Gallardo C, Elizalde M, Robles A, Gomez C, Bishop R, Health L, Couacy-Hymann E, Fasina FO, Pelayo V, Soler A, Arias M. Molecular diagnosis of African swine fever by a new real-time PCR using universal probe library. Transbound Emerg Dis. 2013;60:48–58. https://doi.org/10.1111/j.1865-1682.2012.01317.x.
Gallardo C, Anchuelo R, Pelayo V, Poudevigne F, Leon T, Nzoussi J, Bishop R, Pérez C, Soler A, Nieto R, Martín H, Arias M. African swine fever virus p72 genotype IX in domestic pigs, congo, 2009. Emerg Infect Dis. 2011;17(8):1556–8. https://doi.org/10.3201/eid1708.101877.
Gallardo C, Fernandez-Pinero J, Pelayo V, Gazaev I, Markowska- Daniel I, Pridotkas G, Nieto R, Fernandez-Pacheco P, Bokhan S, Nevolko O, Drozhzhe Z, Perez C, Soler A, Kolvasov D, Arias M. Genetic variation among African swine fever genotype II viruses, eastern and central Europe. Emerg Infect Dis. 2014;20:1544–7.
Gilford-Gonzalez D, Hanotte D. Domesticating animals in Africa: implications of genetic and archaeological findings. J World Prehist. 2011;24:1–23. https://doi.org/10.1007/s1096-010-9042-2.
Gómez-Villamandos JC, Bautista MJ, Sánchez-Cordón PJ, Carrasco L. Pathology of African swine fever: the role of monocyte-macrophage. Virus Res. 2013;173:140–9.
Gonzalez A, Talavera A, Almendral JM, Vinuela. Hairpin loop structure of African swine fever virus DNA. Nucleic Acids Res. 1986;14:6835–44.
Haines FJ, Hofmann MA, King DP, Drew TW, Crooke HR. Development and validation of a multiplex, real-time RT PCR assay for the simultaneous detection of classical and African swine fever viruses. PLoS One. 2013;7(7):e71019.
Heinz FX, Collett MS, Purcell RH, Gould EA, Howard CR, Houghton M, Moormann JM, Rice CM, Thiel H-J. Family flaviviridae. In: van Regenmortel MHV, Fauquet CM, Bishop DHL, Carstens EB, Estes MK, Lemon SM, Maniloff J, Mayo MA, McGeoch DJ, Pringle CR, Wickner RB, editors. Virus taxonomy. Seventh report of the International Committee on Taxonomy of Viruses. San Diego, CA: Academic Press; 2000. p. 859–78.
Hess WR. African Swine Fever: a reassessment. Adv Vet Sci Comp Med. 1981;25:39–67.
Howey EB, O'Donnell V, de Carvalho Ferreira HC, Borca MV, Arzt J. Pathogenesis of highly virulent African swine fever virus in domestic pigs exposed via intraoropharyngeal, intranasopharyngeal, and intramuscular inoculation, and by direct contact with infected pigs. Virus Res. 2013;178(2):328–39. https://doi.org/10.2016/j.virusres.2013.09.024.
Kleiboeker SB, Scoles GA. Pathogenesis of African swine fever virus in Ornithodoros ticks. Anim Health Res Rev. 2001;2(2):121–8.
Kouakou KV, Michaud V, Biego HG, HPG G, Kouakou AV, Mossoun AM, Awuni JA, Minoungou GL, Aplogan GL, Awoumé FK, Albina E, Lancelot R, Couacy-Hymann E. African and classical swine fever situation in Ivory-Coast and neighboring countries, 2008-2013. Acta Trop. 2017;166:241–8.
Kuznar J, Salas ML, Vinuela E. DNA-dependent RNA polymerase in African swine fever virus. Virology. 1980;101:169–75.
Le Potier M-F, Marcé C. Nouvelle avancée de la Peste Porcine Africaine aux frontières de l'Europe: la Biélorussie atteinte. African swine fever is in the vicinity of Europe: first case notified in Belarus. Bull Epid Santé Animal Alim. 2013;58:23–4.

Leblanc N, Corley M, Fernandez Pinero J, Gallardo C, Massembe C, Okurut AR, Health L, Van Heerden J, Sanhez-Vizcaino JM, Stahl K, Belak S. Development of a suspension Microarray for the genotyping of African swine fever virus targeting the SNPs in the C-terminal end of the p72 gene region of the genome. Transbound Emerg Dis. 2012;60(4):378–83.

Levathes LE. When China ruled the seas. The treasure fleet of the Dragon Throne, 1405–1433. New York: Oxford University Press; 1994.

Lubisi BA, Bastos AD, Dwarka RM, Vosloo W. Intra-genotypic resolution of African swine fever viruses from an East African domestic pig cycle: a combined p72-CVR approach. Virus Genes. 2007;35:729–35. https://doi.org/10.1007/s11262-007-0148-2.

McMenamy J, Hjertner B, McNeillya F, Uttenthal A, Gallardo C, Adair B, Allana G. Sensitive detection of African swine fever virus using real-time PCR with a 5'conjugated minor groove binder probe. J Virol Methods. 2010;168:141–6.

Mebus CA. African swine fever. Adv Virus Res. 1988;35:251–69.

Michaud V, Randriamparany T, Albina E. Comprehensive phylogenetic reconstructions of African swine fever virus: proposal for a new Classification and molecular dating of the virus. PLos One. 2013;8(7):e69662. https://doi.org/10.1371/journal.pone.0069662.

Misinzo G, Gwandu FB, Biseko EZ, Kulaya NB, Mdimi L, Kwavi DE, Sikombe CD, Makange M, Madege MJ. *Neisseria denitrificans* restriction endonuclease digestion distinguishes African swine fever viruses of eastern from southern African origin circulating in Tanzania. Res Opin Anim Vet Sci. 2014;4(4):212–7.

Montgomery RE. On a form of swine fever occurring in British East Africa (Kenya Colony). J Comp Pathol. 1921;34:159–91.

Nix RJ, Gallardo C, Hutchings G, Blanco E, Dixon LK. Molecular epidemiology of African swine fever virus studied by analysis of four variable genome regions. Arch Virol. 2006;151:2475–94.

Odemuyiwa SO, Adebayo IA, Ammerlaan W, Ajuwape ATP, Alaka OO, Oyedele OI, Soyelu KO, Olaleye DO, Otesile EB, Muller CP. An outbreak of African swine fever in Nigeria: virus isolation and molecular characterization of the VP72 gene of a first isolate from West Africa. Virus Genes. 2000;20:139–42.

OIE. African swine fever in Burkina Faso; 2004. http://www.oie.int/eng/info/hebdo/AIS17.HTM. Consulted 24 Sept 2012.

OIE (World Organisation for Animal Health). Report archive; 2014. Available at http://www.oie.int/wahis_2/public/wahid.php/Diseaseinformation/reportarchive. Accessed 24 Jan 2014.

Oura C. Overview of African swine fever. In: Kahn CM, Line S, Aiello SE, editors. The Merck veterinary manual. 10. Whitehouse Station, NJ: Merck and Co; 2013. Available at: http://www.merckvetmanual.com/mvm/generalized_conditions/african_swine_fever/overview_of_african_swine_fever.html. Accessed 15 Oct 2015.

Oura CAL, Powell PP, Anderson E, Parkhouse RME. The pathogenesis of African swine fever in the resistant bushpig. J Gen Virol. 1998;79:1439–43.

Penrith ML. African swine fever. Onderstepoort J Vet Res. 2009;76:91–5.

Penrith ML, Vosloo W. Review of African swine fever: transmission, spread and control. J S Afr Vet Assoc. 2009;80(2):58–62.

Penrith ML, Thompson GR, Bastos ADS. African swine fever. In: Coetzer JAW, Tustin RC, editors. Infectious diseases of livestock, vol. 3. 2nd ed. Cape Twon: Oxford University Press; 2004. p. 1088–119.

Plowright W, Parker J, Pierce MA. African swine fever virus in ticks (Ornithodoros moubata Murray) collected from animal burrows in Tanzania. Nature. 1969;221:1071–3.

Plowright W, Perry CT, Pierce MA, Parker J. Experimental infection of the argasid tick, Ornithodoros moubata porcinus, with African swine fever virus. Arch Gesamte Virusforsch. 1970a;31:33–50.

Plowright W, Perry CT, Pierce MA. Transovarial infection with African swine fever virus in the argasid tick, Ornithodoros moubata porcinus, Walton. Res Vet Sci. 1970b;11:582–4.

Plowright W, Perry CT, Greig A. Sexual transmission of African swine fever virus in the tick, Ornithodoros moubata porcinus, Walton. Res Vet Sci. 1974;17:106–13.

Plowright W, Thompson GR, Nesser JA. African swine fever. In: Coetzer JAW, Thompson GR, Tustin RC, editors. Infectious diseases of livestock with special reference to Southern Africa. New York: Oxford University Press; 1994. p. 558–99.
Quembo CJ, Jori F, Vosloo W, HEATH L. Genetic characterization of African swine fever virus isolates from soft ticks at the wildlife/domestic interface in Mozambique and identification of a novel genotype. Transbound Emerg Dis. 2017;65(2):1–12. https://doi.org/10.1111/tbed.12700.
Rahimi P, Sohrabi A, Ashrafihelan J, Edalat R, Alamdar M, Masoudi M, Mastofi S, Azadmanesh K. Emergence of African swine fever virus, Northwestern Iran. Emerg Infect Dis. 2010;16:1946–8.
Ravaomanana J, Michaud V, Jori F, Andriatsimahavandy A, Roger F, Albina E, Vial L. First detection of African swine fever virus in Ornithodoros porcinus in Madagascar and new insights into tick distribution and taxonomy. Parasit Vectors. 2010;3:115. https://doi.org/10.1186/1756-3305-3-115. http://www.parasitesandvectors.com/content/3/1/115
Rodriguez JF, Vinuela E. Analysis of the complete nucleotide sequence of African swine fever virus. Virology. 1995;208:249–78.
Rojo G, GarcõÂa-Beato R, Vinuela E, Salas ML, JoseÂ Salas JA. Replication of African swine fever virus DNA in infected cells. Virology. 1999;257:524–36.
Rousset D, Randriamparany T, Maharavo Rahantamalala CY, Randriamahefa N, Zeller H, Rakoto-Andrianarivelo M, Roger F. Introduction de la Peste Porcine Africaine à Madagascar, histoire et leçons d'une émergence. Arch Inst Pasteur Madagascar. 2001;67(1–2):31–3.
Salas ML, Kuznar J, Vinuela E. Polyadenylation, methylation and capping of the RNA synthesized *in vitro* by African swine fever virus. Virology. 1981;113:484–91.
Salas ML, Rey-Campos J, Almendral JM, Talavera A, Vinuela E. Transcription and translation maps of African swine fevervirus. Virology. 1986;152:228–40.
Sanchez-Botija C. Reservorious del virus de la peste porcina Africana. Bull Off Int Epiz. 1963;60:895–9.
Sogo JM, Almendral JM, Talavera A, Vinuela E. Terminal and internal inverted repetitions in African swine fever virus DNA. Virology. 1984;133:271–5.
Takamatsu H, Martins C, Escribano JM, Alonso C, Dixon LK, Salas ML, Revilla Y. Asfariviridae. In: King AMQ, Adams MJ, Carsterns EB, Leikowitz EJ, editors. Virus taxonomy. Ninth report of the ICTV. Oxford: Elsevier; 2011. p. 153–62.
Tignon M, Gallardo C, Iscaro C, Hutet E, Van der Stede Y, Kolbasov D, De Mia GM, Le Potier M-F, Bishop RP, Arias M, Koenen F. Development and inter-laboratory validation study of an improved new real-time PCR assay with internal control for detection and laboratory diagnosis of African swine fever virus. J Virol Methods. 2011;178:161–70.
Vandeputte J, Chappuis G. Classical swine fever: the European experience and a guide for infected areas. Rev Sci Tech. 1999;18:638–47.
Vinuela E. African swine fever virus. Curr Top Microbiol Immunol. 1985;116:151–79.
Wardley RC. African swine fever virus. Arch Virol. 1983;76:73–90.
Wengler G. Family flaviviridae. In: Francki RIB, Fauquet CM, Knudson DL, Brown F, editors. Classification and nomenclature of viruses. Fifth report of the International Committee on Taxonomy of Viruses. Berlin: Springer; 1991. p. 223.
Wesley BD, Tuthill AE. Genome relatedness among African swine fever virus field isolates by restriction endonucleases analysis. Prev Vet Med. 1984;2:53–62.
Wilkinson PJ. African swine fever. In: Manual of standards for diagnostic test and vaccines. 4th ed. Paris: Office International des Epizooties; 2000. p. 189–98.
Wittek R, Moss B. Tandem repeats within the inverted terminal repetition of vaccinia virus DNA. Cell. 1980;21:277–84.
Yanez FJ, Rodriguez JM, Nogal ML, Yuste L, Enriquez C, Rodriguez HE, Vinuela E. Analysis of the complete nucleotide sequence of African swine fever virus. Virology. 1995;208:249–78.

Chapter 17
Avian Influenza

C. A. Meseko and D. O. Oluwayelu

Abstract Avian influenza (AI) is a viral disease of birds that can be transmitted to other animals including humans. Outbreaks of highly pathogenic avian influenza (HPAI) subtype H5N1 occurred in poultry in Africa for the first time in 2006 following its emergence in Asia. Subsequent spread of the infection, multiple introductions, and fatal transmission to humans globally and in some African countries and connected regions underscore the importance of this transboundary animal disease to livestock, livelihoods, and public health. The prevention and control of future AI outbreaks in order to mitigate its socioeconomic and public health impact cannot be overemphasized. Repeated incursions of HPAI into Africa and intermingling of multiple susceptible hosts could drive emergence of pathogens at the interface between humans and animals in a shared environment. Intersectoral synergy for the control and mitigation of avian influenza based on the principles of One Health is therefore desirable. The need to understand the nature of the disease, its epidemiology, distribution, applied surveillance strategies, diagnosis, and measures for its control are germane and discussed in this chapter as it applies to Sahelian Africa and connected regions. Basic knowledge of influenza virus and its impact on human and animal health would help to better prepare the next generation of professionals to cope with the rapidly evolving challenges posed by this infectious disease of global reckoning.

Keywords Avian influenza · Control · Epidemiology · One Health · Sahelian Africa · Socio-economics

C. A. Meseko (✉)
Regional Centre for Animal Influenza and Transboundary Animal Diseases, National Veterinary Research Institute, Vom, Nigeria

Institute for Virus Diagnostic, Friedrich-Loeffler-Institut, Federal Research Institute for Animal Health, Island of Riems, Germany

D. O. Oluwayelu
Faculty of Veterinary Medicine, Department of Veterinary Microbiology, Influenza Research and Diagnostic Laboratory, University of Ibadan, Ibadan, Nigeria

M. Kardjadj et al. (eds.), *Transboundary Animal Diseases in Sahelian Africa and Connected Regions*, https://doi.org/10.1007/978-3-030-25385-1_17

Disease Description and History

Avian influenza viruses (AIV) belong to the family Orthomyxoviridae in the genus *Influenza A virus*. They are enveloped spherical viruses with negative-sense, single-stranded RNA genome, which consists of eight segments. These code for at least 11 proteins (Fig. 17.1; Swayne and Suarez 2000; Capua and Alexander 2003). AIV can be differentiated into subtypes on the basis of their surface hemagglutinin (HA) and neuraminidase (NA) glycoproteins. Until now, 16 HA and 9 NA subtypes are known for AIV in waterfowls which can occur in manifold combinations (Alexander 2007). Recently, the 17th and 18th HA and 10th NA subtypes were described in bats (Tong et al. 2013). Due to the segmented genome, reassortment of the surface protein as well as segments coding for the "inner" proteins can be exchanged when two different influenza viruses infect one host cell at the same time (thus generating viruses with possibly new biological characteristics, different pathogenicity, or different host specificity). Reassortment between two influenza A viruses (IAVs) that differ in all eight segments can give rise to 256 distinct genotypes (Phipps et al. 2017).

In addition to IAVs, the genera influenza B, C, and D viruses belong to the family Orthomyxoviridae. Others are Thogotovirus, Quaranjavirus, and Isavirus associated with arthropods and fish. While IAVs can infect a broad range of avian and mammalian animals including humans, B and C are human associated while C and D were detected in swine and bovine hosts (Chiapponi et al. 2016). The nomenclature of influenza virus types A, B, C, and D includes information on the antigenic type of the virus (based on the antigenic specificity of the nucleoprotein, e.g., A); the host of origin (for strains isolated from nonhuman sources), e.g., chicken; geographical origin, e.g., Ghana, strain number, e.g., 15VIR2588-10, year of isolation, e.g., 2015; and subtypes based on H and N proteins, e.g., H5N1. Hence, AIV isolated in Ghana in 2015 is typically named (A/chicken/Ghana/15VIR2588-2015/H5N1) (Tassoni et al. 2016; Louten 2016).

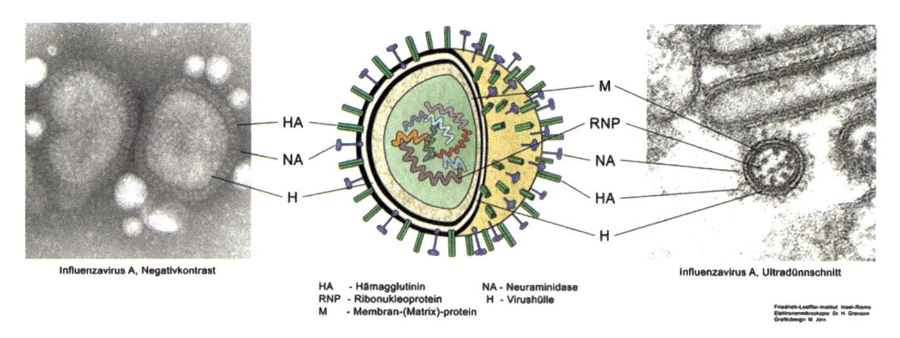

Fig. 17.1 Schematic structure (center), negative contrast (left), and ultra-thin section (right) of influenza A virus showing HA (hemagglutinin), NA (neuraminidase), M (Matrix), RNP (ribonucleoprotein) and H (envelope). Image reproduced with kind permission of Friedrich-Loeffler-Institut, Insel Riems, Germany

Biologically, AIVs are differentiated into low pathogenic AIV (LPAIV) and highly pathogenic AIV (HPAIV) with regard to the severity of the illness and the mortality rate especially in gallinaceae birds. HPAIV and LPAIV also bear markers in the amino acid sequence of the HA gene cleavage site that are determinants of either high pathogenicity (multiple basic amino acids like PQRRRK*G) or low pathogenicity (lacking in multiple basic amino acid, e.g., PQRETR*G) (Luczo et al. 2015). Different cellular enzymes (proteases) are able to process the HA precursor protein into HA I and HA II proteins depending on the HA cleavage site motif and therefore induce local or systemic infections (Klenk and Garten 1994).

HPAIV infections are notifiable in every case while LPAIV of subtypes H5 and H7 are also important because of their capability to mutate to the HPAI form by acquiring additional basic amino acids in the HA cleavage site during passages in avian hosts. These mutations occur because most RNA viruses lack a genetic proofreading mechanism; hence, small errors that occur when the virus replicates go undetected, uncorrected, and accumulate (Sanjuán and Domingo-Calap 2016). The OIE definition of HPAI at the molecular level is based on the presence of these multiple basic amino acids: arginine (R), lysine (K), or glutamine (E) at the HA cleavage site. Other genomic markers of virulence are also being identified and characterized, particularly the PB2 protein. Because of the proclivity of a low virulent H5 or H7 becoming virulent by mutation in poultry hosts, all H5 and H7 viruses are notifiable. Independent of the pathogenicity, the most important AIV subtypes that have been transmitted from birds to humans since 1996 are H5, H7, and H9 which are also of global public health concern (Alexander and Brown 2000). In Sahelian and sub-Saharan Africa, H5 is the most important subtype of AIV in combination with N1, N2, N3, and N8 clinically manifesting as either LPAI or HPAI and has been frequently reported since 2006 (Joannis et al. 2008; Couacy-Hymann et al. 2009; Coker et al. 2014; Monne et al. 2015, Abdelwhab et al. 2016). In addition, AIV subtypes H7 and H9 have also been detected in wild birds and poultry in Algeria, Egypt, Morocco, and Burkina Faso (Aly et al. 2010; Osman et al. 2015; Zecchin et al. 2017).

In its severe form, HPAIV, formerly described as "classical fowl plague," is highly infectious and primarily affects poultry. Early occurrence of an unknown disease was reported in Italy as far back as 1878 by Perroncito (Lupiani and Reddy 2009). In 1901, the causative agent was shown to be ultra-filterable (i.e., virus) and in 1955 the virus was demonstrated to be closely related to mammalian influenza viruses (Alexander and Brown 2000). The term "fowl plague" was substituted in 1981 with highly pathogenic avian influenza (HPAI). Throughout the 1900s, few outbreaks of HPAI in poultry were recorded across the world but since 2003, HPAI H5N1 panzootic, which presumably arose before 1997 in Southern China, has caused serious economic and social consequences (Lupiani and Reddy 2009).

When newly emerged Asian lineage of A/H5N1 (derived from archetype A/Gs/Gd/1/96 virus) caused infection and massive die-offs in migratory water fowls and poultry in 2005 (Xu et al. 1999), there were concerns that the virus could be introduced to Africa and connected regions through poultry trade or migratory waterfowls from Asia and Europe (Liu et al. 2005). Predictably, the first outbreak occurred in north-central Nigeria in January 2006 (Joannis et al. 2006). The

Table 17.1 Countries of Sahelian Africa and year of first reported HPAI H5 and H7 as on September 20, 2017

Year	Subtype	Species	Country	Deaths and culling of poultry/other birds
2006	H5N1	Poultry	Niger	530 Deaths, 21,504 depopulated
2016	H5N1	Poultry		88,290 Deaths
2016	H5N1	Poultry	Sudan	87,370 Deaths, 107,327 depopulated
2006	H5N1	Poultry	Burkina Faso	123 Deaths, 7 depopulated
2016	H5N1	Poultry		152,936 Deaths, 16,886 depopulated

Source: ProMED (https://www.promedmail.org/eafr) OIE (http://www.oie.int/en/animal-health-in-the-world/update-on-avian-influenza/)

outbreaks observed in Africa were thus an expansion of the Euro-Asian spread of HPAI that started in 2003 in Asia and entered Europe for the first time at the end of 2005. Subsequently, it spread to other countries in West and North Africa including Egypt (where it has since become endemic), Niger, Cameroon, Sudan, etc. (Table 17.1). The devastating impact of AIV in Sahelian Africa and connected regions, risks of repeated incursions, and potential for emergence of pandemic strains require responses that would strengthen human and veterinary surveillance for early detection and control (Breiman et al. 2007).

Geographical Distribution

Southern China has been proposed as an epicenter for the generation of influenza pandemics and H5N1 genetic diversity. It is the country of origin of H5N1 HPAI strains and ongoing source of emergent and reemergent HPAI viruses (Shortridge and Stuart-Harris 1982; Gutiérrez et al. 2009). Before 2003, outbreaks of HPAI in poultry had been rare events. However, since 2004 a devastating HPAI H5N1 epizootic, which presumably arose before 1997 in Southern China and subsequently caused numerous outbreaks in poultry farms as well as in live bird markets in Hong Kong, emerged and has spread to East and Middle Asia, Europe, and Africa, causing enormous losses to the poultry population (Kalthoff et al. 2010). The global spread of HPAIV H5N1 clade 2.2 from Asia to Europe and Africa began in 2003–2005 following outbreak in migratory birds at Qinghai Lake in China. The virus is believed to have spread from China to other regions through the transport of poultry and bird migration in successive waves that occurred in 2003, 2005, 2006, and 2015 (Gutiérrez et al. 2009; Monne et al. 2015). In addition, outbreaks of HPAI that resulted from circulating LPAI H5, H7, and H9 viruses have been reported in poultry worldwide (Capua and Marangon 2000; Bulaga et al. 2003; Capua and Alexander 2007; Iqbal et al. 2009; Snoeck et al. 2011). More recently in 2016–2017, novel strains of H5 in the Gs/GD/96 lineage (mainly H5N8) have spread in poultry and wild birds across Europe, parts of Asia, the Middle East, and West Africa and also extended, for the first time, to Eastern and Southern Africa—Uganda, Congo, Zimbabwe, and South Africa (FAO 2017) (Fig. 17.2).

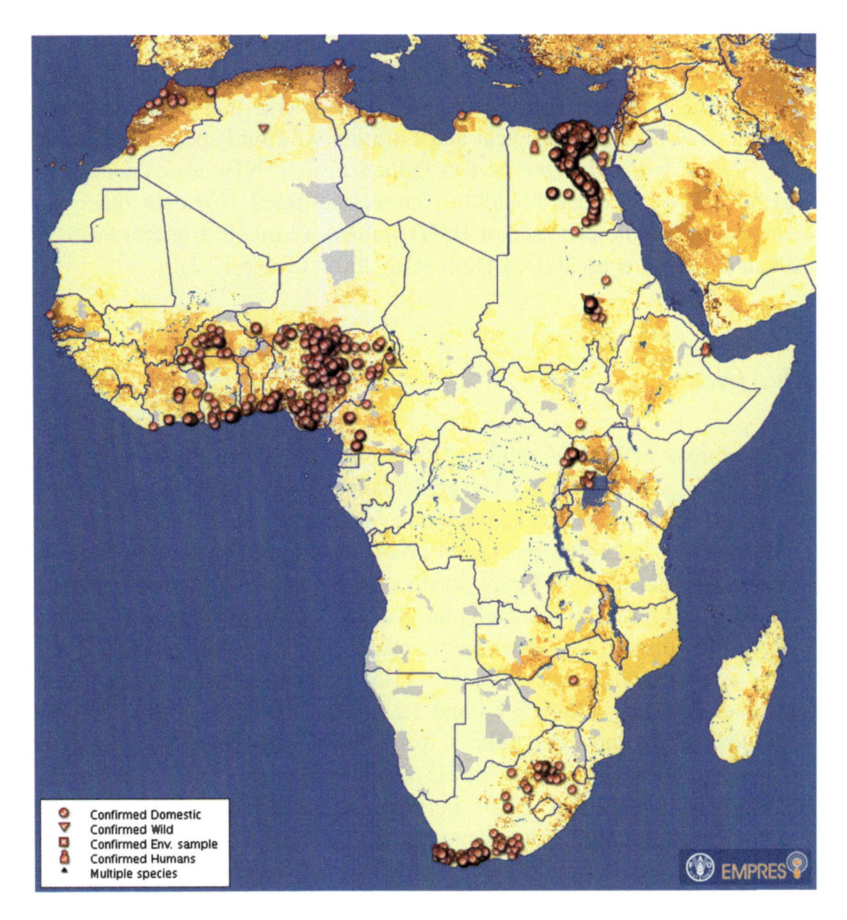

Fig. 17.2 Geographical distribution of highly pathogenic avian influenza in domestic poultry, wild birds, and humans (indicated with symbols) in Sahelian Africa and connected regions. Observation date from 01/01/2016 to 30/11/2017. Map created from FAO-Empress-I. http://empres-i.fao.org/ei3gmapcomp/?MAPREQUEST=ef1695db-7d71-4e8e-8328-06e2b29f919d8 (Accessed 30 November, 2017)

Sahelian Africa

Since its first report in 1996, HPAIV H5N1 prototype virus A/Goose/Guangdong/1/96 (A/Gs/Gd/1/96), which emerged in China in 1996, has undergone various genetic reassortments producing many genotypes/clades and spreading into many countries including those of Africa and the Middle East. Subsequently, multiple introductions of different clades to countries in Africa were reported (Ducatez et al. 2006; Fusaro et al. 2009; Cattoli et al. 2009). The HPAI H5N1 scourge has not spared the Sahelian African territory which is a semiarid tropical savannah ecoregion in Africa that forms the transitional zone between the Sahara Desert to the north and the more humid savannah belt to the south. The Sahel Belt stretches across Senegal, Mauritania,

Mali, Burkina Faso, Niger, northeastern Nigeria, Chad, and Sudan (http://www.newworldencyclopedia.org/entry/Sahel). In early surveillance studies, Gaidet et al. (2007) detected avian influenza virus (AIV) by reverse transcriptase-polymerase chain reaction in both Afro-tropical and Eurasian wild bird species in Mauritania, Senegal, Mali, Chad, and Niger; neither influenza A (H5N1) viruses nor any highly pathogenic AIV were detected in all the samples collected. However, other workers have since reported the presence of H5N1 influenza virus in Burkina Faso, Niger, and Sudan (Ducatez et al. 2007; Cattoli et al. 2009) (Table 17.1).

North Africa

The first outbreak of HPAIV subtype H5N1 in North Africa occurred in Egyptian poultry in 2006 (Aly et al. 2006). The epidemiological role of Egypt in the transmission of influenza A viruses was highlighted by El-Zoghby et al. (2013) who reported that Egypt acts as a bridge between Europe, Asia, and Africa, and millions of migrating birds pass through Egypt on their flights annually particularly in winter seasons where the northern Nile Delta lakes act as a major refuge for a multitude of bird species. The disease has since become enzootic in the country, making Egypt one of the few countries with clade 2.2.1 HPAI H5N1 virus that has diversified into two distinct subclades designated 2.2.1.1 and 2.2.1.2. Viruses in clade 2.2.1.1 circulated in vaccinated poultry, including turkeys, from 2007 to 2014 despite intensive blanket vaccination and using over 20 diverse H5 vaccines (Aly et al. 2006; Abdel-Moneim et al. 2009; Abdelwhab et al. 2016). Further, an outbreak of HPAIV of the H5N1 subtype was diagnosed in a backyard poultry farm in eastern Libya in 2014 (Kammon et al. 2015). By the end of 2016 and later in 2017, H5N8 clade 2.2.2.4 had also been introduced to the North African countries of Egypt and Tunisia (OIE 2017) (Table 17.2). In addition, LPAI virus subtype H9N2 has been reported to circulate in Egypt, Libya, Tunisia, and Morocco (Monne et al. 2008; El-Zoghby et al. 2012; Kammon et al. 2015; El-Houadfi et al. 2016). Also, LPAI H7N7 has been recorded in a black kite in Egypt (Aly et al. 2010).

Arabian Peninsula

The first detection of HPAI H5N1 in the Middle East was recorded in Turkey in a flock of backyard turkeys in October 2005 (Williams and Peterson 2009). By early 2006, some Middle Eastern countries including Kuwait, Israel, West Bank/Gaza Strip, Iran, Iraq, and Jordan had detected HPAI H5N1 virus from samples of dead poultry or dead wild birds (Kalthoff et al. 2010). In addition, H9N2 infections were reported in the Middle East and Asia causing widespread outbreaks in commercial chickens in Iran, Saudi Arabia, Pakistan, China, Korea, the United Arab Emirates, Israel, Jordan, Kuwait, Lebanon, Oman, and Iraq (reviewed in Alexander 2007; Aamir et al. 2007; Perk et al. 2009).

Table 17.2 Countries of North Africa and year of first reported HPAI H5 and H7 as on September 20, 2017

Year	Subtype	Species	Country	Deaths and culling of poultry/other birds
2006–2017	H5N1	Poultry	Egypt[a]	≥1,085,080 Deaths, ≥8.9 million depopulated[b]
2016	H5N8	Wild bird (common coot)		2 Deaths
2017	H7N1	Migratory birds	Algeria	≥1800 Deaths
2014	H5N1	Poultry	Libya[a]	23 Deaths, 39 depopulated
2016	H5N8	Wild birds	Tunisia	30 Deaths

[a]Countries with human infection of avian influenza and fatalities
[b]As on July 2008 but situation is continuous (endemic)
Source: ProMED (https://www.promedmail.org/eafr) OIE (http://www.oie.int/en/animal-health-in-the-world/update-on-avian-influenza/)

Sub-Sahelian Africa

Low pathogenic avian influenza (LPAI) and HPAI viruses including subtypes H5, H7, and H9 are reportable diseases of major concern to the poultry industry worldwide (Swayne et al. 2011). By early 2006, HPAI H5N1 virus had been detected from samples of dead poultry or dead wild birds in seven African countries (Nigeria, Niger, Cameroon, Burkina Faso, Sudan, Cote d'Ivoire, and Djibouti) (Kalthoff et al. 2010) (Table 17.3). Between May and December 2007, outbreaks had also been reported in Ghana, Togo, and Benin Republic (Cattoli et al. 2009). Specifically, in West Africa, outbreaks of HPAI caused by H5N1 viruses belonging to genetic clades 2.2, 2.2.1, 2.2.2, and 2.2.3 occurred from about February 2006 through July 2008 (Aiki-Raji et al. 2008; Fusaro et al. 2010). This is, however, not comparable to the number of outbreaks and spread of the disease across the globe in the last 10 years (2007–2017) during which HPAI outbreaks have increased in alarming proportion and the impact, in terms of the number of birds involved and the costs of disease control, has dramatically escalated. After 7 years of disappearance of the virus in Nigeria, another wave of HPAI epizootic reported in January 2015 was shown to be caused by a novel clade (2.3.2.1c) of A (H5N1) virus (Monne et al. 2015). By the end of 2016 and later in 2017, H5N8 clade 2.2.2.4 had also been introduced to Nigeria as well as other African countries such as Cameroon, Uganda, Congo, Zimbabwe, and South Africa (OIE 2017) (Table 17.3).

According to the Food and Agricultural Organization (FAO 2015), agroecological drivers of HPAI H5N1 similar to those present in endemic regions of South East Asia are present in Nigeria and West Africa, thus implying an increased risk of disease spread in the region. Traditional trading patterns, both formal and informal, between Nigeria and neighboring countries provide an opportunity for cross-border or inter-regional disease spread. Currently, countries at immediate risk of cross-border infection include Benin, Cameroon, Chad, Ghana, Niger Republic, Togo, as

Table 17.3 Countries of sub-Sahelian Africa and year of first reported HPAI H5 and H7 as on September 20, 2017

Year	Subtype	Species	Country	Deaths and culling of poultry/ other birds
2006	H5N1	Poultry	Nigeria[a]	≥500,000 Deaths, ≥1.8 million depopulated
2009	H5N2	Wild bird		–
2015–2017	H5N1	Poultry		≥265,000 Deaths, ≥2.5 million depopulated
2016	H5N8	Mixed species		15 Deaths, 235 depopulated
2006	H5N1	Poultry	Cameroon	50 Deaths, 8 depopulated
2016	H5N1	Poultry		47,562 Deaths, 85,592 depopulated
2017	H5N8	Peafowl		24 Deaths
2010	H5N1	Poultry	Djibouti[a]	4 Deaths, 18 depopulated
2007	H5N1	Poultry	Ghana	12,884 Deaths, 21,612 depopulated
2016	H5N1	Poultry		38,067 Deaths, 104,059 depopulated
2007	H5N1	Poultry	Togo	4011 Deaths, 4418 depopulated
2016	H5N1	Poultry		11,300 Deaths, 3030 depopulated
2007	H5N1	Poultry	Benin	180 Deaths, 504 depopulated
2006	H5N1	Poultry	Cote d'Ivoire	723 Deaths, 1100 depopulated
2016	H5N1	Poultry		83,135 Deaths, 73,528 depopulated
2017	H5N8	Wild birds-White-winged tern/poultry	Uganda	1200 Deaths
2017	H5N8	Waterfowls/poultry	Congo DR	≥31,000 Deaths and depopulated
2017	H5N8	Poultry	Zimbabwe	7845 Deaths, 75155 depopulated
1961	H5N3	Wild bird (tern)	South Africa	1300 Deaths
2004/2006	H5N2	Ostrich		30,000/7334 Depopulated
2017	H5N8	Poultry/Ostrich		≥160,000 Deaths, 1.5 million depopulated

[a]Countries with human infection of avian influenza and fatalities

Source: ProMED (https://www.promedmail.org/eafr), OIE (http://www.oie.int/en/animal-health-in-the-world/update-on-avian-influenza/)

well as other countries that had reported outbreaks in the past such as Côte d'Ivoire, Djibouti, and the Sudan. These countries need to be prepared to detect and respond in a timely manner to possible incursions of the disease (FAO 2015).

Economic/Public Health Impact

Economic Impact AIVs pose significant and ongoing threat to livestock production and economy not only in Sahelian Africa but globally. Apart from direct mortality, depopulation of birds as a method of eradication, and secondary economic impacts on livestock resources, food, and associated trade/businesses are also adversely affected. The cost of controlling HPAIV including surveillance, laboratory infrastructures, and monies paid as compensation to farmers are overwhelming and impact negatively on the lean resources of developing nations. The HPAI has often been characterized by 100% morbidity in affected flocks and intensively reared species including turkeys, chickens, guinea-fowl, and quail often exhibit 90–100% mortality rates within a few days of infection (Capua and Marangon 2000). Since 2003, millions of birds (mostly chickens) have died or been culled because of HPAI outbreaks, which has resulted in severe impacts on the poultry sector in many countries. By June 2007, HPAI H5N1 had spread from Southeast Asia to 62 other countries, with more than 250 million birds dead or destroyed and an estimated impact of more than US$12 billion (Martins 2012). In Nigeria, the 2006 HPAI H5N1 outbreak especially affected backyard and medium-scale poultry farmers and caused a decline in egg and chicken sales by greater than 80% within 2 weeks after the announcement of the outbreak. Sales of poultry feed also dropped by over 80% with 80% of workers on affected farms and 45% on unaffected farms losing their jobs (UNDP 2006). Over 1.8 million birds were culled and more than 623 million naira were paid as compensation to poultry farmers during the 2006–2007 H5N1 outbreaks in Nigeria (Oladokun et al. 2012). HPAIV outbreaks have significant economic impacts causing widespread disruption to poultry production and trade in Sahelian Africa where a large proportion of the rural population depends on livestock for livelihood. Sub-Saharan Africa has the largest livestock populations in the world and developing countries could account for 70% of world meat production by 2050 (Alexandratos and Bruinsma 2012); hence, decimating diseases like HPAI is a threat to global animal health and food security including international trade.

Public Health Avian influenza virus is one of the most important public health pathogens that have emerged from animal reservoirs. Outbreaks are of concern not only to the poultry industry in which they cause an economically important disease but also to human health (Henning et al. 2013; Gao et al. 2013). While AIVs infect variety of bird hosts including free-living and captive-caged birds, domestic ducks, chickens, turkeys, and other domestic poultry (Căpua and Alexander 2007), they are capable of being transmitted across the human–animal–wildlife interface. The World Health Organization (WHO) realized the potential role of animal influenza on the origin of human pandemics as far back as 1958 and promoted studies on the ecology of the virus even in wild animals (Alexander 1986).

HPAIV is capable of zoonotic infection of human hosts via direct contact with infected birds and a derivative of A/Goose/Guangdong/1/96 (A/Gs/Gd/1/96) strain caused the first direct avian to human transmission and fatal human case of H5N1 in Hong Kong SAR in 1997 (Claas et al. 1998). Subsequent fatal transmissions and spread of HPAIVs to humans have resulted in AI being considered one of the most important animal diseases, if not the most important (Capua and Alexander 2007).

Since the official confirmation of HPAI A(H5N1) in Egypt in February 2006, outbreaks have been reported in at least 1101 commercial poultry farms, and the cumulative number of confirmed human cases of H5N1 infection between 2006 and July 2017 was 359 with 120 deaths (WHO 2017). The first and only human case of HPAI H5N1 infection in Nigeria was reported in January 2007. Other African countries with records of human deaths include Libya and Djibouti (Tables 17.2 and 17.3). Globally, as on July 2017, a total of 859 laboratory confirmed cases of HPAI A(H5N1) with 453 deaths (cumulative proportion of fatal cases: 52.7%) had been reported to the World Health Organization from 16 countries (WHO 2017). Though no death due to HPAIV H7N9 has been reported in Africa, the Asian lineage A/H7N9 that was first reported in China in March 2013 had infected 1557 persons and killed over 622 (40%) by September 2017. It is now ranked as the influenza virus with the highest potential pandemic risk. When it was first reported in 2013 only low pathogenic strains were detected in human, poultry, and environmental specimens, but the virus has now mutated to HPAIV with low susceptibility to antiviral drugs (CDC 2017). HPAI infection in humans manifests in severe respiratory signs and pneumonia which progress to septic shock and multi-organ failure. Most infections due to HPAIV H5N1 and H7N9, the two most important subtypes, have limited human-to-human transmission but are easily spread from poultry to human (Peiris et al. 2007; Gao et al. 2013; Poovorawan et al. 2013). The public health concern with respect to the HPAI H5N1 virus is that it may change into a form that is highly infectious for humans and that can spread easily from person to person. This could mark the start of a global outbreak or pandemic. Exposure to poultry birds in LBMs and backyard production systems with poor biosecurity and hygiene are considered the most important routes of such transmission requiring close monitoring, especially in Asia and Africa (Bulaga et al. 2003; Paul et al. 2011; Aiki-Raji et al. 2015).

Epidemiology of Avian Influenza

Influenza A viruses infect a wide variety of birds including free-living, captive-caged, and domestic poultry. All 16 subtypes of IAV viruses are known to be present in wild bird population especially waterfowls (family Anatidae, order Anseriformes), which include ducks, geese, and swans (Capua and Alexander 2007). The order Charadriiformes including shorebirds, gulls, and terns also harbor influenza virus but of a different gene pool from the Anseriformes (Kawaoka et al. 1988). These two bird orders are thus the most important in the transmission and spread of both LPAI and HPAI. Since 1973, avian influenza viruses have been isolated from over 105 wild bird species and some aquatic and terrestrial animals (Olsen et al. 2006).

The LPAIV gene pools are evolutionarily stable in waterfowls mostly without causing clinical diseases (Olsen et al. 2006). Nevertheless, in the course of antigenic drift (mutation) and antigenic shift (reassortment), changes in the gene may cause hitherto low pathogenic viruses to become highly pathogenic. The precursor of A/Goose/Guangdong/1/96 (A/Gs/Gd/1/96) and subsequent clades that emerged was thought to be an LPAI virus circulating in wild aquatic birds (Alexander 2000). The progenitors are probably H3N8 and H7N1 viruses from Nanchang (China) and H1N1 and H5N3 viruses from Hokkaido (Japan) (Mukhtar et al. 2007). In addition, LPAIVs may cause clinical infection when transmitted to domestic birds like chickens and turkeys. The virus replicates in high concentrations in the intestines of infected wild birds and is voided along with feces, thereby contaminating the environment and transmission to other domestic and feral birds (Webster et al. 1992). Birds infected with AIV shed large quantities of virus in their feces as well as in their saliva and nasal secretions in the first 2 weeks of infection. Infected droppings and secretions from both symptomatic and asymptomatic migratory waterfowls can contaminate water and the environments and ensure secondary transmission and spread (Capua and Alexander 2007; WHO 2007).

The gene pool in aquatic birds provides the genetic diversity required for emergence of pandemic strains especially after reassortment of genes in intermediate hosts like the pig (Ma et al. 2009). For H5N1 HPAI virus, the change in the primary route of transmission from fecal/oral to the respiratory route in land birds, especially minor poultry species such as quails and pheasants, has also been considered significant in the epidemiology of the virus, especially in its transmission to mammals (Perez et al. 2003; Mararova et al. 2003; Humbred et al. 2006).

Following a major outbreak of HPAI H5N1 in 2005 in wild birds in Lake Qinghai, an important breeding site for migratory birds in China (Gilbert et al. 2006), outbreaks along migratory bird routes from South-East Asia and Mongolia through Eastern Europe to Western Europe were subsequently observed. Wild birds were thus suspected to play a role in the long-distance spread of HPAI H5N1 along these migratory routes (Li et al. 2004). The Asian lineage of HPAIV H5 clade 2 was thereafter detected for the first time in sub-Saharan Africa in January 2016, and it was strongly suspected to have followed intercontinental migration of waterfowls from Asia and Europe to Africa (Joannis et al. 2006; Olsen et al. 2006; Gaidet et al. 2008). Infected migratory birds from Asian or European congregating wetlands could have direct interaction with bridge species like domestic ducks or they may contaminate the environment from which the bridge species get infected (Cecchi et al. 2008; Vakuru et al. 2012). Ecological factors including free-range birds in wetlands contribute to the transmission of influenza to domestic poultry flocks including local fowls, chickens, and turkeys. Over 65–80% of poultry holdings in rural Africa are kept as backyard and free-range local birds with minimal or no biosecurity where they sometimes have contact with feral birds (Adene and Oguntade 2006). These birds that may be infected are in turn regularly sold into LBMs where they have greater chances of coming in contact with other species, thereby serving as animate and inanimate fomites (Meseko et al. 2007; Capua and Alexander 2007). Thus, LBMs have been recognized as a major source of avian influenza transmission not only in Asia but also in sub-Saharan Africa (Amonsin et al. 2008; Fusaro et al. 2009; Indriani et al. 2010; Oladokun et al. 2012; Oluwayelu et al. 2015, 2017).

Apart from direct transmission of HPAI from an infected migratory bird to domestic poultry, the presence of terrestrial, free-range local birds in areas frequented by avian influenza virus-infected waterfowl provides the right ecological and epidemiological setting around water bodies, wetlands, and flood plains in many countries in sub-Saharan Africa for LPAIVs to mutate into or reassort with HPAI strains (Cecchi et al. 2008). This potential was revealed when H5N2 LPAI was detected in spur-winged geese (*Plectropterus gambensis*) around the Hadejia-Nguru wetland in northeast Nigeria in 2008 and the virus was thought to be a reassortant between European and African influenza gene pools. Phylogenetic analyses revealed that all of the genes, except the nonstructural (NS) genes, of the LPAI H5N2 viruses were more closely related to genes recently found in wild and domestic birds in Europe (Snoeck et al. 2011). In a previous investigation of wild waterfowl also caught around the Hadejia-Nguru wetland, they were found positive for H5N2 HPAI virus even though they showed no clinical signs (Gaidet et al. 2008). These observations underscore the risk for domestic birds and poultry as wild waterfowls may introduce LPAI viruses into free-range domestic ducks in villages and wetland areas when they mingle with wild or feral birds as was later confirmed by Coker et al. (2014). Egypt has also been suggested as a region bridging Europe, Asia, and Africa in influenza epidemiology as migrating birds pass through the country on their annual flights particularly in winter when the northern Nile Delta lakes act as wintering site for multitudes of bird species (El-Zoghby et al. 2012). The risk of long-distance transmission of AIVs from Asia/Europe to Africa is buttressed by data on three major migratory bird flyways crisscrossing Africa. The flyways include the East Africa-Asia, Atlantic-America, and Black Sea/Mediterranean flyways. Waterfowls, waders, birds of prey, and over two billion songbirds migrate from Europe to sub-Saharan Africa each year with the potential of introducing AIVs if infected (Hahn et al. 2009). Molecular evidence and phylogenetic relatedness of HPAIVs from Asia/Europe show that these regions are epidemiologically linked as illustrated in Fig. 17.5 (Tassoni et al. 2016).

Zoonoses and One Health Sixty-one percent of 1415 species of infectious organisms known to be pathogenic to humans are zoonotic and 75% of emerging infectious diseases among which IAV is important are also zoonotic (Taylor et al. 2001). Influenza A viruses originating from animals especially AIV can directly infect humans or adapt to infect humans following mutation or reassortment of genes (Alexander and Brown 2000). Such viruses have the potential to cause influenza pandemic given that the human population has no previous exposure and hence lacks immunity to it. Pandemic influenza viruses as observed in recent decades are likely to emerge from either poultry or pig IAV that gained pandemic potential through mutations and reassortments with circulating human viruses (Reid et al. 1999, 2004; Smith et al. 2009). The Asian lineage HPAI A/H5N1 virus shows strong zoonotic characteristics, and it was transmitted from birds to different mammalian species including humans (Kalthoff et al. 2010). Similarly, A/H7N9 has been responsible for annual waves of mortality in humans since 2013 in China and with over 90% of cases associated with exposure to poultry birds (CDC 2017). In light of these zoonoses, it is imperative that animal and human health services cooperate and promote joint surveillance of IAV that may have consequences on human and animal health. The One-Health approach to disease investigation would enhance intersectoral collaboration between veterinary and public health services (von Dobschuetz et al. 2015). The need to control the spread of HPAI at its animal reservoir source to decrease the risk of human infections (zoonoses) and potential emergence of a human pandemic strain that may also be re-transmitted to animals (reverse zoonoses) cannot be overemphasized.

Symptoms of Avian Influenza

Avian influenza viruses are mostly shed in feces and respiratory secretions which reflect in the symptoms that may be observed. The LPAIVs may be asymptomatic or show mild gastrointestinal and respiratory signs, while HPAIVs usually cause severe illness in chickens and turkeys. Drop in egg production (layers), reduced feed/water intake, ruffled feathers, and apathy may signal infection in poultry. HPAI may occur suddenly in a flock as peracute or acute disease in its severe form with incubation period as short as 0–2 days. The first observation by farmers or consulting veterinarian is massive and rapid mortality reaching 90–100% within 72 h in poultry birds, chickens and turkeys being most susceptible (Liu et al. 2005; Bouma et al. 2009). In some cases, all ages of birds and all species are found dead within 2 weeks of the onset of clinical signs (Adene et al. 2006).

The massive deaths notwithstanding, AIV is not necessarily severe in all birds, because clinical outcome of infection depends on the species, age, the strain of infecting virus, and tissue tropism (Alexander et al. 1978; Chen et al. 2004). Sometimes few clinical signs are seen before death in susceptible species. Less vulnerable species such as waterfowls, also considered as natural reservoir hosts of avian influenza, are most often asymptomatic (Meseko et al. 2007; Gaidet et al. 2008; Meseko et al. 2010a, b). In this species, the virus replicates in the intestine and is shed via the feco-oral route. However, evolutionary shift in feco-oral route in aquatic birds to respiratory shedding in terrestrial species also contributes to expanding the clinical presentations (Perez et al. 2003). More significantly, the switch from LPAI to HPAI also completely alters the manifestation of avian influenza in birds because of the systemic spread and tissue pantropism often noticed with HPAI.

HPAIVs affect different organs and systems including respiratory, circulatory, reproductive, and nervous presenting varieties of symptoms and signs. During the first HPAI H5N1 outbreak among waterfowls in Lake Qinghai China, the two major clinical signs observed were gastrointestinal and neurological in nature and include diarrhea, tremor, and opisthotonus (Liu et al. 2005). Respiratory and neurologic signs of labored breathing, sneezing, cough, ocular/nasal discharges, torticollis, circling, loss of balance, and head tremors were also recorded in ducks in experimental infections (Sturm-Ramirez et al. 2005; Keawcharoen et al. 2008). These signs are similar to those seen in infection with very virulent, velogenic strains of Newcastle disease virus that presents similar clinical manifestation as HPAI and should always be considered in differential diagnosis of HPAI (Swayne 2013).

The maiden outbreak of HPAI H5N1 in Africa in 2006 presented a variety of clinical signs that included edema which was visible in featherless parts of the head, wattles, and legs. Cyanosis of the comb, subcutaneous hemorrhages of the shanks/feet, greenish diarrhea, labored breathing, and severe dyspnea were manifested (Joannis et al. 2006; Fig. 17.3). Field investigation of many cases in Africa revealed further signs such as difficulty in breathing, sneezing and coughing, drooling saliva, and greenish/brownish/yellow diarrhea. Other signs are nose bleeding, somnolence, recumbency (sternal and lateral), dyspnea, and moist rales. Young birds may be noticed with convulsion, torticollis, and stretched neck. Adult chickens have edema of the head and facial region, extensive cyanosis, and hemorrhagic patches of the combs and wattles. Turkeys appeared most susceptible among the gallinaceous species with fulminating infection. Shortly before death, they are recumbent with swollen paranasal sinuses. Those not recumbent have unsteady gait, with backward motion and moving in cycles—torticollis and ataxia (Adene et al. 2006).

During recurrent outbreaks of HPAI in West Africa (2006–2007 and 2015–2017), poultry farmers and operators of live bird markets in few cases observed transient signs before massive deaths. Some of these signs include respiratory distress, watery feces, cyanotic comb, hemorrhagic shanks/feet, torticollis, and paralysis of the limb (Akanbi et al. 2016). The most significant clinical findings following infection with HPAIV are massive deaths which usually begin among a group of birds in the flock before infection spreads through contact and proximity. The mortality pattern is usually staggered before it finally infects all birds within a house or compartment. In few cases, farmers do not recognize abnormality until 5 days post-infection, and without any intervention, an entire flock of chicken could die within 12 days of introduction of virus to the farm (Yoon et al. 2005).

Lesions/Pathology of Avian Influenza

LPAI may be symptomless and without lesions, while HPAI often kills birds too quickly without premonitory signs or marked pathological lesions being noticed. However, both LPAIV and HPAIV infections may show varieties of pathology associated with tissue tropism. HPAIVs are pantropic, while LPAIV replication is restricted to the respiratory and/or enteric tracts and does not produce lesions outside these systems (Mo et al. 1997).

At postmortem, the most apparent physical signs of infection caused by HPAIV are severe subcutaneous hemorrhages and hyperemia on all the featherless areas of the body including head, feet, and shank (Fig. 17.3). However, many carcasses have good body conformation because HPAI is not a chronic, debilitating disease and many birds may lack macroscopic lesion (Elbers et al. 2004), although inflammatory degenerative and necrotic lesions of the musculoskeletal system including myositis and hemorrhagic ulcers can be seen in few cases. The gross pathological lesions on the respiratory organs and tissues include multifocal hemorrhages, catarrhal tracheitis, cloudy and diphtheritic air sacculitis, pulmonary congestion/edema, and viral pneumonia with the lungs most often severely edematous, congested, and hemorrhagic. In the circulatory system, there

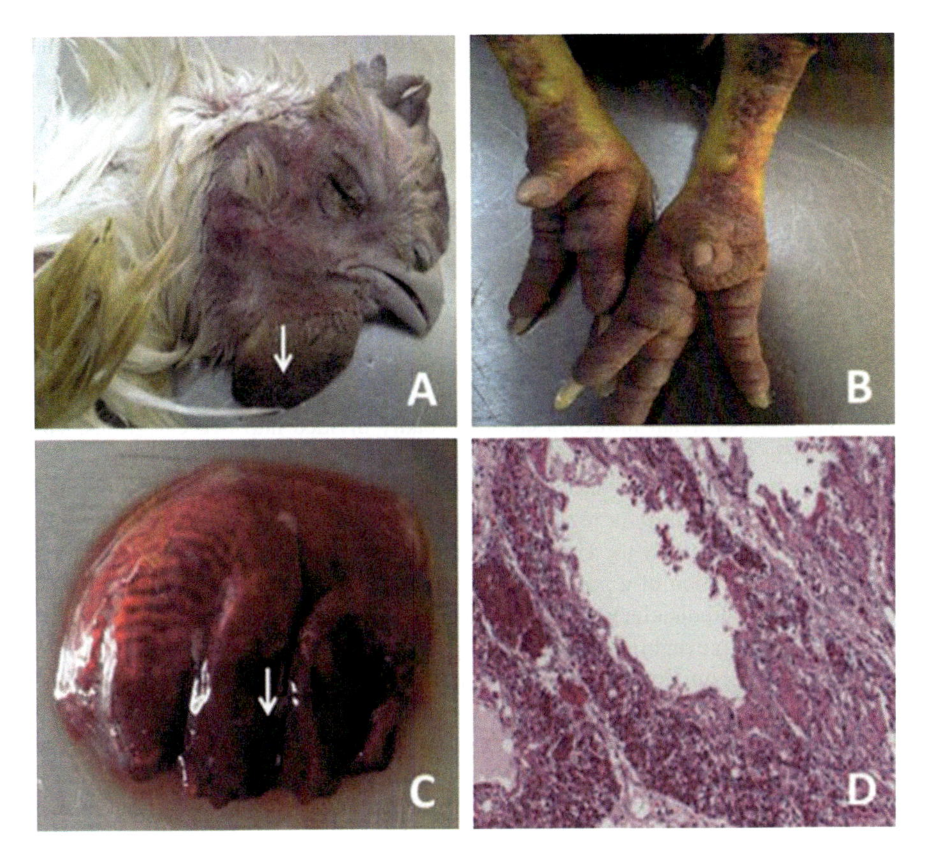

Fig. 17.3 Gross pathological lesions and histopathology (**a**) Swollen comb and wattle with subcutaneous hemorrhages of the face (arrow), (**b**) Shank and feet with diffuse subcutaneous hemorrhages, (**c**) Lung with severe congestion (arrow) and edema, (**d**) Lung with parabronchiolar epithelial necrosis and desquamation, vascular congestion, and expansion of the interstitium by mixed cellular infiltrates (Reproduced from Akanbi et al. 2016)

are sub-epicardial hemorrhages while in the digestive system, ecchymotic hemorrhages in the serosal surface of the intestines and perigastric petechial hemorrhages are widespread. Catarrhal enteritis is also seen in the duodenum with thickened and hemorrhagic cecal tonsils. Hemorrhages on the proventricular mucosa and its margin are also prominent (Teifke et al. 2007).

Early outbreaks of HPAI H5N1 in wild birds in China manifested gross pathological lesions that include pancreatic necrosis, with extensive areas of lytic necrosis. These are consistent with the pathology observed in domestic geese and ducks experimentally infected with H5N1 (Li et al. 2004). In the nervous system, the brain pathology includes glial cell infiltration and perivascular cuffing, which explains the severe neurological signs often manifested (Liu et al. 2005). These severe neurologic signs in ducks are also associated with multifocal viral encephalitis. Apart from hemorrhages in the brain, neuronal/Purkinje cell necrosis in the cerebrum and cerebellum was widespread in field cases. In the reproductive track, oophoritis, eggs binding to the oviduct, and involuted and degenerated ova are seen and the presence of ruptured ova and yolk in the abdominal cavity causes peritonitis and air sacculitis. Among layer birds, ovarian/follicular hemorrhages are also observed and eggs are without shells and appear bleached or whitish. Other viscera including the liver, spleen, and kidney may be congested, hemorrhagic, enlarged, and diffusely necrotic (Adene et al. 2006; Akanbi et al. 2016).

Diagnosis of Avian Influenza

The diagnosis of avian influenza can be broadly divided into two categories, viz. field/clinical diagnosis and laboratory/confirmatory diagnosis. It is important for these two components to work together for an efficient identification and control of HPAI in a given territory or region. This is because the success of any control effort depends on how quickly the virus is detected both on the field and in the laboratory. The rapid declaration of a positive or negative case enables immediate quarantine, depopulation, and disinfection. These steps are critical for effective stamping out of the virus and to prevent infection and spread of HPAI to other premises or territories.

Disease recognition is the first step in HPAI diagnosis; this combines physical observation of clinical signs and anamnesis. HPAI should be suspected when sudden and massive mortality is observed with evidence of biosecurity breach such as the introduction of new birds to farms, equipment, and other fomites (Meseko et al. 2007). It is also possible to be aware of epidemiological scenarios either globally or within the region that may likely serve as early warning. A good example is that prior to 2006 when HPAI H5N1 from Asia spread to Eastern Europe and the Middle East, ecological indicators of migratory waterfowls' flight to the Sahelian region were eventually followed by detection of the virus in poultry in the region. Farmers and consulting veterinarians therefore need to be openly sensitive to observe early warnings of clinical signs and postmortem findings and report field suspicions of sudden and massive deaths for immediate laboratory tests and confirmation. The field diagnosis of the first incursion of Asian lineage HPAI H5N1 into the Sahelian region was achieved based on such strong epidemiological, clinical, and postmortem findings and was confirmed following further laboratory analysis of biological specimens (Adene et al. 2006; Aly et al. 2006; Joannis et al. 2006; Couacy-Hymann et al. 2009).

AIVs can be detected in oropharyngeal, tracheal, and cloacal swabs of live birds. Swabs, tissues, and feces of dead birds are regular diagnostic specimens. When specimens are received in the laboratory, varieties of classical and molecular diagnostic techniques can be applied for the diagnosis of avian influenza. Among these are rapid antigen detection, polymerase chain reaction (PCR), virus isolation in cultures, and serological identification. Other advanced techniques for the characterization of influenza virus such as gene sequencing, biological infectivity studies, vaccine development, and clinical trials may be adopted for enhanced evolutionary analysis and for developing tools, devices, and products for comprehensive control of influenza virus. The diagnostic approach for avian influenza also depends on the facilities and human resources at the disposal of concerned laboratory. In sub-Saharan Africa and connected regions, laboratories were initially only able to carry out basic diagnosis which has improved over time to include advanced classical and molecular diagnostic techniques some of which are discussed below. Whichever diagnostic test method is employed, it is important to bear in mind that HPAI is a zoonotic disease of high consequence requiring adoption of biosecurity and biosafety practices in laboratory procedures (Du et al. 2012).

Rapid Antigen Detection In recent times, it has been possible to detect HPAI on the field or pen side using fecal materials or cloacal swabs. The test is simple and rapid and requires few materials while providing reliability in terms of relative diagnostic sensitivity and relative diagnostic specificity as high as 84.6% and 100%, respectively (Slomka et al. 2012). The rapid antigen test was successfully deployed for rapid identification of HPAI H5N1 in Nigeria during the 2006–2007 epidemics but requires further laboratory confirmation by other techniques (Meseko et al. 2010a, b).

Polymerase Chain Reaction This is a rapid molecular method for the diagnosis of AIV. It is fast, sensitive, and of moderate cost. Reverse transcriptase-polymerase chain reaction (RT-PCR) can be done by conventional or real-time techniques. Both begin with the extraction of nucleic acid from cloacal/tracheal swabs, tissue homogenates, or virus isolates in allantoic fluid or cell culture supernatants. Amplification of nucleic acid (RNA) in a thermocycler and post-amplification identification of positive and negative samples by processing in gel electrophoresis is a required step in conventional RT-PCR (Fig. 17.4). The real-time RT-PCR (RT-qPCR) is an advanced modification where both amplification and identification take place simultaneously and the result can be visualized as florescence curves on a computer monitor in real time. All tests are carried out with variations of primers, probes, and enzymes as is fit for each test or diagnostic procedure of interest (Suarez et al. 2007). It is extensively deployed in confirmatory diagnosis of HPAI cases in wild birds and poultry in African laboratory settings (Joannis et al. 2008; Couacy-Hymann et al. 2009).

Virus Isolation The proof of causative agent is virus isolation as was established in the Koch's postulate (Tabrah 2011). Virus or pathogen isolation is considered in many fields as the gold standard by which other diagnostic techniques are measured. Virus isolation is also described as a necessary method preceding molecular and biological characterization. Specimens from suspected cases of avian influenza which may include cloacal/tracheal swabs and parenchymatous tissues are processed and inoculated in 9–11 day-old embryonated chicken eggs (specific pathogen-free or specific antibody-negative) or in Madin-Darby Canine Kidney (MDCK) or other cell cultures and incubated at 35–37 °C. Embryo mortality and/or cytopathic effects (CPE) is an indication of virus infection. Other tests are required to further identify and quantify the virus of interest (WHO 2002).

Serology Avian influenza virus possesses a cell surface glycoprotein known as hemagglutinin (HA) which is able to agglutinate red blood cells (RBCs). This is used to identify and subtype influenza virus antigens/antibodies by the technique of hemagglutination inhibition (HI) test against homologous or heterologous virus antigens/antibodies as the case may be. Other serological methods include enzyme-linked immunosorbent assay (ELISA) and virus neutralization test (VNT). ELISA techniques are solid-phase reactions where antigen generic for influenza A viruses or subtype-specific antigens are bound to a microtiter plate. Antibody with specificity for the antigen binds and can be visualized generally by an enzyme-linked anti-antibody and a subsequent color reaction. Various modifications of the ELISA technique to identify antibodies or antigens are described as sandwich, capture, competition, direct, and indirect ELISAs. The VNT is a serological test used in laboratories with capacity to maintain cell cultures. An antibody-positive sample that is incubated with the respective virus and neutralizes its infectivity can be verified microscopically by reduction in cytopathic effect caused by the virus or visualized by immunofluorescence or other methods (WHO 2002). Although the agar gel immunodiffusion (AGID) test is being replaced by more sensitive techniques, it is still useful in low-resource laboratories. The principle is that homologous virus antigen can react with antibody-positive serum to form a line of precipitation in an agar gel plate and therefore can identify antibodies. This method was used extensively in the early days of HPAI diagnosis in Nigeria (Joannis et al. 2006).

Virus Characterization Avian influenza can be characterized by biological and molecular methods. Biological characterization is carried out by inoculating susceptible animal models with virus usually through intravenous or intracerebral routes and monitoring clinical outcomes of morbidity and mortality. When a virus isolate is injected intravenously into ten (6 weeks old) specific pathogen-free (SPF) chickens, they are clinically examined daily for 10 days. The scoring consists of (0) if normal, (1) if sick, (2) if severely sick, and (3) if dead (dead animals are scored as (3) at each of the remaining days after death). The intravenous pathogenicity index (IVPI) is the average score per bird per observation over the 10-day period. An index of 3.0 means all birds died within 24 h, and an index of 0.00 means no birds was sick during the observation period. The IVPI is used as clinical assessment of the virulence of AIVs and expression of their pathogenicity, which is able to differentiate between the very virulent HPAIV (IVPI > 1.2) and LPAIV, as well as host susceptibility.

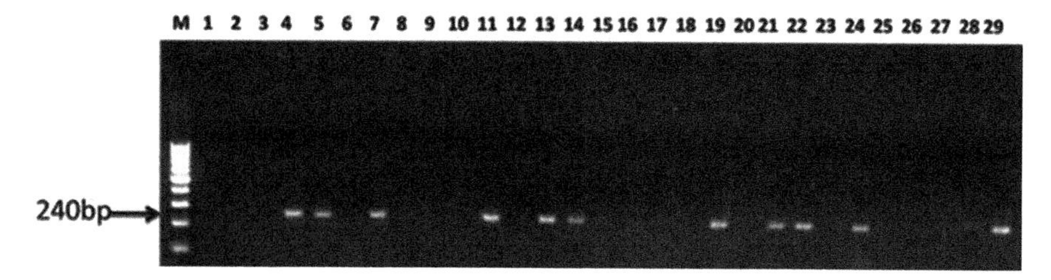

Fig. 17.4 Gel electrophoresis of HA gene of HPAI H5N1, obtained during field outbreaks in Nigeria in 2015. The electrophoresis lanes are numbered 1–29 including negative (1) and positive (29) controls. The DNA marker (M) shows positive reaction at 240 bp (MA-gene). (Reproduced from Akanbi et al. 2016)

Molecular methods like RT-PCR assays and genome sequencing are useful for studying virus genomics and evolution. They enable researchers to identify subtype and pathogenicity of the virus, reassortment events, and changes in molecular markers, e.g., of virulence or drug resistance. Nucleotide sequencing can demonstrate the presence of multiple basic amino acids at the cleavage site of the HA gene as previously described in Section "Disease Description and History". Both biological and molecular characterizations have been adopted by the OIE in description of notifiable HPAI. Gene sequencing and phylogenetic analyses (Fig. 17.5) also increase knowledge of influenza evolutionary diversity, molecular epidemiology, forecasting, and options for control. With bioinformatics tools, the origin, spread, relationship, and biodiversity of AIVs can be diagrammatically presented in phylogenetic trees. For example, in Fig. 17.5, multiple introduction of HPAIV clade

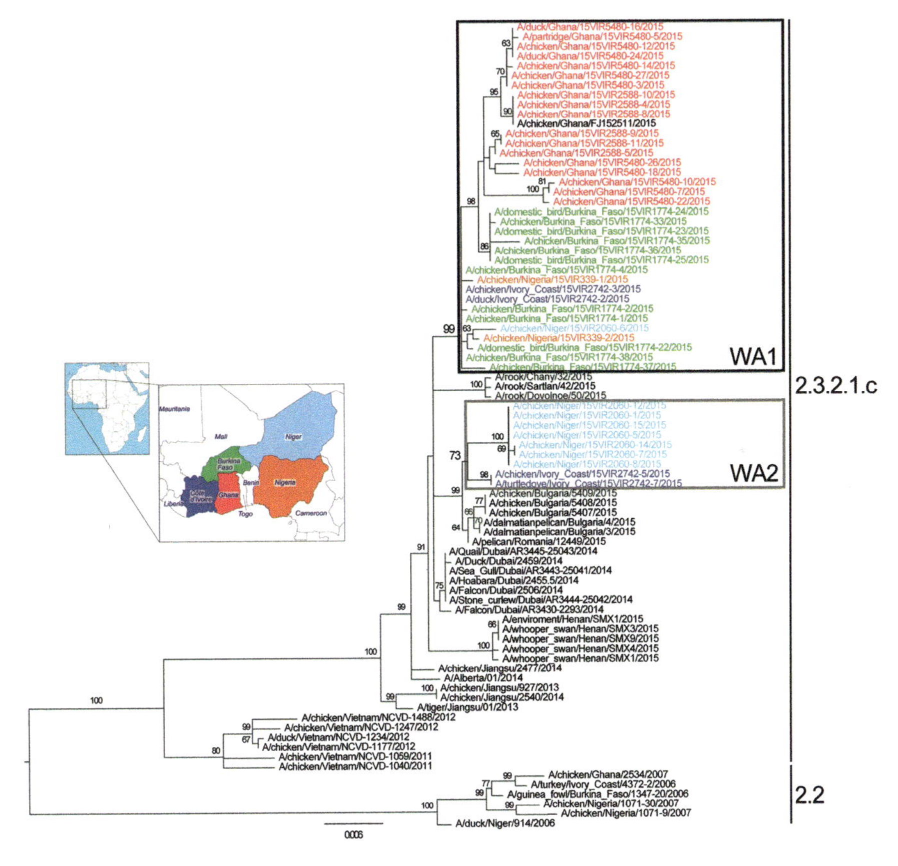

Fig. 17.5 Phylogenetic tree of hemagglutinin (HA) gene segment of highly pathogenic avian influenza (H5N1) viruses obtained from West Africa, 2015, showing the widespread distribution of two subgroups of clade 2.3.2.1c. (Reproduced from Tassoni et al. 2016)

2.3.2.1c designated West Africa (WA) 1 and 2 was described during the 2015 epizootics in West Africa (Tassoni et al. 2016).

Application of Prevention and Control/Surveillance and Control Strategies

The peculiar ecology and epidemiology of AIVs previously described require that preventive measures and control strategies should include biosecurity, surveillance, and depopulation. In exceptional cases, vaccination may also be adopted. These measures take into account extremely complex scenarios: the characteristics of the poultry-producing sector in its entirety, the eco-epidemiological situation in the country, the response capacity of the veterinary infrastructure, and the available resources. These features must also be integrated with the sociocultural environment, including those linked to traditional rearing of birds as livelihood, subsistence backyard farming, and commercial or in few cases recreational purposes (Capua and Alexander 2007).

Depopulation and Decontamination Globally, an enormous number of poultry have died from direct infection with AIV, and countless numbers of poultry flocks at risk have been depopulated as a measure to contain the virus and prevent further spread (Feare 2007). Depopulation and decontamination as first-line measures for controlling HPAI have been implemented in many countries in Sahelian and sub-Saharan Africa where millions of poultry had been destroyed due to mortality or culling (Joannis et al. 2008; Otte et al. 2008). However, without credible, adequate, and rapidly implemented compensation programs linked to depopulation efforts, farmers lack incentive to report bird die-offs and may attempt to salvage the situation by transporting such birds to unaffected areas and LBMs, thereby spreading AIV (FAO 2005).

Biosecurity Biosecurity is defined as the implementation of measures to reduce the risk of introduction and dissemination of disease agents (FAO/AVSF/DAH 2005; FAO 2008). Although ways of classifying these measures may vary, they all refer to the same basic principles of bio-exclusion (i.e., preventing infectious agents from entering the farm) and bio-containment (i.e., preventing infectious agents from exiting) (Charisis 2008) and are implemented via segregation to raise barriers to infectious diseases, cleaning, and disinfection (FAO/AVSF/DAH 2005; FAO 2008; Charisis 2008). Biosecurity measures for preventing HPAI outbreaks should include separation of poultry by age and species, not raising multi-age birds, or raising one species instead of several (Cristalli and Capua 2007; Oluwayelu et al. 2015), since species mixing increases HPAI H5N1 virus transmission because of the risk of introducing asymptomatic reservoirs (Henning et al. 2009). Perimeter fencing, cleaning, and disinfection of surroundings, equipment, people (footwear, foot dips, handwashing), and buildings have also proven to be effective in interrupting potential HPAI H5N1 spread. It is important to eliminate contact between domestic and wild/feral birds, and restriction of movement of people/visitors can limit the risk of

HPAIV introduction into flocks (reviewed in Conan et al. 2012). The continual movement of birds into, through and out of LBMs (like a melting pot), as well as attempts to sell infected dead or dying birds provide opportunity for the introduction, entrenchment, and dissemination of AIVs (Amonsin et al. 2008; Indriani et al. 2010). These practices as well as other marketing practices that make LBMs to become high-risk environments providing excellent prospects for transmission of AI infection from birds to humans and other animals in the markets should therefore be discouraged. Other risky practices in LBMs include the tradition of keeping different species of birds and sick and healthy birds in the same cages or in close proximity. Bird slaughter, evisceration, and processing of raw poultry meat without personal protective equipment (PPE) are also a risk factor to be monitored. Further, interventions to reduce market-based disease transmission such as routine cleaning and disinfection to decontaminate surfaces, daily disposal and removal of waste from the market to eliminate AIV reservoirs, segregation of poultry-related activities into zones to limit virus spread, as well as periodic market rest days with thorough cleaning (Bulaga et al. 2003; Mullaney 2003; WHO 2004; Trock et al. 2008) should be adopted for implementation in LBMs across Sahelian Africa and connected regions where the tradition of slaughtering live fowls in the market is a common practice.

Surveillance and Monitoring To forestall introduction or emergence of new strain (s) of avian influenza virus that can constitute potential threat to animals and public health, there is a need for continuous surveillance of AIVs and IAVs by both human and animal health authorities. Routine surveillance, whether passive or active, is essential for detecting new cases of HPAI H5N1 outbreak for early warning and rapid activation of control measures (Biswas et al. 2009; Brown et al. 2015). In addition, continuous surveillance for AIVs is important in order to prevent outbreaks and possibly identify potential carriers and reservoirs of the virus that can be included in future surveillance programs (Coker et al. 2014). These monitoring activities should include serological, virological, and genetic/molecular surveillance in order to achieve timely detection of the emergence of reassortant/variant strains. Molecular epidemiology is also critical for monitoring genetic and antigenic changes occurring in influenza viruses that circulate among humans and animals (Smith et al. 2006, 2009). In addition, participatory epidemiology which is the systematic use of participatory approaches and methods to improve understanding of diseases and options for animal disease control (Catley et al. 2012), as well as the use of risk mapping to support targeted surveillance and decision making for prevention and control of avian influenza, is important and should be critical components of any program for early detection, reporting, and surveillance of avian influenza (Stevens et al. 2010).

Further, in order to provide a basis for the design of realistic and sustainable risk-based surveillance and control strategies thereby stemming the resurgence of HPAI in Sahelian Africa and connected regions, a better knowledge of the interactions between the local drivers of the disease as well as innovative approaches for integration of poultry production and marketing systems into risk assessment frameworks is required. Novel tools such as social network and value chain analyses

(Klovdahl 1985; Humphrey and Napier 2005) should therefore be applied to identify points and weak links in the disease transmission chain that can be the foci of future mitigation efforts.

Vaccination Vaccination of poultry does not prevent a new infection but can reduce the virus output and may be used for poultry around outbreak zones. According to Capua et al. (2004), optimal vaccination with currently available AI vaccines when selected properly and administered correctly will not only protect against clinical signs and mortality but also reduce the levels and duration of virus excretion. The major disadvantages of vaccination are: (a) the distribution of the virus by apparently healthy animals and (b) the increased opportunity for virus mutations/reassortment. The three categories of strategies proposed for vaccination by the FAO are (1) response to an outbreak, employing perifocal vaccination (ring vaccination) or vaccination only of domestic poultry at high risk, in combination with the destruction of infected domestic poultry; (2) vaccination in response to a "trigger," upon the detection of the disease by surveillance studies, in areas where biosecurity is difficult to implement (e.g., areas with high density of poultry farms); and (3) Preemptive baseline vaccination of chickens and other avian species when the risk of infection is high and/or the consequences of infection are very serious (FAO 2012). When used properly, especially in addition to, rather than instead of other measures such as increased biosecurity and stamping out, vaccines may be a powerful tool in the eradication of AI infections (Ellis et al. 2004). In endemic infected countries where vaccination is currently being practiced, it has been proposed that the following aspects should be considered if vaccination is to remain an important tool in the control plan: (1) matching vaccine strains to currently circulating strains; (2) matching challenge strains to currently circulating strains; (3) maintaining high vaccination coverage; (4) ensuring vaccine efficacy not only in a laboratory setting but also in the field; and (5) evaluating vaccine efficacy on an annual basis (Kayali et al. 2016).

Conclusion: A Regional Strategy for Surveillance and Control of AIV

The risk of continuous introduction and circulation of avian influenza viruses in Sahelian Africa and connected regions (North Africa, Arabian Peninsula, and sub-Sahelian Africa north of the Equator) not only constitutes a threat to animal and human health but also negatively impacts the local economy and livelihood through depletion of livestock resources, income, job losses/unemployment, and aggravation of poverty. It is therefore imperative that constant efforts be made to monitor emergence and evolution of AIVs across these regions and to implement sustainable control strategies to improve animal and public health. To mitigate the impacts of avian influenza in these regions, a strategy of elimination by depopulation and decontamination combined with permanent or continuous risk-based disease surveillance and strengthening of veterinary infrastructures/services for early detection and response is proposed. While authorities of countries with recurrent or endemic status of HPAIVs may be reluctant in adopting blanket vaccination as a control option, research and development (R&D) of vaccines should not be discouraged. Regulated vaccination may also be adopted on case-by-case basis because vaccination alone cannot eliminate the virus and is meant to be part of an integrated control program appropriate to the local situation (FAO 2013a, b). This is with consideration for the peculiar socioeconomic, geographic, cultural, and political circumstances relevant to each country. To ensure effectiveness of the surveillance programs, social network and poultry value chain analyses should be incorporated as tools for identification of possible risk factors and critical points in the affected areas and in the production system that can be targeted for interventions. In addition, it is recommended to establish or revitalize existing avian influenza information and early warning-early response (EWER) systems in each country through the combined efforts of research, educational, and industrial institutions. Further, although biosecurity is recognized and practiced as the first line of defense against disease transmission in intensive and commercial holdings, it should also be introduced and strictly enforced in smallholder, family-owned, or backyard poultry farms, in LBMs, and in the environment in the context of One Health.

References

Aamir UB, Wernery U, Ilyushina N, Webster RG. Characterization of avian H9N2 influenza viruses from United Arab Emirates 2000–2003. J Virol. 2007;361:45–55.

Abdel-Moneim AS, Shany SA, Fereidouni SR, Eid BT, El-Kady MF, Starick E, Harder T, Keil GM. Sequence diversity of the haemagglutinin open reading frame of recent highly pathogenic avian influenza H5N1 isolates from Egypt. Arch Virol. 2009;154:1559–62.

Abdelwhab EM, Hassan MK, Abdel-Moneim AS, Naguib MM, Mostafa A, Hussein IT, Arafa A, Erfan AM, Kilany WH, Agour MG, et al. Introduction and enzootic of A/H5N1 in Egypt: virus evolution, pathogenicity and vaccine efficacy ten years on. Infect Genet Evol. 2016;40:80–90.

Adene DF, Oguntade AE. The structure and importance of the commercial and village based poultry industry in Nigeria. Rome: FAO; 2006.

Adene DF, Wakawa AM, Abdu PA, et al. Clinico-pathological and husbandry features associated with the maiden diagnosis of avian influenza in Nigeria. Nig Vet J. 2006;27(1):32–8.

Aiki-Raji CO, Aguilar PV, Kwon YK, Goetz S, Suarez DL, Jethra AI, et al. Phylogenetics and pathogenesis of early avian influenza viruses (H5N1). Nig Emerg Infect Dis. 2008;14(11):1753–5.

Aiki-Raji CO, Adebiyi AI, Agbajelola VI, et al. Surveillance for low pathogenic avian influenza viruses in live-bird markets in Oyo and Ogun States, Nigeria. Asian Pac J Trop Dis. 2015;5 (5):369–73.

Akanbi BO, Meseko CA, Odita CI, Shittu I, Rimfa AG, Ugbe D, Pam L, Gado DA, Olawuyi KA, Mohammed SB, Kyauta II, Bankole NO, Ndahi W, Joannis TM, Ahmed MS, Okewole PA, Shamaki D. Epidemiology and clinicopathological manifestation of resurgent highly pathogenic avian influenza (H5N1) virus in Nigeria, 2015. Niger Vet J. 2016;37(3):175–86.

Alexander DJ. Avian influenza-historical aspects. In: Proceedings of the second international symposium on avian influenza; 1986. p. 4–13.

Alexander DJ. A review of avian influenza in different bird species. Vet Microbiol. 2000;74 (1–2):3–13.

Alexander DJ. An overview of the epidemiology of avian influenza. Vaccine. 2007;25(30):5637–44.

Alexander DJ, Brown IH. Recent zoonoses caused by influenza A viruses. Rev Sci Tech. 2000;19:197–225.

Alexander DJ, Allan WH, Parsons D, Parsons G. The pathogenicity of four avian influenza viruses for chickens, turkeys and ducks. Res Vet Sci. 1978;24:242–7.

Alexandratos N, Bruinsma J. World agriculture towards 2030/2050: the 2012 revision. ESA working paper No. 12-03. FAO: Rome; 2012.

Aly MM, Arafa A, Hassan MK. Epidemiological findings of outbreaks of disease caused by highly pathogenic H5N1 avian influenza virus in poultry in Egypt during 2006. Avian Dis. 2006;52 (2):269–77.

Aly MM, Arafa A, Kilany WH, Sleim AA, Hassan MK. Isolation of a low pathogenic avian influenza virus (H7N7) from a black kite (*Milvus migrans*) in Egypt in 2005. Avian Dis. 2010;54:457–60.

Amonsin A, Choatrakol C, Lapkuntod J, Tantilertcharoen R, Thanawongnuwech R, Suradhat S, et al. Influenza virus (H5N1) in live bird markets and food markets, Thailand. Emerg Infect Dis. 2008;14(11):1739–42.

Biswas PK, Christensen JP, Ahmed SS, Das A, Rahman MH, Barua H, et al. Risk for infection with highly pathogenic avian influenza virus (H5N1) in backyard chickens, Bangladesh. Emerg Infect Dis. 2009;15:1931–6.

Bouma A, Claassen I, Natih K, Klinkenberg D, Donnelly CA, et al. Estimation of transmission parameters of H5N1 avian influenza virus in chickens. PLoS Pathog. 2009;5(1):e1000281. https://doi.org/10.1371/journal.ppat.1000281.

Breiman RF, Nasidi A, Katz MA, Njenga MK, Vertefeuille J. Preparedness for highly pathogenic avian influenza pandemic in Africa. Emerg Infect Dis. 2007;13(10):1453–8. https://doi.org/10.3201/eid1310.070400.

Brown M, Moore L, McMahon B, Powell D, LaBute M, Hyman JM, et al. Constructing rigorous and broad biosurveillance network for detecting emerging zoonotic outbreaks. PLoS One. 2015;10(5):e0124037. https://doi.org/10.1371/journal.pone.0124037.

Bulaga LL, Garber L, Senne DA, Myers TJ, Good R, Wainwright S, et al. Epidemiologic and surveillance studies on avian influenza in live-bird markets in New York and New Jersey, 2001. Avian Dis. 2003;47:996–1001.

Capua I, Alexander DJ. Human health implications of avian influenza viruses and paramyxoviruses. Eur J Clin Microbiol Infect Dis. 2003;23:1–6.

Capua I, Alexander DJ. Avian influenza infections in birds—a moving target. Influenza Other Respir Viruses. 2007;1:11–8.

Capua I, Marangon S. The avian influenza epidemic in Italy, 1999–2000: a review. Avian Pathol. 2000;29:289–94.

Capua I, Terregino C, Cattoli G, Toffan A. Increased resistance of vaccinated turkeys to experimental infection with an H7N3 low pathogenicity avian influenza virus. Avian Pathol. 2004;33:158–63.

Catley A, Alders RG, Wood JLN. Participatory epidemiology: approaches, methods, experiences. Vet J. 2012;191(2):151–60.

Cattoli G, Monne I, Fusaro A, Joannis TM, Lombin LH, et al. Highly pathogenic avian influenza virus subtype H5N1 in Africa: a comprehensive phylogenetic analysis and molecular characterization of isolates. PLoS One. 2009;4(3):e4842. https://doi.org/10.1371/journal.pone.0004842.

Cecchi G, Ilemobade A, Le Brun Y, Hogerwerf L, Slingenbergh J. Agro-ecological features of the introduction and spread of the highly pathogenic avian influenza (HPAI) H5N1 in northern Nigeria. Geospat Health. 2008;3(1):7–16.

Centers for Disease Control (CDC). Morbidity mortality weekly report update on A/H7N9; 2017 Sept 8. www.cdc.gov/mmwr/volume/66/wr/mm6635a2.htm

Charisis N. Avian influenza biosecurity: a key for animal and human protection. Vet Ital. 2008;44 (4):657–69.

Chen H, Deng G, Li Z, Tian G, Li Y, Jiao P, Zhang L, Liu Z, Webster RG, Yu K. The evolution of H5N1 influenza viruses in ducks in southern China. Proc Natl Acad Sci USA. 2004;101 (28):10452–7.

Chiapponi C, Faccini S, De Mattia A, et al. Detection of influenza D virus among swine and cattle, Italy. Emerg Infect Dis. 2016;22(2):352–4.
Claas EC, de Jong JC, van Beek R, Rimmelzwaan GF, Osterhaus AD. Human influenza virus A/HongKong/156/97(H5N1) infection. Vaccine. 1998;16:977–8.
Coker T, Meseko C, Odaibo G, Olaleye DOO. Circulation of the low pathogenic avian influenza subtype H5N2 virus in ducks at a live bird market in Ibadan, Nigeria. Infect Dis Poverty. 2014;3(1):38. https://doi.org/10.1186/2049-9957-3-38.
Conan A, Goutard FL, Sorn S, Vong S. Biosecurity measures for backyard poultry in developing countries: a systematic review. BMC Vet Res. 2012;8:240. https://doi.org/10.1186/1746-6148-8-240.
Couacy-Hymann E, Danho T, Keita D, Bodjo SC, Kouakou C, Koffi YM, Beudje F, Tripodi A, de Benedictis P, Cattoli G. The first specific detection of a highly pathogenic avian influenza virus (H5N1) in ivory coast. Zoonoses Public Health. 2009;56:10–5.
Cristalli A, Capua I. Practical problems in controlling H5N1 high pathogenicity avian influenza at village level in Vietnam and introduction of biosecurity measures. Avian Dis. 2007;51(1 Suppl):461–2.
Du L, Li Y, Gao J, Zhou Y, Jiang S. Potential strategies and biosafety protocols used for dual-use research on highly pathogenic influenza viruses. Rev Med Virol. 2012;22(6):412–9.
Ducatez MF, Olinger CM, Owoade AA, De Landtsheer S, Ammerlaan W, Niesters HG, Osterhaus AD, Fouchier RA, Muller CP. Avian flu: multiple introductions of H5N1 in Nigeria. Nature. 2006;442:37.
Ducatez MF, Tarnagda Z, Tahita MC, Sow A, De Landtsheer S, Londt BZ, Brown IH, Osterhaus ADME, Fouchier RAM, et al. Genetic characterization of HPAI (H5N1) viruses from poultry and wild vultures, Burkina Faso. Emerg Infect Dis. 2007;13:611–3.
Elbers AR, Kamps B, Koch G. Performance of gross lesions at postmortem for the detection of outbreaks during the avian influenza A virus (H7N7) epidemic in The Netherlands in 2003. Avian Pathol. 2004;33(4):418–22.
El-Houadfi M, Fellahi S, Nassik S, Guérin J-L, Ducatez MF. First outbreaks and phylogenetic analyses of avian influenza H9N2 viruses isolated from poultry flocks in Morocco. Virol J. 2016;13:140. https://doi.org/10.1186/s12985-016-0596-1.
Ellis TM, Leung CYHC, Chow MKW, et al. Vaccination of chickens against H5N1 avian influenza in the face of an outbreak interrupts virus transmission. Avian Pathol. 2004;33:405–12.
El-Zoghby EF, Arafa A, Hassan MK, Aly MM, Selim A, Kilany WH, Selim U, Nasef S, Aggor MG, Abdelwhab EM, Hafez MH. Isolation of H9N2 avian influenza virus from bobwhite quail (*Colinus virginianus*) in Egypt. Arch Virol. 2012;157(6):1167–72. https://doi.org/10.1007/s00705-0121269-z.
El-Zoghby EF, Aly MM, Nasef SA, Hassan MK, Arafa A, Selim AA, Kholousy SG, Kilany WH, Safwat M, Abdelwhab EM, Hafez HM. Surveillance on A/H5N1 virus in domestic poultry and wild birds in Egypt. Virol J. 2013;10:203. https://doi.org/10.1186/1743-422X-10-203.
FAO. Compensation and related financial support policy strategy for avian influenza: emergency recovery and rehabilitation of the poultry sector in Vietnam; 2005. Available from: http://www.fao.org/docs/eims/upload/213835/agal_compensationwp_vietnam_jun05.pdf. Accessed 26 Sept 2017.
FAO. Biosecurity for highly pathogenic avian influenza: issues and options. Rome: FAO; 2008.
FAO. FAO recommendations on the prevention, control and eradication of highly pathogenic avian influenza (HPAI) in Asia, September 2004; 2012. Available from: http://www.fao.org/docs/eims/upload/165186/FAOrecommendationsonHPAI.pdf. Accessed 24 Sept 2017.
FAO. Fifth report of the global programme for the prevention and control of highly pathogenic avian influenza, January 2011–January 2012; 2013a. Available at http://www.fao.org/docrep/017/i3139e/i3139e.pdf. Accessed 24 Sept 2017.
FAO. OFFLU technical meeting: developing guidance on vaccines and vaccination against HPAI from lessons learned 4 to 6 December 2013 in Beijing, China; 2013b. http://www.offlu.net/fileadmin/home/en/meeting-reports/pdf/OFFLU_Beijing_2013/OFFLU_Recommendations_Beijing_Dec_2013_final.pdf. Accessed 13 Sept 2017.

FAO. H5N1 HPAI spread in Nigeria and increased risk for neighbouring countries in West Africa. FAO Empres Watch Vol. 32; 2015 Apr. WWW.FAO.ORG/AG/EMPRES.HTML. Accessed 21 Nov 2017.
FAO. Highly pathogenic H5 avian influenza in 2016 and 2017—observations and future perspectives. FOCUS ON, No. 11; 2017 Nov. Rome.
FAO, AVSF, DAH. Prevention and control of Avian flu in small scale poultry. A guide for veterinary paraprofessionals in Vietnam. FAO, AVSF, DAH: Rome; 2005.
Feare CJ. The spread of avian influenza. Ibis. 2007;149(2):424–5. https://doi.org/10.1111/j.1474-919X.2007.00711.x.
Fusaro A, Joannis T, Monne I, Salviato A, Yakubu B, Meseko C, Oladokun T, Fassina S, Capua I, Cattoli G. Introduction into Nigeria of a distinct genotype of avian influenza virus (H5N1). Emerg Infect Dis. 2009;15(3):445–7.
Fusaro A, Nelson MI, Joannis T, Bertolotti L, Monne I, Salviato A, et al. Evolutionary dynamics of multiple sublineages of H5N1 influenza viruses in Nigeria from 2006 to 2008. J Virol. 2010;84:3239–47.
Gaidet N, Dodman T, Caron A, Balança G, Desvaux S, et al. Avian influenza viruses in water birds, Africa. Emerg Infect Dis. 2007;13:626–9.
Gaidet N, Cattoli G, Hammoumi S, Newman SH, Hagemeijer W, Takekawa JY, Cappelle J, Dodman T, Joannis T, Gil P, Monne I, Fusaro A, Capua I, Manu S, Micheloni P, Ottosson U, Mshelbwala JH. Evidence of infection by H5N2 highly pathogenic avian influenza viruses in healthy wild waterfowl. PLoS Pathog. 2008;4(8):e1000127. https://doi.org/10.1371/journal.ppat.1000127.
Gao R, Cao B, Hu Y, Feng Z, Wang D, Hu W, et al. Human infection with a novel avian-origin influenza A (H7N9) virus. N Engl J Med. 2013;368(20):1888–97.
Gilbert M, Xiao X, Domenech J, Lubroth J, Martin V, Slingenbergh J. Anatidae migration in the western palearctic and spread of highly pathogenic avian influenza H5N1 virus. Emerg Infect Dis. 2006;12(11):1650–6.
Gutiérrez RA, Naughtin MJ, Horm SV, San S, Buchy P. A(H5N1) virus evolution in South East Asia. Viruses. 2009;1(3):335–61.
Hahn S, Bauer S, Liech F. The natural link between Europe and Africa: 2.1 billion birds on migration. Oikos. 2009;118:624–6.
Henning KA, Henning J, Morton J, Long NT, Ha NT, Meers J. Farm- and flock-level risk factors associated with highly pathogenic avian influenza outbreaks on small holder duck and chicken farms in the Mekong Delta of Viet Nam. Prev Vet Med. 2009;91(2–4):179–88.
Henning J, Bett B, Okike I, Abdu P, Perry B. Incidence of highly pathogenic avian influenza H5N1 in Nigeria, 2005–2008. Transbound Emerg Dis. 2013;60(3):222–30.
Humbred J, Guan Y, Webster RG. Comparison of the replication of influenza A viruses in Chinese ring-necked pheasants and chukar partridges. J Virol. 2006;80:2151–61.
Humphrey J, Napier L. The value chain approach as a tool for assessing distributional impact of standards on livestock markets: guidelines for planning a programme and designing case studies. In: FAO AGA/ESC initiative on market exclusion. Brighton: Institute of Development Studies; 2005.
Indriani R, Samaan G, Gultom A, Loth L, Indryani S, Adjid R, et al. Environmental sampling for avian influenza virus A (H5N1) in live-bird markets, Indonesia. Emerg Infect Dis. 2010;16 (12):1889–95.
Iqbal M, Yaqub T, Reddy K, McCauley JW. Novel genotypes of H9N2 influenza A viruses isolated from poultry in Pakistan containing NS genes similar to highly pathogenic H7N3 and H5N1 viruses. PLoS One. 2009;4:e5788.
Joannis TM, Lombin LH, de Benedictis P, Cattoli G, Capua I. Confirmation of H5N1 avian influenza in Africa. Vet Rec. 2006;158(9):309–10.
Joannis TM, Meseko CA, Oladokun AT, et al. Serologic and virological surveillance of avian influenza in Nigeria, 2006–7. Euro Surveill. 2008;13:42.
Kalthoff D, Globig A, Beer M. (Highly pathogenic) avian influenza as a zoonotic agent. Vet Microbiol. 2010;140(3–4):237–45.

Kammon A, Heidari A, Dayhum A, Eldaghayes I, Sharif M, Monne I, et al. Characterization of avian influenza and newcastle disease viruses from poultry in Libya. Avian Dis. 2015;59:422–30.
Kawaoka Y, Chambers TM, Sladen WL, Webster RG. Is the gene pool of influenza viruses in shorebirds and gulls different from that in wild ducks? Virology. 1988;163:247–50.
Kayali G, Kandeil A, El-Shesheny R, Kayed AS, Maatouq AM, Cai Z, McKenzie PP, Webby RJ, El Refaey S, Kandeel A, Ali MA. Avian influenza A(H5N1) virus in Egypt. Emerg Infect Dis. 2016;22(3):379–88.
Keawcharoen J, van Riel D, van Amerongen G, Bestebroer T, Beyer WE, van Lavieren R, Osterhaus ADME, Fouchier RAM, Kuiken T. Wild ducks as long-distance vectors of highly pathogenic avian influenza virus (H5N1). Emerg Infect Dis. 2008;14(4):600–7.
Klenk H, Garten W. Host cell proteases controlling virus pathogenicity. Trends Microbiol. 1994;2 (2):39–43.
Klovdahl AS. Social networks and the spread of infectious diseases: the AIDS example. Soc Sci Med. 1985;21(11):1203–16.
Li KS, et al. Genesis of a highly pathogenic and potentially pandemic H5N1 influenza virus in eastern Asia. Nature. 2004;430(6996):209–13.
Liu J, Xiao H, Lei F, Zhu Q, Qin K, Zhang XW, Zhang XL, Zhao D, Wang G, Feng Y, Ma J, Liu W, Wang J, Gao GF. Highly pathogenic H5N1 influenza virus infection in migratory birds. Science. 2005;309(5738):1206. https://doi.org/10.1126/science.1115273.
Louten J. Chapter 10—Influenza viruses. In: Essential human virology. Boston: Academic Press; 2016. p. 171–91. isbn:9780128009475. https://doi.org/10.1016/B978-0-12-800947-5.00010-7.
Luczo JM, Stambas J, Durr PA, et al. Molecular pathogenesis of H5 highly pathogenic avian influenza: the role of the haemagglutinin cleavage site motif. Rev Med Virol. 2015;25:406–30.
Lupiani B, Reddy SM. The history of avian influenza. Comp Immunol Microbiol Infect Dis. 2009;32:311–23.
Ma W, Kahn RE, Richt JA. The pig as a mixing vessel for influenza viruses: human and veterinary implications. J Mol Genet Med. 2009;3(1):158–66.
Mararova NV, Ozaki H, Kida H, Webster RG, Perez DR. Replication and transmission of influenza viruses in Japanese quail. Virology. 2003;310:8–15.
Martins NRS. An overview on avian influenza. Braz J Poult Sci. 2012;14(2):71–87.
Meseko CA, Oladokun AT, Shehu B. An outbreak of highly pathogenic avian influenza (HPAI) in a mixed farm by the introduction of a water fowl. Niger Vet J. 2007;28(3):67–9.
Meseko CA, Oladokun AT, Ekong PS, Fasina FO, Shittu IA, Sulaiman LK, Egbuji AN, Solomon P, Ularamu HG, Joannis TM. Rapid antigen detection in the diagnosis of highly pathogenic avian influenza (H5N1) virus in Nigeria. Diagn Microbiol Infect Dis. 2010a;68(2):163–5.
Meseko CA, Oladokun AT, Solomon P, Yakubu B. Detection of highly pathogenic avian influenza (H5N1) in apparently healthy ducks (anas sparsa sparsa) in live bird markets, Nigeria. Niger Vet J. 2010b;31(2):164–9.
Mo IP, Brugh M, Fletcher OJ, Rowland GN, Swayne DE. Comparative pathology of chickens experimentally inoculated with avian influenza viruses of low and high pathogenicity. Avian Dis. 1997;41(1):125–36.
Monne I, Ormelli S, Salviato A, De Battisti C, Bettini F, Salomoni A, et al. Development and validation of a one-step real-time PCR assay for simultaneous detection of subtype H5, H7, and H9 avian influenza viruses. J Clin Microbiol. 2008;46:1769–73.
Monne I, Meseko C, Joannis TM, et al. Highly pathogenic avian influenza A (H5N1) virus in poultry, Nigeria, 2015. Emerg Infect Dis. 2015;21(7):1275–7.
Mukhtar MM, Rasool ST, Song D, Zhu C, Hao Q, Zhu Y, Wu J. Origin of highly pathogenic H5N1 avian influenza virus in China and genetic characterization of donor and recipient viruses. J Gen Virol. 2007;88:3094–9.
Mullaney R. Live-bird market closure activities in the north-eastern United States. Avian Dis. 2003;47(3 Suppl):1096–8.
OIE. Updates on avian influenza; 2017. http://www.oie.int/en/animal-health-in-the-world/update-on-avian-influenza/. Accessed 26 Sept 2017.

Oladokun AT, Meseko CA, Ighodalo E, John B, Ekong PS. Effect of intervention on the control of highly pathogenic avian influenza in Nigeria. Pan Afr Med J. 2012;13:14.. http://www.panafrican-med-journal.com/content/article/13/14/full/

Olsen B, Munster VJ, Wallensten A, Waldenstrom J, Osterhaus ADME, Fouchier RAM. Global patterns of influenza A virus in wild birds. Science. 2006;31:2384–8.

Oluwayelu DO, Aiki-Raji CO, Adigun OT, Olofintuyi OK, Adebiyi AI. Serological survey for avian influenza in Turkeys in three states of Southwest Nigeria. Influenza Res Treat. 2015. https://doi.org/10.1155/2015/787890.

Oluwayelu DO, Omolanwa A, Adebiyi AI, Aiki-Raji CO. Flock-based surveillance for low pathogenic avian influenza virus in commercial breeders and layers, southwest Nigeria. Afr J Infect Dis. 2017;11(1):44–9.

Osman N, Sultan S, Ahmed A, et al. Molecular epidemiology of avian influenza virus and incidence of H5 and H9 virus subtypes among poultry in Egypt in 2009–2011. Acta Virol. 2015;59 (1):27–32.

Otte J, Hinrichs J, Rushton J, Roland-Holst D, Zilberman D. Impacts of avian influenza virus on animal production in developing countries. CAB Rev Perspect Agric Vet Sci Nutr Nat Res. 2008;3(080):1–18.

Paul M, Wongnarkpet S, Gasqui P, Poolkhet C, Thongratsakul S, Ducrot C, Roger F. Risk factors for highly pathogenic avian influenza (HPAI) H5N1 infection in backyard chicken farms, Thailand. Acta Trop. 2011;118(3):209–16.

Peiris JSM, de Jong MD, Guan Y. Avian influenza virus (H5N1): a threat to human health. Clin Microbiol Rev. 2007;20(2):243–67.

Perez DR, Webby RJ, Webster RG. Land-based birds as potential disseminators of avian/mammalian reassortant influenza A viruses. Avian Dis. 2003;47:1114–7.

Perk S, Golender N, Banet C, Shihmanter E, Pirak M, Tendler Y, Lipkind M, Panish A. Phylogenetic analysis of hemagglutinin, neuraminidase, and nucleoprotein genes of H9N2 avian influenza viruses isolated in Israel during the 2000–2005 epizootic. Comp Immunol Microbiol Infect Dis. 2009;32:221–38.

Phipps KL, Marshall N, Tao H, Danzy S, Onuoha N, Steel J, Lowen AC. Seasonal H3N2 and 2009 pandemic H1N1 influenza A viruses re-assort efficiently but produce attenuated progeny. J Virol; 2017. https://doi.org/10.1128/JVI.00830-17.

Poovorawan Y, Pyungporn S, Prachayangprecha S, Makkoch J. Global alert to avian influenza virus infection: from H5N1 to H7N9. Pathog Glob Health. 2013;107(5):217–23.

Reid AH, et al. Origin and evolution of the 1918 'Spanish' influenza virus hemagglutinin gene. Proc Natl Acad Sci USA. 1999;96:1651–6.

Reid AH, Taubenberger JK, Fanning TG. Evidence of an absence: the genetic origins of the 1918 pandemic influenza virus. Nat Rev Microbiol. 2004;2:909–14.

Sanjuán R, Domingo-Calap P. Mechanisms of viral mutation. Cell Mol Life Sci. 2016;73 (23):4433–48. https://doi.org/10.1007/s00018-016-2299-6.

Shortridge KF, Stuart-Harris CH. An influenza epicentre. Lancet. 1982;2:812–3.

Slomka MJ, To TL, Tong HH, Coward VJ, Mawhinney IC, Banks J, Brown IH. Evaluation of lateral flow devices for identification of infected poultry by testing swab and feather specimens during H5N1 highly pathogenic avian influenza outbreaks in Vietnam. Influenza Other Respir Viruses. 2012;6(5):318–27.

Smith GJD, Fan XH, Wang J, Li KS, Qin K, Zhang JX, Vijaykrishna D, Cheung CL, Huang K, Rayner JM, et al. Emergence and predominance of an H5N1 influenza variant in China. Proc Natl Acad Sci USA. 2006;103(45):16936–41.

Smith GJD, Vijaykrishna D, Bahl J, Lycett SJ, Worobey M, Pybus OG, Ma SK, Cheung CL, Raghwani J, Bhatt S, et al. Origins and evolutionary genomics of the 2009 swine-origin H1N1 influenza A epidemic. Nature. 2009;459(7250):1122–5.

Snoeck CJ, Adeyanju AT, De Landtsheer S, Ottosson U, Manu S, Hagemeijer W, Mundkur T, Muller CP. Reassortant low-pathogenic avian influenza H5N2 viruses in African wild birds. J Virol. 2011;92:1172–83.

Stevens KB, Costard S, Métras R, Theuri W, Hendrickx S, Pfeiffer DU. Risk mapping for HPAI H5N1 in Africa: improving surveillance for virulent bird flu. Final report and risk maps. Under early detection, reporting and surveillance—avian influenza in Africa project. USAID; 2010. p. 48.
Sturm-Ramirez KM, Hulse-Post DJ, Govorkova EA, Humberd J, Seiler P, Puthavathana P, et al. Are ducks contributing to the endemicity of highly pathogenic H5N1 infl uenza virus in Asia? J Virol. 2005;79:11269–79.
Suarez DL, Das A, Ellis E. Review of rapid molecular diagnostic tools for avian influenza virus. Avian Dis. 2007;51:201–8.
Swayne DE. Chapter 6: Avian influenza. In: Diseases of poultry. 13th ed. Ames, IA: Wiley.
Swayne DE, Suarez DL. Highly pathogenic avian influenza. Rev Sci Tech. 2000;19(2):463–82.
Swayne DE, Pavade G, Hamilton K, Vallat B, Miyagishima K. Assessment of national strategies for control of high-pathogenicity avian influenza and low-pathogenicity notifiable avian influenza, with emphasis on vaccines and vaccination. Rev Sci Tech. 2011;30(3):83970.
Tabrah FL. Koch's postulates, carnivorous cows, and tuberculosis today. Hawaii Med J. 2011;70 (7):144–8.
Tassoni L, Fusaro A, Milani A, Lemey P, Adongo Awuni J, Sedor VB, et al. Genetically different highly pathogenic avian influenza A(H5N1) viruses in West Africa, 2015. Emerg Infect Dis. 2016;22(12):2132–6. https://doi.org/10.3201/eid2212.160578.
Taylor LH, Latham SM, Woolhouse MEJ. Risk factors for human disease emergence. Philos Trans R Soc Lond B Biol Sci. 2001;356(1411):983–9. https://doi.org/10.1098/rstb.2001.0888.
Teifke JP, Klopfleisch R, Globig A, Starick E, Hoffmann B, Wolf PU, Beer M, Mettenleiter TC, Harder T. Pathology of natural infections by H5N1 highly pathogenic avian influenza virus in mute (*Cygnus olor*) and whooper (*Cygnus cygnus*) swans. Vet Pathol. 2007;44(2):137–43.
Tong S, Zhu X, Li Y, Shi M, Zhang J, et al. New world bats harbor diverse influenza A viruses. PLoS Pathog. 2013;9(10):e1003657. https://doi.org/10.1371/journal.ppat.1003657.
Trock SC, Gaeta M, Gonzalez A, Pederson JC, Senne DA. Evaluation of routine depopulation, cleaning, and disinfection procedures in the live-bird markets, New York. Avian Dis. 2008;52 (1):160–2.
United Nations Development Program – Abuja. Socio-economic impact of avian influenza in Nigeria; 2006 Jul. Available from http://www.un-nigeria.org/docs/socioecon_ai.pdf. Accessed 24 Sept 2017.
Vakuru CT, Manu SA, Ahmed GI., Junaidu K, Newman S, Nyager J, Iwar, VN, Mshelbwala, GM, Joannis, T, Maina, JA, Apeverga, PT. Situation-based survey of avian influenza viruses in possible "bridge" species of wild and domestic birds in Nigeria. Influenza Res Treat. 2012; 2012:567601. Available from http://dx.doi/org/10.1155/2012/56601.
von Dobschuetz S, de Nardi M, Harris KA, Munoz O, Breed AC, Wieland B, Dauphin G, Lubroth J, KDC S, FLURISK Consortium. Influenza surveillance in animals: what is our capacity to detect emerging influenza viruses with zoonotic potential? Epidemiol Infect. 2015;143(10):2187–204.
Webster RG, Bean WJ, Gorman OT, Chambers TM, Kawaoka Y. Evolution and ecology of influenza A viruses. Microbiol Rev. 1992;56(1):152–79.
WHO. Manual on animal influenza diagnosis and surveillance. WHO/CDS/CSR/NCS/2002.5 Rev. 1; 2002. http://www.who.int/csr/resources/publications/influenza/whocdscsrncs20025rev.pdf. Accessed 13 Sept 2017.
WHO. Healthy marketplaces in the Western Pacific Region: guiding future action, applying a settings approach to the promotion of health in marketplaces. Geneva: World Health Organization; 2004. [Online]. Available from http://www.wpro.who.int/publications/pub_9290611707/en/. Accessed 26 Sept 2017.
WHO. Review of latest available evidence on potential transmission of avian influenza (H5N1) through water and sewage and ways to reduce the risks to human health. Water, Sanitation and Health. WHO/SDE/WSH/06.1; 2007. Available from http://apps.who.int/iris/bitstream/10665/204275/1/WHO_SDE_WSH_06.1_eng.pdf. Accessed 26 Sept 2017.

WHO. Cumulative number of confirmed human cases for avian influenza A(H5N1) reported to WHO, 2003–2017; 2017. www.who.int/entity/influenza/human_animal.../2017_05_16_tableH5N1.pdf?ua=1. Accessed 29 Sept 2017.
Williams RAJ, Peterson AT. Ecology and geography of avian influenza (HPAI H5N1) transmission in the Middle East and northeastern Africa. Int J Health Geogr. 2009;8:47. https://doi.org/10.1186/1476-072X-8-47.
Xu X, Subbarao EK, cox NJ, et al. Genetic characterization of the pathogenic influenza A/Goose/Guangdong/1/96 (H5N1) virus: similarity of its haemagglutinin gene to those of H5N1 viruses from the 1997 outbreaks in Hong Kong. Virology. 1999;261:15–9.
Yoon H, Park CK, Nam HM, Wee SH. Virus spread pattern within infected chicken farms using regression model: the 2003-2004 HPAI epidemic in the Republic of Korea. J Vet Med. 2005;52(10):428–31.
Zecchin B, Minoungou G, Fusaro A, Moctar S, Ouedraogo-Kaboré A, Schivo A, et al. Influenza A (H9N2) virus, Burkina Faso. Emerg Infect Dis. 2017;23(12):2118–9. https://doi.org/10.3201/eid2312.171294.

Chapter 18
Newcastle Disease

Giovanni Cattoli and William G. Dundon

Abstract Newcastle disease is a devastating and economically important disease caused by virulent avian paramyxovirus type 1 viruses, commonly known as Newcastle disease virus (NDV), which affects many domestic and wild avian species globally. Despite the availability of cheap and effective vaccines (if correctly administered), the disease continues to threaten the livelihoods and food security of millions of people worldwide. This chapter concentrates specifically on the description and circulation of NDV and its impact on poultry in the Arabian peninsula and the Northern, Sahel and sub-Sahel regions of Africa. Topics discussed include a brief history of the disease and its economic impact, disease description (e.g., clinical signs and lesions), distribution of viral genotypes within the regions discussed, epidemiology (e.g., seasonality, host range and transmission), disease diagnosis, control, and vaccination. Concluding remarks on the current and future challenges for low-income countries in the control of this important disease are provided.

Keywords Newcastle disease · Avian paramyxovirus type 1 · NDV genotype · NDV vaccine

Disease Description and History

Etiology and Disease Description

Newcastle disease (ND) is a contagious disease of poultry caused by virulent avian paramyxovirus type 1 (APMV-1) viruses, which belong to the family *Paramyxoviridae*

G. Cattoli (✉) · W. G. Dundon
Animal Production and Health Laboratory, Joint FAO/IAEA Division of Nuclear Techniques in Food and Agriculture, Department of Nuclear Sciences and Applications, International Atomic Energy Agency, Seibersdorf, Austria
e-mail: g.cattoli@iaea.org; w.dundon@iaea.org

M. Kardjadj et al. (eds.), *Transboundary Animal Diseases in Sahelian Africa and Connected Regions*, https://doi.org/10.1007/978-3-030-25385-1_18

in the *Avulavirus* genus (ICTV 2009).[1] The virus has a negative, single-stranded RNA genome of approximately 15 kb composed of 6 genes encoding for 6 structural proteins: fusion (F), nucleoprotein (NP), matrix (M), phosphoprotein (P), RNA polymerase (L), and hemagglutinin-neuraminidase (HN). Proteins V and W are encoded by RNA editing of the P protein.

One of the first attempts to classify the different types of APMV-1 viruses circulating in birds was based on the determination of viral virulence using chicken embryos. This resulted in classification of APMV-1 into three categories of viruses referred to as lentogenic, mesogenic, and velogenic depending on the rapidity by which they caused embryo mortality (i.e., >90 h; 60–90 h; <60 h, respectively) (Alexander 2008). This categorization has also been used to describe the capacity of a given virus to induce clinical disease and mortality in chickens in the field. APMV-1 can also be referred to as of low-, moderate-, or highly virulent for poultry. Under experimental conditions, a pathogenicity index for an APMV-1 isolated in the laboratory can be calculated in day-old chicks to assess the virulence of the virus. Internationally recognized and described in the *Manual of Diagnostic Tests and Vaccines for Terrestrial Animals* (OIE 2017), the intracerebral pathogenicity index (ICPI) can provide a score from 0.0 (avirulent) to a maximum value of 2.0 (virulent). According to OIE, APMV-1 viruses with an ICPI score of 0.7 or greater are considered as Newcastle disease viruses and, as such, should be officially notified (OIE 2017). Thus, all the ND viruses (NDV) are APMV-1 but not all the APMV-1 viruses are NDV.

The most important molecular mechanism controlling the virulence of APMV-1—although probably not the only one—relies on the cleavability of the viral F protein (de Leeuw et al. 2005). The F protein is responsible for the fusion between the virus and the host cell membrane and is essential for viral entry and replication in host cells. Similar to the hemagglutinin glycoprotein of type A influenza viruses, the F protein is produced as a precursor F_0 that must be cleaved by host cellular proteases in order to be activated. The virulence of the virus is directly related to the presence of multiple basic amino acids (Arginine—R and Lysine—K) at the F_0 cleavage site which allow for the proteolytic cleavage of the F protein by proteases present in a wide range of tissues and organs, thereby resulting in the infection of multiple organs. In contrast, viruses of low or moderate virulence for poultry do not possess multiple basic amino acids at the F_0 cleavage site and the fusion protein can only be activated by trypsin-like enzymes. This restricts the infection to tissues where these trypsin-like enzymes are present, such as the digestive or respiratory tracts. It is therefore possible to infer the virulence of APMV-1s based on the nucleotide sequence and the deduced amino acid sequence of the segment encompassing the cleavage site of the F. This virulence criterion has been

[1]Currently, ICTV reported three genera, named Orthoavulavirus, Metaavulavirus, and Paraavulavirus, within a new subfamily Avulavirinae of the family Paramyxoviridae (ICTV 2019). Based on the last report, APMV-1 viruses (and ND viruses) are identified in genus Avian orthoavulavirus 1 (AOAV-1). In this chapter, the commonly known denomination ND and APMV-1 is maintained.

internationally recognized and is used in parallel or as an alternative to the in vivo tests (ICPI) to define and notify ND outbreaks (OIE 2017).

Antigenically, APMV-1 viruses are considered homogeneous and a single serotype is described. Genetically, these viruses present great variability and a number of classification systems based on the nucleotide sequence of the gene encoding for the F protein have been developed (Ballagi-Pordány et al. 1996; Aldous et al. 2003; Czeglédi et al. 2006; Diel et al. 2012). According to the guidelines of the most recent classification and nomenclature proposed by Diel et al. (2012), two classes—Class I and Class II—and 19 distinct genotypes can be described (Diel et al. 2012; Dimitrov et al. 2016a).

Class I APMV-1 consists of a single genotype in which avirulent strains isolated from wild birds (mainly) and poultry are grouped. To date, a single exception has been reported in the scientific literature concerning one outbreak in Ireland caused by a virulent Class I APMV-1 (Alexander et al. 1992). Class II APMV-1 consists of virulent and avirulent viruses isolated globally from a large variety of bird species, both wild and domestic. At the time of writing this chapter,[2] 18 distinct genotypes (I–XVIII) and several subgenotypes have been described in Class II. A comprehensive overview on the currently described APMV-1 genotypes has been recently published (Dimitrov et al. 2016a).

History

Although the first clinical reports of a disease affecting poultry with clinical signs resembling those of ND date back to the end of the nineteenth/beginning of the twentieth century in Scotland and South Africa (cited in Abolnik 2017), the first outbreaks ever reported of confirmed ND occurred in Java, Indonesia, in 1926 and in Newcastle upon Tyne, England, in 1927 (Alexander 2009). In Africa, the earliest official diagnoses of ND were reported in 1945 and 1946 in South Africa and Mozambique, although it is likely that the virus was already circulating in the 1930s (Mapaco et al. 2016; Abolnik 2017). A few years later in 1952, Nigeria made its first official detection (Shittu et al. 2016a). Almost in the same period the disease was reported in Sudan and Uganda in 1951 and 1955, respectively (George 1992; Khalafalla et al. 1992). In Egypt, ND was first officially identified in 1948 (Orabi et al. 2017). At that time, the virus was most likely also circulating in other North African countries as indicated by publications on ND investigations and vaccination in Tunisia (Cordier et al. 1950a, b). Subsequently, prevalence studies conducted in unvaccinated chickens in Sahelian countries such as Mauritania and Niger (1990) and countries in other regions such as Morocco (1992), Benin, and Cameroon (1992) indicated the extensive circulation of ND virus (NDV) (reviewed in Awan et al. 1994).

[2]Currently, 2019, genetic classification was reviewed by a group of experts and at least 20 Class II genotypes have been identified (Dimitrov et al. 2019).

Economic Impact

ND is one of the most important infectious diseases of poultry, causing high mortality in chickens and other gallinaceous species and huge financial losses to the poultry economies (both industrial and backyard) worldwide. In sub-Saharan Africa, it is well recognized that backyard chickens represent an extremely important source of protein for people and contribute to the nutrition and income of the rural poor, particularly women and children (Sambo et al. 2015; Aboe et al. 2006; IAEA 2002; Tadelle et al. 2003). For example, in Uganda, 80% of poultry production is based on the traditional free-range village system, 74% in Zambia, and up to 90% in Malawi. In Ethiopia almost 99% of the national egg and poultry meat production is from backyard chickens (Tadelle et al. 2003). Considering that poultry is one of the main sources of animal protein in developed and developing countries, ND epidemics can have serious consequences on food security, particularly in rural economies of developing countries in Africa and Asia. In these areas of the world, ND is identified as the major single constraint to rural poultry development (Awan et al. 1994) and poultry industry. In village chickens, NDV alone can be responsible for nearly 80% of poultry mortalities in some African countries (Miguel et al. 2013). A study conducted in Nigeria, probably the largest poultry economy in sub-Saharan Africa, estimated total chicken deaths attributed to ND to be approximately 25.5 million heads with a reduction of 26.5 million units in egg production. The financial burden of ND in Nigeria is estimated in 8.9 billion Naira (approximately 25 million USD) (Fadiga et al. 2013).

Geographical Distribution, Description, and Distribution of Newcastle Disease Virus Genotypes

Geographical Distribution

Outbreaks of ND have been reported worldwide and currently only a few, often geographically isolated countries have never experienced the disease (http://www.oie.int). To date, ND can be considered endemic in Africa, likely affecting poultry in all of the African countries with a few exceptions (i.e., Seychelles). NDV is also circulating in the Arabian peninsula as indicated by reports from Saudi Arabia, Qatar, and Oman (Abu Elzein et al. 1999; Haroun et al. 2015; Al Shekaili et al. 2015) and the OIE notifications in the recent past from Bahrain, Kuwait, Qatar, Oman, and Saudi Arabia (http://www.oie.int/wahis_2/public/wahid.php/Diseaseinformation/statusdetail). A study in Africa identified areas characterized by low altitudes, wet forest biomes, and high poultry and human density as the most favorable for ND maintenance and spread in rural poultry (Miguel et al. 2013). According to AU-IBAR (http://www.au-ibar.org/newcastle-disease), ND was reported by 31 African Union countries in 2011. Based on a recent consultation of the OIE website (http://www.oie.int/wahis_2/public/wahid.

php/Diseaseinformation/statuslist, accessed on 09/26/2017), 36 African countries suspected or clinically reported the disease. However, based on one investigation conducted among African Union (AU) Member States between 2000 and 2011, it appeared that about 60% of AU countries did not submit "ND outbreak reports consistently enough to be able to conduct meaningful data analyses" to OIE (Gardner and Alders 2014). It is worth noting that clinical disease caused by NDV can be indistinguishable from highly pathogenic avian influenza (HPAI) in the field. Thus, it would be expected that the increased poultry surveillance and testing triggered by the HPAI incursions in Africa over the past decade would have been reflected in an increased official reporting of ND, at least in the HPAI-infected countries. However, this appears not to be the case: comparing the OIE notifications for ND before and after the HPAI incursions in 2006 in AU countries, only Burkina Faso demonstrated a significant increase in outbreak reporting (Gardner and Alders 2014). On the other hand, it seems that the increased HPAI surveillance implemented in many countries of the African continent has indeed brought new data to the scientific community on the NDV circulation in Africa, with several new virus genotypes being discovered and characterized (see Paragraph "Description and distribution of NDV genotypes"). Taking together, these data and observations suggest that ND in Africa is most likely under-surveyed and, even more likely, underreported.

Description and Distribution of Newcastle Disease Virus Genotypes

During the last decade, several studies have reported on the circulation of multiple and, sometimes novel, APMV-1 genotypes in Africa. In this chapter, a description of the APMV-1 genotypes identified to date and the information on the distribution of each genotype in the African continent and Arabian peninsula are provided. The genotype nomenclature is based on the publications of Diel et al. (2012) and Dimitrov et al. (2016a).

In the Africa and the Arabian peninsula, only Class II ND viruses have been reported to date. Of the 18 Class II genotypes described globally, 13 have been reported in Africa (i.e., genotypes I–VIII, XI, XIII, XIV, XVII, and XVIII). There are no recent reports of genotype VIII at a global level and it is suspected that they no longer circulate in poultry. The situation is similar for genotype IV, although a recent report from Morocco appears to indicate its presence in the region (El Kanthour et al. 2017). Viruses of genotype VI, an antigenic and host variant of APMV-1 that has been reported to infect pigeons and doves in many countries, have also been reported in Africa and in the Arabian peninsula. Similarly, virulent viruses of genotype VII, responsible for several epidemics in Eurasia, and genotype XIII are believed to be the dominant virulent genotype in poultry in Northern, Southern, and Eastern African countries. Interestingly, some of the most recently discovered genotypes are restricted to certain regions of Africa only. This is the case for the virulent viruses of genotype XI, described in Madagascar only (Maminiaina et al. 2010), and virulent genotypes XIV, XVII, and XVIII. The latter three genotypes are the dominant genotypes responsible for the recent poultry outbreaks in West and Central African countries (Cattoli et al. 2010; de Almeida et al. 2013; Snoeck et al. 2013a).

Despite NDV is extensively circulating in Northern Africa and the Arabian peninsula, there is a paucity of information on the genotypes affecting these regions. The dominant ND genotypes in Egypt, Libya, Tunisia, and the Arabian peninsula is probably genotype VII, as indicated in recent publications from Egypt, Libya, and Qatar and F gene sequence submissions to public databases from Tunisia (Mohamed et al. 2011; Radwan et al. 2013; Kammon et al. 2015; Saad et al. 2017; Haroun et al. 2015). In fact, a personal communication from Dr. A Ghram (Pasteur Institute, Tunis, Tunisia) has confirmed the presence of genotype VII in Tunisia. Genotype IV was reported in Morocco (El Khantour et al. 2017) and genotypes II and VI in Egypt (Radwan et al. 2013; Mansour et al. 2017). A table summarizing the geographical distribution of genotypes is presented (Table 18.1). More details for each of the genotypes are provided below and in Figs. 18.1 and 18.2.

Table 18.1 NDV Class II genotypes reported in the Sahel, North Africa, the Arabian peninsula, and sub-Sahel regions

Region	Genotype	Country
Sahel	II	Burkina Faso
	IV	Sudan
	VI	Sudan
	VII	Sudan
	XIV	Burkina Faso, Mali, Mauritania, Niger
	XVII	Burkina Faso, Niger
	XVIII	Mali, Mauritania
North Africa	II	Egypt
	IV	Morocco
	VI	Egypt
	VII	Egypt, Libya, Tunisia
Arabian peninsula	VI	UAE, Saudi Arabia, Kuwait
	VII	UAE, Saudi Arabia, Qatar
Sub Sahel	I	Cameroon, Nigeria
	II	Cameroon, Ethiopia, Nigeria
	III	Nigeria
	IV	Nigeria
	VI	Ethiopia, Nigeria
	XIV	Benin, Cameroon, Cote d'Ivoire, Nigeria
	XVII	Benin, Cameroon, Central African Republic
	XVIII	Cote d'Ivoire, Nigeria, Togo

For details on each genotype and its distribution, refer to Sect. "Description and Distribution of Newcastle Disease Virus Genotypes"

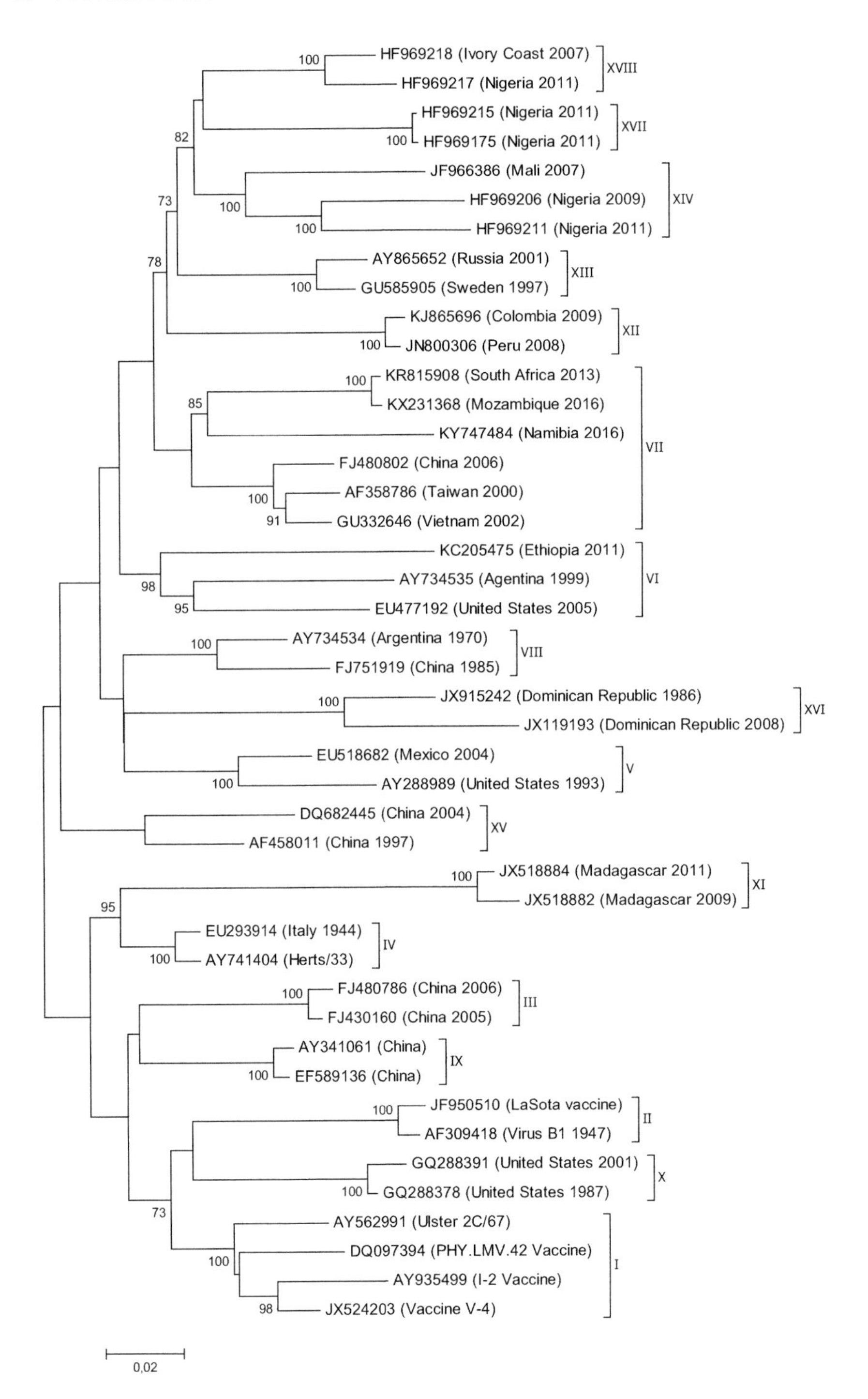

Fig. 18.1 Maximum likelihood analysis using the MEGA6 software of the full F gene nucleotide sequence (1662 bp) of representatives of the eighteen (I–XVIII) of Class II APMV-1s. The numbers

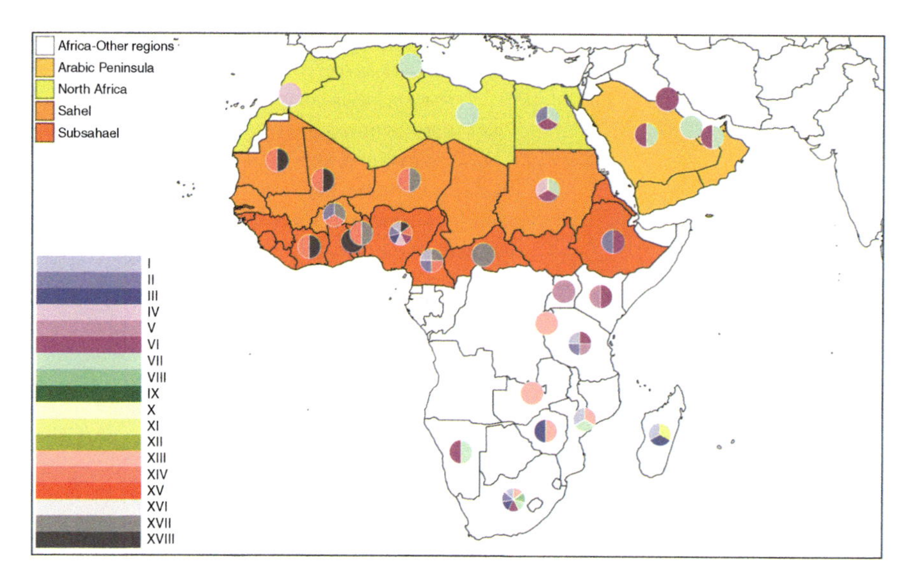

Fig. 18.2 Distribution of Class II NDV genotypes in Africa and the Arabian peninsula. Colors of countries represent the distinct regions targeted in this chapter. The legend in the bottom-left corner of the map indicates the colors assigned to the NDV genotypes (from genotype I–XVIII). Colors in the circles indicate the past or current presence of one or multiple genotypes in a given country

Class I This class consists of a single genotype and its viruses are circulating mainly in the wild bird populations in Eurasia and America. To date, there are no reports of these Class I viruses in the African continent. This might be explained by the fact that the surveillance undertaken in wild bird species in Africa uses molecular techniques (e.g., PCR) which are not capable of efficiently detecting the genome of Class I viruses (Cappelle et al. 2015). All but one of the viruses included in this class are considered avirulent for poultry (Aldous et al. 2003).

Class II Genotype I This genotype contains some virulent but mainly avirulent viruses identified throughout the world (Dimitrov et al. 2016a). Viruses of genotype I include the Queensland V-4, Ulster 2C/67, and I-2 vaccine strains, all of which are widely used in vaccine production and application in Africa. Isolates with F gene sequences almost identical to the V-4/I-2 strains were circulating in South Africa and Tanzania in the 1990s, Cameroon in 2007 and 2011, and Mozambique and Nigeria in 2005 (Abolnik et al. 2004; Yongolo et al. 2011; Snoeck et al. 2009, 2013a, b; Fringe et al. 2012). Genetically similar viruses are likely to be circulating in many other African countries given the widespread use of these genotype I live vaccines on the continent. Genotype I lentogenic viruses have been reported in healthy wild birds in

---

Fig. 18.1 (continued) indicate the bootstrap values calculated from 500 bootstrap replicates. The different genotypes are numbered according to Diel et al. (2012)

Madagascar, while genotype I viruses not related to the vaccine strains were detected in spur-winged geese (*Plectropterus gambensis*) in Nigeria (Snoeck et al. 2013a).

Class II Genotype II This genotype includes virulent, such as the Texas GB/48 strain, and avirulent viruses which cause no, or mild, respiratory disease in chickens. The genotype includes some of the most common vaccine strains used for the production of inactivated and live vaccines, namely Beaudette/45, La Sota/46, and B1/47 strains. Mesogenic NDVs of this genotype (i.e., Komarov or Roakin strain) are also used as vaccine strains in areas where virulent NDV is endemic, particularly in some countries of Asia and Africa. Viruses related to genotype II vaccine strains have been identified in Burkina Faso, Cameroon, Egypt, Ethiopia, Nigeria, Tanzania, and South Africa (Abolnik et al. 2004; Snoeck et al. 2009, 2013a; Yongolo et al. 2011; Fentie et al. 2014a, b; Radwan et al. 2013; Saad et al. 2017). As stated previously for genotype I, genetically similar viruses are likely to be circulating in many other countries of the region given the widespread use of these live vaccines on the continent. Interestingly, virulent genotype II viruses responsible for outbreaks in chickens were reported in Egypt in 2005–2006 (Mohamed et al. 2011; Radwan et al. 2013). These viruses were almost identical to historical strains isolated more than 60 years ago in the USA (chicken/USA/TX/GB/1948 and chicken/USA/New Jersey-Roakin/1946) (Dimitrov et al. 2016b).

Class II Genotype III Genotype III includes the mesogenic vaccine strain Mukteswar/chicken/India/1940, widely used in Southeast Asia and suspected to be the progenitor of the recent genotype III viruses circulating in the field. Virulent strains isolated in several areas of the world are also included in this genotype, i.e., isolates from Australia (1932) and Japan (Miyadera/51). In Africa, this genotype was reported in Zimbabwe and South Africa some decades ago (Bwala et al. 2009; Dimitrov et al. 2016a). More recently, a virus isolated from a chicken in Nigeria in 2006 revealed a genetic sequence almost identical (3 nt difference) to strain Mukteswar (Snoeck et al. 2009). De Almeida et al. (2009) reported the circulation of genotype III in Madagascar at the time of the first ND epidemics in the country, just after the Second World War.

Class II Genotype IV Virulent ND viruses isolated decades ago (1933–1989) in chickens and quails in Europe, Asia, and Africa are included in this genotype. The UK strain Herts/33, frequently used as a reference strain for challenge studies, is also in genotype IV. Reports have indicated the presence of genotype IV in Nigeria (1973 and 1980, Shittu et al. 2016a) and in Sudan in the past (1991, Dimitrov et al. 2016a). Some authors have hypothesized that this genotype is no longer circulating in poultry (Dimitrov et al. 2016a). However, a report from India demonstrated the circulation of this genotype in chickens and free-range pigeons in the southern part of India (Tirumurugaan et al. 2011). Also of note is one recent paper from Morocco which suggests the presence of this genotype in the country (El Khantour et al. 2017). Interestingly, the genotype XI circulating in poultry in Madagascar is thought to have derived from an ancestor close to genotype IV introduced onto the island in the 1950s and is possibly the result of a self-contained evolution due to the geographical and ecological characteristics of this island (Maminiaina et al. 2010).

Class II Genotype V Virulent viruses belonging to this genotype were responsible for the second ND panzootic that likely originated in Asia in the 1960s. The genotype V virus introduction into Europe and North America in the 1970s was associated with the importation of infected psittacine birds (Ballagi-Pordány et al. 1996). This genotype also contains one isolate from a human, isolated from a poultry worker in 1973 (Aldous et al. 2003). Virulent viruses of this genotype have been detected in wild birds in Canada and the USA since 1995 (i.e., cormorants, pelicans, gulls) (Dimitrov et al. 2016a). In sub-Saharan Africa, genotype V caused several outbreaks in free-range poultry in Tanzania in the 1990s (Yongolo et al. 2011), and it was also reported in poultry in live birds markets across Uganda and in Kenya in 2010 and 2011 (Byarugaba et al. 2014). To our knowledge, no further outbreaks in poultry caused by this genotype were reported in the region in the last decade.

Class II Genotype VI Viruses belonging to this genotype are distributed worldwide and based on the molecular marker for pathogenicity; they are all virulent for poultry. Phenotypically, though, the genotype VI includes velogenic and mesogenic strains. Genetically they are quite variable and ten distinct subgenotypes have been described to date (VIa–VIj, Ganar et al. 2017; Sabra et al. 2017). Genotype VI viruses are predominantly detected in bird species of the family *Columbidae*, such as pigeons and doves throughout the world. Genotype VI viruses responsible for the pigeon panzootic that reached Europe in 1981 (also known as the third ND panzootic) demonstrated distinct antigenic variation when typed with monoclonal antibodies (Alexander 2008). For this reason, strains included in this genotype are also named as pigeon paramyxoviruses type 1 (PPMV-1). PPMV-1 was reported in pigeons and doves in South Africa in 1980 and reemerged in 2002, when it was isolated in layer chickens suffering with respiratory disease (Abolnik et al. 2004). An outbreak of ND in pigeons with high mortality and morbidity, likely caused by PPMV-1, also occurred in Sudan in 1982 (Eisa and Omer 1984). This genotype is probably extensively circulating in the pigeon/dove population of Africa and the Arabian peninsula and reports throughout the last two decades revealed its presence in Kuwait, Saudi Arabia, and the United Arab Emirates (Lomniczi et al. 1998; Abu Elzein et al. 1999; Aldous et al. 2003, 2004); South Africa (Abolnik et al. 2008); and Tanzania (Yongolo et al. 2011), Egypt (Mansour et al. 2017; Sabra et al. 2017), Ethiopia (de Almeida et al. 2013; Fentie et al. 2014a, b), Namibia (Molini et al. 2018), Nigeria, and Kenya (Van Borm et al. 2012; Snoeck et al. 2013b). The detection of this genotype in feral and domestic birds indicates that the circulation of PPMV-1 in the "pigeon" reservoir poses a continuous threat for poultry and endangered wild bird species (Abolnik et al. 2008).

Class II Genotype VII Genotype VII represents a large group of viruses virulent for chickens and responsible for major epizootics of ND in poultry over the last two decades in Asia and Europe. These viruses are also circulating in South America and Africa. Genetically, it is a rather heterogeneous group and several subgenotypes—VIIa to VIIk—have been described to date (Dimitrov et al. 2016a; Molini et al. 2017). Viruses belonging to genotype VII emerged in the early 1990s in East Asia and Europe where they became the dominant virulent viruses circulating in poultry (Lomniczi et al. 1998). Around the same time (1991–1993), genotype VII viruses caused outbreaks in poultry in South Africa and Mozambique (Herczeg et al. 1999; Abolnik 2017). This genotype is still causing huge poultry losses worldwide; in Africa it is actively circulating mainly in the southern part of the continent, as recently demonstrated in Mozambique (Mapaco et al. 2016), South Africa (Abolnik 2017; Abolnik et al. 2018), and Namibia (Molini et al. 2017). Viruses of genotype VII were also responsible for ND outbreaks in Sudan in 2003–2006 (Hassan et al. 2010), Libya in 2013 (Kammon et al. 2015), and Egypt in 2011 to present (Radwan et al. 2013; Saad et al. 2017). There is also evidence of genotype VII viruses in Tunisia in 2013 from GenBank submissions KX228393 and KU175357 and personal communications (A. Ghram). Viruses isolated in ostriches and poultry in the United Arab Emirates in 1999 and 2014, Qatar in 2008, and Saudi Arabia in 2000 also belonged to this genotype (Aldous et al. 2003; Haroun et al. 2015; Fuller et al. 2017).

Class II Genotype VIII All the viruses belonging to this genotype are classified as virulent, based on the cleavage site of the F gene. This group of viruses no longer appears to be circulating in poultry; however, in the past (from 1960 to 2000) it was detected in Latin America, Asia, Europe, and South Africa where no genotype VIII viruses have been isolated since 2000 (Dimitrov et al. 2016a; Abolnik 2017).

Class II Genotype IX Based on the amino acid sequence of the F gene cleavage site, genotype IX viruses are classified as virulent. They have been mainly isolated from poultry and apparently healthy wild birds in China between 1985 and 2011 (Dimitrov et al. 2016a). This genotype has not been reported in Africa and the Arabian peninsula to date.

Class II Genotype X At the time of writing, this genotype has not been reported in Africa and in the Arabian peninsula. It appears to be restricted to North and South America where genotype X viruses have been isolated mainly in wild birds. Viruses in this genotype are classified as avirulent based on the amino acid sequence of the F gene cleavage site (Dimitrov et al. 2016a).

Class II Genotype XI This genotype appears to be geographically restricted to Madagascar. The genotype XI viruses described to date are all virulent based on the amino acid sequence of the F gene cleavage site and were detected both in poultry and in wild birds (Maminiaina et al. 2010; Cappelle et al. 2015). Viruses belonging to this genotype likely share a common ancestor with genotype IV viruses (see above).

Class II Genotype XII Viruses belonging to this genotype are all virulent and, to date, they have been detected in poultry in Latin America and China. There are no reports of this genotype in Africa (Dimitrov et al. 2016a).

Class II Genotype XIII Viruses in this genotype are all classified as virulent based on the amino acid sequence of the F gene cleavage site. To date, they have been isolated in Bangladesh, Russia, Sweden, India, Iran, and Pakistan (Diel et al. 2012; de Almeida et al. 2013). According to the new classification proposed by Diel et al.

(2012), recent studies indicated the presence of this genotype in Africa too, namely in Burundi, Mozambique, South Africa, Zambia, and Zimbabwe (Snoeck et al. 2013a; Abolnik et al. 2017).

Class II Genotype XIV The viruses of this genotype are all virulent strains based on the amino acid sequence of the F gene cleavage site and ICPI tests performed on some isolated strains (Cattoli et al. 2010; de Almeida et al. 2013). Interestingly, this genotype appears to be among the dominant genotypes in West Africa: genotype XIV viruses have been reported in poultry in Benin, Burkina Faso, Cameroon, Côte d'Ivoire, Mauritania, Niger, and Nigeria (Cattoli et al. 2010; de Almeida et al. 2013; Snoeck et al. 2013a). In Mali and Mauritania, genotype XIV has also been detected in wild bird species, suggesting a potential epidemiological role of wild birds in the maintenance and local spread of ND (Cappelle et al. 2015).

Class II Genotype XV This genotype includes viruses identified in poultry in China between 1997 and 2004 and which are believed to be recombinant viruses that are not maintained in poultry or wild birds (Diel et al. 2012; Dimitrov et al. 2016a). There are no reports of this genotype in Africa and in the Arabian peninsula.

Class II Genotype XVI All genotype XVI viruses are classified virulent based on the amino acid sequence of the F gene cleavage site. The ICPI test performed on one strain confirmed the velogenic pathotype. Their geographical distribution appears to be restricted to Central America and the Caribbean region (Dimitrov et al. 2016a).

Class II Genotype XVII All of the viruses belonging to this genotype are virulent based on the amino acid sequence of the F gene cleavage site. Similar to genotypes XIV and XVIII, this genotype circulates in West and Central Africa. It has been reported in poultry in Benin, Burkina Faso, Cameroon, Central African Republic Côte d'Ivoire, Niger, and Nigeria (Cattoli et al. 2010; de Almeida et al. 2013; Snoeck et al. 2013a; Shittu et al. 2016b).

Class II Genotype XVIII Similar to genotypes XIV and XVII, genotype XVIII viruses are all classified as virulent and circulate in poultry in West African countries, including Côte d'Ivoire, Mali, Mauritania, Nigeria, and Togo (Snoeck et al. 2013a). Snoeck et al. (2013b) reported this genotype in one wild bird species in Côte d'Ivoire.

Epidemiology

Seasonality Generally speaking, ND infections can occur at any time of the year. Based on the ND reports to the African Union Inter-African Bureau for Animal Resources (AU-IBAR), in Africa there "appears to be no temporal trend for ND occurrence on the continent, suggesting the lack of seasonality for the risk factors that determine occurrence and maintenance of the disease" (http://www.au-ibar.org/newcastle-disease). Nevertheless, several scientific reports from Africa suggest that certain periods of the year might be at higher risk for ND outbreaks in rural poultry.

In the semiarid and arid regions of Cameroon and Chad, two epidemic peaks can be observed: one during the intense heat of March and April and the second in the cold dry months of December and January (Awa and Achukwi 2010). Similarly, two peaks were described in Zambia: during the hot, dry season (September/October) and the hot, humid season (January/March) (Awan et al. 1994). In Mauritania, ND outbreaks apparently increase particularly during the hot season beginning in March (Awan et al. 1994). In Uganda, the hot and dry seasons of the year see the highest incidences of outbreaks in rural poultry (Awan et al. 1994). It appears that the apparent seasonality of ND outbreaks is well known among farmers in Uganda. In fact, "just before the dry season sets in, farmers panic and start selling off their stock. This usually triggers the spread of disease" (George 1992). In southeastern Nigeria, the ND outbreak peak has been reported during the dry, harmattan season (November–February) and another marginal peak during the height of the rainy season (June–July) (Shittu et al. 2016a). A recent meta-analysis on the occurrence of ND in African backyard poultry confirmed that ND epidemics are peaking during the dry season at the continental level (Miguel et al. 2013).

Host Range Virtually, all species of birds can be considered susceptible to infection with APMV-1; a comprehensive literature review published 30 years ago (Kaleta and Baldauf 1988) concluded that at least 241 species from 27 orders of birds are susceptible to natural or experimental NDV infection. All species of commercially reared poultry are susceptible to virulent NDV, although ducks can be more resistant to the disease. A milder disease than that seen in chickens has been reported in some commercially reared species such as ostrich and pheasants, with variation in disease severity depending on the age of the animals infected (Alexander 2000). Reported observations from infected ostrich farms in South Africa indicate that birds susceptible to clinical disease are usually of young age, poor performers, and with concurrent bacterial infections or on unbalanced diet (Verwoerd et al. 1997).

APMV-1 can frequently be detected in wild birds species, mainly viruses of low virulence for chickens belonging to Class I and to genotypes I, II, and VI in Class II. However, virulent NDV can also be isolated from wild bird species where they can cause subclinical infections or clinical disease and mortality. The possibility of virulent NDV infection in wild birds became apparent during outbreaks in Canada and the USA in the 1990s, when mortality was reported in double-crested cormorants (*Phalacrocorax auritus*). On the African continent, both avirulent and virulent APMV-1 were detected in clinically healthy wild birds belonging to 45 different species, i.e., anatids, waders, passerines, rails, ciconiiformes, and gulls (Cappelle et al. 2015). In this monitoring study, performed on samples collected from more than 9000 wild birds in four African countries (Madagascar, Mali, Mauritania, and Zimbabwe), the circulation of NDV occurred all year round with no apparent seasonality. The only exception was in Mali, where a higher infection rate was revealed during the dry season for two consecutive years. Genetically, some of the viruses detected in wild birds were related to those responsible for outbreaks in poultry suggesting a potential role of wild birds in the maintenance of NDV in different regions of Africa and the possibility for these wild species to transmit and spread the disease to poultry (Cappelle et al. 2015). This is particularly apparent for genotype VI viruses, which primarily infect pigeons and doves, but which have been detected in poultry in South Africa (Abolnik et al. 2008) and Ethiopia (Damena et al. 2016) confirming cross-species transmission.

Transmission The most common mechanism for viral transmission is believed to be through the inhalation or ingestion of infectious virus particles and/or contact with mucous membranes, such as conjunctiva. Virus replicating in the respiratory tract can in fact be shed via large or small droplets containing infectious particles. Gastrointestinal infections promote the excretion of the virus through feces and infection can then be transmitted directly or indirectly through the ingestion of contaminated feed and water. Inhalation of contaminated dust can also be a mechanism of transmission. The significance of vertical transmission has been debated for a long time and is still unclear (Alexander 2009). Infection of laying hens by virulent NDV may result in the infection and death of embryos during incubation and the cessation of egg laying by the diseased birds. NDV can penetrate the intact egg shell (as well as cracked and broken shells) once laid while contaminated feces on the shell can also be a source of infection for embryos and hatched chicks (Alexander 2008).

Source and Spread of Newcastle Disease Virus For virulent NDV, clinically diseased domestic birds are unequivocally the main source of infection, although the virus can also be transmitted by subclinically infected birds. This can happen in cases of infection of species that are more resistant to the disease, such as ducks or adult ostriches, or during infections of a vaccinated poultry population. In fact, vaccinated birds can be protected from the clinical disease but still shed infectious viruses after challenge with virulent NDV strains (Dortmans et al. 2012; Fentie et al. 2014b; Dimitrov et al. 2017). On rare occasions, virulent NDV have originated from viruses of low virulence for poultry: this was described for viruses isolated in Ireland in 1990 and Australia in 1998 (Alexander 2009). As there is evidence that virulent NDV can be present in clinically healthy wild birds (see previous section—Host range), these species can also represent a source of infection for poultry. This is very relevant in sub-Saharan Africa where scavenging, free-ranging poultry is predominant. Feces excreted by infected birds may contain a large amount of infectious virus that can survive for prolonged period of time in the environment. Therefore, material, products, and feed contaminated with feces from infected birds can represent a source of infection (Awan et al. 1994). Similarly, carcasses of infected birds can be a source of virus, and it has been observed in rural areas farmers dumping dead animals in the common garbage or in open areas outside the farm (Antipas et al. 2012).

The ND outbreaks that occurred in the UK in 1984 were believed to be spread by feed contaminated by infected feral pigeons (Alexander et al. 1985). In rural areas of Africa, infected and diseased chickens are slaughtered and eaten by farmers. During slaughtering, it is usual practice to throw viscera into the field to feed other animals including poultry, thus contributing to the spread of infection (Awan et al. 1994). Movement of live, infected birds during the incubation period of the disease (or infected vaccinated birds) and contaminated poultry products or feed represent the major mechanism for the secondary spread of the NDV. Live bird markets are hot spots for ND transmission and spread. Farmers try to sell their chickens when they start showing clinical signs of disease or just before the periods considered at higher risk of infection, i.e., at the beginning of the dry season (Awan et al. 1994).

Movement of people and equipment from farm to farm or between farms and markets can represent another method for secondary spread. Airborne spread has been discussed and considered as a potential mechanism for spread in some outbreaks; however, its relevance is questionable particularly in rural poultry systems characterized by a low-density poultry population unable to generate a sufficiently dense aerosol (Awan et al. 1994; Alexander 2009). In addition, contaminated hatcheries can be a source of infection and hot spots for disease spread. NDV can survive in water for a long period of time; thus water surfaces such as village ponds contaminated by infected poultry or wild birds can be a source of infection, particularly for free-ranging, backyard chickens (Awan et al. 1994).

Clinical Signs and Lesions

ND viruses have been historically grouped into distinct pathotypes—velogenic, mesogenic, lentogenic—based on the clinical signs they can induce in chickens (Table 18.2). The velogenic strains are responsible for high and acute mortalities in chicken flocks. Mortality can suddenly appear in the absence of other evident clinical signs, similar to highly pathogenic avian influenza infections. In other cases, signs commonly observed during infection with virulent viruses are weakness, lethargy, prostration, edema around the eyes and the head, respiratory signs, nervous signs, green diarrhea, and a sudden drop in egg production in laying birds. The clinical

Table 18.2 NDV pathotypes and associated clinical features

Pathotype				
	Velogenic			
	Viscerotropic	Neurotropic	Mesogenic	Lentogenic
Pathogenicity test				
MDT	<60	<60	60–90	>90
ICPI	>1.5	>1.5	0.7–1.5	<0.7
*Clinical signs***				
Gastro-enteric signs	+++	–	–	–
Respiratory signs	–	+++	++	+*
Nervous signs	++*_	+++	++*	–
Drop in egg production	+++	+++	++*	+*
Morbidity	+++	+++	++	+*
Mortality	+++ (100%)	+++ (50–100%)	+	+*

Modified from Terregino and Capua (2009a) and Cattoli et al. (2011)
MDT Mean Death Time, *ICPI* IntraCerebral Pathogenicity Index
+++ severe, ++ intermediate, + mild
* Observed in young or immunocompromised chickens, ** clinical signs observed in chickens

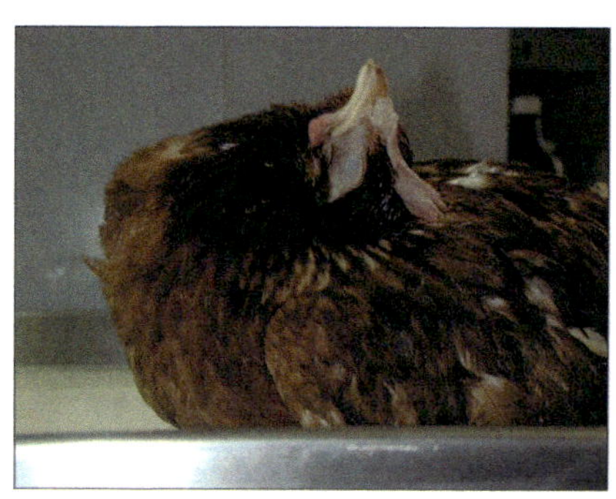

Fig. 18.3 Layer hens naturally infected with velogenic neurotropic NDV exhibiting nervous signs (torticollis and paresis). (Courtesy of C. Terregino, IZSVe, Padova, Italy)

signs may also depend on the strain involved in the infection: in fact, velogenic strains can be either neurotropic (causing predominantly neurologic signs such as head twitch, tremors, paralysis, and opisthotonus; Fig. 18.3) or viscerotropic (causing predominantly gastrointestinal signs such as severe enteritis with diarrhea). In turkeys, some velogenic viruses can cause less severe disease compared to chickens. Turkeys infected with these strains may survive longer but, as a result, also shed viruses for longer periods (Piacenti et al. 2006; Terregino and Capua 2009a). In ostriches, clinical signs are mainly reported in young birds (5–9 months of age) and mostly include nervous signs such as torticollis, tremors, total paralysis, and death in approximately 30% of the infected birds. Pigeons and doves are frequently infected with virulent viruses of genotype VI, the so-called PPMV-1. Young birds may show high morbidity and mortality, with mainly nervous signs. In adult Columbiformes, PPMV-1 can cause less severe disease or even subclinical infections. ND caused by virulent strains has also been described in some game bird species such as partridges and pheasants and the clinical signs are similar to those reported in chickens. In young birds, the mortality can be very high and, in laying birds, a sudden drop in egg production can be seen. Ducks and geese are considered to be more resistant to virulent NDV infections compared to chickens. Nevertheless, mild to severe disease has been reported in these species on rare occasions, with gastro-enteric and nervous signs mainly being observed (Terregino and Capua 2009a).

Diffuse hemorrhages are frequently the only clear lesions observed during necropsy of chickens and other bird species infected with velogenic viscerotropic strains (Fig.18.4). Hemorrhages are often localized in the gastrointestinal tract, including the proventriculum, ceca, and small intestine. Necrotic-hemorrhagic lesions are also frequently observed in the lymphatic intestinal tissue and in the spleen, which appears enlarged and edematous. Edema and hemorrhages can be observed in the ovaries in addition to yolk peritonitis and misshaped eggs. Velogenic neurotropic viruses cause minimal gross lesions (Terregino and Capua 2009a; Cattoli et al. 2011).

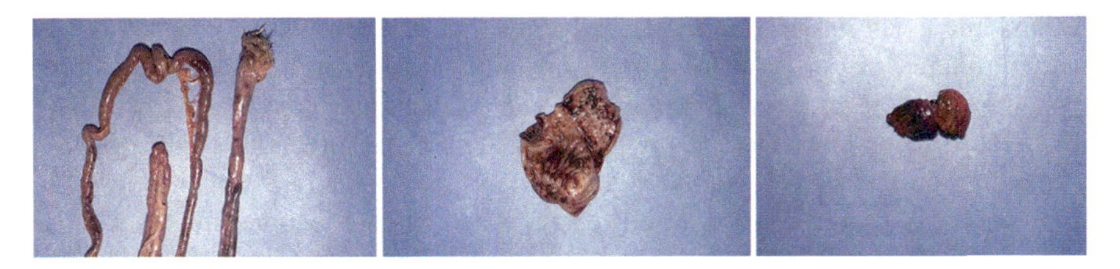

Fig. 18.4 Left: Guinea fowl infected with vNDV, intestine exhibiting necrotic-hemorrhagic lesions of lymphatic tissue through serosal wall. Center: Pheasant infected with velogenic NDV (vNDV), hemorrhages in the proventriculus. Right: Guinea fowl infected with vNDV exhibiting pneumonia. (Courtesy of C. Terregino, IZSVe, Padova, Italy)

In chickens, mesogenic strains can cause respiratory disease, a marked drop in egg production that may last several weeks, and decreased feed consumption (Alexander 2008). Mortality is usually lower, unless the infection is complicated by other pathogens or factors that may exacerbate the disease. Generally speaking, lentogenic strains cause mild or no disease. In young unvaccinated chickens, lentogenic infections can cause respiratory disease and some mortality. Mesogenic and lentogenic viruses produce minimal, if any, gross lesions (Cattoli et al. 2011).

The clinical signs and lesions in chickens experimentally infected with some of the dominant virulent NDV genotypes circulating in Africa, such as genotypes VII, XIV, and XVII, have been described (Susta et al. 2015). Clinical signs and lesions were consistent with those caused by velogenic viscerotropic ND strains and were similar irrespective of the genotype. All the infected birds died within 4 days post-infection, shortly after the onset of clinical signs such as ruffled plumage, conjunctivitis, and prostration. Gross lesions included severe conjunctivitis, visceral hemorrhages, splenic necrosis, thymic hemorrhages, and atrophy (Susta et al. 2015).

Diagnosis

Clinical and Differential Diagnosis

Given the highly contagious nature of this transboundary animal disease and the mandatory notification of its occurrence according to international (OIE) and national regulations, it is essential to be able to detect and confirm the infection as early as possible in order to implement control measures that will reduce losses and viral spread. The first step in raising suspicion of ND within a flock is based on clinical diagnosis. Unfortunately, the clinical signs and gross lesions described in the previous section are only suggestive since none of them can be considered specific for ND. In cases of acute mortality, severe depression, nervous, gastrointestinal, and respiratory signs, a range of other etiologies should be considered as part of differential diagnosis. In addition to managerial errors, such as deprivation of feed, water, or air, and poisoning, other avian infectious diseases should be included in the differential diagnosis. Acute viral diseases that need to be considered are highly pathogenic avian influenza (HPAI), infectious bursal disease (IBD) caused by the very virulent viruses, the acute form of infectious laryngotracheitis (ILT) and fowl pox (FP), and respiratory diseases caused by infectious bronchitis virus (IBV). Bacterial diseases may include fowl cholera, septicemic infections by *E. coli*, and depending on the species involved (e.g., turkeys, psittacine birds, or pigeons) also salmonellosis, ornithosis, Pacheco's disease, and viral infections by other avian paramyxoviruses.

To further complicate the clinical and differential diagnoses of ND, it should be noted that birds can be co-infected by multiple pathogens. In fact, in several ND endemic areas of the African continent, the above listed pathogens are also circulating, such as HPAI of the H5 subtype or H9N2 low pathogenicity avian influenza (LPAI) viruses, IB, IBD, FP viruses, and salmonella. The collection of suitable samples and their quick submission to the diagnostic veterinary laboratory are therefore essential to confirm the suspected disease.

Sample Collection

The respiratory and digestive tracts are the main replication sites for NDV as well as for the majority of the other pathogens listed above; thus, the correct samples for virus detection should include oro/pharyngeal or tracheal swabs, cloacal swabs, and internal organs such as trachea, lungs, intestines, or feces. In cases where virus isolation is required, samples should be properly packaged to avoid leakage and the cold chain should be maintained in order for longer transportation to preserve the viability of the virus. Blood should be collected without anticoagulant to obtain serum for antibody detection. The brachial vein under the wing is the preferred site for blood collection.

Laboratory Diagnosis

Serology The detection of NDV-specific antibodies is commonly performed to evaluate the exposure of susceptible birds to vaccines or field viruses. It is useful for screening large animal populations and for the implementation of post-vaccination monitoring to evaluate the immune status of the population and the vaccination coverage. However, it should be kept in mind that it is not possible to obtain specific information on the virus strain birds are exposed to using serological tests (e.g., serological tests cannot distinguish between infection by velogenic, mesogenic, or lentogenic field strains or live attenuated vaccine). The standard techniques to detect NDV antibodies are the enzyme-linked immunosorbent assay (ELISA) and the hemagglutination inhibition (HI) test. These tests and the exact procedure are reviewed and described in several manuals (Alexander 2008; OIE 2017; Terregino and Capua 2009b).

Virus Detection and Characterization To date, polymerase chain reaction (PCR)-based molecular methods are probably the tests most commonly used in laboratories for the early and rapid detection of NDV. The first documented attempt to identify the virus by reverse transcription (RT)-PCR was in 1991 (Jestin and Jestin 1991). Originally, the RT-PCR-based tests were applied on viruses previously cultured in embryonated eggs. Then, following the progress made in molecular technologies more sensitive tests became available based on classical (i.e., gel-based) RT-PCR and real-time RT-PCR which can be directly applied on clinical specimens for screening and diagnostic purposes. In general terms, these tests are highly specific and sensitive with detection limits which may vary from 10 to 1000 egg infectious dose 50 (EID_{50}) (Cattoli et al. 2011). Given the extensive circulation of both APMV-1 strains of low pathogenicity and ND vaccine strains in poultry, the detection of APMV-1 viruses in a flock without further characterization provides information of limited use. The main advantages of these PCR-based tests are that they not only enable the detection, but also the rapid genetic characterization of APMV-1 detected in a flock. In fact, PCR-based techniques and genetic sequencing allow for the rapid determination of the amino acid sequence of the F protein cleavage site, thereby determining pathotype of the NDV (i.e., virulent or avirulent strain) and, consequently, allowing for a rapid notification of ND according to national and international regulations (OIE 2017).

Gene sequence data are also used for phylogenetic studies and for APMV-1 genotyping. PCR-based assays can also be multiplexed, enabling for the rapid and cost-effective detection of multiple pathogens in one single test. The molecular techniques applied for the detection and characterization of APMV-1 viruses and the confirmation of NDV are reviewed and described in several publications (Cattoli and Monne 2009; Cattoli et al. 2011; OIE 2017). The high genetic variability of APMV-1 and the emergence of novel variants and genotypes represent one of the main constraints for the application of molecular techniques. For example, the majority of the PCR-based tests applied for the detection of Class II APMV-1 strains causing ND are not capable of detecting Class I APMV-1 viruses (Kim et al. 2008; Fuller et al. 2009). In addition, for Class II viruses, point mutations or deletions in the genes targeted by these molecular tests may result in decreased sensitivity of the validated tests and in false-negative results (Kim et al. 2008; Cattoli et al. 2009; Fuller et al. 2009).

Virus isolation in specific pathogen-free (SPF) embryonated chicken eggs or in embryonated eggs from ND serum antibody-negative (SAN) layers or cell culture remains the prescribed test for international trade (OIE 2017) and has some advantages over molecular tests. Indeed, the sensitivity of virus isolation is not affected by the genetic variability of the virus and it allows for the replication of the virus in the laboratory to be used for further antigenic or vaccine challenge studies and for the production of novel vaccines or antigens to be used in serological assays. Virus isolates are also necessary for the categorization of the virus as lentogenic, mesogenic, and velogenic and for the pathotype determination using the ICPI. Virus isolation techniques require the presence of viable viruses in the sample that are capable of replicating in cell culture. Thus, these techniques can generate false-negative results in cases in which viruses have been inactivated due to a variety of reasons (e.g., sample degradation, exposure of samples to high temperatures, or active disinfectants). Virus isolation and characterization procedures have been described in detail in previous manuals (Terregino and Capua 2009b; OIE 2017).

Control of Newcastle Disease

As for other infectious diseases, the prevention and control of ND has the main objectives of preventing susceptible birds from becoming infected or reducing the number of susceptible animals in the population by vaccination. The first objective can be achieved through the implementation of strict biosecurity measures, particularly at the farm level, which would prevent the virus from coming in contact with susceptible birds or reduce the risk of the virus escaping from the infected farm (biocontainment). These activities include the control of the movement of animals and poultry products and movement restrictions and quarantine implementation for imported birds. Depending on national veterinary policies, an ND control program may target the eradication of the disease through compulsory culling of infected and in-contact birds, destruction of the related commodities, movement restrictions, cleaning and disinfection of the infected or at-risk premises, and restrictions or ban on trade and animal exhibitions. In some cases, control and eradication strategies include the so-called preventive or prophylactic vaccination of the susceptible population. In this case, birds are vaccinated even in the absence of outbreaks in order to create an "immunological barrier" to reduce the risk of the introduction of the virus or its spread. In other cases, strategies may include "ring" vaccination around the site of the confirmed outbreaks to establish a buffer zone and slow down the spread of the virus, thus giving time to implement and conduct other control measures, surveillance, and monitoring activities. With the emergence of highly pathogenic avian influenza epizootics worldwide in the last two decades, it became evident that there was no "universal" control policy and that the prevention and control of transboundary animal diseases should be tailored according to the epidemiological situation and the technical, economical, and human resources of the country. Given the global spread of APMV-1 viruses and their endemic circulation in poultry in many countries, the control of ND relies on vaccination as the main tool to prevent mortalities and production losses. However, it should always be kept in mind that an effective control of ND can be achieved through the organized combination of early detection (i.e., application of rapid and specific diagnostic assays), implementation of biosecurity measures, the application of direct control strategies (i.e., culling and movement restrictions), and effective vaccination.

ND Vaccines

The first vaccines for ND which became available in the USA in 1945 were based on inactivated viruses. Initially, they were not extensively adopted by the poultry industry, mainly because of their cost and poor effectiveness demonstrated in some outbreaks. A few years later in 1948, the first live attenuated vaccines were licensed in America, based on natural or attenuated mesogenic or lentogenic strains (reviewed in Alexander 2008; Dimitrov et al. 2017). The vast majority of the inactivated and live ND vaccines marketed in the last 50 or 60 years to the present day contain the same lentogenic or mesogenic vaccine strains belonging to Class II genotype I (e.g., Queensland V-4, Ulster 2C/67 and I-2), genotype II (e.g., lentogenic Beaudette/45, La Sota/46, VG/GA, strain F and B1/47, or mesogenic Komarov and Roakin), and genotype III (e.g., mesogenic Mukteswar strain). Today, the ND vaccines available on the market are either inactivated, live, or vectored vaccines.

Inactivated Vaccines

The seed virus of these vaccines is inactivated with formalin or beta-propiolactone and adjuvanted mainly with either aluminum hydroxide or oil emulsion. These vaccines are safer, do not cause adverse reactions in vaccinated birds, and are easier to store compared to live vaccines. The inactivated vaccine strain can also be incorporated in multivalent vaccines together with other viral and bacterial antigens, for example, infectious bronchitis virus (IBV), infectious bursal disease virus (IBDV), *Pasteurella multocida,* and *Salmonella pullorum.* Inactivated vaccines usually elicit good and long-lasting humoral antibody responses. The immune response provoked by these vaccines is less affected by maternally derived antibodies (MDA) compared to live vaccines (Alexander 2008; Awa et al. 2009); therefore, they are frequently administered to day-old chicks at the hatchery level. Inactivated vaccines have the disadvantage of requiring individual administration either subcutaneously or intramuscularly. In some countries, a withdrawal period before vaccinated birds can be slaughtered for human consumption is required. Inactivation procedures and individual administration can make these vaccines more expensive to produce and to administer compared to live vaccines. Birds vaccinated with inactivated vaccines tend to shed larger amounts of viruses when challenged with field strains compared to birds vaccinated with live vaccines (Miller et al. 2009). Inactivated multivalent vaccines are produced in Africa and tested in both village chickens and poultry farms (Awa et al. 2009).

Live Vaccines

Live vaccine viruses can be lentogenic or mesogenic and the immune responses provoked by these vaccines are directly related to the residual pathogenicity of the vaccine virus used. Because of the greater virulence of mesogenic strains, they are usually only administered as a secondary vaccine in birds already primed with lentogenic strains. In addition, mesogenic vaccines may predispose poultry flocks to secondary infections (Visnuvinayagam et al. 2015). These vaccines are rarely applied by the poultry industry in countries where the infectious pressure is low or moderate, but are still used in some countries where virulent field viruses circulate endemically (Dey et al. 2014). Lentogenic viruses, such as La Sota strain, may also have residual pathogenicity for birds resulting in a higher level of neutralizing antibodies and, at the same time, some adverse reactions, particularly in turkey (Abdul-Aziz and Arp 1983). For this reason, La Sota-based live vaccines are frequently used in countries where virulent field viruses are extensively circulating. Countries where the infectious pressure is low or moderate may opt for less virulent lentogenic strains, such as VG/GA or B1.

Live vaccines have some major advantages. Lentogenic live vaccines can be administered to birds by mass-application methods, such as aerosol sprays or through drinking water, reducing the time for vaccination and making operating costs cheaper (Siccardi 1966). Also, live vaccines stimulate humoral and mucosal immunity and protection occurs earlier compared to inactivated vaccines. The live vaccine strain replicates in and is shed by the vaccinated birds, thereby spreading within the flock and making the vaccine exposure of susceptible birds more homogeneous (Ahlers et al. 1999). Generally speaking, the immune response to live vaccines can be negatively affected by the presence of preexisting immunity, particularly by the presence of MDA. Also, these vaccines are sensitive to storage and transport conditions: for example, if the cold chain is not fully maintained vaccine efficacy can be reduced.

In sub-Saharan Africa, the majority of poultry is reared under extensive, free-ranging rural conditions posing additional challenges to the control of ND by vaccination. In this regard, the development during the 1980s/1990s of easy-to-administer, heat-tolerant live vaccines has made ND vaccination of rural poultry much more practical in both developed and developing countries. The Queensland V4 and I2 strains are the most common thermostable vaccine strains used in Africa and Asia for ND vaccination. The V4-based vaccine has become commercially available to the poultry industry and farmers since the 1990s. As well as the V4 vaccine, the I2 vaccine was also developed by the University of Queensland as part of a project aiming at improving village chicken ND vaccination. The master seed is held in Australia and available free of charge to developing countries (Spreadbrow 2015). In one recent study, it was demonstrated that NDV thermostability is associated with the hemagglutinin-neuraminidase (HN) virus protein, opening new possibilities for the development of novel thermostable, engineered virus strains (Wen et al. 2016).

Several reports based on experimental and field challenges indicate that live thermostable vaccines based on the V4 or I2 strains, if properly administered, are effective and can protect village chickens from clinical disease and mortality caused by virulent NDV strains circulating in Africa (Foster et al. 1999; Ahlers et al. 1999; Illango et al. 2005; Susta et al. 2015; Fentie et al. 2014b). Although these thermostable vaccines can be administered via drinking water, studies conducted on African village chickens recommended eyedrop administration to achieve better protective humoral immunity with clinical protection ranging from 70 to 80% (Sagild and Spalatin 1982; Foster et al. 1999; Ahlers et al. 1999; Illango et al. 2005).

Very often, live vaccines are marketed in vials or lyophilized tablets containing 500 or 1000 doses. This may represent a limiting factor for vaccination campaigns of backyard poultry, where 30–50 chickens are the average size of a backyard flock (Lal et al. 2014). Novel formulations of freeze-dried or nano-encapsulated vaccine tablets based on I2 or La Sota strain have enabled the preparation of thermostable and cost-effective vaccines in tablets containing 10 or 50 doses, thus reducing waste and costs (Wambura 2011; Lal et al. 2014).

Vectored Vaccines

Progress in molecular biology and biotechnology has contributed to the development of novel, recombinant ND vaccines using avian viruses as vectors. Fowl pox (FPV) and Turkey herpes viruses (HVT) have been used to develop vectored vaccines expressing the major immunogenic ND antigens, such as the F and the HN proteins. HVT-based vaccines can also confer optimal clinical protection in the presence of MDA. NDV has also been used as a vector for foreign genes to create multivalent recombinant poultry vaccines, for example, bivalent vaccines against NDV and IBDV or avian influenza virus (AIV) or IBV. In some mass vaccinations, as is the cases for HVT-vectored vaccines, the vaccine can be administered in ovo or in day-old chicks at the hatchery level. However, it should be taken into account that the presence of preexisting immunity to the vector or to the vectored foreign genes may reduce the efficacy of the vectored vaccines. Advantages and limitations of vectored and other novel vaccines for ND are comprehensively reviewed in recent publications (Choi 2017; Dimitrov et al. 2017).

Newcastle Disease Vaccination

Despite the fact that ND vaccines have been available for the last 60 years and are extensively applied globally, the effective control of ND remains a serious challenge. The disease continues to threaten both the poultry industry and the rural poultry sector causing huge economic losses and food security concerns. Although ND in Africa is well known among farmers and the benefit of vaccination in village chickens has been demonstrated (Knueppel et al. 2010), it appears that ND vaccination is not always practiced in poultry flocks, particularly in the rural sector, even in regions experiencing extensive circulation of virulent field viruses (Aboe et al. 2006; Awa et al. 2009; Kouakou et al. 2015). Thus, limited vaccine coverage and the presence of fully susceptible poultry populations is one of the main reasons for the continuous occurrence of ND outbreaks in Africa and elsewhere.

Mortality and severe drop in egg production caused by ND infections in vaccinated poultry flocks have been reported worldwide, including Africa. For example, ND outbreaks in vaccinated commercial and rural chickens were described in South Africa (Bwala et al. 2009), Ethiopia (Fentie et al. 2014a), and Eritrea (Mihreteab et al. 2017). The use of vaccines of poor quality or that have been inadequately stored and administered may explain vaccine failures in many instances. To effectively control ND, the vaccination program implemented should ensure adequate and homogeneous vaccination coverage. It has been estimated that at least 85% of a flock should be properly vaccinated and respond accordingly in order to achieve effective herd immunity (van Boven et al. 2008). Humoral immune response is an important mechanism of protection following vaccination, and hemagglutination inhibition (HI) antibody levels higher than 16 or 32 are necessary to protect birds in the field (Kapczynski et al. 2013). In mass vaccination practices using aerosolized vaccines or vaccines in drinking water, the vaccine intake may greatly differ between individuals and this may result in heterogeneous immune responses within a flock (Alexander et al. 2004; Dortmans et al. 2012). Furthermore, mass vaccination effectiveness may be influenced by the quality of the water, the ambient temperature, the type and quality of the equipment used to distribute the vaccines, and the instruments to generate aerosols. In one study in Sudan, the antibody titers of groups of chickens vaccinated with La Sota live vaccine differed based on the type of water used to reconstitute the vaccine, with the bottled water giving the best results compared to tap water, artesian and shallow well water, and surface water (Khalil and Khalafalla 2011).

It is believed that intense vaccination pressure contributes to the evolution of ND viruses with the resultant emergence of novel genotypes (Dimitrov et al. 2017). In addition, this genetic variability of circulating viruses raises concerns about the true efficacy of currently used ND vaccines which contain seed virus strains developed decades ago. Antigenic divergence between the vaccine strains and the field NDVs currently circulating may explain the reduced efficacy of ND vaccines (Miller et al. 2009, 2013). Indeed, there is evidence indicating that better clinical protection and reduced viral shedding and transmission are obtained in chickens vaccinated with strains homologous to the challenge viruses. However, similar performance can be obtained with classical (and heterologous) vaccines as long as they provide a strong humoral response in the host—which directly depends on the vaccine quality and antigen content—and if sufficient time is allowed for birds to react immunologically (Dortmans et al. 2012; Miller et al. 2013).

It is important to keep in mind that sterilizing immunity cannot be achieved using the ND vaccines currently available, especially under field conditions. Presently, properly implemented vaccination programs using good-quality vaccines can achieve important goals such as good clinical protection, reduction of economic losses, and secure food and incomes in rural settings. However, it cannot achieve effective control and eradication of the disease. In fact, existing ND vaccines cannot prevent infection and so challenge viruses can still replicate to various extents and be shed in vaccinated birds (Kapczynski et al. 2013; Miller et al. 2009, 2013; Susta et al. 2015; Fentie et al. 2014b).

Conclusions

Despite the fact that ND is a well-known threat for poultry economies worldwide that has been studied by many for the last 90 years, it still remains one of the biggest veterinary health challenges for intensive as well as extensive poultry sectors. Although the use of vaccination has certainly contributed to limiting economic losses due to the disease, at least in intensive farming systems, the control of infection is still an unresolved issue in all affected countries. In many developing countries, the implementation of animal control programs in general is difficult (Msoffe et al. 2010). A lack of trained personnel, resources, and funds are common problems in Africa which result in serious impacts on public health and animal production. It has been reported that most of the financial resources allocated to the implementation of livestock health policies are directed mainly to species considered of higher value such as cattle, while little attention being paid to the control of poultry diseases (Mubamba et al. 2016). This is in direct contrast with several publications identifying poultry as the main animal protein source for rural communities in Africa with over 70% of poultry products coming from village chickens (Hailemichael et al. 2016). For example, in Zambia, ND is classified as a management disease, meaning that its control is entirely under the responsibility of the farmer and ND vaccination has not been subsidized by the government since the 1980s (Musako and Abolnik 2012; Mubamba et al. 2016). The capacity for early recognition and rapid confirmation of a disease is the first requirement for the effective control of transboundary animal diseases such as ND. Implementing biosecurity and biocontainment measures to prevent the introduction of the virus into the farm and its subsequent spillover should be the priority of an effective animal infectious disease control strategy, particularly in the industrial and semi-industrial poultry sector. However, it should be recognized that such strategies have additional challenges in their application when dealing with village chickens in developing countries. In these cases, the poultry population is dispersed and fragmented; the bird population in the epidemiological units (i.e., villages) is heterogeneous, frequently composed of different species of different ages owned by different smallholders; the movement of animals and animal products is difficult to trace and control. Control by vaccination in village poultry can be hampered by the rapid turnover of birds (e.g., replacing or selling dying birds). Because of this turnover, it has been estimated that by 4 months following vaccination 30% of birds in a given flock will be potentially unvaccinated and thus unprotected (Oakeley 2000). The persistence of fully susceptible birds in the extensive poultry sector facilitates the maintenance of the virus in the environment and represents a constant threat for the intensive poultry sector. The community-based approach, involving and educating community leaders and smallholders, can facilitate access to vaccination practices and should aim at improving the management of poultry husbandry by introducing basic biosecurity practices at the village and farm level (Oakeley 2000; Aboe et al. 2006; McCrindle et al. 2007; Msoffe et al. 2010). The proper training of the staff conducting vaccination is important. In one study from

South Africa, the antibody response to vaccination was much lower in poultry vaccinated by inexperienced volunteers compared to poultry vaccinated by experienced staff (McCrindle et al. 2007). With the current vaccines, the single vaccination practice applied in rural poultry in several countries appears not to be adequate enough in reducing virus spread and mortalities and, therefore, a prime-boost immunization scheme would be more recommendable (Fentie et al. 2014a, b). Novel approaches to vaccine development and vaccination strategies could improve the situation by increasing vaccination coverage and clinical protection of the poultry population and by also reducing viral shedding in order to better control the spread of the infection. As concluded by Oakeley (2000), it should also be taken into account that ND is a major constraint for poultry development in Africa, but it is by no means the only one. National or regional ND control programs should run in parallel with the implementation of sustainable agricultural policies for the development of the poultry economy while also addressing the other health, management, and husbandry problems this livestock sector is suffering from.

Acknowledgements The authors wish to thank Dr. Calogero Terregino (IZSVe, Padova, Italy) for the pictures provided in Figs. 18.3 and 18.4 and Dr. Ivancho Naletoski (Joint FAO/IAEA Division, Vienna, Austria) for the creation of the map in Fig. 18.2.

References

Abdul-Aziz TA, Arp LH. Pathology of the trachea in turkeys exposed by aerosol to lentogenic strains of Newcastle disease virus. Avian Dis. 1983;27(4):1002–11.

Aboe PA, Boa-Amponsem K, Okantah SA, Butler EA, Dorward PT, Bryant MJ. Free-range village chickens on the Accra Plains, Ghana: their husbandry and productivity. Trop Anim Health Prod. 2006;38(3):235–48.

Abolnik C. History of Newcastle disease in South Africa. Onderstepoort J Vet Res. 2017;84(1): e1–7.

Abolnik C, Horner RF, Maharaj R, Viljoen GJ. Characterization of a pigeon paramyxovirus (PPMV-1) isolated from chickens in South Africa. Onderstepoort J Vet Res. 2004;71 (2):157–60.

Abolnik C, Gerdes GH, Kitching J, Swanepoel S, Romito M, Bisschop SP. Characterization of pigeon paramyxoviruses (Newcastle disease virus) isolated in South Africa from 2001 to 2006. Onderstepoort J Vet Res. 2008;75:147–52.

Abolnik C, Mubamba C, Dautu G, Gummow B. Complete genome sequence of a Newcastle disease genotype XIII virus isolated from indigenous chickens in Zambia. Genome Announc. 2017;5 (34):e00841-17.

Abolnik C, Mubamba C, Wandrag DBR, Horner R, Gummow B, Gautu G, Bisschop SPR. Tracing the origins of genotype VIIh Newcastle disease in southern Africa. Transbound Emerg Dis. 2018;65(2):e393–403.

Abu Elzein EM, Manvell R, Alexander D, Alafaleq AI. Pigeon paramyxovirus-1 (P-group) as the cause of severe outbreaks in fancy *Columba livia* in Saudi Arabia. Zentralbl Veterinarmed B. 1999;46(10):689–92.

Ahlers C, Hüttner K, Pfeiffer D. Comparison between a live and an inactivated vaccine against Newcastle disease in village chickens. A field study in northern Malawi. Trop Anim Health Prod. 1999;31(3):167–74.

Aldous EW, Mynn JK, Banks J, Alexander DJ. A molecular epidemiological study of avian paramyxovirus type 1 (Newcastle disease virus) isolates by phylogenetic analysis of a partial nucleotide sequence of the fusion protein gene. Avian Pathol. 2003;32(3):239–56.
Aldous EW, Fuller CM, Mynn JK, Alexander DJ. A molecular epidemiological investigation of isolates of the variant avian paramyxovirus type 1 virus (PPMV-1) responsible for the 1978 to present panzootic in pigeons. Avian Pathol. 2004;33(2):258–69.
Alexander DJ. Newcastle disease in ostriches (*Struthio camelus*) - a review. Avian Pathol. 2000;29 (2):95–100.
Alexander DJ. Newcastle disease and other avian paramyxoviridae infections. In: Saif YM, editor. Diseases of poultry. 12th ed. Ames, IA: Iowa State/University Press; 2008. p. 541–69.
Alexander DJ. Ecology and epidemiology of Newcastle disease. In: Capua I, Alexander DJ, editors. Avian influenza and Newcastle disease. Milan: Springer; 2009. p. 19–26.
Alexander DJ, Wilson GW, Russell PH, Lister SA, Parsons G. Newcastle disease outbreaks in fowl in Great Britain during 1984. Vet Rec. 1985;117(17):429–34.
Alexander DJ, Campbell G, Manvell RJ, Collins MS, Parsons G, McNulty MS. Characterisation of an antigenically unusual virus responsible for two outbreaks of Newcastle disease in the Republic of Ireland in 1990. Vet Rec. 1992;130:65–8.
Alexander DJ, Bell JG, Alders RG. Technology review: Newcastle disease. Rome: Food and Agriculture Organization of the United Nations; 2004.
Antipas BB, Bidjeh K, Youssouf ML. Epidemiology of Newcastle disease and its economic impact in Chad. Eur J Exp Biol. 2012;2(6):2286–92.. http://www.imedpub.com/articles/epidemiology-of-newcastle-disease-and-its-economic-impact-in-chad.pdf
Awa DN, Achukwi MD. Livestock pathology in the central African region: some epidemiological considerations and control strategies. Anim Health Res Rev. 2010;11(2):235–44.
Awa DN, Ngo Tama AC, Njoya A, Jumbo SD, Mefomdjo P. The potential role of an inactivated thermostable vaccine in the control of Newcastle disease in traditionally free-roaming poultry in central and West Africa. Trop Anim Health Prod. 2009;41(3):285–90.
Awan MA, Otte MJ, James AD. The epidemiology of Newcastle disease in rural poultry: a review. Avian Pathol. 1994;23(3):405–23.
Ballagi-Pordány A, Wehmann E, Herczeg J, Belák S, Lomniczi B. Identification and grouping of Newcastle disease virus strains by restriction site analysis of a region from the F gene. Arch Virol. 1996;141(2):243–61.
Bwala DG, Abolnik C, van Wyk A, Cornelius E, Bisschop SP. Efficacy of a genotype 2 Newcastle disease vaccine (Avinew) against challenge with highly virulent genotypes 5d and 3d. J S Afr Vet Assoc. 2009;80(3):174–8.
Byarugaba DK, Mugimba KK, Omony JB, Okitwi M, Wanyana A, Otim MO, Kirunda H, Nakavuma JL, Teillaud A, Paul MC, Ducatez MF. High pathogenicity and low genetic evolution of avian paramyxovirus type I (Newcastle disease virus) isolated from live bird markets in Uganda. Virol J. 2014;11:173.
Cappelle J, Caron A, Servan De Almeida R, Gil P, Pedrono M, Mundava J, Fofana B, Balança G, Dakouo M, Ould El Mamy AB, Abolnik C, Maminiaina OF, Cumming GS, De Visscher MN, Albina E, Chevalier V, Gaidet N. Empirical analysis suggests continuous and homogeneous circulation of Newcastle disease virus in a wide range of wild bird species in Africa. Epidemiol Infect. 2015;143(6):1292–303.
Cattoli G, Monne I. Molecular diagnosis of Newcastle disease virus. In: Capua I, Alexander DJ, editors. Avian influenza and Newcastle disease. Milan: Springer; 2009. p. 127–32.
Cattoli G, De Battisti C, Marciano S, Ormelli S, Monne I, Terregino C, Capua I. False-negative results of a validated real-time PCR protocol for diagnosis of Newcastle disease due to genetic variability of the matrix gene. J Clin Microbiol. 2009;47(11):3791–2.
Cattoli G, Fusaro A, Monne I, Molia S, Le Menach A, Maregeya B, Nchare A, Bangana I, Maina AG, Koffi JN, Thiam H, Bezeid OE, Salviato A, Nisi R, Terregino C, Capua I. Emergence of a new genetic lineage of Newcastle disease virus in west and Central Africa—implications for diagnosis and control. Vet Microbiol. 2010;142(3–4):168–76.
Cattoli G, Susta L, Terregino C, Brown C. Newcastle disease: a review of field recognition and current methods of laboratory detection. J Vet Diagn Investig. 2011;23(4):637–56.

Choi KS. Newcastle disease virus vectored vaccines as bivalent or antigen delivery vaccines. Clin Exp Vaccine Res. 2017;6(2):72–82.
Cordier G, Clavieras J, Ounais A. Research on Newcastle disease virus in Tunisia. Ann Inst Pasteur. 1950a;78(2):242–61.
Cordier G, Clavieras J, Ounais A. Vaccination against Newcastle disease in Tunisia. Ann Inst Pasteur. 1950b;78(3):302–6.
Czeglédi A, Ujvári D, Somogyi E, Wehmann E, Werner O, Lomniczi B. Third genome size category of avian paramyxovirus serotype 1 (Newcastle disease virus) and evolutionary implications. Virus Res. 2006;120(1–2):36–48.
Damena D, Fusaro A, Sombo M, Belaineh R, Heidari A, Kebede A, Kidane M, Chaka H. Characterization of Newcastle disease virus isolates obtained from outbreak cases in commercial chickens and wild pigeons in Ethiopia. Springerplus. 2016;5:476.
de Almeida R, Maminiaina OF, Gil P, Hammoumi S, Molia S, Chevalier V, Koko M, Andriamanivo HR, Traoré A, Samaké K, Diarra A, Grillet C, Martinez D, Albina E. Africa, a reservoir of new virulent strains of Newcastle disease virus? Vaccine. 2009;27(24):3127–9.
de Almeida RS, Hammoumi S, Gil P, Briand FX, Molia S, Gaidet N, Cappelle J, Chevalier V, Balança G, Traoré A, Grillet C, Maminiaina OF, Guendouz S, Dakouo M, Samaké K, Bezeid Oel M, Diarra A, Chaka H, Goutard F, Thompson P, Martinez D, Jestin V, Albina E. New avian paramyxoviruses type I strains identified in Africa provide new outcomes for phylogeny reconstruction and genotype classification. PLoS One. 2013;8(10):e76413.
de Leeuw OS, Koch G, Hartog L, Ravenshorst N, Peeters BP. Virulence of Newcastle disease virus is determined by the cleavage site of the fusion protein and by both the stem region and globular head of the haemagglutinin-neuraminidase protein. J Gen Virol. 2005;86(6):1759–69.
Dey S, Chellappa MM, Gaikwad S, Kataria JM, Vakharia VN. Genotype characterization of commonly used Newcastle disease virus vaccine strains of India. PLoS One. 2014;9(6):e98869.
Diel DG, da Silva LH, Liu H, Wang Z, Miller PJ, Afonso CL. Genetic diversity of avian paramyxovirus type 1: proposal for a unified nomenclature and classification system of Newcastle disease virus genotypes. Infect Genet Evol. 2012;12(8):1770–9.
Dimitrov KM, Ramey AM, Qiu X, Bahl J, Afonso CL. Temporal, geographic, and host distribution of avian paramyxovirus 1 (Newcastle disease virus). Infect Genet Evol. 2016a;39:22–34.
Dimitrov KM, Lee DH, Williams-Coplin D, Olivier TL, Miller PJ, Afonso CL. Newcastle disease viruses causing recent outbreaks worldwide show unexpectedly high genetic similarity to historical virulent isolates from the 1940s. J Clin Microbiol. 2016b;54(5):1228–35.
Dimitrov KM, Afonso CL, Yu Q, Miller PJ. Newcastle disease vaccines-a solved problem or a continuous challenge? Vet Microbiol. 2017;206:126–36.
Dimitrov KM, Abolnik C, Afonso CL, Albina E, Bahl J, Berg M, Briand FX, Brown IH, Choi KS, Chvala I, Diel DG, Durr PA, Ferreira HL, Fusaro A, Gil P, Goujgoulova GV, Grund C, Hicks JT, Joannis TM, Torchetti MK, Kolosov S, Lambrecht B, Lewis NS, Liu H, Liu H, McCullough S, Miller PJ, Monne I, Muller CP, Munir M, Reischak D, Sabra M, Samal SK, Servan de Almeida R, Shittu I, Snoeck CJ, Suarez DL, Van Borm S, Wang Z, Wong FYK. Updated unified phylogenetic classification system and revised nomenclature for Newcastle disease virus. Infect Genet Evol. 2019;74:103917.
Dortmans JC, Peeters BP, Koch G. Newcastle disease virus outbreaks: vaccine mismatch or inadequate application? Vet Microbiol. 2012;160(1–2):17–22.
Eisa M, Omer EA. A natural outbreak of Newcastle disease in pigeons in the Sudan. Vet Rec. 1984;114(12):297.
El Khantour A, Darkaoui S, Tatár-Kis T, Mató T, Essalah-Bennani A, Cazaban C, Palya V. Immunity elicited by a Turkey Herpesvirus-vectored Newcastle disease vaccine in Turkey against challenge with a recent genotype IV Newcastle disease virus field strain. Avian Dis. 2017;61(3):378–86.
Fadiga M, Jost C, Ihedioha J. 2013. Financial costs of disease burden, morbidity and mortality from priority livestock diseases in Nigeria. https://cgspace.cgiar.org/bitstream/handle/10568/33418/ResearchReport_33.pdf?sequence=2.

Fentie T, Heidari A, Aiello R, Kassa T, Capua I, Cattoli G, Sahle M. Molecular characterization of Newcastle disease viruses isolated from rural chicken in Northwest Ethiopia reveals the circulation of three distinct genotypes in the country. Trop Anim Health Prod. 2014a;46 (2):299–304.

Fentie T, Dadi K, Kassa T, Sahle M, Cattoli G. Effect of vaccination on transmission characteristics of highly virulent Newcastle disease virus in experimentally infected chickens. Avian Pathol. 2014b;43(5):420–6.

Foster HA, Chitukuro HR, Tuppa E, Mwanjala T, Kusila C. Thermostable Newcastle disease vaccines in Tanzania. Vet Microbiol. 1999;68(1–2):127–30.

Fringe R, Bosman AM, Ebersohn K, Bisschop S, Abolnik C, Venter E. Molecular characterisation of Newcastle disease virus isolates from different geographical regions in Mozambique in 2005. Onderstepoort J Vet Res. 2012;79(1):E1–7.

Fuller CM, Collins MS, Alexander DJ. Development of a real-time reverse-transcription PCR for the detection and simultaneous pathotyping of Newcastle disease virus isolates using a novel probe. Arch Virol. 2009;154(6):929–37.

Fuller C, Löndt B, Dimitrov KM, Lewis N, van Boheemen S, Fouchier R, Coven F, Goujgoulova G, Haddas R, Brown I. An Epizootiological report of the re-emergence and spread of a lineage of virulent Newcastle disease virus into Eastern Europe. Transbound Emerg Dis. 2017;64 (3):1001–7.

Ganar K, Das M, Raut AA, Mishra A, Kumar S. Emergence of a deviating genotype VI pigeon paramyxovirus type-1 isolated from India. Arch Virol. 2017;162:2169–74.

Gardner E, Alders R. Livestock risks and opportunities: Newcastle disease and Avian influenza in Africa. GRF Davos Planet@Risk, Vol 2, No 4 (2014): special issue on one health (part II/II). 2014. http://vet.tufts.edu/wp-content/uploads/Gardner-MCM-case-study-pub-article.pdf.

George MM. Epidemiology of Newcastle disease and the need to vaccinate local chickens in Uganda. In: Spradbrow PB, editors. Newcastle disease in village chickens, control with Thermostable Oral vaccines. Proceedings, international workshop held in Kuala Lumpur, Malaysia, 6–10 October 1991. Canberra: Centre for International Agricultural Research (ACIAR); 1992. pp. 155–158.

Hailemichael A, Gebremedhin B, Gizaw S, Tegegne A. Analysis of village poultry value chain in Ethiopia: implications for action research and development. LIVES working paper 10. Nairobi: International Livestock Research Institute (ILRI); 2016. https://cgspace.cgiar.org/handle/10568/71088.

Haroun M, Mohran KA, Hassan MM, Abdulla NM. Molecular pathotyping and phylogenesis of the first Newcastle disease virus strain isolated from backyard chickens in Qatar. Trop Anim Health Prod. 2015;47(1):13–9.

Hassan W, Khair SA, Mochotlhoane B, Abolnik C. Newcastle disease outbreaks in the Sudan from 2003 to 2006 were caused by viruses of genotype 5d. Virus Genes. 2010;40(1):106–10.

Herczeg J, Wehmann E, Bragg RR, Travassos Dias PM, Hadjiev G, Werner O, Lomniczi B. Two novel genetic groups (VIIb and VIII) responsible for recent Newcastle disease outbreaks in southern Africa, one (VIIb) of which reached southern Europe. Arch Virol. 1999;144 (11):2087–99.

IAEA. Characteristics and parameters of family poultry production in Africa. Results of a FAO/IAEA Co-ordinated Research Programme on Assessment of the effectiveness of vaccination strategies against Newcastle disease and Gumboro disease using immunoassay-based technologies for increasing farmyard poultry production in Africa. Vienna: IAEA; 2002.

ICTV. International Committee of Taxonomy of Viruses, Ninth Report; 2009 Taxonomy Release. https://talk.ictvonline.org/ictv-reports/ictv_9th_report/negative-sense-rna-viruses-2011/w/negrna_viruses/199/paramyxoviridae.

ICTV International Committee on Taxonomy of Viruses. Virus Taxonomy: 2018b Release. 2019. Available at https://talk.ictvonline.org/taxonomy/.

Illango J, Olaho-Mukani W, Mukiibi-Muka G, Abila PP, Etoori A. Immunogenicity of a locally produced Newcastle disease I-2 thermostable vaccine in chickens in Uganda. Trop Anim Health Prod. 2005;37(1):25–31.

Jestin V, Jestin A. Detection of Newcastle disease virus RNA in infected allantoic fluids by in vitro enzymatic amplification (PCR). Arch Virol. 1991;118(3–4):151–61.

Kaleta EF, Baldauf C. Newcastle disease in free-living and pet birds. In: Alexander DJ, editor. Newcastle disease. Boston, MA: Kluwer Academic; 1988. p. 197–246.

Kammon A, Heidari A, Dayhum A, Eldaghayes I, Sharif M, Monne I, Cattoli G, Asheg A, Farhat M, Kraim E. Characterization of avian influenza and Newcastle disease viruses from poultry in Libya. Avian Dis. 2015;59(3):422–30.

Kapczynski DR, Afonso CL, Miller PJ. Immune responses of poultry to Newcastle disease virus. Dev Comp Immunol. 2013;41(3):447–53.

Khalafalla AI, Fadol MA, Hameid OA, Hussein YA, el Nur M. Pathogenic properties of Newcastle disease virus isolates in the Sudan. Acta Vet Hung. 1992;40(4):329–33.

Khalil AA, Khalafalla AI. Analysis and effect of water sources used as diluents on Newcastle disease vaccine efficacy in chickens in the Sudan. Trop Anim Health Prod. 2011;43(2):295–7.

Kim LM, Suarez DL, Afonso CL. Detection of a broad range of class I and II Newcastle disease viruses using a multiplex real-time reverse transcription polymerase chain reaction assay. J Vet Diagn Investig. 2008;20(4):414–25.

Knueppel D, Cardona C, Msoffe P, Demment M, Kaiser L. Impact of vaccination against chicken Newcastle disease on food intake and food security in rural households in Tanzania. Food Nutr Bull. 2010;31(3):436–45.

Kouakou AV, Kouakou V, Kouakou C, Godji P, Kouassi AL, Krou HA, Langeois Q, Webby RJ, Ducatez MF, Couacy-Hymann E. Prevalence of Newcastle disease virus and infectious bronchitis virus in avian influenza negative birds from live bird markets and backyard and commercial farms in Ivory-Coast. Res Vet Sci. 2015;102:83–8.

Lal M, Zhu C, McClurkan C, Koelle DM, Miller P, Afonso C, Donadeu M, Dungu B, Chen D. Development of a low-dose fast-dissolving tablet formulation of Newcastle disease vaccine for low-cost backyard poultry immunisation. Vet Rec. 2014;174(20):504.

Lomniczi B, Wehmann E, Herczeg J, Ballagi-Pordány A, Kaleta EF, Werner O, Meulemans G, Jorgensen PH, Manté AP, Gielkens AL, Capua I, Damoser J. Newcastle disease outbreaks in recent years in western Europe were caused by an old (VI) and a novel genotype (VII). Arch Virol. 1998;143(1):49–64.

Maminiaina OF, Gil P, Briand FX, Albina E, Keita D, Andriamanivo HR, Chevalier V, Lancelot R, Martinez D, Rakotondravao R, Rajaonarison JJ, Koko M, Andriantsimahavandy AA, Jestin V, Servan de Almeida R. Newcastle disease virus in Madagascar: identification of an original genotype possibly deriving from a died out ancestor of genotype IV. PLoS One. 2010;5(11): e13987.

Mansour SMG, Mohamed FF, Eid AAM, Mor SK, Goyal SM. Co-circulation of paramyxo- and influenza viruses in pigeons in Egypt. Avian Pathol. 2017;46(4):367–75.

Mapaco LP, Monjane IV, Nhamusso AE, Viljoen GJ, Dundon WG, Achá SJ. Phylogenetic analysis of Newcastle disease viruses isolated from commercial poultry in Mozambique (2011–2016). Virus Genes. 2016;52(5):748–53.

McCrindle CM, Bisschop SP, Modise K. Evaluation of the application of a thermostable Newcastle disease vaccine by community volunteers in the north West Province of South Africa. J S Afr Vet Assoc. 2007;78(3):158–62.

Miguel E, Grosbois V, Berthouly-Salazar C, Caron A, Cappelle J, Roger F. A meta-analysis of observational epidemiological studies of Newcastle disease in African agro-systems, 1980–2009. Epidemiol Infect. 2013;141(6):1117–33.

Mihreteab B, Gide B, Nguse F, Petros Y, Simon Y. Outbreak investigation of Newcastle disease virus from vaccinated chickens in Eritrea. Afr J Biotechnol. 2017;16:1717–23.

Miller PJ, Estevez C, Yu Q, Suarez DL, King DJ. Comparison of viral shedding following vaccination with inactivated and live Newcastle disease vaccines formulated with wild-type and recombinant viruses. Avian Dis. 2009;53(1):39–49.

Miller PJ, Afonso CL, El Attrache J, Dorsey KM, Courtney SC, Guo Z, Kapczynski DR. Effects of Newcastle disease virus vaccine antibodies on the shedding and transmission of challenge viruses. Dev Comp Immunol. 2013;41(4):505–13.

Mohamed MH, Kumar S, Paldurai A, Samal SK. Sequence analysis of fusion protein gene of Newcastle disease virus isolated from outbreaks in Egypt during 2006. Virol J. 2011;8:237.

Molini U, Aikukutu G, Khaiseb S, Cattoli G, Dundon WG. First genetic characterization of Newcastle disease viruses from Namibia: identification of a novel VIIk subgenotype. Arch Virol. 2017;162(8):2427–31.

Molini U, Aikukutu G, Khaiseb S, Cattoli G, Dundon WG. Phylogenetic analysis of pigeon paramyxoviruses type-1 identified in mourning collared doves (*Streptopelia decipiens*), Namibia. J Wildlife Dis. 2018;54(3):601–6.

Msoffe PL, Bunn D, Muhairwa AP, Mtambo MM, Mwamhehe H, Msago A, Mlozi MR, Cardona CJ. Implementing poultry vaccination and biosecurity at the village level in Tanzania: a social strategy to promote health in free-range poultry populations. Trop Anim Health Prod. 2010;42 (2):253–63.

Mubamba C, Ramsay G, Abolnik C, Dautu G, Gummow B. A retrospective study and predictive modelling of Newcastle disease trends among rural poultry of eastern Zambia. Prev Vet Med. 2016;133:97–107.

Musako C, Abolnik C. Determination of the seroprevalence of Newcastle disease virus (avian paramyxovirus type 1) in Zambian backyard chicken flocks. Onderstepoort J Vet Res. 2012;79 (1):E1–4.

Oakeley RD. The limitations of a feed/water based heat-stable vaccine delivery system for Newcastle disease-control strategies for backyard poultry flocks in sub-Saharan Africa. Prev Vet Med. 2000;47(4):271–9.

OIE. 2017. Manual of diagnostic tests and vaccines for terrestrial animals. http://www.oie.int/international-standard-setting/terrestrial-manual/access-online/.

Orabi A, Hussein A, Saleh AA, El-Magd MA, Munir M. Evolutionary insights into the fusion protein of Newcastle disease virus isolated from vaccinated chickens in 2016 in Egypt. Arch Virol. 2017;162(10):3069–79. https://doi.org/10.1007/s00705-017-3483-1.

Piacenti AM, King DJ, Seal BS, Zhang J, Brown CC. Pathogenesis of Newcastle disease in commercial and specific pathogen-free turkeys experimentally infected with isolates of different virulence. Vet Pathol. 2006;43(2):168–78.

Radwan MM, Darwish SF, El-Sabagh IM, El-Sanousi AA, Shalaby MA. Isolation and molecular characterization of Newcastle disease virus genotypes II and VIId in Egypt between 2011 and 2012. Virus Genes. 2013;47(2):311–6.

Saad AM, Samy A, Soliman MA, Arafa A, Zanaty A, Hassan MK, Sultan AH, Bazid AI, Hussein AH. Genotypic and pathogenic characterization of genotype VII Newcastle disease viruses isolated from commercial farms in Egypt and evaluation of heterologous antibody responses. Arch Virol. 2017;162(7):1985–94.

Sabra M, Dimitrov KM, Goraichuk IV, Wajid A, Sharma P, Williams-Coplin D, Basharat A, Rehmani SF, Muzyka DV, Miller PJ, Afonso CL. Phylogenetic assessment reveals continuous evolution and circulation of pigeon-derived virulent avian avulaviruses 1 in Eastern Europe, Asia, and Africa. BMC Vet Res. 2017;13(1):291.

Sagild IK, Spalatin J. Newcastle disease vaccination with the V4 strain in Malawi: laboratory and field studies. Avian Dis. 1982;26(3):625–8.

Sambo E, Bettridge J, Dessie T, Amare A, Habte T, Wigley P, Christley RM. Participatory evaluation of chicken health and production constraints in Ethiopia. Prev Vet Med. 2015;118 (1):117–27.

Shekaili TA, Clough H, Ganapathy K, Baylis M. Sero-surveillance and risk factors for avian influenza and Newcastle disease virus in backyard poultry in Oman. Prev Vet Med. 2015;122 (1–2):145–53.

Shittu I, Joannis TM, Odaibo GN, Olaleye OD. Newcastle disease in Nigeria: epizootiology and current knowledge of circulating genotypes. Virus. 2016a;27(4):329–39.

Shittu I, Sharma P, Joannis TM, Volkening JD, Odaibo GN, Olaleye DO, Williams-Coplin D, Solomon P, Abolnik C, Miller PJ, Dimitrov KM, Afonso CL. Complete genome sequence of a genotype XVII Newcastle disease virus, isolated from an apparently healthy domestic duck in Nigeria. Genome Announc. 2016b;4:e01716–15.

Siccardi FJ. Effect of vaccination during an outbreak of Newcastle disease on a broiler-breeder chicken farm in Nigeria. Avian Dis. 1966;10(4):422–7.
Snoeck CJ, Ducatez MF, Owoade AA, Faleke OO, Alkali BR, Tahita MC, Tarnagda Z, Ouedraogo JB, Maikano I, Mbah PO, Kremer JR, Muller CP. Newcastle disease virus in West Africa: new virulent strains identified in non-commercial farms. Arch Virol. 2009;154(1):47–54.
Snoeck CJ, Owoade AA, Couacy-Hymann E, Alkali BR, Okwen MP, Adeyanju AT, Komoyo GF, Nakouné E, Le Faou A, Muller CP. High genetic diversity of Newcastle disease virus in poultry in west and Central Africa: cocirculation of genotype XIV and newly defined genotypes XVII and XVIII. J Clin Microbiol. 2013a;51(7):2250–60.
Snoeck CJ, Adeyanju AT, Owoade AA, Couacy-Hymann E, Alkali BR, Ottosson U, Muller CP. Genetic diversity of Newcastle disease virus in wild birds and pigeons in West Africa. Appl Environ Microbiol. 2013b;79(24):7867–74.
Spreadbrow PB. Thermostable vaccines in the control of Newcastle disease in village chickens. 2015. https://kyeemafoundation.org/about-us/our-history/history-of-the-i-2-nd-vaccine/.
Susta L, Jones ME, Cattoli G, Cardenas-Garcia S, Miller PJ, Brown CC, Afonso CL. Pathologic characterization of genotypes XIV and XVII Newcastle disease viruses and efficacy of classical vaccination on specific pathogen-free birds. Vet Pathol. 2015;52(1):120–31.
Tadelle D, Million T, Alemu Y, Peters KJ. Village chicken production systems in Ethiopia: 2. Use patterns and performance valuation and chicken products and socio-economic functions of chicken. Livestock research for rural development. 2003;15:10. http://www.lrrd.org/lrrd15/1/tadeb151.htm.
Terregino C, Capua I. Clinical traits and pathology of Newcastle disease infection and guidelines for farm visit and differential diagnosis. In: Capua I, Alexander DJ, editors. Avian influenza and Newcastle disease. Milan: Springer; 2009a. p. 113–22.
Terregino C, Capua I. Conventional diagnosis of Newcastle disease virus infection. In: Capua I, Alexander DJ, editors. Avian influenza and Newcastle disease. Milan: Springer; 2009b. p. 123–5.
Tirumurugaan KG, Kapgate S, Vinupriya MK, Vijayarani K, Kumanan K, Elankumaran S. Genotypic and pathotypic characterization of Newcastle disease viruses from India. PLoS One. 2011;6(12):e28414.
Van Borm S, Obishakin E, Joannis T, Lambrecht B, van den Berg T. Further evidence for the widespread co-circulation of lineages 4b and 7 velogenic Newcastle disease viruses in rural Nigeria. Avian Pathol. 2012;41(4):377–82.
van Boven M, Bouma A, Fabri TH, Katsma E, Hartog L, Koch G. Herd immunity to Newcastle disease virus in poultry by vaccination. Avian Pathol. 2008;37(1):1–5.
Verwoerd DJ, Gerdes GH, Olivier A, Williams R. Experimental infection of vaccinated slaughter ostriches with virulent Newcastle disease virus. Onderstepoort J Vet Res. 1997;64(3):213–6.
Visnuvinayagam S, Thangavel K, Lalitha N, Malmarugan S, Sukumar K. Assessment of the pathogenicity of cell-culture-adapted Newcastle disease virus strain Komarov. Braz J Microbiol. 2015;46(3):861–5.
Wambura PN. Formulation of novel nano-encapsulated Newcastle disease vaccine tablets for vaccination of village chickens. Trop Anim Health Prod. 2011;43(1):165–9.
Wen G, Hu X, Zhao K, Wang H, Zhang Z, Zhang T, Yang J, Luo Q, Zhang R, Pan Z, Shao H, Yu Q. Molecular basis for the thermostability of Newcastle disease virus. Sci Rep. 2016;6:22492.
Yongolo MG, Christensen H, Handberg K, Minga U, Olsen JE. On the origin and diversity of Newcastle disease virus in Tanzania. Onderstepoort J Vet Res. 2011;78(1):312.

Part III
Bacterial Diseases

Chapter 19
Tuberculosis: A Transboundary Animal Disease in Sahel Africa and Its Connected Regions

Lilian Akudo Okeke

Abstract Tuberculosis is one of the transboundary animal diseases that can spread extremely rapidly, irrespective of national or international borders. Tuberculosis is a zoonotic disease and a poverty-related disease that affects both humans and animals, and it is caused by members of the *Mycobacterium tuberculosis* complex (MTC). Globally, tuberculosis (TB) is a major public health problem with about 9 million people around the world effected by tuberculosis (TB), of which nearly 2 million people died with or from the disease. Tuberculosis is a major cause of death due to infectious diseases, competing with the human immunodeficiency virus (HIV). The burden of tuberculosis in Sahelian Africa has increased tremendously as a result of continuing poverty, political instability, and threat by violence, and these have impeded greatly the progress in implementing effective TB control measures. The majority of the people in the Sahelian region are by tradition and culture semi-nomads and are into farming and raising livestock in a system of transhumance as a way of utilizing the Sahel and sustaining themselves; hence there is that close contact between humans and their animals which translates to a high likelihood of zoonotic transmission between animal and humans. *Mycobacterium tuberculosis* affects humans mainly, while *Mycobacterium bovis* affects a wide range of host species including humans and wild and domestic animals, for example, camel and cattle. This chapter reviews tuberculosis in cattle and other domestic animals like sheep and goats in the Sahel Africa and its connected regions while drivers, history, epidemiology, clinical signs, diagnosis, and modes of transmission were reviewed and methods for the prevention and control of tuberculosis in cattle and other domestic animals like goats and sheep were highlighted.

Keywords Tuberculosis · Transboundary · Zoonosis · Sahel Africa

---

L. A. Okeke (✉)
National Veterinary Research Institute, Vom, Plateau State, Nigeria

African Field Epidemiology Network, Abuja, Nigeria

M. Kardjadj et al. (eds.), *Transboundary Animal Diseases in Sahelian Africa and Connected Regions*, https://doi.org/10.1007/978-3-030-25385-1_19

History of Tuberculosis

Before recorded history, tuberculosis and other species of mycobacteria have been in existence (James 1999). Mycobacteria belongs to the genus of Actinobacteria under the family Mycobacteriaceae (Ryan and Ray 2004). The Actinobacteria genus includes pathogens known to cause serious diseases in vertebrate animals, including tuberculosis and leprosy (McMurray 1996). However, the Latin prefix "myco" means both fungus and wax, which is the "waxy" compounds in the cell wall (McMurray 1996). This attribute makes *Mycobacterium* acid-fast in nature (Bhamidi 2009).

Tuberculosis in cattle and other domestic animals like goats and sheep is caused by *Mycobacterium tuberculosis* complex (MTC), specifically *M. bovis* and *M. caprae* species (Pavlik et al. 2002; Prodinger et al. 2002; Erler et al. 2004). Tuberculosis is characterized by an accelerating development of specific granulomatous lesions of tubercles in affected tissues like the spleen, lymph nodes, and lungs (Omer et al. 1995). Mycobacterium also affects cattle, birds, frogs, fish, small reptiles, and frogs. There has been a risk of transmission of *Mycobacterium bovis* from animals to humans through the ingestion of milk and dairy products by humans (McMurray 1996). However, due to eradication of infected cattle and pasteurization of milk bovine tuberculosis is rarely found in the USA and sub-Saharan Africa. Mycobacterium is easily destroyed by sunlight and ultraviolet irradiation but can survive in the dust and air for weeks and months (McMurray 1996).

During the eighteenth century, many researchers assumed that the days of tuberculosis as a threat to the US population had passed and an incidence of 20,000 new cases per year was slowly declining, even though tuberculosis was still the leading infectious cause of death globally (McMurray 1996); however, the situation in the nineteenth century changed dramatically. After the nineteenth century, the incidence of tuberculosis slightly increased and kept increasing primarily due to the acquired immunodeficiency syndrome (AIDS) epidemic according to Control of Communicable Diseases and Prevention (2007). At the same time, multiple drug resistance strains of *M. tuberculosis* were on the increase.

Geographical Distribution and Economic/Public Health Importance

Bovine tuberculosis (BTB) is one of the transboundary diseases and an infectious disease caused by *Mycobacterium bovis* a member of the *Mycobacterium tuberculosis* complex (MTC) (Ameen et al. 2008). *Mycobacterium tuberculosis* complex species that affect cattle primarily are the *Mycobacterium bovis* and *Mycobacterium caprae* but also affect goats, camels, horses, pigs, dogs, and cats among other animals including human beings (Erler et al. 2004; Prodinger et al. 2002; Thoen and Steel 2009). In the Sahel Africa, tuberculosis in goats and sheep is primarily caused by *M. caprae* species which has a serious socioeconomic impact as comparable as *M. bovis* and poses a public health threat to the livelihoods of livestock farmers. Tuberculosis in animals can lead to reduction in the quality and quantity of meat, milk and diary products, livestock products like hides, skins, and fibers, and animal power for transport and traction.

According to Corbett et al. (2003), worldwide, tuberculosis caused about 2 million deaths and about 9 million new cases had been reported annually, with the sub-Saharan Africa having the highest annual risk of infection with TB, probably due to upsurge of HIV/AIDS pandemic, globalization, land encroachment, and climate change. Globally, *M. bovis* accounts for 3.1% of all human TB cases (Ameen et al. 2008). The global TB burden as a result of *M. bovis* and *M. caprae* in Sahelian Africa is not known. This may be due to the fact that human TB due to *M. bovis* and *M. caprae* cannot be differentiated from that due to *M. tuberculosis* with respect to clinical signs and symptoms and pathological and radiological features (Ameen et al. 2008).

Tuberculosis as reported by Thoen and Steel (2009) contributes largely to the cause of human morbidity and mortality in many developing countries including sub-Saharan Africa. Human tuberculosis (HTB) is caused by *Mycobacterium tuberculosis* but in some cases caused by *M. bovis* from ingestion of contaminated milk. In developed countries, before the advent of control and elimination of bovine tuberculosis (BTB) in which one of the strategy for control was the introduction of milk pasteurization, zoonotic infection with *M. bovis* was a major cause of HTB (Cosivi et al. 1998). In geographic regions where tuberculosis is prevalent in animals, human tuberculosis cases due to *M. bovis* and *M. caprae* occurred (Thoen and Steel 2009) resulting from the ingestion of contaminated unpasteurized milk, raw meat, or improperly cooked meat and also by inhaling cough spray from infected livestock (Ayele et al. 2004).

Tuberculosis (TB) in animals is regarded as a poverty-related zoonosis with little or no attention given to it (WHO 2006). TB in animals has a serious economic impact on livestock, thereby reducing quality and productivity in livestock (Müller et al. 2008). In addition, it can persist in wildlife reservoirs and thus affect the complete ecosystem (Müller et al. 2008). As reported by WHO (2006, 2008), TB is still prevalent in the developed countries; however, it is commonly found prevalent in developing countries, with insufficient financial and human resources to control the disease (Zinsstag et al. 2006).

Tuberculosis in animals is still widespread in the Sahel Africa and its connected regions. The Sahel region of Africa is greatly involved in animal production with countries like Mali, Mauritania, Somalia, Ethiopia, Pakistan, Nigeria, Kenya, Niger, Egypt, and the United Arab Emirates being among the main cattle producing countries and also with published reports documenting the presence of tuberculosis (Müller et al. 2008; Anita 2014). In the Sahel Africa, TB caused by *M. bovis* and *M. caprae* species in humans seems to be especially prevalent among the pastoralists (Cadmus et al. 2006), who herd their cattle across the borders of the country. Pastoralists use milk which they do not usually boil from their cattle for food. TB in humans as a result of *M. bovis* and *M. caprae* species is becoming a serious public health issue in developing countries like Sahel Africa as humans and animals share the same micro-environment and dwelling premises especially in rural areas. The close contacts between humans and animals are increasing everyday globally due to increase in population density and growth especially in low-income developing countries where livestock production is their mainstay and offers a pathway out of poverty (WHO 2006).

In areas without infectious wildlife, nomadic movement which involves the movement of cattle, goat, and sheep from one route to another route remains a possible means for transmission of tuberculosis (WHO 2006). Factors that influence the routes by which cattle, sheep, and goats become infected include age, environment, and local farming practices. Infection via the alimentary (ingestion) route can also be seen in young calves ingesting milk from tuberculosis dams, although mesenteric (intestinal) lesions are relatively rare in countries with advanced control programs (Pavlik et al. 2002). Factors that influence direct transmission via the respiratory route include high stocking density and substantial cattle, sheep, and goat movement. However, zoonotic diseases affect livestock and humans with significant adverse effects on animal productivity and the health of the population especially among the poor and who are more vulnerable to the diseases (Pavlik et al. 2002). According to Corbett et al. (2003), emerging or reemerging animal and human TB caused by pathogenic bacteria of the family of *Mycobacterium tuberculosis* complex is becoming widely spread and affects the livestock industries and human health in the Sahelian African countries (Ayele et al. 2004).

Tuberculosis in cattle, sheep, and goats affects their health, reduces profitability, impacts negatively on international trade, and can alter genetic improvement toward desirable traits in animals (Boland et al. 2010). It also has a negative impact on the welfare of families who are into farming. Transmission from animals to humans and humans to animals still occurs and is considered a public health risk and of high concern, despite control measures and policies instituted by the government which includes herd testing, pasteurization, effective meat inspection, health surveillance, and BCG vaccination (Moda et al. 1996; Smith et al. 2004). Although there is the risk of transmission from humans to animals, some more recent opinion considers this risk to be negligible (Torgerson and Torgerson 2010). Hence, TB control is currently more focused on the implications of international trade. Despite the efforts made toward the eradication of TB through sustained and costly implementation of various control and eradication programs beginning from the nineteenth century, TB has not been eradicated in the Sahelian Africa and its connected regions.

Epidemiology

Tuberculosis caused by *Mycobacterium species* has a worldwide distribution (OIE 2007). In several countries, TB remains a major and costly infectious disease of cattle and other domesticated and wild animal populations which include badgers, possums, deer, goats, sheep, camelids, etc. (Pollock and Neill 2002; Carslake et al. 2011). According to OIE (2011) and Cousins and Roberts (2001), TB in cattle, sheep, and goats is of socioeconomic or public health importance in developing countries and of significance to the international trade of animals and animal products.

A key to understanding bovine TB epidemiology can be seen in the relationship between infection and disease (TB) and the relationship between disease and transmission. Biological, social, and environmental factors are known to influence both transmission and susceptibility of the host (WHO 2014).

Zoonotic tuberculosis is the result of TB infection among domestic species. These domestic animals that play a significant role in the epidemiology of zoonotic tuberculosis include goats, sheep, pigs, farmed deer, and camel, while cases in horses, cats, and dogs are rare (Anita 2014). The occurrence and the distribution of *Mycobacterium* infection in cattle and sheep from a global point of view is low but varies greatly in the husbandry system of the Sahelian African countries. TB infection in animals like goats can be the result of co-grazing with cattle herds infected by TB (Anita 2014).

In pastoralist communities of the Sahelian Africa and its connected regions, animals like camels provide meat and milk and serve as draft power for transportation of goods especially for long-distance journey. The constant close connection between pastoralists and their animals promotes the bidirectional transmission of tuberculosis from animals to humans. The existence of TB and how widespread and significant TB is in camel population is still largely unknown; however, research has shown its presence in the United Arab Emirates, Egypt, Mauritania, Somalia, Pakistan, and Niger as reported by Anita (2014).

Zoonotic TB is the result of the adaptability of Mycobacterium species in different hosts (Radostitis et al. 2000). Zoonotic TB Infection due to *M. bovis* was once a major problem in developed countries, but following the implementation of eradication programs and policies such as test-and-slaughter policy and milk pasteurization, the incidence drastically reduced (Cousins and Roberts 2001). Despite these efforts, surveillance and control activities for bovine tuberculosis are often inadequate and unavailable. Approximately 85% of the cattle population and 82% of the human population of the Sahelian Africa are in regions where surveillance and control activities for bovine tuberculosis are either inadequate or unavailable, thereby making the burden and the epidemiologic and public health impact of the infection among animals and humans greatly unknown (Cosivi et al. 1998). With the emergence and reemergence of drug-resistant strains of *Mycobacterium* species and the rise in TB and HIV/AIDS co-infection, assessing the burden of this disease and its current situation in Africa are further complicated (WHO 2011).

Transmission

M. bovis and *M. caprae* can be transmitted by the inhalation of aerosols, by ingestion of any animal product infected with the disease, or through breaks in the skin. However, the routes of transmission vary between species. Cattle, goats, and sheep serve as reservoirs for tuberculosis. In the absence of main maintenance hosts, populations of spillover hosts do not harbor *Mycobacterium* indefinitely, but may transmit the infection between their members (or to other species) for a time. However, due to high population density, some spillover hosts can become reservoirs for TB (Etter et al. 2006).

Cattle excretes *M. bovis* and *M. caprae* in milk, respiratory secretions, feces, and sometimes in the urine, vaginal secretions, or semen. A larger number of organisms may be shed in the late stages of infection. According to Menzies and Neill (2000), carriers include asymptomatic and anergic carriers. In most cases, *M. bovis* and *M. caprae* are transmitted between cattle via aerosols as a result of close contact among cattle. Animals like sheep and goats become infected when they ingest the organism, as seen in calves that nurse from infected cows. It is important to note that not all infected cows transmit the disease to their young calves (Menzies and Neill 2000). Transmission of any of the *Mycobacterium* species can be either direct through close association between animals and humans or indirect from exposure to viable bacteria in a contaminated environment like pasture, feed, and housing. Movement of cattle, sheep, and goats during co-grazing can facilitate most transmission in areas without infectious wildlife (Barlow et al. 1997). Risk factors like age, environment, and practice of local farming are likely to influence the routes by which cattle, sheep, and goats become infected. Other factors that promote direct transmission via the respiratory route include natural cattle, sheep, and goat behavior and high stocking density.

Clinical Signs

In animals like cattle, sheep, and goats, granulomatous lesions are seen mainly in the lungs, lymph nodes, and spleen (Anita 2014). Often affected animals may remain for so many years without showing any clinical signs. However, some of the classical signs include difficulty in breathing, persistent and painful cough which may develop during chronic stage accompanied by hyperpnea and dyspnea, and drastic change in the animal's body condition, and the animal gradually emaciates with dull and sunken eyeballs (Anita 2014) (Plate 19.1).

Diagnosis

Diagnosis of TB in live cattle in the field involves the use of tuberculin skin test especially among herds of cattle. The tuberculin skin test is conducted by injecting tuberculin intradermally into the cattle. A positive is shown by a delayed swelling around the site of injection and this is otherwise known as hypersensitivity reaction. Bovine tuberculin can be used alone when performing the tuberculin skin test or a comparative test can be employed to distinguish reactions to *M. bovis* from other

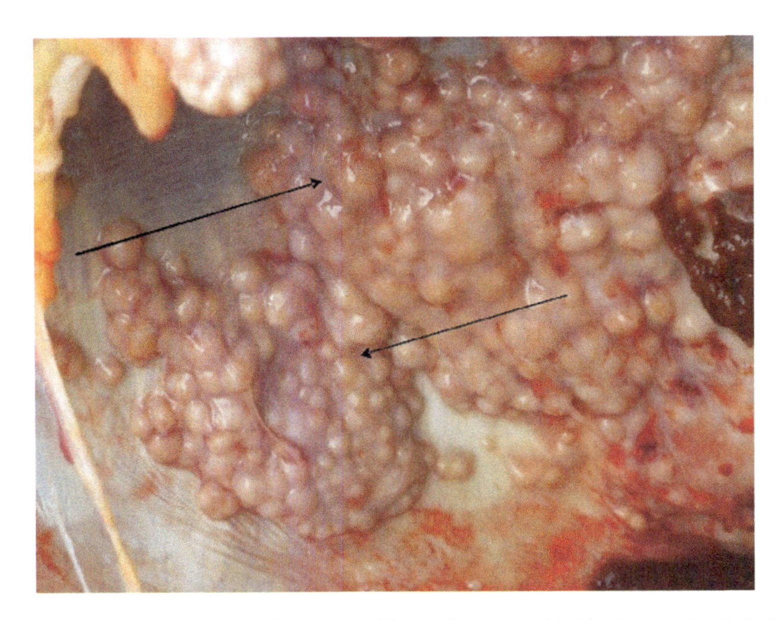

Plate 19.1 Picture showing small discreet grayish nodules (arrows) indicative of tubercle lesions in the lungs of a cow

environmental *Mycobacterium* species (Cousins and Florisson 2005). The use of tuberculin skin test for detection of TB infection is constrained by test sensitivity. A reduction in testing frequency results in increased prevalence and a reduced ability to detect disease. To compound this, a substantial proportion of cattle, sheep, and goats were never tested for TB (Mitchell et al. 2005), and there is also a period of reduced reactivity in infected animals following an initial tuberculin test, the precise duration of which has not been determined.

Laboratory Tests

However, a presumptive diagnosis of TB is done using the acid-fast bacilli (AFB) test whereby direct smears from clinical samples or tissues may be stained with immunoperoxidase technique, Zeihl-Neelson stain, or a fluorescent acid-fast stain (Cousins and Florisson 2005).

For confirmation of TB infection in animals and the differentiation of the causative agent Mycobacterium species, culture test is conducted and usually done by the isolation of *M. bovis* and *M. caprae* on selective media such as Lowenstein-Jensen medium and incubated for a period of 8 weeks (Ochei and Kolhatker 2008). The cultural characteristics of the organism isolated can be confirmed with biochemical tests or the use of polymerase chain reaction (PCR) assay which is quicker but very expensive.

Other molecular techniques that have been applied for TB detection include (a) genetic fingerprinting techniques, for example, spoligotyping; (b) lymphocyte proliferation and gamma interferon assays (a blood test that measures cellular immunity); (c) enzyme-linked immunosorbent assays (ELISA) which measure antibody titers to *M. bovis* and *M. caprae*; and (d) GeneXpert MTB which is a cartridge-based automated diagnostic test that can identify *Mycobacterium tuberculosis* (MTB) complex including *M. bovis* and *M. caprae* and resistance to rifampicin (RIF). With additional financial support provided by the US National Institute of Health (NIH) and OIE (2011), the GeneXpert MTB technique was jointly developed by the laboratory of Professor David Alland at the University of Medicine and Dentistry of New Jersey (UMDNJ) (WHO 2013) in collaboration with the Foundation for Innovative New Diagnostic.

According to the World Health Organization (WHO 2010), the GeneXpert MTB/RIF was endorsed for use in TB endemic countries as categorized by WHO, and its major milestones for global TB diagnosis were also declared after an 18 months rigorous assessment of its effectiveness in TB, MCR-TB, and TB/HIV co-infection on the field (Small and Pai 2010). The GeneXpert test has the propensity to revolutionize TB diagnosis (Van Rie et al. 2010). Deoxyribonucleic acid (DNA) sequences specific for *M. bovis* and *M. caprae* and rifampicin resistance can be detected by the GeneXpert/RIF using polymerase chain reaction technique and relies on the Cepheid GeneXpert system, a platform used for rapid and simple nucleic acid amplification tests (NAAT) (Van Rie et al. 2010). This is achieved by the purification and concentration of the isolate genomic materials or testing of lymph nodes and other tissues from the captured bacteria by sonication and subsequently amplifying the genomic DNA by PCR. Molecular beacons in a real-time format using fluorescent probes were used to identify all the clinically relevant rifampicin resistance-inducing mutations in the RNA polymerase beta (rpo ß) gene in the *M. bovis* and *M. caprae* genome. Xpert TB replaced smear microscopy with a pooled sensitivity of 88% and specificity of 98%. However, sensitivity was only 67% and specificity 98%, when Xpert TB was employed for use as an add-on in cases of negative smear microscopy (Steingart 2013).

Another important molecular technique is the loop-mediated isothermal amplification (LAMP) which is known as a single-tube technique for the amplification of deoxyribonucleic acid (Mori et al. 2001). LAMP is a very cheap isothermal which eradicates the need for expensive thermocyclers used in conventional polymerase chain reaction and can serve as an alternative technique in the future for detection of certain diseases. LAMP technique can be combined with a reverse transcription step to allow for the detection of ribonucleic acid (RNA). LAMP can be used as a simple screening technique on the field or can be used by clinicians at the point of care (Sen and Ashbolt 2010). According to MaCarthur George (2009), in low- and middle-income countries, LAMP may be useful for infectious diseases diagnosis. Major advantages of LAMP include: (1) it is a relatively new DNA amplification technique, (2) simple to use, and (3) relatively inexpensive. In LAMP method, either two or three sets of primers or a polymerase with high strand displacement activity plus a replication activity are used to amplify the target sequence at a constant temperature of 60–65 °C. Four different primers are employed to identify six different regions on the target gene, and this increases the specificity of LAMP. However, as a result of the specific nature of the action of these primers, the quantity of DNA produced in LAMP is substantially higher than the amplification produced by PCR. Research has shown that LAMP is widely being used for detecting infectious diseases such as tuberculosis (Geojith et al. 2011). However, in developing countries, the use of LAMP technique for the detection of other common pathogens is yet to be extensively validated (Small and Pai 2010).

Spoligotyping is a new method developed recently and used for the detection and typing of *M. bovis* and *M. caprae* bacteria simultaneously (Andrea et al. 2005). This method involves the amplification of a highly polymorphic direct repeat locus in the *M. bovis* and *M. caprae* genome by polymerase chain reaction (PCR). Results can be obtained from a *M. bovis* and *M. caprae* culture within 1 day. Features of spoligotyping include: (1) it is very rapid, (2) it can detect causative bacteria, and (3) it provides epidemiologic information on strain identities (Andrea et al. 2005). Spoligotyping is very useful in the surveillance of tuberculosis transmission and in interventions to prevent further spread of this disease especially when implemented in a clinic setting.

Latent Infection

In human TB epidemiology, the concept of "latent" infection is well supported and is defined as where the pathogen resides long term in the host which may or may not be detectable (Manabe and Bishai 2000). However, where infection is relatively common, productive (transmitting) infection is rare. Latent TB is currently regarded as a spectrum of pathogen burden and host immune control (Sridhar et al. 2011), and it is seen to occur when the pathogen is forced into a state of allegedly nonreplicating persistence by a host response (Palmer and Water 2006). However, current researches suggest that these interactions between the pathogen and the host response are largely pathogen driven (Bold and Ernst 2009; Ehlers 2010; Sasindran and Torrelles 2011).

As reported by Brites and Gagneux (2011), the ability of pathogens to establish latent infections enhances the transmission of TB. This ability may have evolved as a means of adaptation for persistence when population densities were low. Prolonged latency period as a result of reactivation may provide an opportunity for these pathogens to overtake an entire generation to access new vulnerable. *M. tuberculosis* has an enduring phenotypic feature, which could allow it to resist the antibacterial resistance of the host and to adapt to long-term survival within the host (Keren et al. 2011) and possibly even the environment. Though the extent of latency and reactivation applicable to cattle and other animal populations is not yet known (Van Rhijn et al. 2008), the process of latency and reactivation has been reported to operate in cattle and possibly other animal populations (Pollock and Neill 2002). The significance of a latent infection is that it can, under normal conditions, reactivate to full-blown TB. The physical nature of the animal during latency determines what strategies to be employed such as post-exposure vaccination and what more effective method to be employed for diagnosis in order to detect the disease at the latent phase. For instance, the disease may not be detectable by current diagnostics methods, if it is truly dormant and hidden from the immune system.

Based on analysis of the size and distribution of lesions in cattle, Fritsche et al. (2004) suggests that disease exacerbation could occur following periods of lesion dormancy, though relatively poor test sensitivity might explain such observations. The concept of latency in cattle can be further illustrated, when cattle with no physical lesions shown on any of the affected organs are positive for TB using IFN test (Neill et al. 1994; Monaghan et al. 1994) or culture test (Cassidy et al. 1999).

Prevention and Control/Adopted Surveillance and Control Strategies

One of the control strategies for the control of tuberculosis in cattle, sheep, and goats and other infected wildlife is the test-and-slaughter or test-and-segregation method (OIE 2007). Other control measures include quarantining of infected herds and sanitation and disinfection. To eliminate cattle that may shed the organism thereby increasing the risk of transmission among herds, affected herds are re-tested periodically by employing the use of tuberculin skin test, otherwise known as the Mantoux test. It is recommended by OIE (2007) that infected herds should be quarantined and animals that have been in contact with reactors should also be traced.

Test-and-slaughter technique is the best method to eradicate tuberculosis from domesticated animals (OIE 2007), though with very negative economic impact. Sanitation and disinfection is another method of prevention that involves the use of 5% phenol iodine solutions with a high concentration of available iodine, glutaraldehyde, and formaldehyde. The use of 1% sodium hypochlorite with a long contact time is also effective in an environment with low concentration of organic material (Cousins and Roberts 2001). Moist heat of 121 °C (250 °F) for a minimum of 15 min has been employed for destroying *M. bovis* in animal products (Cousins and Roberts 2001).

Another method that has shown to produce positive result toward the eradication of TB is culling to reduce the population density thereby decreasing transmission, though the occurrence of *Mycobacterium species* in wildlife reservoir hosts impedes these eradication efforts. Effects of culling include increase in the scattering of the remaining animals and restrictions of supplemental feeding areas (Lees 2004). To mitigate against the effect of culling, fencing around the hay storage areas can prevent access to wildlife. In addition, the interactions between wildlife and domesticated animals can be reduced by implementing biosecurity measures on farms (Cousins and Roberts 2001).

Adequate control measures in order to reduce the incidence in animals and humans should be adopted. The measures may include:

1. Intensified public education for the awareness about the public health implication of zoonotic TB.
2. Active surveillance for TB in goats, cattle, sheep, and other domestic animals at international borders should be instituted.
3. Test-and-slaughter policy should be adopted in order to improve animal and human health.
4. Government should enforce the policy of compulsory tuberculin skin testing of dairy animals such as cattle, twice a year, and destruction of all dairy animals positive for tuberculin skin test and secreting acid-fast bacilli, with full compensation to the owners.

Conclusion

In conclusion, the pastoralist nature of the people of the Sahelian Africa and its connected regions contributes significantly to the transmission of tuberculosis within this region and across borders. Reducing tuberculosis in the Sahelian region requires a long-term TB control and public health strategy which requires enormous capital in terms of finance and manpower.

References

Ameen SA, Adedeji OS, Raheem AK, Leigh OO, Rafiu TA, Ige AO. Current status of bovine tuberculosis in Oyo state. Middle-East J Sci Res. 2008;3(4):207–10.

Andrea G, Alessandra B, Giulia M, Anna DE, Lidia C, Gian PN, Lidia G, Giulio F, Jan DA, Dick VS, Mauro M, Fabio F. Spoligotying and *Mycobacterium tuberculosis*. Emerg Infect Dis. 2005;11:1242–8.

Anita LM. Zoonotic aspects of tuberculosis: disease of the past or re-emerging zoonosis. In: Zoonosis infections affecting humans and animals: focus in public health. Dordrecht: Springer; 2014. p. 891–914.

Ayele WY, Neill SD, Zinsstag J, Pavlik I. Bovine tuberculosis an old disease but a new threat to Africa. Int J Tuberc Lung Dis. 2004;8:924–37.

Barlow ND, Kean JM, Hickling G, Livingstone PG, Robson AB. A simulation model for the spread of bovine tuberculosis within New Zealand cattle herds. Prev Vet Med. 1997;32(1–2):57–75.

Bhamidi S. Mycobacterial cell wall Arabinogalacten. In: Bacterial polysaccharides: current innovations and future trends. Wymondham, UK: Caister Academic; 2009.

Boland F, Kelly GE, Good M, More SJ. Bovine tuberculosis and milk production in infected dairy herds in Ireland. Prev Vet Med. 2010;93(2–3):153–61.

Bold TD, Ernst JD. Who benefits from granulomas, mycobacteria or host? Cell. 2009;136(1):17.

Brites D, Gagneux S. Old and new selective pressures on *Mycobacterium tuberculosis*. Infect Genet Evol. 2011;12(4):678–85.

Cadmus S, Palmer S, Okker M, Dale J, Gover K, Smith W, Jahan K, Hewison RG, Gordon SV. Molecular analysis of human and bovine tubercle bacilli from local setting in Nigeria. J Clin Microbiol. 2006;39:222–7.

Carslake D, Grant W, Green LE, Cave J, Greaves J, Keeling M, McEldowney J, Weldegebriel H, Medley GF. Endemic cattle diseases: comparative epidemiology and governance. Philos Trans R Soc Lond Ser B Biol Sci. 2011;366(1573):1975–86.

Cassidy JP, Bryson DG, Pollock JM, Evans RT, Forster F, Neill SD. Lesions in cattle exposed to *Mycobacterium bovis*-inoculated calves. J Comp Pathol. 1999;121(4):321–37.

Control of Communicable Diseases and Prevention. Tuberculosis. In: Manual of tuberculosis. 18th ed; 2007.

Corbett EL, Watt CJ, Walker N. The growing burden of tuberculosis: global trends and interactions with the HIV epidemic. Arch Intern Med. 2003;163(9):1009–21.

Cosivi O, Grange JM, Daborn CJ, Raviglione MC, Fujikura T, Cousin D, Robinson RA, Huchzermeyer HE, de Kantor AK, Meslin FX. Tuberculosis due to *Mycobacterium bovis* in developing countries. Emerg Infect Dis. 1998;4:59–70.

Cousins DV, Florisson NA. A review of tests available for use in the diagnosis of tuberculosis in non-bovine species. Rev Sci Tech. 2005;24:1039–59.

Cousins DV, Roberts JL. Australia's campaign to eradicate bovine tuberculosis: the battle for freedom and beyond. Tuberculosis (Edinburge). 2001;81(1–2):5–15.

Ehlers S. TB or not TB? Fishing for molecules making permissive granulomas. Cell Host Microbe. 2010;7(1):6–8.

Erler W, Martin G, Sachase K, Naumann L, Kablau D, Beer J, Bartos M, Nagy G, Vetnic Z, Zolnr-Dove M, Pavlik I. Molecular fingerprinting of *Mycobacterium bovis* subsp. *caprae* isolates from Central Europe. J Clin Microbiol. 2004;42:2234–8.

Etter E, Donado P, Jori F, Caron A, Goutard F, Roger F. Risk analysis and bovine tuberculosis; a re-emerging zoonosis. Ann NY Acad Sci. 2006;1081:61–73.

Fritsche A, Engel R, Bulb D, Zellweger JP. Mycobacterium bovis tuberculosis: from animal to man and back. Int J Tuberc Lung Dis. 2004;8:903–4.

Geogie M. Global health diagnostics: research development and regulation: workshop report. London: Academy of Medical Sciences; 2009.

Geojith G, Dhanasekaran S, Chandran SP, Kenneth J. Efficacy of loop mediated isothermal amplification (LAMP) assay for the laboratory identification of *Mycobacterium tuberculosis* isolates in a resource limited setting. J Microbiol Methods. 2011;84(1):71–3.

James H. History of tuberculosis. In: The champion; 1999. p. 636–734.

Keren I, Minami S, Rubin E, Lewis K. Characterization and transcriptome analysis of *Mycobacterium tuberculosis* persisters. Microbiology. 2011;2(3):e00100-11.

Lees VW. Learning from outbreak of bovine tuberculosis near Riding Mountain National Park: application to a foreign animal disease outbreak. Can Vet J. 2004;45:28–34.

Manabe YC, Bishai WR. Latent *Mycobacterium tuberculosis*-persistence, patience, and winning by waiting. Nat Med. 2000;6(12):1327–9.

McMurray DN. 'Mycobacteria and Nocardia' in Baron. In: Baron's medical microbiology. 4th ed. Galveston, TX: University of Texas Medical Branch; 1996.

Menzies FD, Neill SD. Cattle-to-cattle transmission of bovine tuberculosis. Vet J. 2000;158:245–6.

Mitchell A, Bourn D, Mawdsley J, Wint W, Clifton-Hadley R, Gilbert M. Characteristics of cattle movements in Britain: an analysis of records from the cattle tracing system. Anim Sci. 2005;80 (3):265–73.

Moda G, Daborn CJ, Grange JM, Cosivi O. The zoonotic importance of *Mycobacterium bovis*. Int J Tuberc Lung Dis. 1996;77(2):103–8.

Monaghan ML, Doherty ML, Collins JD, Kazda JF, Quinn PJ. The tuberculin tests. Vet Microbiol. 1994;40(1–2):111–24.

Mori Y, Nagamne KN, Notomi T. Detection of loop-mediated isothermal amplification reaction by turbidity derived from magnesium pyrophosphate formation. Biochem Biophys Res Commun. 2001;289:150–4.

Müller B, Steiner B, Bonfoh B, Fané A, Smith NH, Zinsstag J. Molecular characterization of *Mycobacterium bovis* isolated from cattle slaughtered at the Bamako abattoir in Mali. BMC Vet Res. 2008;4:26.

Neill SD, Pollock JM, Bryson DB, Hanna J. Pathogenesis of *Mycobacterium bovis* infection in cattle. Vet Microbiol. 1994;40:41–52.

Ochei J, Kolhatker A. Pathogenecity of mycaobacteria. In: Medical laboratory science. 7th ed. - London: McGraw-Hill; 2008. p. 724–9.

OIE. 2011. http://www.oie.int/animal-health-in-the-world/oie-listed-diseases-2011.

OIE, World Organization for Animal Health. Bovine tuberculosis. In: World animal health information database (WAHID). 2007. http://www.oie.int/wahid.prod/public.php?page=diseasestatuslistanddiseaseid=32.

Omer EE, Sanosi SM, Greafer CJ. Tuberculin sensitivity in Sudanese population in contact with cattle. Bull Anim Health Prod Afr. 1995;28:91–6.

Palmer MV, Water WR. Advances in bovine tuberculosis diagnosis and pathogenesis: what policy makers need to know. Vet Microbiol. 2006;112:181–90.

Pavlik I, Bures F, Janovsky P, Pecenka P, Bartos M, Dvarska L, Mattors L, Kremer K, Van Soolinger D. The last outbreak of bovine tuberculosis in cattle in the Czech Republic in 1995 was caused by *Mycobacterium bovis* subspecies *caprae.* Vet Med. 2002;47:251–63.

Pollock JM, Neill SD. *Mycobacterium bovis* infection and tuberculosis in cattle. Vet J. 2002;163:115–27.

Prodinger WM, Eigentler A, Allerberger F, Schonbauer M, Glawiachnig W. Infection of red deer, cattle, and humans with *Mycobacterium bovis* subsp. caprae in Western Austria. J Clin Microbiol. 2002;40:2270–2.

Radostitis OM, Gray GC, Blood DC, Hinchelift KW. A textbook of disease of cattle, sheep, pig, goat and horses. In: Veterinary medicine. 9th ed. London: Harcourt; 2000. p. 909–18.

Ryan KJ, Ray CG. Sherris medical microbiology. 4th ed. New York: McGraw Hill; 2004.

Sasindran SJ, Torrelles JB. *Mycobacterium tuberculosis* infection and inflammation: what is beneficial for the host and for the bacterium? Front Microbiol. 2011;2:2.

Sen K, Ashbolt NK. Environmental microbiology: current technology and water application. Wymondham, UK: Caister Academic; 2010.

Small PM, Pai M. TB diagnosis-time for a game change. N Engl J Med. 2010;363:1000–71.

Smith RM, Drobniewski F, Gibson A, Montague JD, Logan MN, Hunt D, Hewinson G, Salmon RL, O'Neill B. *Mycobacterium bovis* infection, United Kingdom. Emerg Infect Dis. 2004;10 (3):539–41.

Sridhar S, Pollock K, Lalvani A. Redefining latent tuberculosis. Future Microbiol. 2011;6:1021–35.

Steingart KR. Xpert MTB/RIF assay for pulmonary tuberculosis and rifampicin resistance in adults. Cochrane Database Syst Rev. 2013;1:CD009593.

Thoen CO, Steel JH. Regional and country status reports. Part 2. *Mycobacterium bovis* infections in animals and humans. Ames, IA: IOWA State University Press; 2009. p. 167–345.

Torgerson PR, Torgerson DJ. Public health and bovine tuberculosis: what's all the fuss about? Trends Microbiol. 2010;18(2):67–72.

Van Rhijn I, Godfroid J, Michel A, Rutten V. Bovine tuberculosis as a model for human tuberculosis: advantages over small animal models. Microbes Infect. 2008;10(7):711–5.

Van Rie A, Page-Shipp L, Scott L, Scanne I, Stevens W. Xpert MTB/RIF for point-of-care diagnosis of TB in high-HIV burden resource limited countries. Expert Rev Mol Diagn. 2010;10:937–46.

WHO. 2010. WHO endorses new rapid tuberculosis test.

WHO. 2011. http://www.who.int/tb/publications/global_report/2011/gtbr11_full.pdf.

WHO. 2013. WHO monitoring of Xpert MTB/RIF rollout.

World Health Organization. 2006. Global tuberculosis control report – Annex 1 profiles of high-burden countries (PDF).

World Health Organization. 2014. Childhood TB: training toolkit. www.who.int/tb/challengestraining_manual.

World Health Organization. WHO report 2008: global tuberculosis control.
Zinsstag J, Kazwala RR, Cadmus I, Ayanwale I. *Mycobacterium bovis* in Africa. In: Thoen CO, Steele FH, Gilsdorf MJ, editors. *Mycobacterium bovis* infection in animals and humans. 2nd ed. Ames, IA: IOWA State University Press; 2006. p. 199–210.

Chapter 20
Contagious Bovine Pleuropneumonia

William Amanfu

Abstract Lung sickness or contagious bovine pleuropneumonia (CBPP) is a pneumonic transboundary animal disease that specifically affects cattle (*Bos taurus*, *Bos indicus*) and occasionally water buffalo (*Bubalus bubalis*). The African buffalo (*Syncerus caffer*) is not affected by the disease. CBPP is caused by *Mycoplasma mycoides mycoides* (Mmm) which phylogenetically belongs to the Mycoplasma cluster or group which are pathogens of ruminants. The acute to subacute disease is characterized by pleuropneumonia and severe pleural effusion, with sequestra formation being a predominant feature in the chronic disease. Contagious bovine pleuropneumonia is one of the major diseases affecting cattle in Africa now that rinderpest has been eradicated from the continent. It is partly responsible for food security deficits in areas where the disease occurs. The most important impact of the disease is the effect on meat and milk production and on crop production whereby oxen are used for traction of farm inputs or products and land preparation (plowing) for food crop production. The control or eradication of the disease continues to suffer. This arises out of lack of financial and human resources to support the control of the disease. The decline in performance of the veterinary services due to many factors has been experienced in many countries in Africa. Lack of a credible livestock compensation policy in affected countries has hindered prompt reporting of the disease in many countries in Africa. Often, this results in farmers attempting to treat affected cattle with antibiotics—mostly tetracyclines. The present epidemiology of CBPP in parts of Africa affected by disease demands that preemptive steps are undertaken to protect countries in Southern Africa free from the disease from being infected, thus affecting the lucrative beef industry in that region and negatively affecting people's livelihoods in these countries. At present, CBPP control using vaccination (T1/44 and T1/SR) has been carried out in an uncoordinated manner, with the result that the disease is still prevalent in parts of the continent.

W. Amanfu (✉)
International Veterinary Consultant, Arts Center, Accra, Ghana

M. Kardjadj et al. (eds.), *Transboundary Animal Diseases in Sahelian Africa and Connected Regions*, https://doi.org/10.1007/978-3-030-25385-1_20

Keywords Contagious bovine pleuropneumonia · Fibrinous · Hemorrhagic septicemia · Vaccination · *Pasteurella multocida* · Surveillance · Transboundary animal disease

Introduction

Lung sickness or contagious bovine pleuropneumonia (CBPP) is a pneumonic transboundary animal disease that specifically affects cattle—both humped and humpless. The disease is caused by *Mycoplasma mycoides mycoides* (Mmm) which phylogenetically is a member of the Mycoplasma cluster or group which are pathogens of ruminants. The acute to subacute disease is characterized by pleuropneumonia and severe pleural effusion, with sequestra formation being a predominant feature in the chronic disease. CBPP is one of the major diseases affecting cattle and is partly responsible for food security deficits in areas where the disease occurs. CBPP is very contagious with a mortality rate of up to 50% or more in susceptible cattle populations and causes significant economic losses. The disease is partly responsible for food security deficits in areas where it occurs and significantly affects peoples' livelihoods.

A comprehensive historical account of CBPP disease was provided by Provost et al. (1987). According to these authors, the initial accounts of the disease were described by Gallo in 1550. In 1843, CBPP was introduced into the USA through a dairy cow purchased off an English ship. The spread of the disease was so fast in the USA that by 1884, CBPP had become so widespread that the Federal Government established the Bureau of Animal Industries to combat the disease. In 1887, an intensive campaign to prevent and control animal diseases such as CBPP by quarantine and slaughter began in the USA.

According to records, CBPP was introduced into South Africa by a Friesian bull imported from the Netherlands in 1854. Encountering a naïve CBPP cattle population, the introduction of the disease spread rapidly throughout Africa south of the Sahara and within 2 years caused the death of over 100,000 cattle in South Africa alone (Thiaucourt et al. 2004). CBPP was eradicated from South Africa in 1924, Zimbabwe in 1904, Botswana in 1939 (reappeared in 1995), and Australia in 1972. South America and the Island of Madagascar have never been contaminated by the disease.

Contagious bovine pleuropneumonia is suspected to occur occasionally in the Middle East and possibly parts of Asia where hemorrhagic septicemia caused by strains of *Pasteurella multocida* (B: 2) is prevalent. This disease has symptoms and gross pathology similar to acute phase of CBPP. In Africa, CBPP is found in an area south of the Sahara Desert, from the Tropic of Cancer in the North to the Tropic of Capricorn in the South. Cattle production in Africa is seriously hampered by the presence of CBPP and together with foot and mouth disease is regarded as major impediments to cattle production in Africa, now that rinderpest has been eradicated globally.

The Etiology of CBPP

The organism *Mycoplasma mycoides* subsp. *mycoides* (Mmm) which causes CBPP is a pleomorphic bacterium that lacks a protective cell wall. This feature of the Mycoplasmas explains why antibiotics directed at the cell wall, are ineffective against this bacterium. Primary isolation of this organism is effected in broth media that must contain 10–20% equine serum, glucose, fresh yeast extracts, and antimicrobials to inhibit bacterial and fungal growth.

Initial growth of Mmm in liquid medium is relatively slow and a minimum of 24–48 h are usually required to observe growth in liquid cultures. Cultures in liquid medium often yield the typical whitish cloud floating in the medium, often dispersed by agitation. Solid medium is prepared by the inclusion of 1% agar to the broth medium and grown under microaerophilic conditions. The Mycoplasma *mycoides* subsp. *mycoides* colonies have the “fried egg” appearance typical of mycoplasma colonies. The term pleuropneumonia-like organisms (PPLO) originally ascribed to mycoplasma isolates is based on the typical morphology of Mmm colonies on solid agar medium (Fig. 20.1). The colony size rarely exceeds 1 mm and is easily visualized using a stereomicroscope.

The CBPP agent belongs to the *Mycoplasma mycoides* cluster which consists of six pathogenic mycoplasma species that cause disease in ruminants (Cottew et al. 1987). This cluster shares many genotypic and phenotypic characteristics. The *M. mycoides* cluster comprises five recognized bacteria: *Mycoplasma mycoides* subsp. *mycoides* Small Colony (MmmSC), *M. mycoides* subsp. *mycoides* Large Colony (MmmLC), *M. mycoides* subsp. *capri* (Mmc), *Mycoplasma capricolum* subsp. *capricolum* (Mcc), and *M. capricolum* subsp. *capripneumoniae* (Mccp). With the re-designation of

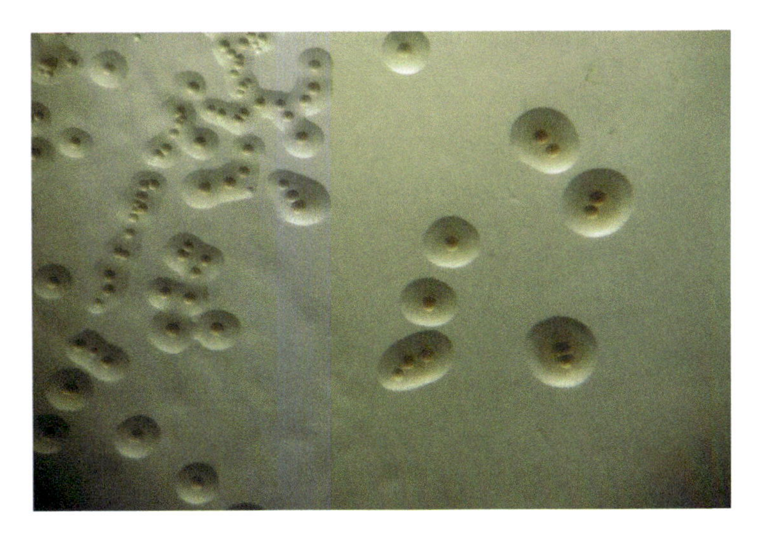

Fig. 20.1 Mmm on solid agar medium. Note the typical fried egg morphology. Source: F. Thiaucourt-Montpellier, France

MmmLC as Mmc, MmmSC can now be designated as Mycoplasma mycoides subsp. mycoides (Mmm) (Manso-Silvan et al. 2009).

Recent evidence suggests that various genotypes of Mmm can be distinguished by the use of techniques such as restriction analysis of whole DNA (Poumarat and Solsona 1995) or Southern Blotting (March et al. 2000; Cheng et al. 1995). Thus, African and European strains of Mmm can be distinguished by genetic (Vilei et al. 2000) as well as antigenic differences (Goncalves et al. 1998), an indication that the outbreaks occurring since 1980 in France, Portugal, Spain, and Italy were not due to the introduction of strains from Africa, but were more probably due to a resurgence of CBPP from a region or regions where the agent was probably never completely eradicated (Thiaucourt et al. 2004).

The agent of CBPP is susceptible to environmental determinants and survives outside the host for an average of 3 days under hot and humid conditions and for 2 weeks in temperate environments. The organism can persist for more than 10 years in frozen pleural fluid (Thiaucourt et al. 2004). It is highly susceptible (within few minutes) to ultraviolet radiation. Serial passage of Mmm in broth culture, in cattle and embryonated chicken eggs alters its virulence and pathogenicity. This property formed the basis for the development of attenuated vaccine strains of the CBPP agent.

The Geographic Extent of CBPP Disease in Africa

In the early 1970s, the disease situation seemed to be under control with the progressive vaccination of cattle against rinderpest and CBPP using the combined vaccine (Bisec) for the two diseases, especially in West and Central Africa. With the cessation of the use of Bisec and other CBPP vaccines, CBPP had a remarkable resurgence in Africa.

Subregional Epidemiologic Trends of CBPP in Africa

West and Central Africa

The main livestock management system, within the cattle producing countries of the two subregions, is the extensive pastoral system characterized by livestock transhumance system within the Sahelian zones. In Mali, the central delta of River Niger is a major convergence zone for seasonal grazing for a huge number of transhumant cattle herds from Mauritania, Burkina Faso, and Niger. In Chad, the Lake Chad basin plays a similar role for transhumant herds from Cameroon, the Central African Republic, Niger, and Nigeria. These seasonal traditional movements of transhumant cattle herds are accompanied by an important flux of cattle trade movements—more than 500,000 heads yearly (Boubacar Seck, personal Communication: 2003) directed southward throughout the year from the Sahelian zones of Burkina Faso, Mali, Mauritania, Niger, Cameroon, and Chad to coastal countries of Benin, Côte d'Ivoire, Ghana, Liberia, Nigeria, Senegal, and Togo. Such annual movements to grazing zones formed the basis of coordinated vaccination against CBPP and rinderpest in the past.

Within West and Central Africa, cattle movement monitoring by veterinary services is known to be suboptimal due to lack of human, financial, and logistic resources. Livestock industry infrastructures such as markets, abattoirs, and slaughter slabs are usually monitored by veterinary authorities, but CBPP suspected cases or lesions noticed at their level are not often integrated into the data of the national CBPP surveillance system nor does it serve as a trigger for coordinated control of disease monitoring and trace-back. The Gambia reported CBPP outbreaks for the first time in 2012 followed by Senegal in the same year. The last recorded outbreak of the disease in Gambia was in 1971 and CBPP vaccination in that country was consequently stopped in 1987. The new outbreaks formally reported to the OIE in September 2012 constituted the reoccurrence of CBPP after over 40 years of freedom from the disease. Senegal had been free from the disease for about the same period.

Southern Africa

Some countries in the subregion experienced outbreaks of CBPP in previously free countries or parts of countries. For the past three decades, outbreaks of CBPP have occurred in Botswana, (1995), D. R. Congo (1991), Tanzania (1990, 1992, 1994), Rwanda (1994), Northwestern Zambia (1997), and Burundi (1997). Apart from Botswana which succeeded in eradicating the disease in 1997, outbreaks are still continuing in some of the countries indicated. The underlining common factors to these outbreaks in previously free areas were (1) illegal movement of cattle from known infected cattle populations, (2) failure of surveillance systems, and (3) lack of emergency preparedness and early reaction to disease outbreaks. In Botswana, outbreaks of CBPP occurred after over half a century of freedom from the disease (Amanfu et al. 1998a). This outbreak was eradicated by the slaughter of about 320,000 infected and in-contact cattle with compensation and restocking with 70,000 cattle imported from the CBPP-free zone of Namibia (Amanfu et al. 1998b). International cooperation in the eradication of CBPP in Botswana was encouraged between South Africa, Namibia, and others. The technical expertise of Onderstepoort Veterinary Institute of South Africa together with technical assistance from the Food and Agriculture Organization of the United Nations and the commitment of the Government of Botswana, underpinned the eradication of CBPP from Botswana (Amanfu 2009). The successful eradication of CBPP from Botswana stands as testimony to the quality of veterinary services in that country and serves as an example of how political commitment and forthrightness can lead to the successful eradication of an animal disease. In Zambia, CBPP appears to be confined to the Western Province and the influx of CBPP-infected cattle from Angola has contributed to the maintenance of this infected zone.

Eastern Africa

Outbreaks of CBPP have been reported from Ethiopia, Kenya, South Sudan, Sudan, Tanzania, and Uganda. CBPP is enzootic in Rwanda, Burundi, and Tanzania; Mozambique has not reported CBPP, but the southern spread of outbreaks in Tanzania is a major threat to that country.

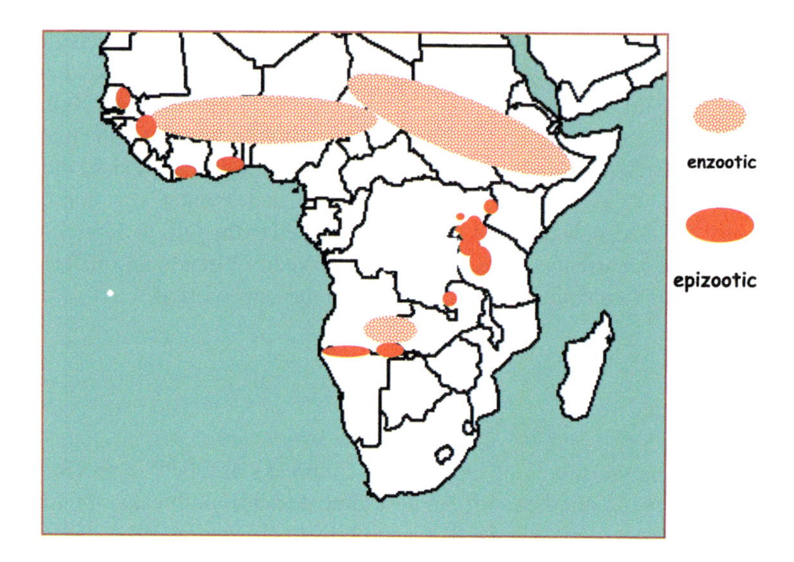

Epidemiology

The three principal factors responsible for the rate of spread of CBPP are (1) intensity of infection, (2) closeness of contact, and (3) the number of susceptible cattle in a herd. Natural transmission of CBPP occurs by droplet infection from cattle with acute clinical disease or from subclinical careers which are actively shedding the organisms to susceptible cattle in close contact. Aerosols containing infected droplets may spread the disease over distances of 20 m or more. Infection of cattle is often observed when large numbers of cattle (1) congregate at watering points on river or lake banks, (2) when cattle are kraaled or tethered together at night, and (3) when infected cattle and susceptible animals are transported together to markets over considerable distances. Thus, direct contact between susceptible and infected cattle appears to be a critical requirement for disease transmission. Mycoplasma mycoides subsp. mycoides may be present in urine of cattle affected by severe disease in the acute phase. Therefore, a urinary tract to nose transmission of Mmm through aerosols of urine may be possible (Masiga et al. 1972). The causative agent of CBPP has been isolated from semen (Goncalves et al. 1998) and detected by preputial washings of infected bulls by PCR technique. Thus, direct sexual transmission and indirect transmission by frozen semen might occur.

Another way in which CBPP disease may be spread or maintained is the development of pulmonary sequestra in clinically recovered animals. These sequestra may persist for several months if not years. The sequestra are made up of necrotic pulmonary tissue which becomes encapsulated by fibrous connective tissue and may contain viable mycoplasma organisms. Such affected cattle are known as "lungers." The role of these lesions in the transmission and persistence of infection is debatable (Windsor and Masiga 1977). Completely closed lesions are not likely to spread infection, but rupture of the sequestra capsule may lead to transmission of CBPP to susceptible animals. Infected calves usually develop carpal and tarsal arthritis. In the outbreaks of CBPP in Botswana, calves aged between 3 and 6 months were found to be affected by full-blown CBPP disease with or without arthritis. In Africa, movement of cattle between herds is common and this is the direct cause of many outbreaks on the continent. These movements can occur through direct market transactions, dowry payments for marriages, cattle congregations at watering points and pastures, and exchange of cattle for breeding or plowing or for carting farm produce. Deliberate and illegal sale of sick CBPP-affected animals to escape control actions in the field, theft of cattle, and civil strife that leads to people moving across borders with their cattle herds could elicit outbreaks of the disease in susceptible cattle populations.

The agent of CBPP has been isolated from the lungs of goats (Kusiluka et al. 2000). However, their role as reservoirs in the spread of CBPP infection is most unlikely. Empirical evidence exists that eradication of the disease from Botswana was by the slaughter of affected and in-contact cattle, without involving sheep and goats which often shared the same kraal and environment with CBPP-affected cattle. From 1997 when CBPP was eradicated, there has been no reoccurrence of the disease anywhere in Botswana.

Pathogenesis of CBPP Disease

The pathogenesis of CBPP disease is poorly understood. Investigations on pathogenesis and other aspects of CBPP disease have been hampered by the difficulties in reproducing the disease in cattle and the cost of cattle to be used as experimental animals in CBPP disease reproduction. Secondly, there is no reliable laboratory animal model that can be used to facilitate research into the various aspects of the disease including research on the production of protective antibodies and the development of requisite studies to standardize test results that correlate antibody response with herd immunity. The pseudo capsule of Mmm organisms made up of the carbohydrate galactan, has been considered as a potential virulence factor as it might play a role in the attachment of the organism to target organs (Gabridge et al. 1985). It may also contribute to the resistance of the organism to phagocytosis and may induce the formation of auto-antibodies to pneumogalactan. Differences in pathogenicity of European and African strains of Mmm have been observed, but it is difficult to achieve conclusive comparisons in such different environmental circumstances. Better insights into the pathogenesis of the disease will be obtained once the characterization of cell types involved in and the cytokines produced during the inflammatory process has been fully elucidated and clarified.

Clinical Signs of the Disease

The incubation period of the disease is generally 3–6 weeks, but may be as long as 6 months. The acute form is characterized by fever, loss of appetite, decreased milk production depression, and rapid labored breathing, which is mostly abdominal. This is followed by coughing, which becomes severe. There is noticeable chest pain and the animal typically faces the direction of the wind. The back is arched with elbows out and the neck extended (Fig. 20.2). Nasal discharges and frothy saliva accumulate around the mouth. Affected animals hide under bushes or prefer to stand in the shade. The animals grunt and coughing is exacerbated by vigorous exercise and percussion of the chest.

The mortality rate from acute CBPP may be as high as 75%, and death usually occurs within 2–3 weeks of the onset of clinical signs, with recovering animals being extremely weak and emaciated. Many become chronic carriers. In the hyperacute form, animals may die suddenly without showing any premonitory signs. In the subacute and chronic forms, clinical signs are milder and may not be detected. Calves up to 6 months may demonstrate infection with CBPP in the form of carpal and tarsal arthritis.

Fig. 20.2 Acute clinical form of CBPP. Note the extended neck and abducted forelimbs of the cow suggestive of respiratory distress. Source: W. Amanfu, Accra-Ghana

Acutely affected animals may die within 1 week as a result of the development of severe serofibrinous pleuropneumonia sometimes with pulmonary infarcts. In many cases, the clinical signs gradually disappear and condition of affected animal may improve rapidly. Recovered animals may harbor sequestra in their lungs in which the infection remains latent. Relapse may be precipitated by stress factors such as long treks in search of pasture and water.

Pathology

Gross

In acute CBPP, there is severe pneumonia with extensive pleural fluid, with the presence of pleural fluid in the chest cavity being a particular characteristic feature of the disease. Up to 30 liters of fluid with fibrin clots may be seen. The lungs show varying degrees of gray or red hepatization (Fig. 20.3). The affected areas of lungs are edematous, varying in coloration from pink to dark red, have a moderately firm consistency, and exude clear or blood-tinged fluid from cut surfaces of the lung. Pleural surfaces over affected areas are thickened, gray to red, and are often covered by friable, yellowish fibrin cast. Lymph nodes especially the bronchial and mediastinal are enlarged and may contain areas of necrosis.

In chronic cases, one or more sequestra which may vary in size from less than 10–300 mm in diameter and more than one lobe is usually present in the lungs. Secondary bacterial infection of sequestra often occurs resulting in a purulent

Fig. 20.3 Acute CBPP: Areas of gray and red hepatization—marbling. Note thickened interlobular septa. Source: W. Amanfu, Accra-Ghana

liquefied content which after rupture of the capsule may escape into the bronchi (Thiaucourt et al. 2004). Lesions in other organs are often seen in acute cases of CBPP. In many cases, a serofibrinous pericarditis with copious amounts of straw-colored exudate occurs. Multiple, small, and whitish infarcts in the kidneys are commonly seen in acute cases of CBPP, as was observed during the Botswana outbreaks of CBPP.

Histopathology

The earliest pulmonary lesions consist of catarrhal bronchiolitis, distension of the lymphatics in the interlobular septa, and thickened alveolar walls. There is proliferation of the cells in lymphatic follicles and an increase in the population of mononuclear cells around bronchioles. There is also lymphatic edema, with distension of subpleural lymphatics (Geering and Amanfu 2002). The necrotic lesion is often demarcated from living tissue by a zone of nuclear debris. Lesions in the lymph nodes are characterized by lymphoid hyperplasia of follicles and accumulation of edema fluid and fibrin in the subcapsular, medullary, and cortical sinuses (Bygrave et al. 1968). Joint lesions in calves consist of hyperemia, edema, and infiltration with lymphocytes and macrophages into the synovial membranes. Thrombosis of lymphatics and blood vessels and fibrin deposits occur on the synovial surfaces of the joint.

Diagnosis

The diagnosis of CBPP is classified into field and laboratory diagnosis. A summary of the various diagnostic features for CBPP has been provided by Geering and Amanfu (2002).

Clinical (Field) Diagnosis

The distinguishing features of clinical diagnosis of CBPP in a cattle herd are characterized by (1) occurrence of respiratory disease in a herd in which there is coughing, dyspnea, and emaciation; (2) the principal respiratory symptoms of fast, difficult, and noisy breathing, discharge from the nose, and coughing especially after exercise; (3) presence of yellowish or straw-colored fluid in the chest cavity; lungs are covered with yellowish fibrinous material that is adhered to the chest wall especially in chronic cases; (4) lungs that do not collapse and are solid, hepatized, or marbled; and (5) sequestra can be seen in the lungs of chronic cases.

Laboratory Diagnosis

The definitive diagnosis/confirmation of CBPP is based on isolation and characterization of the causative agent and/or the finding of specific antigens or antibodies by the appropriate molecular or serological tests, respectively. Lung tissue and affected local lymph nodes should be collected for bacteriologic isolation of causative agent. At least 10 mL of pleural fluid should be collected aseptically. Joint fluid from affected joints of calves should also be collected. Blood samples should be collected from live and in-contact animals, the sera harvested, and stored frozen until tested by the appropriate diagnostic tests. *Mycoplasma mycoides subsp. mmycoides* can be isolated from unpreserved tissue and pleural fluid specimens in suitable mycoplasma media such as Hayflick's or Gourlay's broth containing penicillin and thallium acetate as bacterial inhibitors. Definitive diagnosis can be made with polymerase chain reaction test.

Differential Diagnosis

Acute CBPP lesions must be differentiated from pneumonic pasteurellosis in which *Mannheimia (Pasteurella) haemolytica* and less commonly *Pasteurella multocida* play major roles in the field. Hemorrhagic septicemia often causes throat and brisket edema in affected cattle. This is not a common feature with infection in cattle caused by Mmm. Smears of lung exudates and pleural effusions stained with Giemsa or polychrome methylene blue reveal the typical bipolar staining of the organism. The sequestra of CBPP should not be confused with pulmonary abscesses caused by pyogenic bacteria. Abscess culture and histopathological examinations will assist differentiation from CBPP. Other diseases of importance in differential diagnosis are East Coast fever, bovine tuberculosis, actinobacillosis, traumatic pericarditis, and hydatid cysts.

Control of CBPP

The control of CBPP depends on the epidemiologic situation of the disease in a particular country, the animal production systems for cattle, and the effectiveness of the veterinary services department of that country. Stamping out is the method of choice in CBPP control/eradication, but the technical and financial capacity of many countries in Africa to carry this out effectively is lacking. Another issue to consider is the loss of genetic potential in cattle slaughtered to control CBPP, for example, disease resistance such as trypanotolerance in N'dama cattle. In most parts of Southern Africa where the disease is not present, immediate stamping out of affected region is the preferred method of control. In other parts of Africa with disease endemicity, progressive control by quarantine and vaccination is used. The control of the disease requires an optimum high degree of cooperation between neighboring countries. Cattle producers appear to have no choice, but to treat their animals affected by CBPP with antibiotics. These treatments are frequently suboptimal, leading to higher risks of bacterial antibiotic resistance emergence (Amanfu 2006). The progressive control using attenuated CBPP vaccines is done in most countries affected by the disease. Again, this is suboptimal due to financial and logistic constraints. Research on CBPP vaccines in Kenya, Chad, Senegal, Nigeria, and other countries in Africa coupled with a large multi-donor-funded international campaign known as the Joint Project (JP) 16 resulted in the drastic reduction of clinical disease from most parts of Africa in the 1960s and 1970s (Lubroth et al. 2007).

Types of CBPP Vaccines

Attenuated vaccines against CBPP have been developed by growing the CBPP agent in embryonated chicken eggs and later by subculture in Mycoplasma growth media (Haigh 1997; Sheriff and Piercy 1952). The efficacy of live vaccines is directly related to the primary strains of Mmm used for their production. Attenuated strains of Mmm stimulate the most immunity (Masiga and Windsor 1975), but they also induce the most severe post-vaccination local (Willems') reaction. The live attenuated vaccines currently in use for the control of CBPP are therefore a compromise between virulence, immunogenicity, and safety. In the past, a number of attenuated vaccine strains were developed such as the V5 (Australia), DK32 (Senegal), KH3J (Sudan), and Ben-181 (China). These vaccines were successfully used in the field in affected countries, but in Africa, they were superseded by the T1/44 (originally isolated in Tanzania) and its streptomycin-resistant derivative—T1/SR. Currently, T1/44 and T1/SR are the only Mmm vaccine strains that are available commercially. Broth culture vaccines have been replaced by lyophilized vaccines. A grand parental stock of T1/44 and T1/SR is kept at the African Union Pan African Veterinary Vaccine Centre (AU-PANVAC) at Debre Zeit–Ethiopia. The combined CBPP/rinderpest vaccine (Bisec) was actively used in many parts of West and Central Africa to control the two diseases. The discontinuation of rinderpest vaccination following its eradication and the disuse of Bisec have contributed to the resurgence of CBPP in West Africa (Amanfu 2009).

A communication strategy will have to be put in place to ensure that there is cooperation and that farmers are convinced that CBPP vaccination is perhaps the most cost-effective way to cope with the disease. Special care should be put into designing and perfecting a uniform strategy for cost recovery and price settings (Kairu-Wanyoike et al. 2014). Warnings about possible post-vaccination reactions when using T1/44 for the first time in a herd will have to be clearly indicated as otherwise it could jeopardize future efforts at CBPP control.

Post-Vaccination Monitoring of CBPP

It is critical and very important to specify that in contrast to viral vaccines such as rinderpest or PPR, CBPP vaccines do not induce constant and long-lasting sero-conversions irrespective of the serological test used (Hudson 1968; Gilbert and Windsor 1971). Hence serology is not a good tool to monitor the vaccination campaign efficacies. Herd immunity to CBPP is critical to vaccination efficacy evaluations.

Support Plans for CBPP Prevention and Control

The delay in obtaining funds from the appropriate governmental/non-governmental organization or donor sources is one of the major constraints to rapid response to emergency transboundary animal disease outbreaks such as CBPP, African swine fever, foot and mouth disease, PPR, Newcastle disease, and others. The immediate application of modest funds that are obtained earlier will save major expenditures in the future. Most veterinary services have difficulty getting their governments to set aside funds for animal disease outbreak emergencies because of sheer pressure from other sources that require emergency governmental assistance. However, veterinary services must always have a financial plan for animal disease emergencies.

Compensation

The financial plan should as much as possible include adequate provisions for compensation to owners for any livestock or property destroyed as part of the disease control/eradication campaign. Livestock insurance is uncommon in most countries in Africa. This needs to be looked into to help manage the risks inherent in livestock and poultry production.

Resource Plans

It is essential to prepare a *resource inventory*. This is a compilation of all logistics required to respond to an animal disease emergency such as CBPP. This should include personnel, equipment, and other physical resources (Geering and Amanfu 2002).

References

Amanfu W. The use of antibiotics for CBPP control: the challenges. In: Lubroth J, editor. CBPP control: antibiotics to the rescue. Rome: FAO; 2006. p. 7–11.

Amanfu W. Contagious bovine pleuropneumonia (Lungsickness) in Africa. Onderstepoort J Vet Res. 2009;76:13–7.

Amanfu W, Masupu KV, Adom K, Raborokgwe MV, Bashuriddin JB. An outbreak of contagious bovine pleuropneumonia in Ngamiland district of north-western Botswana. Vet Rec. 1998a;143:46–8.

Amanfu W, Sediadie S, Masupu KV, Benkirane A, Geiger R, Thiaucourt F. Field evaluation of a competitive ELISA for the detection of contagious bovine pleuropneumonia in Botswana. Rev Elev Med Vet Pays Trop. 1998b;51:189–93.

Bygrave AC, Moulton JE, Sheffrine M. Clinical, serological and pathological findings in an outbreak of contagious bovine pleuropneumonia. Bull Epizoot Dis Afr. 1968;16:21–46.

Cheng X, Nicolet J, Poumarat F, Regalla J, Thiaucourt F, Frey J. Insertion element IS1296 in *Mycoplasma mycoides* subsp. small colony identifies a European clonal line distinct from African and Australian strains. Microbiology. 1995;141:3221–8.

Cottew GS, Breard A, Damassa AJ, Erno H, Leach RH, Lefevre PC, Rodwell AW, Smith GR. Taxonomy of the *Mycoplasma mycoides* cluster. Isr J Med Sci. 1987;23:632–5.

Gabridge MG, Chandler DKF, Daniels MJ. Pathogenicity factors in mycoplasmas and spiroplasmas. In: Razin S, Barile MF, editors. The mycoplasmas, vol. IV. Orlando, FL: Academic Press; 1985.

Geering WA, Amanfu W. Preparation of contagious bovine pleuropneumonia contingency plans. In: FAO animal health manuals series, vol. 14. Rome: FAO; 2002.

Gilbert FR, Windsor RS. The immunizing dose of T1 strain mycoplasma mycoides against contagious bovine pleuropneumonia. Trop Anim Health Prod. 1971;3:71–6.

Goncalves R, Regalla J, Nicolet J, Frey J, Nicholas R, Bashiruddin J, De Santis P, Goncalves AP. Antigen heterogeneity among *Mycoplasma mycoides* subsp. *mycoides* SC isolates: discrimination of major surface proteins. Vet Microbiol. 1998;63:13–28.

Haigh AJ. The role of private industry in the transfer of vaccine technology to developing countries. In: Mowat N, Rweyemamu MM, editors. Vaccine manual. The production and quality control of veterinary vaccines for use in developing countries, FAO animal health and production series, vol. 53. Rome: FAO; 1997. p. 165–9.

Hudson JR. Contagious bovine pleuropneumonia. Experiments on the susceptibility and protection by vaccination of different types of cattle. Aust Vet J. 1968;44:83–9.

Kairu-Wanyoike SW, Kaitibie S, Heffernan C, Taylor NM, Gitau GK, Kiara H, McKeever D. Willingness to pay for contagious bovine pleuropneumonia vaccination in Narok South District of Kenya. Prev Vet Med. 2014;115:130–42.

Kusiluka LJ, Ojeniyi B, Friis NF, Kazwala RR, Kokotovic B. Mycoplasmas isolated from the respiratory tract of cattle and goats in Tanzania. Acta Vet Scand. 2000;41:299–309.

Lubroth J, Rweyemamu M, Viljoen G, Diallo A, Dungu B, Amanfu W. Veterinary vaccines and their use in developing countries. Res Sci Tech Off Int Epiz. 2007;26(1):179–201.

Manso-Silvan L, Vilei EM, Sachse K, Djordjevic SP, Thiaucourt F, Frey J. *Mycoplasma leachii* sp. nov. as a new species designation for *Mycoplasma* sp. bovine group 7 of Leach, and reclassification of *Mycoplasma mycoides* subsp. mycoides LC as a serovar of *Mycoplasma mycoides* subsp. capri. Int J Syst Evol Microbiol. 2009;59(6):1353–8.

March JB, Clare J, Brodlie M. Characterization of strains of mycoplasma mycoides small colony type isolated from recent outbreaks of contagious bovine pleuropneumonia in Botswana and Tanzania. Evidence for a new biotype. J Clin Microbiol. 2000;38:1419–25.

Masiga WN, Windsor RS. Immunity to contagious bovine pleuropneumonia. Vet Rec. 1975;97:350–1.

Masiga WN, Windsor RS, Read WCS. A new mode of spread of contagious bovine pleuropneumonia in Africa. Vet Rec. 1972;90:247–8.

Poumarat F, Solsona M. Molecular epidemiology of *Mycoplasma mycoides* subsp. *mycoides* biotype small colony, the agent of contagious bovine pleuropneumonia. Vet Microbiol. 1995;47:305–15.
Provost A, Perreau P, Le Breard A, Goff C, Martel JL, Cottew GS. Contagious bovine pleuropneumonia. Rev Sci Tech OIE. 1987;6:625–79.
Sheriff D, Piercy SE. Experiments with avianised strain of the organism of contagious bovine pleuropneumonia. Vet Rec. 1952;64:615–21.
Thiaucourt F, Van der Lugt JJ, Provost A. Contagious bovine pleuropneumonia. In: Coetzer JAW, Tutsin RC, editors. Infectious diseases of livestock, vol. 3. 2nd ed. Cape Town: Oxford University Press; 2004. p. 2045–58.
Vilei EM, Abdo EM, Nicolet J, Botelho A, Goncalves R, Frey J. Genomic and antigenic differences between the European and African/Australian clusters of *Mycoplasma mycoides* subsp. *mycoides* SC. Microbiology. 2000;146:477–86.
Windsor RS, Masiga WN. Investigation into the role of carrier animals in the spread of contagious bovine pleuropneumonia. Res Vet Sci. 1977;23:224–9.

Chapter 21
Contagious Caprine Pleuropneumonia

Lucía Manso-Silván and François Thiaucourt

Abstract Contagious caprine pleuropneumonia (CCPP), caused by *Mycoplasma capricolum* subsp. *capripneumoniae*, is an OIE-listed disease affecting goats and wild ungulate species. CCPP is present in Africa, the Middle East, and Central Asia, but its exact distribution is unknown, particularly in Asia. It is enzootic in the Middle East and East Africa, while it has only been sporadically reported in North and Central Africa and, though suspected, has never been identified in West Africa. In addition, there are very few studies reporting the prevalence and losses induced by CCPP, which are greatly underestimated. This uncertainty over the distribution and impact of CCPP is partly due to the fastidious nature of its etiologic agent, which is difficult to identify, particularly when it circulates in an insidious, mild, or asymptomatic form, favored by the use of antibiotic treatments. However, specific molecular and serological tests are now available for the diagnosis of CCPP, even in the absence of isolation. The main limitation of CCPP surveillance remains the lack of awareness by veterinary services. Vaccines based on inactivated antigens in saponin can induce good protection and their variable quality may now be assessed using a specific ELISA. However, they are very expensive, and there is a paucity of vaccine producers to satisfy their demand. Efforts must urgently be directed to the development of cheaper, quality-controlled vaccines to be extensively used in the field. The global campaign to eradicate "*peste des petits ruminants*" by 2030 may be a great opportunity to target other goat diseases such as CCPP.

Keywords *Mycoplasma capricolum capripneumoniae* · Mccp · Contagious caprine pleuropneumonia · CCPP

L. Manso-Silván (✉) · F. Thiaucourt
CIRAD, UMR ASTRE, Montpellier, France

INRA, UMR1309 ASTRE, Montpellier, France
e-mail: lucia.manso-silvan@cirad.fr; francois.thiaucourt@cirad.fr

M. Kardjadj et al. (eds.), *Transboundary Animal Diseases in Sahelian Africa and Connected Regions*, https://doi.org/10.1007/978-3-030-25385-1_21

Abbreviations

CBPP	contagious bovine pleuropneumonia
CCPP	contagious caprine pleuropneumonia
cELISA	competitive enzyme-linked immunosorbent assay
FAO	Food and Agriculture Organization of the United Nations
Mccp	*Mycoplasma capricolum* subsp. *capripneumoniae*
OIE	World Organisation for Animal Health

History

Contagious caprine pleuropneumonia (CCPP) is a severe infectious disease of goats and wild ruminant species characterized by unilateral sero-fibrinous pleuropneumonia. The history of CCPP may be divided into three periods.

The first period (1873–1890) corresponds to the earliest clinical descriptions of the disease. CCPP was first described by P. Thomas in 1873 in Algeria, where it was known for time immemorial under the name "*Bou-Frida*" (affecting a "single" lung) (Thomas 1873). Thomas thoroughly described the clinical and pathological features of the acute and chronic forms of the disease. He noted a cyclic appearance of CCPP at 5–7 year intervals, triggered by extreme weather conditions (cold and wet winters), and failed to reproduce the disease by experimental inoculation of pleural exudates from sick animals. He then concluded that the disease was due to the climate rather than to an infectious agent, disregarding the understanding by local herdsmen that animals that recovered from the disease acquired a lifelong protection. The next description was provided by D. Hutcheon in 1881, when CCPP was introduced in the "Colony of the Good Hope," South Africa, through the import of Angora goats from Turkey (Hutcheon 1881). Hutcheon immediately understood that the disease was contagious and put in place a control strategy combining the prompt slaughter of all affected goats in a flock and the "preventive inoculation" of the remaining animals (Hutcheon 1889). The inoculation procedure consisted in the subcutaneous injection of infectious material to induce protection. Implementation of this strategy resulted in the swift eradication of the disease from South Africa, well before the identification of its etiologic agent (Hutcheon 1889).

The second period (1890–1976) is marked by confusion over the etiology of CCPP. This confusion was mainly due to the fastidiousness of the CCPP agent in in vitro culture, coupled to its similarity with other, fast-growing mycoplasmas that can induce similar symptoms and lesions. As a result, efforts to isolate the "true" causative organism were unsuccessful, whereas other mycoplasmas were recovered from suspected cases. With the probable exception of a report in Greece, linked to the importation of goats from Turkey (Melanidi and Stylianopoulos 1928), CCPP descriptions made at that time in Europe actually referred to other diseases, particularly to what is now known as the contagious agalactia syndrome (Thiaucourt and

Bolske 1996). In East Africa the disease was known since the beginning of the twentieth century, with first descriptions dating from 1912 (Schellhase 1912). CCPP was also described in India in 1914 (Walker 1914) and 1940 (Longley 1940), but there were no further reports confirming the presence of the disease in Asia. The CCPP agent was finally isolated and characterized in Kenya in 1976 under the name "F38" (MacOwan and Minette 1976).

The third period (1976 until today) is characterized by a progressive recognition of the real distribution and impact of CCPP and a precise characterization of its etiologic agent. The bacterium isolated in 1976 was soon considered its sole causal agent, since it was the only mycoplasma that fulfilled the Koch's postulates for CCPP (MacOwan 1984). Known for many years by the name of its type strain, F38, it was finally designated *Mycoplasma capricolum* subspecies *capripneumoniae* (*Mccp*) in 1993 (Leach et al. 1993). Specific molecular and serological diagnostic tests were then developed (Bascunana et al. 1994; Thiaucourt et al. 1994; Woubit et al. 2004), contributing to a better understanding of the distribution and impact of CCPP. By the end of the twentieth century, CCPP had been demonstrated in North, Central, and East Africa and in the Middle East, though a wider distribution was suspected, particularly in Asia. The presence of CCPP in this continent was demonstrated through reports in East Turkey, Pakistan, Tajikistan, and China between 2009 and 2011 (Amirbekov et al. 2010; Awan et al. 2009; Cetinkaya et al. 2009; Chu et al. 2011a). Molecular epidemiology analyses proved that the disease was not recently introduced but had circulated in Asia for quite some time (Manso-Silvan et al. 2011). Evolutionary analysis based on high-throughput genomic data also showed that Mccp emerged around 300 years ago (Dupuy et al. 2015), indicating that CCPP is a very recent disease. Finally, the understanding that CCPP is not a goat-specific disease, but can also affect wild ungulate species, was demonstrated in 2007 (Arif et al. 2007), and the number of susceptible wildlife species has not ceased to grow since, posing new challenges for wildlife conservation as well as for the prevention of CCPP introduction in free regions.

Geographical Distribution and Socioeconomic Impact

The exact distribution of CCPP is still uncertain. The disease is difficult to detect due to the fastidiousness of the etiologic agent and confusion with other common goat illnesses. Furthermore, there is a lack of awareness about CCPP and no incentives for declaration. Therefore, goat farmers confronted with respiratory disease in their flocks resort directly to antibiotic treatments, disregarding a formal diagnosis. Nevertheless, if we take into consideration the clinical and isolation data cumulated over decades, the contagiousness of the disease, and the uncontrolled movements of nomadic goatherds in affected regions, we may determine that CCPP is present within an area comprised between Tunisia–Niger, Turkey–Tajikistan–China, and Tanzania (Fig. 21.1). This area covers most of the Sahel and connected regions with the exception of West Africa. The prevalence of CCPP in these regions and its

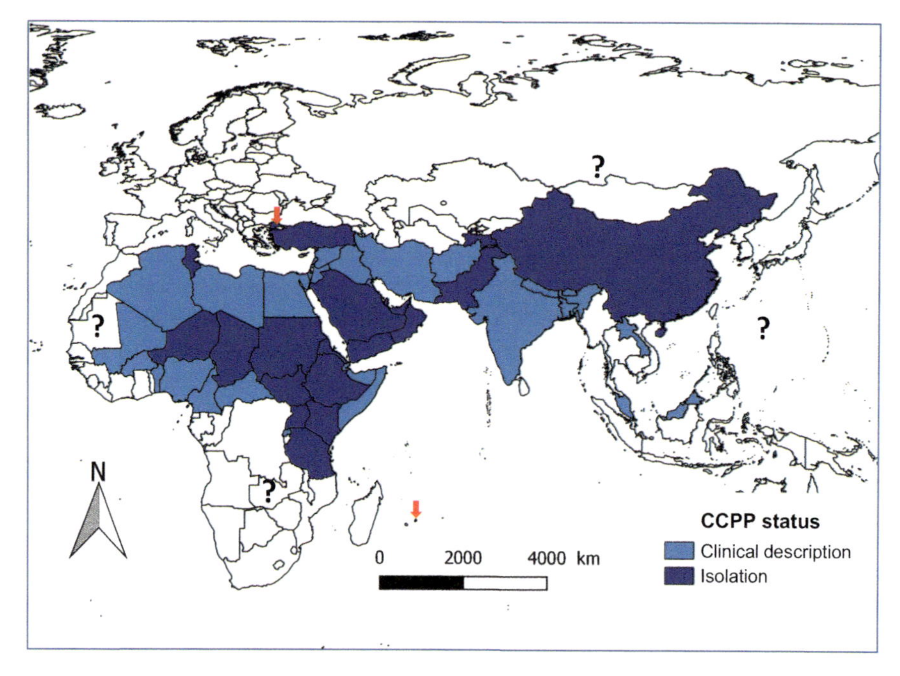

Fig. 21.1 Probable distribution of CCPP: The countries in which the disease has been described and those in which the etiologic agent has been isolated are indicated, with the exception of South Africa, where the disease was introduced and swiftly eradicated before the twentieth century (Hutcheon 1889). The arrows indicate the presence of CCPP in the Thrace region of Turkey and in Mauritius, where Mccp was isolated in 2004 and 2009, respectively (Ozdemir et al. 2005; Srivastava et al. 2010)

extension outside this area are difficult to ascertain, but we know that it is enzootic in the Middle East and in East Africa, whereas it has only been sporadically reported in North and Central Africa and its distribution in Central Asia is not well documented.

The report of CCPP in South Africa dating from 1881 actually referred to animals imported from Turkey (Hutcheon 1881), and the disease has remained a burden in this country and the Arabian peninsula until today. The Middle East is actually considered a reservoir of Mccp strains of different genotypes (Manso-Silvan et al. 2011). This has been explained by the historic importance of Mediterranean trading routes between Turkey, North Africa, and the Arabian peninsula, but also by the frequent importation of animals from diverse origins to the Arabian peninsula during the Muslim feasts, which remains very active to this day.

The current situation in North Africa is somehow uncertain. CCPP was first described in Algeria at the end of the nineteenth century (Thomas 1873), but it was never again reported in this country. The disease was evidenced in Tunisia in 1980 (Perreau et al. 1984) and apparently expanded to Libya. Although no control measures were put in place at that time, there have been no further reports of CCPP in North Africa.

The situation in East Africa is more documented. CCPP was known in Sudan since the beginning of the twentieth century under the name "abu nini" or "moaning disease" (Harbi et al. 1981), although this term also included diseases caused by other mycoplasmas. It is therefore difficult to date the first reports of "real CCPP" in this country. The disease was reported in Tanzania (formerly known as Tanganyika), Kenya, and Eritrea in 1912, 1929, and 1932, respectively (Mettam 1929; Pirani 1932; Schellhase 1912). In Ethiopia, the first reports date from 1982, in the Sudanese border, and 1990 in the border with Somalia (Thiaucourt et al. 1992). CCPP rapidly spread to the Rift Valley and since then it occurs in an enzootic form, as it does in other eastern African countries such as Eritrea, Kenya, Uganda, and Tanzania. Furthermore, CCPP has recently been imported into Mauritius via goats originating from East Africa (Srivastava et al. 2010) and the situation in the Indian Ocean needs to be reevaluated.

When CCPP was first evidenced in Chad in 1987, it was assumed that it had been recently introduced from East Africa (Lefevre et al. 1987), which was supported by recent evolutionary analysis (Dupuy et al. 2015). New outbreaks were detected in the same country in 1994 and in Niger in 1995, but CCPP has not been reported since and the western limits of the distribution of CCPP in Africa are yet to be ascertained (Dupuy et al. 2015).

CCPP never entered America, nor Australia, and the risk of contamination remains minor, as these continents are protected by their isolation and strict bans on the importation of live animals. Europe is also free of CCPP but must be considered at risk since the disease was detected in 2002 and confirmed in 2004 in the Thrace region of Turkey bordering Greece and Bulgaria (Ozdemir et al. 2005). The outbreak was controlled by the use of antibiotics, but no reports followed and the situation in the region needs to be reexamined.

CCPP is included in the list of notifiable diseases of the World Organisation for Animal Health (OIE) because of its very high morbidity and mortality rates causing a major socioeconomic impact, particularly in the arid and semiarid lands of Sahelian Africa and connected regions, where goat farming is of great importance. Goats are essential resources for a large segment of the population in these areas and, above all, for poor communities, which depend on goat farming for sustenance. CCPP produces substantial direct losses resulting from its high mortality, reduced milk and meat yield, and costs of treatment and control. However, the real impact of CCPP in affected regions is difficult to establish, since there are very few studies reporting the prevalence of the disease and the losses induced by it, which have been greatly underestimated. A highly specific ELISA kit for CCPP has recently been made available (Peyraud et al. 2014). This ELISA was used in an international collaborative study to evaluate the prevalence of CCPP in several affected regions of Africa and Asia and is a useful tool for the surveillance and control of CCPP. The value of this ELISA kit was further demonstrated in a cross-sectional survey in Kenya, which confirmed that CCPP is widespread and endemic in the pastoral production systems studied (Kipronoh et al. 2016a) and that regions sharing international boundaries are at a higher risk, emphasizing the need for a unified cross-border approach to CCPP control (Kipronoh et al. 2016b). Finally, participatory epidemiological techniques may represent a very valuable contribution to the surveillance and control of CCPP, in addition to more classical epidemiological approaches based on sero-prevalence analysis, by enhancing their sensitivity and relevance, and encouraging the inclusion of marginalized groups such as small goat farmers and nomadic pastoralists. These participatory approaches have proven to be particularly useful as tools to increase disease awareness by allowing an estimation of the relative incidence and impact of the disease, particularly in the absence of laboratory confirmation. For instance, nomadic pastoralists in Turkana South District, Kenya, ranked CCPP among the most important diseases affecting their goats, together with "peste des petits ruminants" (PPR) and sarcoptic mange (Bett et al. 2009).

Etiology

CCPP is exclusively caused by *Mycoplasma capricolum* subsp. *capripneumoniae* (Mccp). This bacterium belongs to the class *Mollicutes*, order *Mycoplasmatales*, family *Mycoplasmataceae*, and genus *Mycoplasma*. The *Mollicutes*, from the Latin words *mollis* (soft) and *cutis* (skin), trivially referred to as mycoplasmas, comprise the smallest organisms capable of autonomous replication. They are distinguished from other eubacteria by their small size and lack of a cell wall, which renders them pleomorphic and resistant to beta-lactam antibiotics. Though technically Gram-negative, mycoplasmas evolved from Gram-positive bacteria by a process of massive genome reduction involving the loss of many metabolic pathways and adaptation to a commensal or parasitic mode of life.

More precisely, Mccp belongs to the "mycoides cluster" (Table 21.1), a group of very closely related ruminant pathogens that share many genetic and phenotypic features, which has complicated species identification and classification for decades. The taxonomy of the cluster was revised in 2009 based on thorough phylogenetic analysis (Manso-Silvan et al. 2007, 2009). Two sub-clusters were identified, one comprising the *Mycoplasma mycoides* subspecies, and another grouping the *M. capricolum* subspecies and *M. leachii*. With the exception of *M. leachii*, which induces mainly mastitis and arthritis in cattle, all the other group members are associated with OIE-listed diseases. *M. mycoides* subsp. *mycoides* is the agent of contagious bovine pleuropneumonia (CBPP), a disease of cattle presenting many similarities with CCPP. On the other hand, *M. mycoides* subsp. *capri* and *M. capricolum* subsp. *capricolum* are involved in the contagious agalactia syndrome of small ruminants (Thiaucourt and Bolske 1996). This syndrome, also known as MAKePS for mastitis, arthritis, keratoconjunctivitis, pneumonia, and septicemia, implicates additional mycoplasma species (i.e., *Mycoplasma putrefaciens* and *Mycoplasma agalactiae*). The fact that these mycoplasmas can also induce pleuropneumonia lesions has created much confusion and for many years they have erroneously been implicated in the etiology of CCPP, particularly *M. mycoides* subsp. *capri*. However, the infections caused by these mycoplasmas are often associated with other symptoms and lesions affecting several organs, which should allow discrimination from "real CCPP." Mccp was known as "F38 type" since its first

Table 21.1 Taxonomy of the "mycoides cluster"

	Mycoides cluster					
	Mycoides sub-cluster			Capricolum sub-cluster		
1987 taxonomy[a]	*Mycoplasma mycoides* subsp. *capri*	*Mycoplasma mycoides* subsp. *mycoides* LC	*Mycoplasma mycoides* subsp. *mycoides* SC	*Mycoplasma sp.* bovine group 7 of Leach	*Mycoplasma capricolum* subsp. *capricolum*	*Mycoplasma sp.* F38 type
Current taxonomy[b]	*Mycoplasma mycoides* subsp. *capri*		*Mycoplasma mycoides* subsp. *mycoides*	*Mycoplasma leachii*	*Mycoplasma capricolum* subsp. *capricolum*	*Mycoplasma capricolum* subsp. *capripneumoniae*
Associated disease	Contagious agalactia		CBPP	Mastitis, arthritis	Contagious agalactia	CCPP

CBPP contagious bovine pleuropneumonia, *CCPP* contagious caprine pleuropneumonia, *LC* large colony, *SC* small colony

[a]Cottew et al. (1987)

[b]Manso-Silvan et al. (2009)

isolation in Kenya in 1976 (MacOwan and Minette 1976) until 1993 (Leach et al. 1993), when sufficient evidence was gathered to support a subspecies relationship with *M. capricolum* subsp. *capricolum* (Bonnet et al. 1993).

Mccp is a very monomorphic pathogen: there is a single serotype and the genetic diversity of strains is very limited. For this reason, the development of discriminatory genotyping tools for Mccp has been challenging. However, with the advent of high-throughput sequence technologies, large-scale genomic analysis has allowed the fine characterization of Mccp strains (Dupuy et al. 2015). These analyses were also used to date the emergence of Mccp, estimated only at about 300 years ago. The low genetic diversity of Mccp was therefore explained by the fact that it is a very young pathogen that has evolved too recently to allow the accumulation of polymorphisms.

Epidemiology

CCPP affects mostly domestic goats, and there are no differences in susceptibility according to breed, sex, or age. It was long believed to be a goat-specific disease until 2004, when a classical CCPP outbreak was confirmed in wild ruminants kept in a wildlife reserve in Qatar. Since then, there has been a growing number of CCPP reports in both captive and free-ranging wild ungulates, including wild goat (*Capra aegagrus*), Laristan mouflon (*Ovis orientalis laristanica*), Nubian ibex (*Capra ibex nubiana*), gerenuk (*Litocranius walleri*) (Arif et al. 2007), Tibetan antelope (*Pantholops hodgsonii*) (Yu et al. 2013), Arabian oryx (*Oryx leucoryx*) (Chaber et al. 2014), and gazelle species (Nicholas and Churchward 2012). There is no evidence suggesting that wildlife actually acts as CCPP reservoir, since in all the reported cases wild ungulates appear to have been contaminated from domestic goats. More likely, the emergence of CCPP in wildlife has taken place only recently due to increased opportunity for contacts between wild and domestic animals driven by anthropogenic land-use changes and human encroachment into wild habitats. In any case, the fact that CCPP can affect wildlife poses new challenges for surveillance and control and the possible role of these species in the epidemiology of the disease is yet to be elucidated. The strict host specificity of Mccp to the domestic goat had already been questioned when this agent was isolated from sheep in Kenya (Litamoi et al. 1990) and Uganda (Bolske et al. 1995, 1996). However, sheep are certainly less susceptible than goats and are generally refractory to natural and experimental infection (Harbi et al. 1983; Hutcheon 1889). Again, it remains uncertain whether sheep actually play a role as CCPP reservoirs or they just become occasionally infected.

Mccp is very sensitive to physical, chemical, and biological factors and does not persist in the environment, which excludes the possibility of indirect transmission through fomites or animal products. CCPP is exclusively transmitted by the airborne route (via infective droplets excreted through coughing and sneezing) between live animals kept in close contact. However, transmission at a distance of up to 80 m has recently been observed (Lignereux, L., personal communication). Animals in the chronic stage of the disease may present no symptoms and do not display sequestered lesions (which are characteristic of chronic CBPP). However, this does not prevent a long-term carrier status. For example, when CCPP was introduced in South Africa by a shipment of Angora goats from Turkey, no symptoms were observed during the entire journey, which took 7 weeks (Hutcheon 1881).

Harsh climatic conditions and concomitant infections are considered important predisposing factors and have a direct impact on the course of CCPP. Cold winters, as well as abrupt changes in temperature and pluviometry, particularly at the end of the dry season, have been related to disease outbreaks and increased severity. CCPP has also been associated with pox infections, particularly in China (Chu et al. 2011b). The arid and semiarid environment and nomadic farming practices that are characteristic of the Sahel and connected regions play an important role in CCPP transmission, as the disease can easily spread when animals congregate at grazing and watering points.

Symptoms and Lesions

CCPP is a strictly respiratory disease that presents many analogies with CBPP. However, as opposed to CBPP, which affects only adult bovine subjects, CCPP affects caprine and wild ruminants from all age groups (Thiaucourt and Bolske 1996).

The clinical symptoms and pathological features of CCPP were described in detail by Hutcheon (1881) and reviewed by Thiaucourt and Bolske (1996). The incubation of CCPP, before the appearance of the first clinical signs, is of approximately 10 days, though it may extend to over 1 month in some cases. The severity of the clinical signs depends on the disease presentation, from hyperacute to subacute or chronic. The classical clinical picture is observed in the acute form that develops when the etiologic agent infects naïve populations and is associated with morbidity and mortality rates reaching 100 and 80%, respectively. The first signs are reluctance to move, extreme fever (41–43 °C), and anorexia. After 2–3 days of high fever, respiratory signs become apparent. They are characterized by a fast and painful respiration and a violent and productive cough, often accompanied by a grunt. Abortions are frequent in pregnant goats. In the terminal stages, the animals are unable to move. They stand with their forelegs wide apart, neck extended, and mouth open (Fig. 21.2). Drooling and a mucopurulent nasal discharge are frequently observed. Death may quickly follow as a result of respiratory distress or heart failure. Otherwise, the animals recover gradually, though they may still carry and excrete the etiologic agent for months. Subacute or chronic forms of the disease may be observed in enzootic regions or at the end of an epizootic. In these forms, the clinical signs are similar, but milder, and consist in transient, slight fever and occasional coughing, which may only be noticeable following exercise.

Fig. 21.2 Goat with acute CCPP standing with forelegs wide apart and extended neck. Source: F. Thiaucourt. Location: Sultanate of Oman 1990

As for CBPP, CCPP lesions are localized exclusively in the thoracic cavity and often affect a single lung. In both cases, classical lesions consist in severe, exudative pleuropneumonia, associated with sero-fibrinous pleurisy. However, in CCPP there is no thickening of the interlobular septa of the lung, which is characteristic of the interstitial pneumonia observed in CBPP, which may also be found in infections caused by other mycoplasmas of the mycoides cluster. In the acute form of CCPP, large amounts of straw-colored pleural exudate are found in the thoracic cavity, with caseous and fibrinous deposits in the pleura (Fig. 21.3). The affected lung can be completely hepatized and present a fine granular texture with different colors varying from purple to gray. In the chronic form or phase, there is no pleural fluid and pleural adhesions are prominent. The lung tissue presents a black discoloration, but the encapsulated necrotic lesions known as "sequestra" that are characteristic of CBPP are not observed here.

Another distinctive feature of CCPP can be demonstrated under experimental inoculation. The local edematous reactions that are observed in cattle when the CBPP agent is injected subcutaneously do not occur in goats inoculated with Mccp (Hutcheon 1889). This feature also allows its differentiation from other mycoplasmas affecting goats, which induce local reactions and can spread to distant organs resulting in disseminated *disease* and septicemia.

Fig. 21.3 Thoracic cavity of a goat with CCPP showing pleural fluid, fibrin deposits, and lung hepatization. Source: F. Thiaucourt. Location: Ethiopia 1991

Diagnosis

The specific diagnosis of CCPP is essential, since it is an important transboundary disease listed by the OIE as a notifiable disease. The clinical symptoms and lesions can be highly evocative in the acute form of the disease, though the classical presentation is rarely observed in practice due to insidious disease circulation, concomitant infections, or antibiotic treatments. Differential diagnosis with pulmonary diseases such as acute pasteurellosis and other mycoplasmoses presenting respiratory pathology may be required. In any case, laboratory diagnosis is required.

Direct diagnosis by culture and isolation of the etiologic agent is still mandatory for absolute confirmation by the OIE. Moreover, Mccp isolation is essential for the generation of reference culture collections. The preferred samples for isolation are pleural fluid, hepatized lung lesions (ideally in the interface between healthy and diseased tissue), and tracheobronchial and mediastinal lymph nodes from acute disease. Mccp may be grown in rich media such as the modified Hayflick's medium (Thiaucourt et al. 1996) incubated at 37 °C and maximum humidity in anaerobic or microaerophilic conditions. However, it is extremely fastidious and isolation trials are often unsuccessful. Moreover, isolation may be compromised by bacterial overgrowth (including other mycoplasmas), poor sample conservation (including repeated freeze-thawing), and antibiotic treatments. Upon isolation, the different mycoplasmas of the mycoides cluster may be difficult to identify by classical diagnostic techniques, but Mccp presents several characteristic cultural features (Leach et al. 1993). It grows much more slowly than any of the other members and, like *M. mycoides* subsp. *mycoides*, the agent of CBPP, produces characteristic "silky swirls" in broth culture. On solid medium, it produces very small colonies with a marked "fried egg" appearance (Fig. 21.4). A few biochemical tests classically proposed for mycoplasmas may also help orient the diagnosis, but they do not allow the discrimination between members of the mycoides cluster and serological confirmation by growth inhibition or immunofluorescence is required. However, serodiagnosis is often hampered by cross-reactions between Mccp and other cluster members such as Mcc and *M. leachii* (Cottew et al. 1987). In addition, these techniques are time- and labor-consuming, and their sensitivity is low.

Because of the numerous limitations of Mccp isolation and classical diagnosis, the use of molecular methods for the direct detection of this agent in clinical material is a very useful alternative for the confirmation of CCPP outbreaks. The development of specific PCR tests has made Mccp identification much more sensitive and reliable. These assays can be even performed using samples dried on filter paper, which can be easily transported without cold-chain requirements. The first PCR assay for the identification of Mccp was based on the amplification of a 16S rDNA gene fragment, followed by restriction enzyme digestion (Bascunana et al. 1994). A specific PCR test for the direct detection of Mccp was developed in 2004 (Woubit et al. 2004), and it has been extensively validated in the field. This assay was then adapted in order to develop a real-time, quantitative PCR test, which does not require post-PCR processing, thus greatly reducing the risk of contamination (Lorenzon et al. 2008). A one-step, multiplex RT-qPCR assay for the detection of PPR virus, capripoxvirus, and *Pasteurella multocida* has recently been developed for the rapid detection of the individual targeted pathogens, as well as co-infections (Settypalli et al. 2016).

Furthermore, typing tools based on the detection of single-nucleotide polymorphisms allow the discrimination of Mccp strains. The first typing technique was based on polymorphisms observed in the 16S rDNA genes (Pettersson et al. 1998). An approach based on eight locus sequences allowed for a more precise typing of

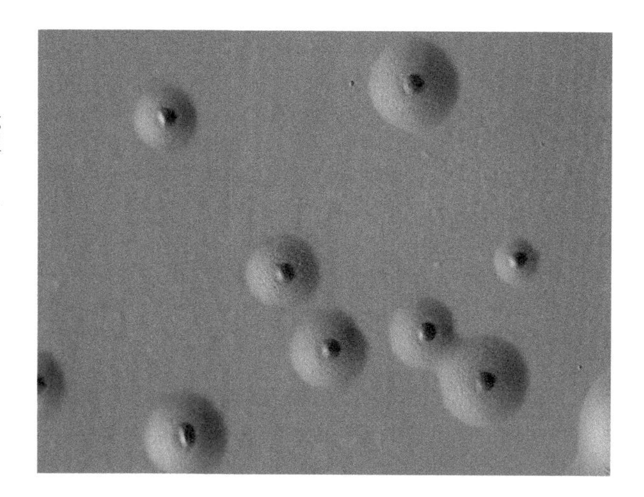

Fig. 21.4 Colonies of *Mycoplasma capricolum* subsp. *capripneumoniae* examined under a dissecting microscope showing a "fried egg" appearance. The rose color of the colony center is due to the reduction of triphenyl-tetrazolium salts included in the agar medium. Source: F. Thiaucourt, CIRAD

strains and the identification of a cluster grouping Asian isolates, demonstrating that CCPP, only recently declared in Central Asia, must have evolved in the continent for quite some time (Manso-Silvan et al. 2011). A large-scale genomic analysis based on high-throughput sequence data recently allowed the development of a highly discriminatory multigene typing system and the definition of six genotyping groups, which showed some correlation with geographic origin (Dupuy et al. 2015). Molecular detection and typing techniques are now available for epidemiological investigations, but the main limitation remains the paucity of Mccp strains or biological samples for analysis.

Several methods are also available for the serological analysis of CCPP. Until the end of the twentieth century, the complement fixation test (CFT) was the method of choice for the detection of antibodies to CCPP, and it is still prescribed by the OIE (2014). The main advantage of this test is that it can be performed quite easily with very little equipment, for example in mobile laboratories. Nevertheless, it lacks specificity. Since it uses crude Mccp antigens, cross-reactions with other mycoplasmas of the mycoides cluster are frequent. In addition, as it detects mostly IgM, its sensitivity may rapidly decline over time and the CFT is therefore more suited to the early detection of cases during an epizootic than to prevalence studies in an enzootic context. Lastly, this technique requires a good laboratory experience and trained personnel.

A specific competitive ELISA (cELISA) for the detection of antibodies to Mccp was developed in 1994 (Thiaucourt et al. 1994) and has recently been commercialized (Peyraud et al. 2014). This assay, based on a monoclonal antibody, is highly specific (99.9%). Although the detection by cELISA is slightly retarded as compared to the CFT, it is also prolonged, so more animals may be detected in the chronic stage of the disease. Given its high specificity, the cELISA can be used for CCPP detection at the herd level. The sensitivity may then be increased by selecting for analysis animals that have shown respiratory symptoms in the previous months. Finally, since CCPP vaccines are made of inactivated, adjuvanted antigens and induce long-lasting serological responses, the cELISA can be used to assess vaccination campaigns. However, in this context, it cannot be used for the detection of outbreaks.

The latex agglutination test, based on latex particles sensitized with polysaccharides secreted by Mccp, provides a rapid confirmation of outbreaks in the field (Rurangirwa et al. 1987b). The test is sensitive in the early stages of the disease, when IgM persists in the serum. However, it lacks specificity, since the same polysaccharide is secreted by *M. capricolum* subsp. *capricolum* and *M. leachii* (Table 21.1) (Bertin et al. 2015).

Treatment, Prevention, and Control

Mycoplasmas are intrinsically resistant to penicillins and other antibiotics acting on cell wall synthesis, and they become rapidly resistant to aminosides such as streptomycin. However, they are sensitive to many other antibiotics that may be used to treat CCPP. The most effective are tetracyclines, macrolides such as erythromycin or tylosin, and fluoroquinolones such as enrofloxacin (El Hassan et al. 1984; Ozdemir et al. 2006). In order to be effective, antibiotic treatments must be implemented as early as possible and a duration of at least 5 days is necessary. The use of long-acting formulations may therefore be essential to guarantee that an appropriate protocol is achieved, particularly in nomadic herds (Thiaucourt et al. 1996). Antibiotherapy results in reduction of clinical symptoms and pathology, but it does not assure complete clearance of the infection, so animals may act as chronic, asymptomatic carriers (El Hassan et al. 1984). Furthermore, since antibioresistance is considered one of the most important threats to human health, there is a global trend to reduce the use of antibiotics. For these reasons, antibiotherapy may only be recommended as part of well-designed eradication strategies, coupled to prophylactic measures.

Since CCPP is exclusively transmitted by direct contact between infected and susceptible animals, standard methods for disease control may be applied. Sanitary measures constitute the most effective strategy in disease-free countries, and the first golden rule is the prohibition of importation of live animals from infected or suspected zones. As shown in South Africa at the end of the nineteenth century, CCPP can be eradicated from an entire country or region by a combination of strategies including drastic stamping-out policies and movement restrictions (Hutcheon 1889). However, restriction of animal movements in regions that practice extensive nomadic livestock herding is not feasible. Furthermore, the massive slaughter of infected herds is unacceptable in most countries where the disease is enzootic. A medical prophylaxis based on the use of vaccines is the preferred alternative to reduce the prevalence and limit the expansion of CCPP in these regions.

Inoculation procedures for the prevention of CCPP are very old. Already in the 1880s, following the introduction of the disease in South Africa, Hutcheon showed that it was possible to "vaccinate" goats by inoculating infected lung extracts subcutaneously (Hutcheon 1889). A high passage culture of Mccp inoculated intratracheally to goats significantly reduced their susceptibility to CCPP infection, but it did not prevent clinical infection (MacOwan and Minette 1978). On the other hand, crude inactivated antigens adjuvanted with either Freund's complete or incomplete adjuvant afforded good protection against CCPP and prevented the development of clinical signs (Rurangirwa et al. 1981). The unsuitable oil adjuvants were then replaced by saponin and these preparations, which conferred good protection after a single injection (Rurangirwa et al. 1987a), are still prescribed by the OIE (2014).

When correctly produced, these vaccines are highly effective and confer relative long-lived immunity, lasting a little over 1 year. The main advantages of these preparations are their safety, thermostability, and compatibility with antibiotic treatments. Since they induce a strong seroconversion, the vaccination campaigns may be monitored by cELISA, though this may hamper the detection of outbreaks (Peyraud et al. 2014). The main drawback of these vaccines is their high production cost. This is due to the fastidious nature of Mccp, which grows slowly in vitro and requires rich and expensive supplements, including important amounts of animal serum. In addition, the complexity of these cultures makes it difficult to purify mycoplasma proteins and compromises the quality of the final product, which may contain suboptimal amounts of antigen, contaminated with medium components such as serum albumin. This can be very problematic, since the current quality control of these vaccines relies essentially on a measure of total protein content. However, the specific CCPP cELISA may also be used to assess their quality by measuring the intensity and duration of the seroconversion induced in vaccinated animals (Peyraud et al. 2014). Indeed, vaccines produced according to the optimal formulation described by Rurangirwa (Rurangirwa et al. 1987a) induced high antibody levels lasting for at least 5 weeks, whereas formulations containing suboptimal amounts of antigen and/or adjuvant generally resulted in either weaker or shorter serological responses (Peyraud et al. 2014).

Furthermore, there is a paucity of vaccine producers and the demand of CCPP vaccines largely exceeds the offer. Kenya and Ethiopia are currently the leading manufacturers of CCPP vaccines. Their goat population is very high (29 million heads in Ethiopia and 25 million in Kenya in 2014, according to FAO stats) and plays an essential role in rural households, as well as in the national economy. A high prevalence of CCPP has been reported in both countries (Asmare et al. 2016; Kipronoh et al. 2016a). The Kenya Veterinary Vaccines Production Institute (KEVEVAPI) declared the production of 3.8 million doses of CCPP vaccines in 2015, of which 0.8 million were exported (OIE-WAHIS) (OIE). In Ethiopia, 3.4 million doses were produced the same year by the National Veterinary Institute (NVI), but none could be exported due to the inability to satisfy the internal demand. These data strongly emphasize the need for increased production of CCPP vaccines.

Conclusion

CCPP is a transboundary disease constituting an important threat to the livelihood of goat farmers in the Sahel and connected regions. Given its high morbidity and mortality, CCPP can spread, irrespective of national borders, and cause serious socioeconomic damage. Consequently, it is included in the list of notifiable diseases of the OIE. However, in spite of its undeniable importance, the exact distribution of CCPP and its economic impact in affected regions are still largely unknown. This may be partly explained by the fastidious nature of Mccp, which may remain unnoticed, confused with other mycoplasmas of the "mycoides cluster" that are often found in goats and can produce similar symptoms and lesions. However, specific molecular and serological methods for the diagnosis of CCPP are now available and should contribute to improved disease surveillance. The main limitation remains the lack of awareness concerning CCPP among veterinary services, which are more concerned with PPR and vector-borne viral diseases.

There is an urgent need for increased CCPP awareness by veterinary services, diagnostic laboratories, and policy-makers in affected and suspected regions, as well as those that may be considered at risk, in order to improve CCPP surveillance and control. Participatory epidemiological techniques may be a useful approach to increase this awareness and may contribute to surveillance systems in addition to more classical epidemiological approaches based on sero-prevalence analysis, which are already available (Peyraud et al. 2014). In any case, it will be essential to take into consideration the transboundary nature of this disease, which calls for a regional, unified cross-border approach to disease surveillance and control.

Another important consideration for the surveillance of CCPP is that it can affect wildlife (Arif et al. 2007). This has been confirmed in several captive and free-ranging wild ruminants in the Arabian peninsula and Tibet but may concern additional species and geographical locations. Therefore, CCPP must be considered a threat to endangered wild ungulates and efforts should be made to avoid contact between these animals and domestic goats, which will not be an easy task in the context of increased human encroachment into wildlife territories. In addition, wild ungulates may constitute a reservoir for pathogenic mycoplasmas, including Mccp. Therefore, the exchange of zoo animals for conservation purposes may contribute to the emergence of mycoplasma diseases and represents a risk of CCPP introduction into disease-free areas.

The lack of awareness regarding CCPP is particularly alarming in North, Central, and West Africa, where it is virtually unknown or disregarded by veterinary services. It is somehow surprising that CCPP has never been detected in West Africa, particularly when we consider the distribution of other contagious diseases of goats such as PPR. Similarly, it seems unlikely that the disease would disappear from North and Central Africa, where there have been no reports since the 1980s and 1990s, respectively. Indeed, the regions where CCPP has been detected and in which efficient eradication programs have not been implemented must be considered infected, even in the absence of subsequent official declarations. Once established, it may subsist in an insidious, mild, or asymptomatic form, most certainly favored by the uncontrolled use of antibiotic treatments. Efforts should be urgently made to define the incidence of CCPP in North and Central Africa and to determine the western limits of its distribution in this continent.

The situation is very different in the Middle East and in East African countries. CCPP is known to be prevalent in these regions and is often acknowledged as an important threat to goat herding, but disease control is limited by the lack of cost-effective vaccines. Although there is a paucity of CCPP vaccine producers at present, thanks to increased CCPP awareness additional manufacturers are becoming interested in the production of these vaccines, particularly in the Middle East. When correctly produced, CCPP vaccines based on Mccp antigens in saponin are highly effective and confer relatively long-lived immunity, but they are extremely expensive. Their high production cost is due to the fastidious nature of Mccp, which requires extremely rich culture medium and grows very poorly. As a result, the quality of commercially available vaccines is variable and deserves to be controlled by cELISA (i.e., analysis of the intensity and duration of seroconversion in vaccinated animals). In the absence of cost-effective prophylactic treatments, CCPP cases are treated with antibiotics, which may contribute to the development of multidrug resistance, particularly when the optimal dose and duration of treatments are not respected. Therefore, new efforts should be urgently directed to the development of cheaper, quality-controlled vaccines.

Owing to its large distribution and insidious nature, the eradication of CCPP from the Sahel and connected regions will be extremely challenging. Sanitary control measures, which are most effective in disease-free areas, cannot be applied here. Drastic stamping-out policies are neither acceptable nor viable in most countries, where there is no governmental compensation. Furthermore, the restriction of animal movements in regions that practice extensive nomadic herding is not feasible. Antibiotic treatments should only be recommended as part of well-designed control strategies, coupled to a medical prophylaxis. Indeed long-acting antibiotics and inactivated vaccines may be injected simultaneously in order to reduce the prevalence and limit the expansion of CCPP in emergency situations. However, a preventive vaccination should always be the preferred strategy to control CCPP in these regions. The challenge will be the production of sufficient amounts of cost-effective vaccines for vaccination to be implemented on a large scale in the field. A multivalent approach combining several goat pathogens may be explored, since the simultaneous immunization against multiple diseases would dramatically reduce vaccination costs. The co-administration of vaccines may be an interesting alternative in the absence of multivalent formulations. An encouraging example may be found in the simultaneous vaccination of cattle against rinderpest and CBPP during the rinderpest eradication campaign (Brown and Taylor 1966); the campaign was very successful and no interference was observed between the two vaccine valences. The combination of CCPP and PPR vaccines would be strategic considering the current global campaign for the eradication of PPR by 2030.

References

Amirbekov M, Murvatulloev S, Ferrari F. Contagious caprine pleuropneumonia detected for the first time in Tajikistan. EMPRES Transbound Anim Dis Bull. 2010;35:20–2.

Arif A, Schulz J, Thiaucourt F, Taha A, Hammer S. Contagious caprine pleuropneumonia outbreak in captive wild ungulates at Al Wabra Wildlife Preservation, State of Qatar. J Zoo Wildl Med. 2007;38(1):93–6.

Asmare K, Abayneh T, Mekuria S, Ayelet G, Sibhat B, Skjerve E, Szonyi B, Wieland B. A meta-analysis of contagious caprine pleuropneumonia (CCPP) in Ethiopia. Acta Trop. 2016;158:231–9. https://doi.org/10.1016/j.actatropica.2016.02.023.

Awan MA, Abbas F, Yasinzai M, Nicholas RA, Babar S, Ayling RD, Attique MA, Ahmed Z, Wadood A, Khan FA. First report on the molecular prevalence of *Mycoplasma capricolum* subspecies *capripneumoniae* (Mccp) in goats the cause of contagious caprine pleuropneumonia (CCPP) in Balochistan province of Pakistan. Mol Biol Rep. 2009;37(7):3401–6. https://doi.org/10.1007/s11033-009-9929-0.

Bascunana CR, Mattsson JG, Bolske G, Johansson KE. Characterization of the 16S rRNA genes from *Mycoplasma* sp. strain F38 and development of an identification system based on PCR. J Bacteriol. 1994;176(9):2577–86.

Bertin C, Pau-Roblot C, Courtois J, Manso-Silvan L, Tardy F, Poumarat F, Citti C, Sirand-Pugnet P, Gaurivaud P, Thiaucourt F. Highly dynamic genomic loci drive the synthesis of two types of capsular or secreted polysaccharides within the *Mycoplasma mycoides* cluster. Appl Environ Microbiol. 2015;81(2):676–87. https://doi.org/10.1128/AEM.02892-14.

Bett B, Jost C, Allport R, Mariner J. Using participatory epidemiological techniques to estimate the relative incidence and impact on livelihoods of livestock diseases amongst nomadic pastoralists in Turkana South District, Kenya. Prev Vet Med. 2009;90(3–4):194–203. https://doi.org/10.1016/j.prevetmed.2009.05.001.

Bolske G, Johansson KE, Heinonen R, Panvuga PA, Twinamasiko E. Contagious caprine pleuropneumonia in Uganda and isolation of *Mycoplasma capricolum* subspecies *capripneumoniae* from goats and sheep. Vet Rec. 1995;137(23):594.

Bolske G, Mattsson JG, Bascunana CR, Bergstrom K, Wesonga H, Johansson KE. Diagnosis of contagious caprine pleuropneumonia by detection and identification of *Mycoplasma capricolum* subsp. *capripneumoniae* by PCR and restriction enzyme analysis. J Clin Microbiol. 1996;34(4):785–91.

Bonnet F, Saillard C, Bove JM, Leach RH, Rose DL, Cottew GS, Tully JG. DNA relatedness between field isolates of *Mycoplasma* F38 group, the agent of contagious caprine pleuropneumonia, and strains of *Mycoplasma capricolum*. Int J Syst Bacteriol. 1993;43(3):597–602.

Brown RD, Taylor WP. Simultaneous vaccination of cattle against rinderpest and contagious bovine pleuropneumonia. Bull Epizoot Dis Afr. 1966;14(2):141–6.

Cetinkaya B, Kalin R, Karahan M, Atil E, Manso-Silvan L, Thiaucourt F. Detection of contagious caprine pleuropneumonia in East Turkey. Rev Sci Tech. 2009;28(3):1037–44.

Chaber AL, Lignereux L, Al Qassimi M, Saegerman C, Manso-Silvan L, Dupuy V, Thiaucourt F. Fatal transmission of contagious caprine pleuropneumonia to an Arabian oryx (*Oryx leucoryx*). Vet Microbiol. 2014;173(1–2):156–9. https://doi.org/10.1016/j.vetmic.2014.07.003.

Chu Y, Gao P, Zhao P, He Y, Liao N, Jackman S, Zhao Y, Birol I, Duan X, Lu Z. Genome sequence of *Mycoplasma capricolum* subsp. *capripneumoniae* strain M1601. J Bacteriol. 2011a;193(21):6098–9. https://doi.org/10.1128/JB.05980-11.

Chu Y, Yan X, Gao P, Zhao P, He Y, Liu J, Lu Z. Molecular detection of a mixed infection of Goatpox virus, Orf virus, and *Mycoplasma capricolum* subsp. *capripneumoniae* in goats. J Vet Diagn Investig. 2011b;23(4):786–9. https://doi.org/10.1177/1040638711407883.

Cottew GS, Breard A, DaMassa AJ, Erno H, Leach RH, Lefevre PC, Rodwell AW, Smith GR. Taxonomy of the *Mycoplasma mycoides* cluster. Isr J Med Sci. 1987;23(6):632–5.

Dupuy V, Verdier A, Thiaucourt F, Manso-Silvan L. A large-scale genomic approach affords unprecedented resolution for the molecular epidemiology and evolutionary history of contagious caprine pleuropneumonia. Vet Res. 2015;46:74. https://doi.org/10.1186/s13567-015-0208-x.

El Hassan SM, Harbi MS, Abu Bakr MI. Treatment of contagious caprine pleuropneumonia. Vet Res Commun. 1984;8(1):65–7.

FAO Food and Agriculture Organization of the United Nations. http://www.fao.org/. Accessed 10 October 2017.

Harbi MS, El Tahir MS, Macowan KJ, Nayil AA. *Mycoplasma* strain F38 and contagious caprine pleuropneumonia in the Sudan. Vet Rec. 1981;108(12):261.

Harbi MS, El Tahir MS, Salim MO, Nayil AA, Mageed IA. Experimental contagious caprine pleuropneumonia. Trop Anim Health Prod. 1983;15(1):51–2.

Hutcheon D. Contagious pleuro-pneumonia in angora goats. Vet J. 1881;13:171–80.

Hutcheon D. Contagious pleuropneumonia in goats at Cape Colony, South Africa. Vet J. 1889;29:399–404.

Kipronoh AK, Ombui JN, Kiara HK, Binepal YS, Gitonga E, Wesonga HO. Prevalence of contagious caprine pleuro-pneumonia in pastoral flocks of goats in the Rift Valley region of Kenya. Trop Anim Health Prod. 2016a;48(1):151–5. https://doi.org/10.1007/s11250-015-0934-0.

Kipronoh KA, Ombui JN, Binepal YS, Wesonga HO, Gitonga EK, Thuranira E, Kiara HK. Risk factors associated with contagious caprine pleuro-pneumonia in goats in pastoral areas in the Rift Valley region of Kenya. Prev Vet Med. 2016b;132:107–12. https://doi.org/10.1016/j.prevetmed.2016.08.011.

Leach RH, Erno H, MacOwan KJ. Proposal for designation of F38-type caprine mycoplasmas as *Mycoplasma capricolum* subsp. *capripneumoniae* subsp. nov. and consequent obligatory relegation of strains currently classified as *M. capricolum* (Tully, Barile, Edward, Theodore, and Erno 1974) to an additional new subspecies, *M. capricolum* subsp. *capricolum* subsp. nov. Int J Syst Bacteriol. 1993;43(3):603–5. https://doi.org/10.1099/00207713-43-3-603.

Lefevre PC, Breard A, Alfarouk I, Buron S. *Mycoplasma* species F 38 isolated in Chad. Vet Rec. 1987;121(24):575–6.
Litamoi JK, Wanyangu SW, Simam PK. Isolation of *Mycoplasma* biotype F38 from sheep in Kenya. Trop Anim Health Prod. 1990;22(4):260–2.
Longley E. Contagious pleuro-pneumonia of goats. Ind J Vet Sci Anim Hus. 1940;10:127–97.
Lorenzon S, Manso-Silvan L, Thiaucourt F. Specific real-time PCR assays for the detection and quantification of *Mycoplasma mycoides* subsp. *mycoides* SC and *Mycoplasma capricolum* subsp. *capripneumoniae*. Mol Cell Probes. 2008;22(5–6):324–8. https://doi.org/10.1016/j.mcp.2008.07.003.
MacOwan KJ. Role of mycoplasma strain F38 in contagious caprine pleuropneumonia. Isr J Med Sci. 1984;20(10):979–81.
MacOwan KJ, Minette JE. A mycoplasma from acute contagious caprine pleuropneumonia in Kenya. Trop Anim Health Prod. 1976;8(2):91–5.
MacOwan KJ, Minette JE. The effect of high passage *Mycoplasma* strain F38 on the course of contagious caprine pleuropneumonia (CCPP). Trop Anim Health Prod. 1978;10(1):31–5.
Manso-Silvan L, Perrier X, Thiaucourt F. Phylogeny of the *Mycoplasma mycoides* cluster based on analysis of five conserved protein-coding sequences and possible implications for the taxonomy of the group. Int J Syst Evol Microbiol. 2007;57(10):2247–58. https://doi.org/10.1099/ijs.0.64918-0.
Manso-Silvan L, Vilei EM, Sachse K, Djordjevic SP, Thiaucourt F, Frey J. *Mycoplasma leachii* sp. nov. as a new species designation for *Mycoplasma* sp. bovine group 7 of Leach, and reclassification of *Mycoplasma mycoides* subsp. *mycoides* LC as a serovar of *Mycoplasma mycoides* subsp. *capri*. Int J Syst Evol Microbiol. 2009;59(6):1353–8. https://doi.org/10.1099/ijs.0.005546-0.
Manso-Silvan L, Dupuy V, Chu Y, Thiaucourt F. Multi-locus sequence analysis of *Mycoplasma capricolum* subsp. *capripneumoniae* for the molecular epidemiology of contagious caprine pleuropneumonia. Vet Res. 2011;42:86. https://doi.org/10.1186/1297-9716-42-86.
Melanidi C, Stylianopoulos M. La pleuropneumonie contagieuse des chèvres en Grèce. Rev Gén Méd Vét. 1928;37:490–3.
Mettam R. Contagious pleuro-pneumonia of goats in East Africa. In: Paper presented at the Pan African veterinary conference, Pretoria. 1929.
Nicholas R, Churchward C. Contagious caprine pleuropneumonia: new aspects of an old disease. Transbound Emerg Dis. 2012;59(3):189–96. https://doi.org/10.1111/j.1865-1682.2011.01262.x.
OIE. Contagious caprine pleuropneumonia. In: Manual of diagnostic tests and vaccines for terrestrial animals. Paris: World Organisation for Animal Health; 2014.
OIE WAHIS, World Animal Health Information Database. http://www.oie.int/wahis_2/public/wahid.php/Wahidhome/Home. Accessed 09 October 2017.
Ozdemir U, Ozdemir E, March JB, Churchward C, Nicholas RA. Contagious caprine pleuropneumonia in the Thrace region of Turkey. Vet Rec. 2005;156(9):286–7.
Ozdemir U, Loria GR, Godinho KS, Samson R, Rowan TG, Churchward C, Ayling RD, Nicholas RA. Effect of danofloxacin (Advocin A180) on goats affected with contagious caprine pleuropneumonia. Trop Anim Health Prod. 2006;38(7–8):533–40.
Perreau P, Breard A, Le Goff C. Experimental infection in goats caused by mycoplasma strain F.38 (contagious caprine pleuropneumonia). Ann Microbiol. 1984;135A(1):119–24.
Pettersson B, Bolske G, Thiaucourt F, Uhlen M, Johansson KE. Molecular evolution of *Mycoplasma capricolum* subsp. *capripneumoniae* strains, based on polymorphisms in the 16S rRNA genes. J Bacteriol. 1998;180(9):2350–8.
Peyraud A, Poumarat F, Tardy F, Manso-Silvan L, Hamroev K, Tilloev T, Amirbekov M, Tounkara K, Bodjo C, Wesonga H, Nkando IG, Jenberie S, Yami M, Cardinale E, Meenowa D, Jaumally MR, Yaqub T, Shabbir MZ, Mukhtar N, Halimi M, Ziay GM, Schauwers W, Noori H, Rajabi AM, Ostrowski S, Thiaucourt F. An international collaborative study to determine the prevalence of contagious caprine pleuropneumonia by monoclonal antibody-based cELISA. BMC Vet Res. 2014;10:48. https://doi.org/10.1186/1746-6148-10-48.

Pirani A. Sulla pleropolmonite infettivo-contagiosa delle capre in Eritrea. Profilassi. 1932;5:170–5.
Rurangirwa FR, Masiga WN, Muthomi E. Immunity to contagious caprine pleuropneumonia caused by F-38 strain of *Mycoplasma*. Vet Rec. 1981;109(14):310.
Rurangirwa FR, McGuire TC, Kibor A, Chema S. An inactivated vaccine for contagious caprine pleuropneumonia. Vet Rec. 1987a;121(17):397–400.
Rurangirwa FR, McGuire TC, Kibor A, Chema S. A latex agglutination test for field diagnosis of contagious caprine pleuropneumonia. Vet Rec. 1987b;121(9):191–3.
Schellhase W. Ein Beitrag zur Kenntnis der ansteckenden Lungenbrustfellentzundung der Ziegen in Deutsch-Ostafrica. Zeit Infect Haustiere. 1912;12:70–83.
Settypalli TB, Lamien CE, Spergser J, Lelenta M, Wade A, Gelaye E, Loitsch A, Minoungou G, Thiaucourt F, Diallo A. One-step multiplex RT-qPCR assay for the detection of Peste des petits ruminants virus, Capripoxvirus, *Pasteurella multocida* and *Mycoplasma capricolum* subspecies (ssp.) *capripneumoniae*. PLoS One. 2016;11(4):e0153688. https://doi.org/10.1371/journal.pone.0153688.
Srivastava AK, Meenowa D, Barden G, Salguero FJ, Churchward C, Nicholas RA. Contagious caprine pleuropneumonia in Mauritius. Vet Rec. 2010;167(8):304–5. https://doi.org/10.1136/vr.c3816.
Thiaucourt F, Bolske G. Contagious caprine pleuropneumonia and other pulmonary mycoplasmoses of sheep and goats. Rev Sci Tech. 1996;15(4):1397–414.
Thiaucourt F, Breard A, Lefevre PC, Mebratu GY. Contagious caprine pleuropneumonia in Ethiopia. Vet Rec. 1992;131(25–26):585.
Thiaucourt F, Bolske G, Libeau G, Le Goff C, Lefevre PC. The use of monoclonal antibodies in the diagnosis of contagious caprine pleuropneumonia (CCPP). Vet Microbiol. 1994;41(3):191–203.
Thiaucourt F, Bolske G, Leneguersh B, Smith D, Wesonga H. Diagnosis and control of contagious caprine pleuropneumonia. Rev Sci Tech. 1996;15(4):1415–29.
Thomas P. In: Jourdan A, editor. Rapport médical sur le Bou Frida. Algiers: Publication du gouvernement général civil de l'Algérie; 1873.
Walker GK. Pleuro-pneumonia of goats in the Kangra district, Punjab, India. J Comp Pathol. 1914;27:68–71.
Woubit S, Lorenzon S, Peyraud A, Manso-Silvan L, Thiaucourt F. A specific PCR for the identification of *Mycoplasma capricolum* subsp. *capripneumoniae*, the causative agent of contagious caprine pleuropneumonia (CCPP). Vet Microbiol. 2004;104(1–2):125–32. https://doi.org/10.1016/j.vetmic.2004.08.006.
Yu Z, Wang T, Sun H, Xia Z, Zhang K, Chu D, Xu Y, Xin Y, Xu W, Cheng K, Zheng X, Huang G, Zhao Y, Yang S, Gao Y, Xia X. Contagious caprine pleuropneumonia in endangered Tibetan antelope, China, 2012. Emerg Infect Dis. 2013;19(12):2051–3. https://doi.org/10.3201/eid1912.130067.

Chapter 22
Cowdriosis/Heartwater

Frédéric Stachurski, Arona Gueye, and Nathalie Vachiéry

Abstract Heartwater (or cowdriosis) is a tick-borne disease caused by *Ehrlichia ruminantium*, an obligatory intracellular bacterium of the order Rickettsiales, transmitted by several ticks of the genus *Amblyomma*. The organism is genetically highly variable which prevented until now the development of efficient vaccines. The disease is enzootic in sub-Sahelian Africa and in some Caribbean islands. It affects domestic and wild ruminants, the susceptibility to cowdriosis varying greatly between breeds and species: African wildlife shows mainly asymptomatic infections; local cattle breeds are generally protected due to enzootic stability; and introduced cattle breeds and small ruminants, even in enzootic regions, are usually susceptible to heartwater and can suffer high mortality rates. Cowdriosis is characterized by a sudden and acute fever followed by nervous, respiratory, and gastrointestinal symptoms and by hydrothorax and hydropericardium during postmortem examination. In West Africa, the only vector is *Amblyomma variegatum*, present in areas where pluviometry is higher than 500 mm. Therefore, animals of a high proportion of the Sahelian region are usually not infested by the tick and not infected by the bacterium. They are thus susceptible when introduced in southern parts of the Sahel or in the subhumid neighboring areas of the West African countries, for example during transhumance. Tetracyclines are effective drugs to treat heartwater when administered before occurrence of the nervous symptoms. Various vaccines have been tested, and are still developed, but, up to now, none of them showed enough effectiveness against all the field strains of *E. ruminantium* to allow its marketing. Prevention is therefore mainly achieved by drastic vector control or, on the contrary, acquisition of enzootic stability following tick infestation combined with tetracycline treatment as soon as hyperthermia occurs.

F. Stachurski (✉) · N. Vachiéry
CIRAD, UMR ASTRE, Montpellier, France

ASTRE, CIRAD, INRA, University of Montpellier, Montpellier, France
e-mail: frederic.stachurski@cirad.fr

A. Gueye
Cité ISRA, Marinas Bel Air, Dakar, Senegal
e-mail: arogueye@orange.sn

M. Kardjadj et al. (eds.), *Transboundary Animal Diseases in Sahelian Africa and Connected Regions*, https://doi.org/10.1007/978-3-030-25385-1_22

Keywords *Ehrlichia ruminantium* · Heartwater · Cowdriosis · *Amblyomma variegatum* · Tick-borne disease · Epidemiology · Enzootic stability · Prevention · Treatment

Introduction

Cowdriosis is an infectious, virulent, and noncontagious disease of ruminants due to *Ehrlichia ruminantium* (formerly known as *Cowdria ruminantium*), an obligate intracellular bacterium naturally transmitted by ticks of the genus *Amblyomma* (Camus et al. 1996; Allsopp 2015). Domestic ruminants (cattle, sheep, and goats) as well as wild ruminants (buffaloes, antelopes, Cervidae, etc.) are prone to infection. However, African wildlife appears to be less susceptible, showing mainly asymptomatic infections (Camus et al. 1996). Cowdriosis is responsible for high mortality rates in sheep and goats. In cattle, losses can also be significant in some circumstances, for example in austral or eastern Africa where numerous susceptible cattle of exotic (i.e., nonindigenous) breeds have been imported and are reared. Besides, drastic tick control campaigns were implemented in some of these countries, leading to the disappearance of an enzootic stability situation (see below), which can be disastrous when tick control is later interrupted (Lawrence et al. 1980). Losses are by far lower in West Africa because the cattle reared there are still predominantly of less susceptible local breeds, except in some dairy farms located in peri-urban areas of big cities (Adakal et al. 2013). Moreover, tick control was, for historical reasons, never organized by the national veterinary services in West Africa (Uilenberg 1984), each farmer taking care of his own herd, using generally less expensive and often less efficient methods and products (Adakal et al. 2013).

Cowdriosis is characterized by a sudden and acute fever followed by severe nervous, respiratory, and gastrointestinal symptoms. Postmortem examination is characterized mainly by hydrothorax and hydropericardium, hence the other name of the disease, "heartwater." In Africa, cowdriosis is enzootic in all regions where the vector ticks *Amblyomma* are present (Allsopp 2015). About ten species have been proved, naturally or experimentally, able to transmit *E. ruminantium*. The main vector is *A. variegatum* because of its high efficiency and its wide distribution (Walker and Olwage 1987). But other species, like *A. hebraeum*, *A. pomposum*, *A. lepidum*, or *A. astrion* (Uilenberg 1983), may have locally a dominating role and are sometimes the only vector (*A. hebraeum* in South Africa, *A. astrion* in São Tomé) (Fig. 22.1).

Amblyomma variegatum is the only *E. ruminantium* vector which hugely increased its distribution area outside of its original sub-Saharan cradle. This tick and the transmitted bacterium are now established in some of the Indian Ocean islands (Madagascar, Comoros, La Réunion, Maurice) and in three of the Caribbean islands, Guadeloupe, Marie-Galante, and Antigua (Walker and Olwage 1987). On the other hand, in many other West Indies islands (Martinique, Dominica, St. Kitts &

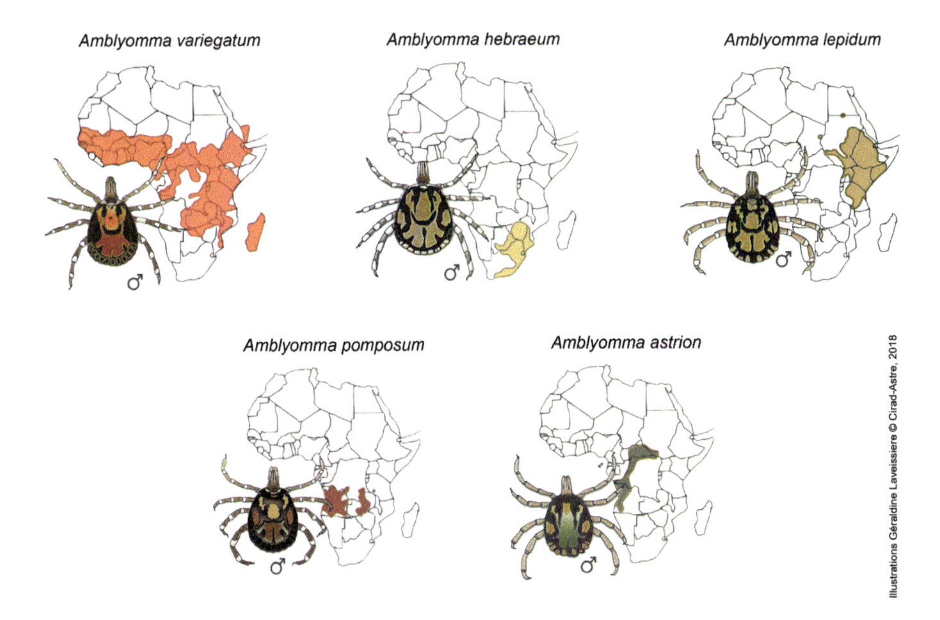

Fig. 22.1 Distribution of the five main vectors of *Ehrlichia ruminantium* in Africa (sources: Walker and Olwage 1987; Walker et al. 2003)

Nevis, etc.), only the tick was introduced, either by cattle or hide trade (legal or not) or by dissemination of immature stages by migratory birds like cattle egret (*Bubulcus ibis*) (Barré et al. 1987; Corn et al. 1993), but not *E. ruminantium*. On some of these islands (Montserrat), *A. variegatum* seems not to have been able to settle; on others, an eradication campaign carried out in the 1990s (Caribbean Amblyomma Program) led to the disappearance of the tick; and others remained infested at the end of the campaign, although generally with low infestation levels (Fig. 22.2). In the south of the Arabian peninsula (Yemen, Oman), the tick was also recorded (Walker and Olwage 1987) but the disease was never reported.

Because the West Indies are so close to the American mainland, and because of the possibility for migrating birds to carry infected ticks, the presence of *A. variegatum* and *E. ruminantium* in the Caribbean is a permanent threat for cattle production of many countries of the continent. The tick could find in many regions in Brazil, Venezuela, the Pacific coast of Peru and Chile, in Colombia and Central American countries, in Mexico, and in the southern USA, the environment and climatic conditions allowing its establishment (Barré et al. 1987; Barré 1997). As some autochthonous *Amblyomma*, *A. maculatum* above all and to a lesser extent ticks of the *A. cajennense* group, have been experimentally shown to be able to transmit *E. ruminantium* (Barré et al. 1987), it is easy to understand that heartwater is a cause for concern for veterinary authorities of the continent (Burridge et al. 2002). Cowdriosis was besides identified as one of the 12 most important transboundary diseases by the US department of homeland security (Roth et al. 2013).

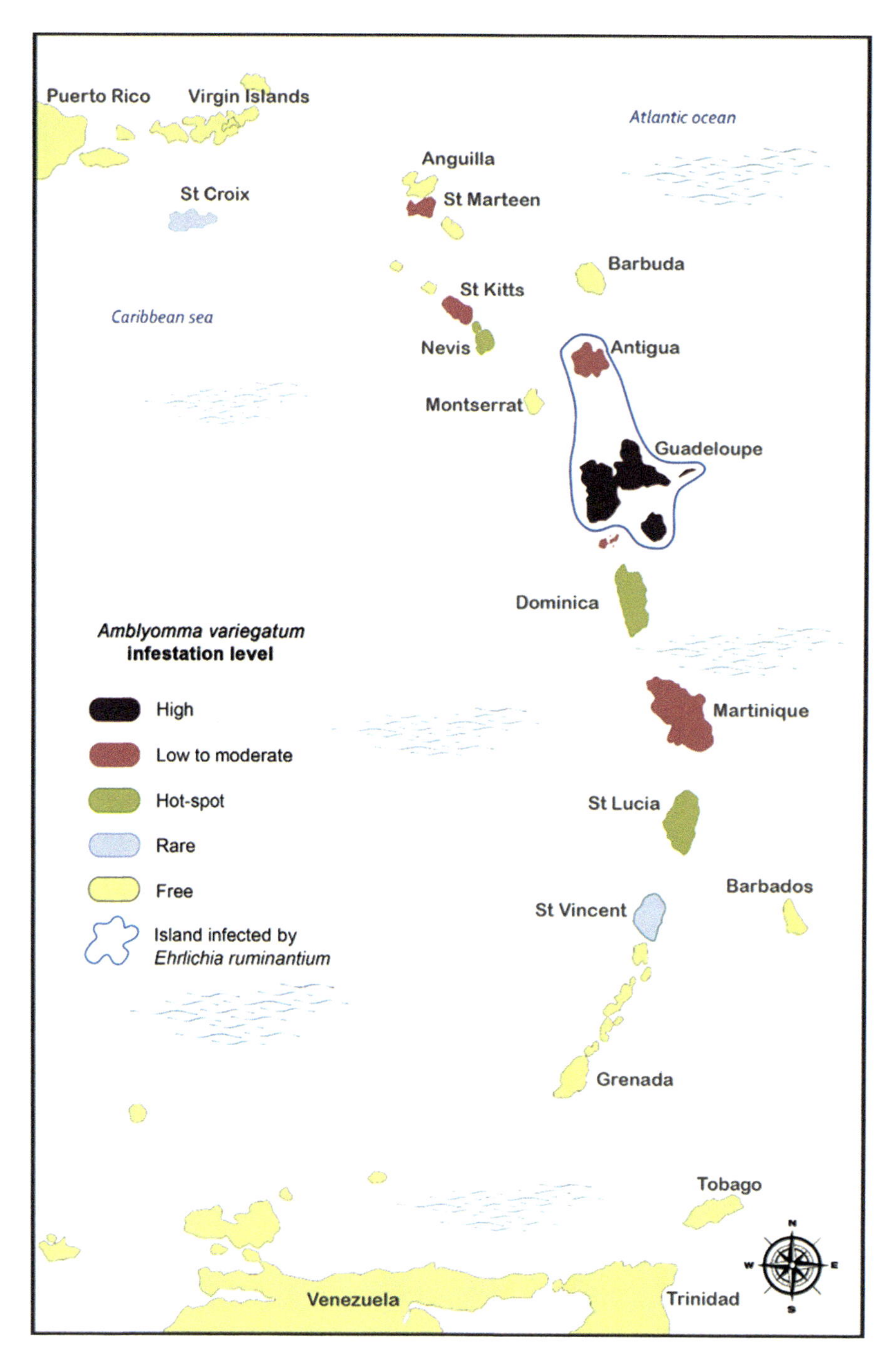

Fig. 22.2 Distribution of *Amblyomma variegatum* in the Caribbean in 2011 (sources: CaribVet network; CIRAD)

In Western and Central Sahel, *A. variegatum* is the only known vector of cowdriosis: the distribution and importance of the disease are therefore intimately related to the tick distribution. It is classically considered that the species cannot persist in regions where the annual rainfall is lower than 500 mm (Morel 1969): *A. variegatum* is thus absent from a high proportion of the Sahel area. Due to the lack of transmission of the cowdriosis pathogen, ruminants of these regions are not immunized against heartwater. They are therefore susceptible to the disease when moved to areas where *A. variegatum* is endemic, i.e., to the most humid southern parts of the Sahel or to the neighboring Sudanian region. In this latter area, *A. variegatum* is very abundant and is the most harmful species, responsible for direct damage (Pegram and Oosterwijk 1990; Stachurski et al. 1993) and indirect losses due to heartwater but also dermatophilosis (Plowright 1956). High mortality rates can thus be observed in Sahelian cattle and small ruminants when temporally or permanently moved to areas where the tick is present. In such susceptible animals, cowdriosis may develop peracute symptoms and death may occur in less than 48 h.

Ehrlichia ruminantium infection might, however, remain unnoticed in regions where *A. variegatum* is abundant, in herds where only cattle of local breeds are reared. Such an "enzootic stability" situation, as previously defined for babesiosis (Mahoney and Ross 1972), may occur because of the usual high *A. variegatum* infestation levels with high infection rates by *E. ruminantium*, and because calves are less susceptible to the disease due to various reasons (see below). On the other hand, small ruminants, even those of local breeds which are infested by the tick for generations, do not benefit from that enzootic stability: losses due to cowdriosis, occurring mainly at the beginning of the dry season, may affect a high proportion of lambs and kids born during the previous months.

Ehrlichia ruminantium, the pathogen of heartwater, was identified by Cowdry in 1925, but few studies have been carried out on that bacterium in the following decades because it was neither possible to cultivate it in vitro nor to use laboratory animal models. The situation changed in the 1970s when strains virulent for mice were isolated (du Plessis and Kumm 1971) and especially when *E. ruminantium* cell culture was developed (Bezuidenhout et al. 1985). Various experimental vaccines (inactivated, attenuated, and recombinant) were afterward tested (Allsopp 2015) which generally give good protection against homologous challenge. They have nevertheless a lower effectiveness to protect against the heterologous field strains, which impairs their use. This is mainly due to high genetic variability of the pathogen with sometimes low cross-immunity (Jongejan et al. 1991a), while many strains cohabit in limited areas (Adakal et al. 2010b). The control of the disease in areas without enzootic stability is therefore difficult because there is no commercial vaccine and because the cost of efficient tick control methods might exceed the value of the animals. Fortunately, antibiotics of the tetracycline family are efficient if given properly and rapidly, as soon as fever is noticed.

Although it is rarely done, cowdriosis is a notifiable disease to OIE (=World Organisation for Animal Health). A human case, possibly due to *E. ruminantium*, which has also been proved to grow in human cell cultures (Totté et al. 1993), was recorded many years ago (Allsopp et al. 2005). But there has been no further record from that time onward to confirm the zoonotic potential of heartwater.

Etiology

Ehrlichia ruminantium is an obligate intracellular bacterium which infects endothelial cells of the blood vessels, neutrophils, and macrophages. Based on the characterization of its 16S RNA, *Cowdria ruminantium* was renamed in the early 2000s and clustered in the genus *Ehrlichia*, belonging to the *Anaplasmataceae* family (Dumler et al. 2001).

During the febrile phase of the disease, the bacterium is circulating in the bloodstream or present in the neutrophils (Prozesky and Du Plessis 1987). Isolation of *E. ruminantium* in ruminant endothelial cell cultures is possible using blood collected from infected animals during hyperthermia. Pioneered in South Africa in 1985 and considerably simplified since then, this technique allowed to identify two phases in the development cycle. In vacuoles of the cytoplasm of the infected cells, there are the "reticulate bodies," the non-infective form, which multiply by binary fission and develop into morulas. When these morulas are mature, cell lysis is observed and the "elementary bodies," the free infectious form, are released (Jongejan et al. 1991b; Marcelino et al. 2015). In in vitro cell culture conditions, the cycle of *E. ruminantium* lasts 4–6 days according to the strain (Marcelino et al. 2015).

Ehrlichia ruminantium development cycle is still not fully understood, in ticks as well as in ruminant hosts. In ruminants, local multiplication occurs initially in reticulo-endothelial cells and macrophages present where the infected tick inoculates the bacterium. The infection of these cells is made easier because of the immunomodulation action of the numerous salivary molecules inoculated during the blood meal of the tick (Rodrigues et al. 2017). The infected cells are drained to the nearest lymph node; the bacterium is then disseminated via the blood to the whole body and infects the endothelial cells of the various blood vessels where further multiplications occur (Prozesky and Du Plessis 1987). In the tick, following the infecting blood meal, *E. ruminantium* infects cells of the gut, crosses then the wall of the digestive tract and, carried by the hemolymph, colonizes the salivary glands. This maturation occurs during the molting of the tick. Once the new tick stage is attached to its host, *E. ruminantium* multiplies during a few days in the acini of the salivary glands before it is inoculated to the vertebrate (Prozesky and Du Plessis 1987). Because of this high multiplication during the first days of the tick blood meal, a single infected tick is enough to cause the disease.

The recent advances in molecular biology allowed the characterization of numerous *E. ruminantium* strains by using typing methods based on one or several genes or on whole genome sequencing. *Ehrlichia ruminantium* genetic characterization was first implemented by targeting a single conserved gene or region (i.e., a fragment of the *pCS20* region) or a single polymorphic gene (*map-1* gene, encoding for a Major Antigenic Protein) which was amplified by PCR and sequenced. MLST (multi-locus sequence typing) or MLVA (multi-locus variable analysis) techniques were also developed based on the amplification of multiple targeted genes (Adakal et al. 2009; Pilet et al. 2012). A high genetic diversity was observed within very restricted geographical areas in Burkina Faso (Adakal et al. 2010b), Madagascar (Raliniaina et al. 2010), or Guadeloupe (Vachiéry et al. 2008), where more than 10 different genotypes were identified in areas of 25–30 km^2. Moreover, strains from the same genotypes were detected in many different areas, from Southern Africa toward Western Africa and the Caribbean. A recent phylogeography study on strains from worldwide areas using MLST showed that the strains are clustered into two main genetic groups, one including only strains of West Africa and the other including strains from the Caribbean, the Indian Ocean, and Southern and Western Africa. This study also highlighted the introduction routes of *E. ruminantium* strains from Western Africa toward West Indies and from Mozambique toward the Indian Ocean islands (Cangi et al. 2016).

Some *E. ruminantium* reference strains isolated in vitro were fully sequenced: Gardel strain originating from Guadeloupe and Welgevonden strain from South Africa (Collins et al. 2005; Frutos et al. 2006). The analysis of full genomes allowed to explain genome plasticity and the repeated occurrence of recombination between various strains. This gene recombination was confirmed by the genetic characterization of worldwide strains using MLST (Cangi et al. 2016). Recombination plays thus an important role in the high genetic diversity observed. However, despite these advances, no relationship could so far be observed between genetic groups and protection within strains belonging to these groups. It is therefore not yet possible to identify the optimal strains to include in a regional vaccine, based on their genotype, without carrying out cross-protection experiments to evaluate their protection ability against other strains.

Epidemiology

Epidemiology of cowdriosis is closely linked to the distribution and life cycle of its vectors, the *Amblyomma* ticks. In eastern parts of the Sahel, the epidemiological situation is rather complex because several vectors infest the ruminants and because the climatic zones are not as clearly delineated as in West Africa. In western and central parts of the Sahelian area, and particularly to the south of these regions which are the areas where the tick is present (heartwater is strictly speaking not a Sahelian disease), the only vector is *A. variegatum*. This tick is a three-host tick, each stage (larvae, nymphs, adults; Fig. 22.3) infesting a new host where it takes a blood meal before it drops off for molting (larvae, nymphs) or egg laying (females). The feeding period lasts 5–8 days for larvae and nymphs and 7–14 days for females (Hoogstraal 1956; Yonow 1995). Males can remain attached for 2–3 months, mating successively with several females.

In equatorial regions infested by the tick where there is no real dry season and where rains occur more or less all year round (coast of Gulf of Guinea, West Indies, east coast of Madagascar, etc.), the three stages infest simultaneously and practically permanently their hosts. There is therefore no season more favorable than others to *E. ruminantium* transmission (Camus et al. 1996).

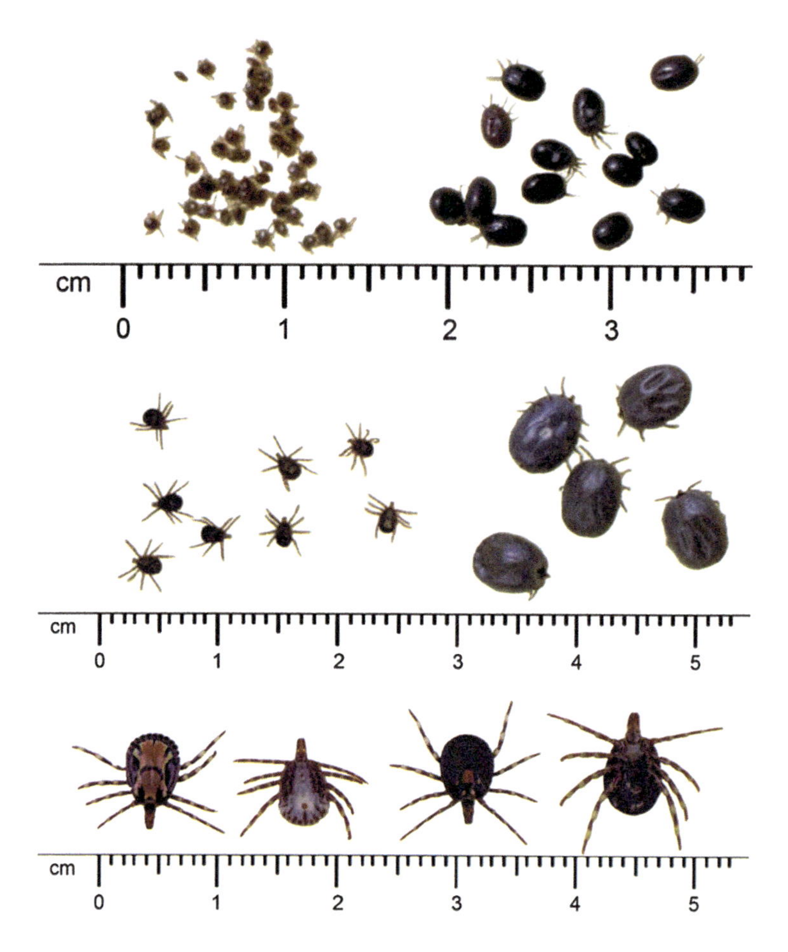

Fig. 22.3 The three stages of *Amblyomma variegatum*. From top to bottom: unfed and engorged larvae, unfed and engorged nymphs, unfed adults (dorsal and ventral faces of a male, dorsal and ventral faces of a female) (photos: F. Stachurski)

On the contrary, in the Sudanian area and in the neighboring southern part of Sahelian area, climate is characterized by a well-defined and a yearly succession of dry and rainy seasons. In that case, *A. variegatum* life cycle and cowdriosis epidemiology are very dependent on rains. Adult ticks infest ruminants during the first part of the rainy season, larvae hatching from eggs laid in that period parasitize their hosts at the end of the same rainy season, and nymphs resulting from engorged larvae infest animals at the beginning of the following dry season. In a favorable environment, i.e., when humidity and temperature are favorable, unfed adults may survive up to 18 months and unfed nymphs up to 12 months waiting for a host. On the other hand, eggs and larvae are very prone to desiccation and die quickly at the beginning of the dry season, when ground humidity becomes too low to allow survival of these stages (Barré and Garris 1989; Pegram and Banda 1990).

As the incubation period of the eggs is quite long and may last about 100 days, *A. variegatum* can establish itself only in regions where humidity is high enough. It is thus usually considered that the tick is absent when annual rainfall is lower than 500 mm (Morel 1969; Petney et al. 1987). But survival of eggs and larvae is probably more dependent on the duration of the rainy season and therefore on the humidity at ground level than on the amount of annual rainfall. It is consequently possible to find live larvae during the dry season in well-sheltered places such as gallery forests. It explains also that *A. variegatum* and cowdriosis are observed in some Sahelian regions where rainfall is lower than 500 mm, like the Niayes in Senegal. In that area, humidity remains high during a large part of the year because of shallow ground water and numerous ponds. One of us (AG) even observed *A. variegatum* ticks along Senegal River in a region where rainfall is lower than 400 mm but where ground humidity remains high enough, allowing survival of the most sensitive life stages. Tick populations are able to establish in such areas when Sahelian ruminants, moved during the rainy season for transhumance in regions where the pastures are infested by the tick, come back with attached ticks. If these drop in areas with favorable environment and microclimate, they can survive. It is then possible to observe cases of heartwater in sites a priori unsuitable for tick survival and on animals which never move from those regions to areas where the vector is endemic.

Infected ticks are the main reservoir of *E. ruminantium*, tick survival being apparently not impaired by the presence of the bacterium. Transmission of *E. ruminantium* by the vector is transstadial: a tick acquires the pathogen during its blood meal and the next or the two next stages can transmit it (Camus et al. 1996). As there is no transovarial transmission, from female toward the next generation, only nymphs and adults are vectors. In south Sahel and in the neighboring Sudanian area, there are therefore two periods during which cowdriosis outbreaks can occur: the beginning of the rainy season, during the infestation peak by adult ticks, and the beginning of the dry season, when the ruminants are infested by the nymphs (see above).

It is mainly during the dry season that cowdriosis is observed in small ruminants within enzootic regions, because goats and sheep are, in that period, infested in 2–3 months by tens or hundreds of nymphs. Considering the known infection rates of that stage (Mahan et al. 1998b), it is likely that all animals are infected by *E. ruminantium* in that period. On the other hand, small ruminants' infestation by adult ticks during the rainy season is low (only few ticks per head on average, and a high proportion which are not parasitized by adult *A. variegatum*; see Fig. 22.4). Consequently, only some sheep and goats are infected by the pathogen at that period.

The situation is quite different for cattle. During the dry season, their infestation by nymphs is very high, adult animals being sometimes infested by more than 1000 nymphs, and during the rainy season, cattle are infested by tens to hundreds adult *A. variegatum* (Stachurski et al. 1993). Calves are infested by adult *A. variegatum* from their birth, even if they are kept all day round near the night paddock and are not brought to the pasture with the other cattle (Fig. 22.5). These latter capture the adult ticks, which first attach to the inter-digital areas, when grazing. The ticks later

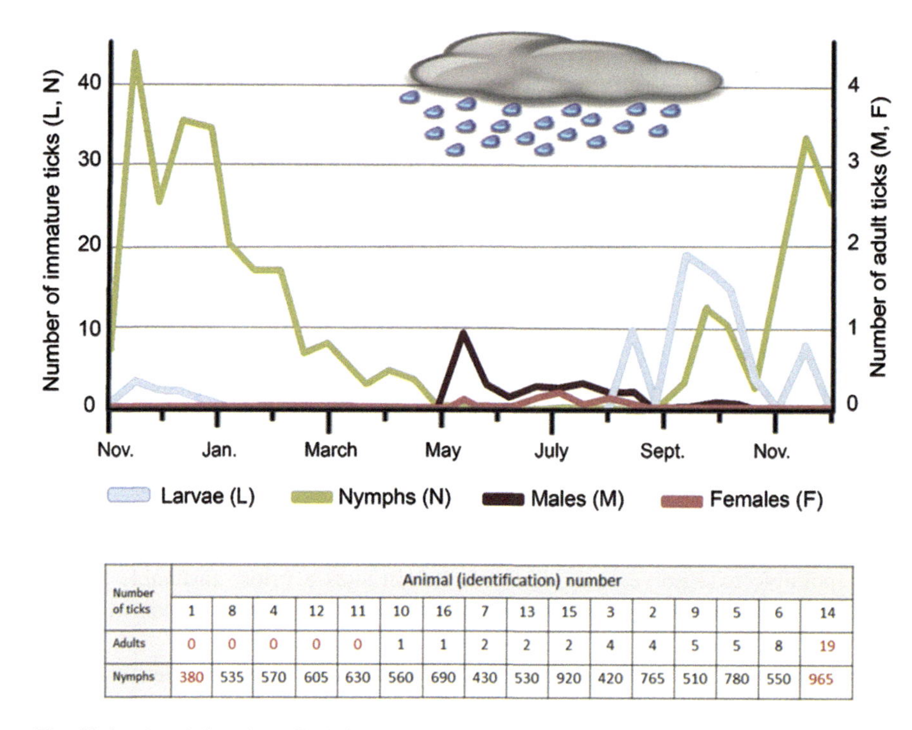

Number of ticks	Animal (identification) number															
	1	8	4	12	11	10	16	7	13	15	3	2	9	5	6	14
Adults	0	0	0	0	0	1	1	2	2	2	4	4	5	5	8	19
Nymphs	380	535	570	605	630	560	690	430	530	920	420	765	510	780	550	965

Fig. 22.4 Mean infestation of Djallonke ewes by the three stages of *Amblyomma variegatum* in a village near Bobo-Dioulasso, Burkina Faso. The table shows the annual cumulative numbers of nymph and adult ticks for each of the 16 monitored sheep

reach the predilection attachment sites when animals are laying down, mainly during the night (Stachurski 2000). *Amblyomma variegatum* may then move to infest other animals in the kraal: such movement can easily occur from the dam to its calf lying down nearby.

The infection rate of the vector ticks by *E. ruminantium* is high. That of adult ticks generally exceeds 10% (Faburay et al. 2007b; Molia et al. 2008; Cangi 2017) although it was sometimes found to be lower than 5% (Adakal et al. 2010a). That of the nymphs, for *A. variegatum* (Adakal et al. 2010a) as well as for *A. hebraeum* (Mahan et al. 1998b), is about 3%. Consequently, cattle are infected by cowdriosis in the dry season as well as in the rainy season and therefore attain enzootic stability. Calves are infected when they are young and have a low susceptibility to the disease. They can thus acquire a protective immunity which is later on recurrently boosted due to high nymph and adult infestations. On the other hand, sheep do not live in enzootic stability: some lambs, mainly those born just before or during the rainy season, lose their neonatal tolerance to heartwater before they are infested by infected ticks. They are thus susceptible to cowdriosis when infested by nymphs in the next dry season, and a high proportion of these animals may die. The difference between small ruminants and cattle might be due to their size since *A. variegatum* adults are attracted toward larger animals which produce more efficient stimuli

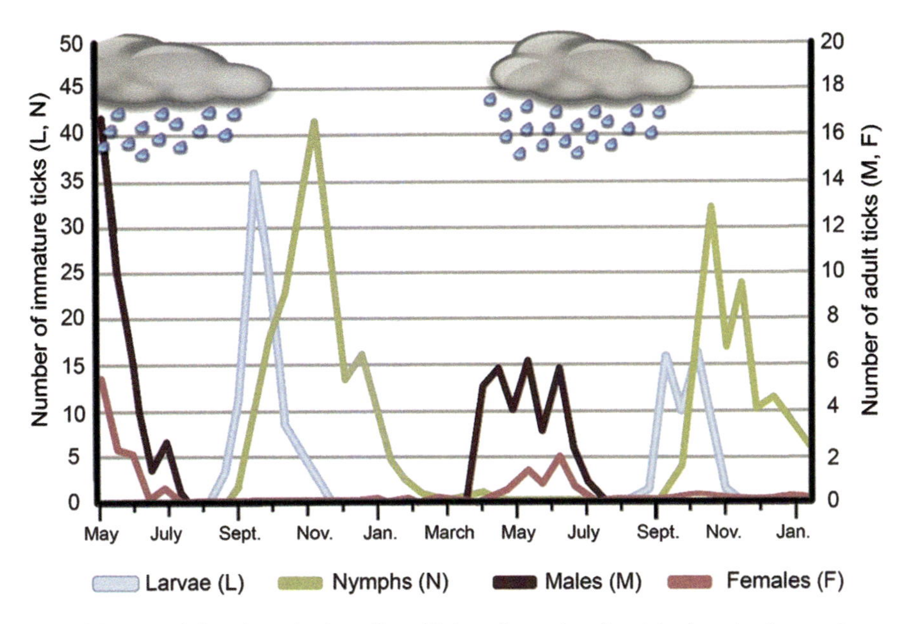

Fig. 22.5 Mean infestation of calves (from birth to 8 months of age) in three herds reared near Bobo-Dioulasso, Burkina Faso, by the three stages of *Amblyomma variegatum*

(carbon dioxide emitted by expiration, ground vibrations due to animal movements, etc.) (Barré 1989; Meltzer 1996).

According to the authors, the main cause for the protection allowing young ruminants to overcome early *E. ruminantium* infection is variable. For some, although maternal antibodies are not protective, there is transmission of a kind of maternal immunity, probably cellular, to the newborn calves. The colostrum of cows reared in enzootic areas could thus protect calves, even those of dams bred in heartwater-free areas (Deem et al. 1996a). This colostrum might also contain the bacterium which could thus quickly infect the calves, some of them being even infected in utero (Deem et al. 1996b). For other authors, the main protection is due to an innate resistance (or tolerance) allowing all young ruminants to overcome early *E. ruminantium*, as well as other tick-borne pathogens, infection (Uilenberg 1995). This innate resistance is more efficient in indigenous ruminant breeds but is also observed in exotic animals. The importance and duration of the protection may be variable according to ruminant species (it could be more efficient and persistent in cattle, allowing the setting up of enzootic stability), but the data regarding this issue are still fragmented and unclear. Finally, the centuries-old cohabitation of local breeds and *E. ruminantium* led to a natural selection of less-susceptible ruminant populations, which is not only reflected by a better calf innate resistance (Camus et al. 1996).

In animals contaminated by an infected tick, bacteremia appears 1–2 days before the fever onset. Ruminants which overcome the infection, and those which were already immunized, develop (or boost) a cellular-mediated protective immunity which lasts probably more than 2 years (Stewart 1987) and does not prevent ruminants to be chronic carriers of *E. ruminantium* for weeks or months (Andrew and Norval 1989). These asymptomatic carriers are an important source of infection for the ticks, more efficient than animals dying of a subacute or acute form of cowdriosis. But not all of the infesting *Amblyomma* ticks acquire infection on such animals because bacteremia is not continuous but shows transitory "eclipses." This allows, however, a sufficiently high infection rate of the ticks to ensure maintenance of enzootic stability in cattle. Once animals have overcome a first *E. ruminantium* infection, they are permanently immunized unless they are infected by a strain completely different from those usually present in the area where they are raised and against which there is a lack of cross-immunity.

Cowdriosis is thus constantly present in herds bred in areas where *A. variegatum* is established. Local cattle do not suffer, or only marginally, from the infection because of enzootic stability due to regular infestations by vector ticks. On the other hand, high mortality might be observed in sheep and goats of these areas, mainly because of their low infestation by the adult ticks during the rainy season. But what of the ruminants living in the drier areas of the Sahel where *A. variegatum* is not established? They are of course not immunized against *E. ruminantium* when they are young and are therefore susceptible to the disease when they are infected later on. This infection may occur when animals are, during the dry season, i.e., during the nymphal infestation period, brought for transhumance in more humid regions where the vector tick is present. A mortality rate of 35% was thus observed in sheep brought from northern Senegal to pastures located near the Atlantic coast of the country, and death reached 20% for adult Sahelian sheep brought from the north to the south of Burkina Faso at the beginning of the dry season, the mortality being only stopped by the systematic treatment of the surviving sheep with oxytetracycline (see the "treatment" section below). Such Sahelian sheep introductions are nevertheless regularly made by farmers of the subhumid Sudanian area of the country in order to try to increase the format of the local Djallonke sheep (a West African dwarf breed), and regularly, a high mortality is observed in these animals, except for those kept in town where the infestation risk is lower. Such mortalities due to cowdriosis are naturally also observed in exotic (European or American) cattle, sheep, or goat breeds introduced in Africa with the purpose of increasing production.

Cowdriosis could also be observed in ruminants moved from one enzootic region to another where they might be infected by completely different *E. ruminantium* strains. But such a situation has never been reported, which is not surprising given that tens of strains are circulating permanently in enzootic regions (Adakal et al. 2010b). The "transboundary" aspect of cowdriosis concerns thus mainly movements of ruminants between Sahelian and non-Sahelian regions, whether these are in the same or in a neighboring country.

The situation could change with the current climatic changes. If rainfall (and/or the length of the rainy season) would decrease in the south of the Sahel and the northern Sudan zone, *A. variegatum* could no more survive there and cattle reared in that areas would lose their immunity against cowdriosis. On the contrary, if rainfall (and/or the length of the rainy season) would increase in areas where it reaches presently 400–500 mm, the tick, introduced by ruminants infested further south, could become established. Some of these ticks would certainly be infected by *E. ruminantium*. Ruminants of these areas would then face a new disease, deaths would be noticed, and several years would certainly be needed before the situation would become stabilized and the infestation of the animals reaches a level allowing enzootic stability.

At the border of the distribution area of a tick, where the climatic conditions are not very favorable to parasite survival, the population size usually highly fluctuates from 1 year to the next (Uilenberg 1995): the infestation of the animals is therefore often too low and young ruminants are thus often not infected when they are resistant/tolerant to the disease. Although information is missing on this matter, it is moreover likely that some of the ruminants living in the Sahelian area, in regions where the annual rainfall fluctuates around 500 mm, are in that situation: they graze pastures where *Amblyomma variegatum* can survive but where tick density is most probably low for some years or some biotopes do not allow tick survival. Some of these ruminants are neither in an early age nor regularly infected, and some of them should show symptoms and even die when they contract *E. ruminantium* infection. This lack of enzootic stability is probably more or less offset by natural selection since these ruminant populations are living for a long time near regions where the vector tick and the pathogen are enzootic.

Symptoms and Lesions

They were described by numerous authors, from Alexander (1931) to Allsopp (2015). The natural incubation period is 2–3, sometimes 4, weeks. As mentioned above, because of the multiplication of the bacterium in the acini of the tick salivary glands during a few days after attachment, a single infected tick is enough to transmit the pathogen and provoke the disease. Following experimental injection of infected blood or of cell cultures of *E. ruminantium*, the disease can occur more quickly, in less than 10 days.

Several forms of cowdriosis are described, the severity of symptoms and lesions depending greatly on species, breed, age of the infected ruminant, as well as the *E. ruminantium* strain. In the peracute form, death can occur within 24 h without other symptoms than hyperthermia and perhaps some modification of the behavior (see below). In that case, antibodies might even not be detected if serum is collected for a diagnostic purpose. In the acute and subacute forms, the first symptom is the modification of behavior and the deterioration of general condition: the ruminant lies down longer in the morning, stands apart, has a ruffled hair coat, eats less, and has a fast respiration. Fever reaches rapidly 41 °C and sometimes 42 °C. The production of antibodies, which starts generally with the onset of pyrexia, does not prevent death because humoral immunity is not protective (Stewart 1987).

After this period of general weakness and prostration, breathing difficulties and nervous symptoms are usually observed. These latter (hypersensitivity, lateral recumbent position, pedaling, convulsions) occur during a few minutes alternating with periods of quiet prostration. In the acute form, the animal dies within 3–6 days and recovery usually does not occur once the nervous symptoms are observed. Unusual forms are sometimes observed, with diarrhea and without nervous symptoms, or with low hyperthermia. In the subacute form, the modification of the general condition (prostration, loss of appetite, and therefore loss of weight) might be the only symptom observed. Within 7–10 days, if no nervous symptoms occurred, the animal may recover and the symptoms disappear. A mild form may finally be observed with only transient pyrexia and even cases without any symptom in immune animals or resistant breeds. Animals with such mild forms are often unnoticed but are very important for the epidemiology (see above).

The forms described here are observed mainly with small ruminants. With cattle, the symptoms are similar but diarrhea occurs more frequently.

No lesion is pathognomonic although hydropericardium and hydrothorax (transudation of yellow fluid in the cavities) are often observed (Fig. 22.6), but they might also occur in other diseases. Other symptoms like pulmonary edema or intestinal congestion may also be observed. These lesions may lack in the peracute form.

The cause of the symptoms and lesions remains doubtful. They are generally attributed to increased vascular permeability which may induce liquid effusion in the cavities, responsible for heart and respiratory dysfunctions, and in the brain, leading to cerebral edema which would cause the nervous symptoms. But it is not known whether the cause for such a vascular permeability increase is a toxin or another mechanism (Camus et al. 1996).

Diagnosis

Cowdriosis is generally suspected if nervous symptoms, high fever, or unexplained death is observed when *A. variegatum* adults or nymphs (or other vectors in eastern and southern Africa) infest the ruminants.

Parasitologic Diagnosis

The definite diagnosis in case of death is the observation of *E. ruminantium* clusters in the vascular endothelial cells. They are searched for in the capillaries of the cortex, because this organ is easier to examine, but they may also be observed in the capillaries of the lung or in the endothelium of arteries like aorta. Brain smears, more accurately "brain crush smears," are prepared and stained as described in box 1 (Camus et al. 1996). Although the bacterium is relatively quickly damaged in the organs at ambient temperature after death, it is still possible to observe it in an animal body after 12–24 h. When the head is stored in a fridge, samples may be successfully examined several days or weeks after the death.

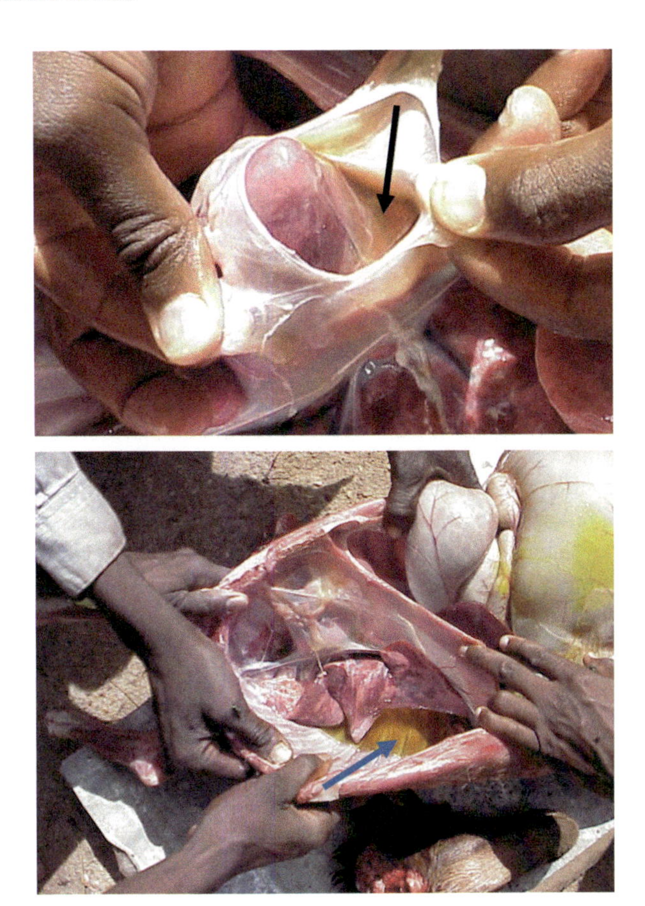

Fig. 22.6 Hydropericardium and hydrothorax following experimental infection of a sheep (photos: F. Stachurski)

To prepare cortex samples in order to diagnose cowdriosis, it is necessary to:

- Remove the skin of the head and open the cranium (with a machete, for example);
- Free the cortex from the meninges; then take superficially a small piece of gray matter (the size of a pinhead or of a millet seed; Fig. 22.7); alternatively, a piece of cerebellar cortex may be more easily be collected with a curette or a teaspoon through the *foramen occipitale* once the head has been removed from the body (Schreuder 1980);
- This cortex piece is put down near an end of a microscope slide and crushed down with the opposite end of another slide.

(continued)

- The brain fragment is then stretched, both slides remaining pressed against one another and being pushed in opposite directions as far as possible, as otherwise the gray matter is not sufficiently spread out thinly and the capillaries not correctly arranged.
- After 2–3 min of air drying and 3–5 min of fixation with methanol, the sample is stained.
- A rapid coloration kit (RAL 555) or 30 min Giemsa staining may be used; in South Africa, the preferred method is to stain tissue sections with an immunoperoxidase-labeled polyclonal antibody against *E. ruminantium*, followed by counterstaining with hematoxylin (Allsopp 2015).

Under the microscope, capillaries are first located with a low-power lens. Then, *E. ruminantium* is looked for with a high-power lens in the cytoplasm of the endothelial cells, close to the nucleus (Fig. 22.8). They appear as bunch of grapes of reddish-purple organisms (nuclei are dark blue), generally coccoid but sometimes annular or horseshoe-shaped. The microorganisms of the same clusters have roughly the same size, which can vary between clusters. It is sometimes difficult to find the clusters and a prolonged search may be needed. The method is highly dependent on the quality of the staining and on the experience of the observer: people not used to identifying the *E. ruminantium* colonies may miss the diagnosis. Moreover, *E. ruminantium* clusters should be differentiated from other parasites like *Babesia bovis* which is responsible for cerebral babesiosis.
Be careful: as nervous symptoms are also observed in animal infected with rabies, it is necessary to be very cautious to prepare the "brain crush smear"!

Serological Diagnosis

Two ELISA methods, detecting antibodies against the immunodominant membrane protein of *E. ruminantium*, MAP-1, are available: an indirect ELISA test using a fraction (MAP 1-B) of the protein (Van Vliet et al. 1995; Mondry et al. 1998; Semu et al. 2001) and a competitive ELISA using a recombinant *map-1* gene cloned in a baculovirus and a monoclonal antibody (Katz et al. 1997). The specificity of these tests was improved but some cross-reactions still occur with *E. canis*, *E. chaffeensis* (responsible for a human ehrlichiosis but also infecting ruminants), and Panola Mountain *Ehrlichia* (Semu et al. 2001; Sayler et al. 2016). In small ruminants, the MAP 1-B ELISA is able to detect antibodies during 4–6 months after infection whereas antibodies disappear rapidly in cattle although they can remain immune for years: the lack of antibodies in cattle does not mean that animals are not protected

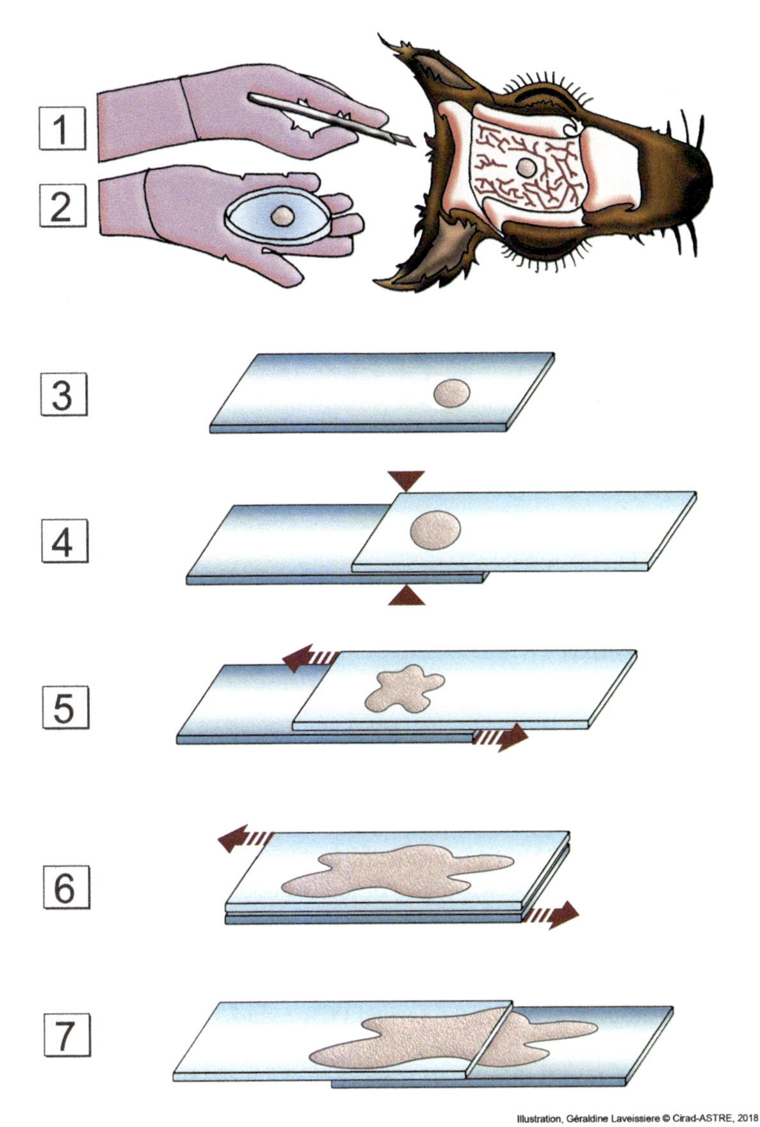

Fig. 22.7 Preparation of "brain crush smears" from the cortex of a sheep

and have not been infected by the pathogen during the previous months. Therefore, the ELISA test does not allow to determine the status of a region or a herd. More accurately, if positive animals are detected, it is sure that *E. ruminantium* is circulating in the area whereas the reverse is not certain: lack of positive cattle does not mean that the pathogen did not infect the examined animals (Semu et al. 2001).

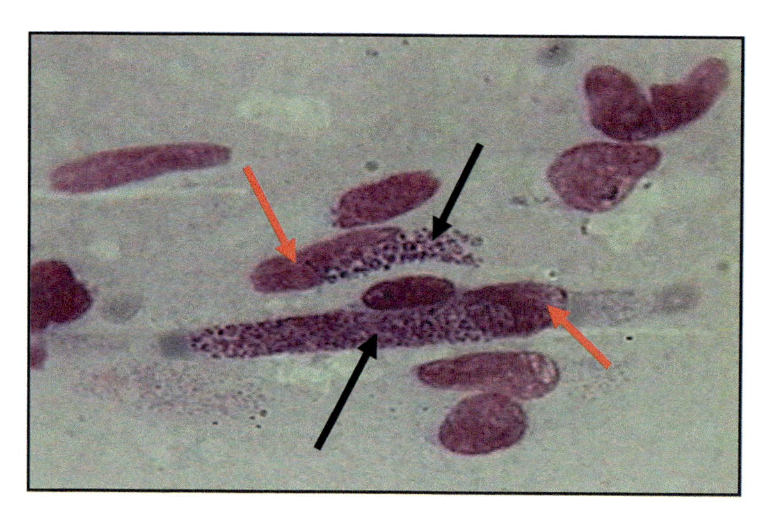

Fig. 22.8 *Ehrlichia ruminantium* clusters (black arrows) near the nuclei (red arrows) of capillary endothelial cells (photo: N. Vachiéry)

Molecular Diagnosis

Various techniques were developed to detect the presence of *E. ruminantium* in ruminant samples. The method used so far by the OIE reference laboratory was a nested PCR targeting a DNA fragment specific to the bacterium and named *pCS20*. The detection limit is 1–6 organisms per sample (blood, organ, or tick), either stored in ethanol or deep-frozen (Martinez et al. 2004). New primers including universal nucleotides even improved the sensibility of the method, allowing the detection of field strains which were non-detected with the previous primers (Molia et al. 2008). The use of nested PCR presents, however, some drawbacks: the method is time-consuming (1.5 day to implement the whole analysis), and there is high risk of contamination, particularly during the second PCR round.

Several real-time PCR methods were recently developed which allowed a reduction of the abovementioned drawbacks. One of these method is as sensitive as the nested PCR (possible detection of 7 organisms per sample) and detects 15 reference strains. Its specificity is, however, lower and cross-reactions occur with *E. canis* and *E. chaffeensis* (Steyn et al. 2008, 2010). Lastly, a new real-time PCR targeting another fragment of the *pCS20* gene was developed. Its sensitivity is at least equivalent to the previous one, the specificity and reproducibility are good, and it detects at least 16 *E. ruminantium* strains of various geographical origins (Cangi et al. 2017).

Whatever the molecular method used, the detection of *E. ruminantium* in sick ruminant blood is possible from 1 to 2 days before the onset of pyrexia to the end of the febrile period. On the other hand, no pathogen detection is possible in blood of asymptomatic animals.

Treatment

Tetracyclines, in particular oxytetracycline (Van Amstel and Oberem 1987) and doxycycline (Immelman and Dreyer 1982), are efficient to treat *E. ruminantium* infections if the treatment is administered quickly, as soon as fever is observed and before nervous symptoms occur. A single intramuscular injection of a long-acting oxytetracycline (LAO) formulation is enough to stop the disease and cure the ruminant which will therefore be immunized and can become asymptomatic carrier of the pathogen, in the same way as animals recovering without treatment. The dose of LAO to administer is 20 mg/kg body weight, i.e., 1 mL of a 20% solution per 10 kg. It is necessary to take into account the withdrawal period of the drug which prevents the consumption of treated animals. If the treatment is implemented too late, the ruminant will nevertheless die rapidly (Van Amstel and Oberem 1987) and the meat will not be edible. It is therefore necessary to determine whether recovery is possible before treating and to inform the farmer. Treatment is sometimes impossible when too sensitive breeds are infected by very virulent strains of *E. ruminantium*.

Vaccination

An immunization method called "infection and treatment" was developed in South Africa. Although sometimes considered as a vaccine, this method consists in an intravenous injection of cryopreserved blood containing virulent *E. ruminantium* organisms of the Ball 3 strain. The temperature of the inoculated ruminants is then monitored daily and animals are treated with antibiotics at the onset of the hyperthermia (Van der Merwe 1987). The animals are thus protected, at least against strains close to Ball 3, and can graze infested pasture where natural infection will consequently maintain the immune status. This method is, however, dangerous since some animals might die of a peracute form before they are treated. It is also not very convenient because the infected blood has to be kept until use in liquid nitrogen without cold chain disruption to prevent the bacterium to lose virulence, which is sometimes difficult in remote areas. Moreover, as the infected blood has to be regularly produced in "donor" sheep, it may contain other pathogens, including virus.

Inactivated vaccines were developed and tested in laboratory and field trials (Martinez 1997; Mahan et al. 1998a; Marcelino et al. 2015). They were efficient toward the homologous strain, with 80–100% protection, but inadequate to protect from all the strains present in the field, even when emulsified in oily adjuvant improving the immune response (Faburay et al. 2007a; Adakal et al. 2010b). A large-scale pre-industrial method of production was, however, developed (bacterium multiplication, purification, chemical inactivation, or lysis) which would allow a rapid marketing if one (or some) strain(s) enabling a good protection would be identified (Marcelino et al. 2006; Peixoto et al. 2007). The improvement of these inactivated vaccines depends probably on the elaboration of a vaccine cocktail of several *E. ruminantium* strains which would cover the antigenic diversity of an endemic area. This involves the determination of the strain diversity in each region as well as isolation of the circulating strains, in order to elaborate a regional multivalent vaccine. But as the knowledge of the genotype is not enough to know whether there is cross-protection between two strains, it is difficult to choose the suitable vaccine strains.

Attenuated strains were obtained from some *E. ruminantium* strains (Senegal, Gardel, Welgevonden) after numerous passages in cell culture (Jongejan et al. 1993; Martinez 1997; Zweygarth et al. 2005). As for inactivated vaccines, these attenuated vaccines are efficient to protect from homologous challenge, but less effective when used in the field, due to the high pathogenic diversity of the circulating strains. Moreover, numerous drawbacks limit or complicate their use: they have to be stored in liquid nitrogen until use, which prevents their use in the field; they have to be injected intravenously, and as for all the live attenuated vaccines, there is perhaps a risk of reversion to virulence in the vaccinated ruminants or in the ticks infesting them, especially since the attenuation mechanisms are so far not identified.

The last vaccine type currently developed are recombinant vaccines (Pretorius et al. 2010). The first one was made from MAP-1 glycosylated recombinant MAP-1 protein and induced immune responses in sheep but challenge was not performed. The second one was made of DNA/boost recombinant protein Erum 2510 which induced protection against homologous challenge. The use of recombinant vaccines has the advantage to include various antigens against the various *E. ruminantium* strains circulating in a region (Faburay et al. 2017). Finally, identification of *E. ruminantium* effectors acting on host cells is now considered in order to develop innovative therapeutic methods based on the blocking of host cell/bacterium interactions.

Other Prophylactic Methods

Prophylaxis by tick control methods aiming at stopping transmission of *E. ruminantium* is very difficult to implement because it is virtually impossible to completely prevent *Amblyomma variegatum* infestation, particularly by nymphs, even when acaricides are applied fortnightly or weekly. Considering the high infection rates of the vectors, attachment of only a few ticks will ensure infection of the hosts. Only ruminants permanently kept away from infested pastures and hay (dairy cattle in stalls, sheep in town, etc.) might remain unaffected by cowdriosis in enzootic areas.

The best prophylaxis is therefore the maintenance of enzootic stability which is achieved by regular infestation of the ruminants by enough ticks to enable them to acquire and maintain a protective immunity against cowdriosis (FAO 1990). The control of *A. variegatum* nymphs is useless because these ticks do not cause direct quantifiable losses (Stachurski et al. 1993). Treatment to prevent *A. variegatum* adult infestation is on the other hand necessary since these ticks cause high direct losses (Pegram et al. 1989; Stachurski et al. 1993). Tick control should, however, not aim at complete elimination of the ticks but at reducing infestation to a level allowing *E. ruminantium* transmission to all animals while limiting the direct losses. Such a strategy is sometimes difficult to accept for livestock owners and for animal health authorities and often difficult to achieve. In Sahelian infested areas, tick control should thus only be performed during the three first months of the rainy season at intervals depending on the re-infestation kinetics and on the residual effects of the used products.

It is by far more difficult to prevent clinical cowdriosis in ruminants introduced from free regions toward enzootic areas, in animals living in regions (including a significant part of the Sahel), where *A. variegatum* infestation is too low to ensure early infection of all the animals, or in sheep and goats losing their innate resistance before they are infested by infected ticks. How to favor the acquisition of protective immunity during the first infection by *E. ruminantium* without inducing serious symptoms? The "infection and treatment" method described above could be replaced, where there is no possibility to produce and store infected blood, by natural infection caused by ticks. But the monitoring of ruminants grazing infested pastures is still more inconvenient and may last longer (weeks and even months) because the onset of fever and disease is by far more uncertain than following infected blood inoculation. But if the animals are introduced in enzootic areas during the nymphal infestation peak, i.e., at the beginning of the dry season, or, as far as cattle are concerned (since sheep and goats are not highly infested by *A. variegatum* adults, as previously mentioned), at the beginning of the rainy season, during the adult infestation peak, they most likely will be rapidly infected by *E. ruminantium*. It is thus possible to daily monitor the animals and to treat them with antibiotics as soon as general condition is affected and pyrexia is noticed. It is also possible to systematically treat the ruminants with long-acting oxytetracycline at predetermined intervals taking into account the natural incubation period of the disease. Purnell (1987) suggested thus to treat at days 7, 14, and 21 after arrival of the animals in an infested area in order to prevent cowdriosis symptoms without hindering acquisition of immunity. When this method is implemented, tick control by acaricide should be prohibited during the entire period of antibiotic treatment in order to increase the infection probability. This method is, however, uncertain and costly, particularly for low-value small ruminants.

Acknowledgements The authors thank Gerrit Uilenberg for fruitful discussions regarding various issues of this chapter, mainly the epidemiology of cowdriosis. Gerrit Uilenberg also helped to translate the manuscript in English.

References

Adakal H, Meyer DF, Carasco-Lacombe C, Pinarello V, Allègre F, Huber K, Stachurski F, Morand S, Martinez D, Lefrançois T, Vachiéry N, Frutos R. MLST scheme of *Ehrlichia ruminantium*: genomic stasis and recombination in strains from Burkina-Faso. Infect Genet Evol. 2009;9(6):1320–8. https://doi.org/10.1016/j.meegid.2009.08.003.

Adakal H, Gavotte L, Stachurski F, Konkobo M, Henri H, Zoungrana S, Huber K, Vachiéry N, Martinez D, Morand S, Frutos R. Clonal origin of emerging populations of *Ehrlichia ruminantium* in Burkina Faso. Infect Genet Evol. 2010a;10(7):903–12. https://doi.org/10.1016/j.meegid.2010.05.011.

Adakal H, Stachurski F, Konkobo M, Zoungrana S, Meyer DF, Pinarello V, Aprelon R, Marcelino I, Alves PM, Martinez D, Lefrançois T, Vachiéry N. Efficiency of inactivated vaccines against heartwater in Burkina Faso: impact of *Ehrlichia ruminantium* genetic diversity. Vaccine. 2010b;28(29):4573–80. https://doi.org/10.1016/j.vaccine.2010.04.087.

Adakal H, Stachurski F, Chevillon C. Tick control practices in Burkina Faso and acaricide resistance survey in *Rhipicephalus (Boophilus) geigyi* (Acari: Ixodidae). Exp Appl Acarol. 2013;59(4):483–91. https://doi.org/10.1007/s10493-012-9610-5.

Alexander RA. Heartwater: the present state of our knowledge of the disease. Pretoria: S. A. Report of the Director of Veterinary Services and Animal Industry, Part I; 1931. pp. 89–150.

Allsopp BA. Heartwater - *Ehrlichia ruminantium* infection. Revue scientifique et technique de l'Office international des Epizooties. 2015;34(2):557–68.

Allsopp MTEP, Louw M, Meyer EC. *Ehrlichia ruminantium*: an emerging human pathogen? Ann N Y Acad Sci. 2005;1063(1):358–60. https://doi.org/10.1196/annals.1355.060.

Andrew HR, Norval RAI. The carrier status of sheep, cattle and African buffalo recovered from heartwater. Vet Parasitol. 1989;34(3):261–6. https://doi.org/10.1016/0304-4017(89)90056-3.

Barré N. Biologie et écologie de la tique *Amblyomma variegatum* (Acarina: Ixodina) en Guadeloupe (Antilles Françaises). PhD Thesis, Université de Paris-Sud-Orsay, Orsay; 1989. p. 268.

Barré N. Les tiques des ruminants dans les Petites Antilles: biologie, importance économique, principes de lutte. INRA Prod Anim. 1997;10(1):111–9.

Barré N, Garris GI. Biology and ecology of *Amblyomma variegatum* (Acari: Ixodidae) in the Caribbean: implications for a regional eradication program. J Agric Entomol. 1989;7(1):1–9.

Barré N, Uilenberg G, Morel PC, Camus E. Danger of introducing heartwater onto the American mainland: potential role of indigenous and exotic *Amblyomma* ticks. Onderstepoort J Vet Res. 1987;54:405–17.

Bezuidenhout JD, Paterson CL, Barnard BJ. In vitro cultivation of *Cowdria ruminantium*. Onderstepoort J Vet Res. 1985;52(2):113–20.

Burridge MJ, Simmons LA, Peter TF, Mahan SM. Increasing risks of introduction of heartwater onto the American mainland associated with animal movements. Ann N Y Acad Sci. 2002;969:269–74.

Camus E, Barré N, Martinez D, Uilenberg G. Heartwater (cowdriosis), a review. Office International des Epizooties (OIE), Paris; 1996. p. 177.

Cangi N. New molecular high throughput methods for *Ehrlichia ruminantium* tick screening and characterization of strain genetic structure in Mozambique and at worldwide scale. PhD Thesis, Université des Antilles, Pointe-à-Pitre, Guadeloupe; 2017. p. 163.

Cangi N, Gordon JL, Bournez L, Pinarello V, Aprelon R, Huber K, Lefrançois T, Neves L, Meyer DF, Vachiéry N. Recombination is a major driving force of genetic diversity in the Anaplasmataceae *Ehrlichia ruminantium*. Front Cell Infect Microbiol. 2016;6:111. https://doi.org/10.3389/fcimb.2016.00111.

Cangi N, Pinarello V, Bournez L, Lefrançois T, Albina E, Neves L, Vachiéry N. Efficient high-throughput molecular method to detect *Ehrlichia ruminantium* in ticks. Parasit Vectors. 2017;10(1):566. https://doi.org/10.1186/s13071-017-2490-0.

Collins NE, Liebenberg J, de Villiers EP, Brayton KA, Louw E, Pretorius A, Faber FE, van Heerden H, Josemans A, van Kleef M, Steyn HC, van Strijp MF, Zweygarth E, Jongejan F, Maillard J-C, Berthier D, Botha M, Joubert F, Corton CH, Thomson NR, Allsopp MT, Allsopp BA. The genome of the heartwater agent *Ehrlichia ruminantium* contains multiple tandem repeats of actively variable copy number. Proc Natl Acad Sci. 2005;102(3):838–43. https://doi.org/10.1073/pnas.0406633102.

Corn JL, Barré N, Thiebot B, Creekmore TE, Garris GI, Nettles VF. Potential role of cattle egrets, *Bubulcus ibis* (Ciconiformes: Ardeidae), in the dissemination of *Amblyomma variegatum* (Acari: Ixodidae) in the Eastern Caribbean. J Med Entomol. 1993;30(6):1029–37.

Deem SL, Donachie PL, Norval RAI. Colostrum from dams living in a heartwater-endemic area influences calfhood immunity to *Cowdria ruminantium*. Vet Parasitol. 1996a;61:133–44.

Deem SL, Norval RAI, Donachie PL, Mahan SM. Demonstration of vertical transmission of *Cowdria ruminantium*, the causative agent of heartwater, from cows to their calves. Vet Parasitol. 1996b;61:119–32.

du Plessis JL, Kumm NAL. The passage of *Cowdria ruminantium* in mice. J S Afr Vet Med Assoc. 1971;42(3):217–21.

Dumler JS, Barbet AF, Bekker CP, Dasch GA, Palmer GH, Ray SC, Rikihisa Y, Rurangirwa FR. Reorganization of genera in the families Rickettsiaceae and Anaplasmataceae in the order Rickettsiales: unification of some species of Ehrlichia with Anaplasma, Cowdria with Ehrlichia and Ehrlichia with Neorickettsia, descriptions of six new species combinations and designation of Ehrlichia equi and 'HGE agent' as subjective synonyms of Ehrlichia phagocytophila. Int J Syst Evol Microbiol. 2001;51(6):2145–65. https://doi.org/10.1099/00207713-51-6-2145.

Faburay B, Geysen D, Ceesay A, Marcelino I, Alves PM, Taoufik A, Postigo M, Bell-Sakyi L, Jongejan F. Immunisation of sheep against hearwater in the Gambia using inactivated and attenuated *Ehrlichia ruminantium* vaccines. Vaccine. 2007a;25:7939–47.

Faburay B, Geysen D, Munstermann S, Taoufik A, Postigo M, Jongejan F. Molecular detection of *Ehrlichia ruminantium* infection in *Amblyomma variegatum* ticks in the Gambia. Exp Appl Acarol. 2007b;42:61–74.

Faburay B, McGill J, Jongejan F. A glycosylated recombinant subunit candidate vaccine consisting of *Ehrlichia ruminantium* major antigenic protein1 induces specific humoral and Th1 type cell responses in sheep. PLoS One. 2017;12(9):e0185495. https://doi.org/10.1371/journal.pone.0185495.

FAO. Report of the FAO expert consultation on revision of strategies for the control of ticks and tick-borne diseases, Rome, 25–29 September 1989. Parassitologia. 1990;32:3–12.

Frutos R, Viari A, Ferraz C, Morgat A, Eychenié S, Kandassamy Y, Chantal I, Bensaid A, Coissac E, Vachiéry N, Demaille J, Martinez D. Comparative genomic analysis of three strains of *Ehrlichia ruminantium* reveals an active process of genome size plasticity. J Bacteriol. 2006;188(7):2533–42. https://doi.org/10.1128/JB.188.7.2533-2542.2006.

Hoogstraal H. African Ixodidae. Vol. 1: ticks of the Sudan. U.S. NaMRU no 3, Research report, Cairo; 1956. p. 1101.

Immelman A, Dreyer G. The use of doxycycline to control heartwater in sheep. J S Afr Vet Assoc. 1982;53(1):23–4.

Jongejan F, Thielemans MJC, Briere C, Uilenberg G. Antigenic diversity of Cowdria ruminantium isolates determined by cross-immunity. Res Vet Sci. 1991a;51(1):24–8. https://doi.org/10.1016/0034-5288(91)90025-J.

Jongejan F, Zandbergen TA, van de Wiel PA, de Groot M, Uilenberg G. The tick-borne rickettsia *Cowdria ruminantium* has a Chlamydia-like developmental cycle. Onderstepoort J Vet Res. 1991b;58(4):227–37.

Jongejan F, Vogel SW, Gueye A, Uilenberg G. Vaccination against heartwater using in vitro attenuated *Cowdria ruminantium* organisms. Revue d'Elevage et de Médecine vétérinaire des Pays tropicaux. 1993;46(1–2):5. https://doi.org/10.19182/remvt.9367.

Katz JB, DeWald R, Dawson JE, Camus E, Martinez D, Mondry R. Development and evaluation of a recombinant antigen, monoclonal antibody-based competitive ELISA for heartwater serodiagnosis. J Vet Diagn Investig. 1997;9(2):130–5.

Lawrence JA, Foggin CM, Norval RAI. The effect of war on the control of the diseases of livestock in Rhodesia (Zimbabwe). Vet Rec. 1980;107:82–5.

Mahan SM, Kumbula D, Burridge MJ, Barbet AF. The inactivated *Cowdria ruminantium* vaccine for heartwater protects against heterologous strains and against laboratory and field tick challenge. Vaccine. 1998a;16(11/12):1203–11.

Mahan SM, Peter TF, Simbi BH, Burridge MJ. PCR detection of *Cowdria ruminantium* infection in ticks and animals from heartwater-endemic regions of Zimbabwe. Ann N Y Acad Sci. 1998b;849:85–7.

Mahoney DF, Ross DR. Epizootiological factors in the control of bovine babesiosis. Aust Vet J. 1972;48:292–8.

Marcelino I, Sousa MFQ, Verissimo C, Cunha AE, Carrondo MJT, Alves PM. Process development for the mass production of *Ehrlichia ruminantium*. Vaccine. 2006;24:1716–25.

Marcelino I, Ventosa M, Pires E, Müller M, Lisacek F, Lefrançois T, Vachiery N, Varela Coelho A. Comparative proteomic profiling of *Ehrlichia ruminantium* pathogenic strain and its high-passaged attenuated strain reveals virulence and attenuation-associated proteins. PLoS One. 2015;10(12):e0145328. https://doi.org/10.1371/journal.pone.0145328.

Martinez D. Analysis of the immune response of ruminants to *Cowdria ruminantium* infection. Development of an inactivated vaccine. PhD Thesis, Faculty of Veterinary Medicine, Universiteit Utrecht, Utrecht; 1997. p. 206.

Martinez D, Vachiéry N, Stachurski F, Kandassamy Y, Raliniaina M, Aprelon R, Gueye A. Nested PCR for detection and genotyping of *Ehrlichia ruminantium*: use in genetic diversity analysis. Ann N Y Acad Sci. 2004;1026(1):106–13. https://doi.org/10.1196/annals.1307.014.

Meltzer MI. A possible explanation of the apparent breed-related resistance in cattle to bont tick (*Amblyomma hebraeum*) infestations. Vet Parasitol. 1996;67:275–9.

Molia S, Frebling M, Vachiéry N, Pinarello V, Petitclerc M, Rousteau A, Martinez D, Lefrançois T. *Amblyomma variegatum* in cattle in Marie Galante, French Antilles: prevalence, control measures, and infection by *Ehrlichia ruminantium*. Vet Parasitol. 2008;153(3–4):338–46. https://doi.org/10.1016/j.vetpar.2008.01.046.

Mondry R, Martinez D, Camus E, Liebisch A, Katz JB, DeWald R, Van Vliet AH, Jongejan F. Validation and comparison of three enzyme-linked immunosorbent assays for the detection of antibodies to *Cowdria ruminantium* infection. Ann N Y Acad Sci. 1998;849(1):262–72. https://doi.org/10.1111/j.1749-6632.1998.tb11058.x.

Morel P.-C. Contribution à la connaissance de la distribution des tiques (Acariens, Ixodidae et Amblyommidae) en Afrique éthiopienne continentale. PhD Thesis, Université de Paris-Sud-Orsay, Orsay; 1969. p. 388.

Pegram RG, Banda DS. Ecology and phenology of cattle ticks in Zambia: development and survival of free-living stages. Exp Appl Acarol. 1990;8:291–301.

Pegram RG, Oosterwijk GPM. The effect of *Amblyomma variegatum* on liveweight gain of cattle in Zambia. Med Vet Entomol. 1990;4:327–30.

Pegram RG, Lemche J, Chizyuka HGB, Sutherst RW, Floyd RB, McCosker PJ. Effect of tick control on liveweight gain of cattle in Central Zambia. Med Vet Entomol. 1989;3:313–20.

Peixoto C, Marcelino I, Amaral AI, Carrondo MJT, Alves PM. Purification by membrane technology of an intracellular *Ehrlichia ruminantium* candidate vaccine against heartwater. Process Biochem. 2007;42(7):1084–9. https://doi.org/10.1016/j.procbio.2007.04.012.

Petney TN, Horak IG, Rechav Y. The ecology of the African vectors of heartwater, with particular reference to *Amblyomma hebraeum* and *A. variegatum*. Onderstepoort J Vet Res. 1987;54:381–95.

Pilet H, Vachiéry N, Berrich M, Bouchouicha R, Durand B, Pruneau L, Pinarello V, Saldana A, Carasco-Lacombe C, Lefrançois T, Meyer DF, Martinez D, Boulouis H-J, Haddad N. A new typing technique for the Rickettsiales *Ehrlichia ruminantium*: multiple-locus variable number tandem repeat analysis. J Microbiol Methods. 2012;88(2):205–11. https://doi.org/10.1016/j.mimet.2011.11.011.

Plowright W. Cutaneous streptothricosis of cattle. I. Introduction and epizootiological features in Nigeria. Vet Rec. 1956;68:350–5.

Pretorius A, Liebenberg J, Louw E, Collins NE, Allsopp BA. Studies of a polymorphic *Ehrlichia ruminantium* gene for use as a component of a recombinant vaccine against heartwater. Vaccine. 2010;28(20):3531–9. https://doi.org/10.1016/j.vaccine.2010.03.017.

Prozesky L, Du Plessis JL. Heartwater. The development and life cycle of *Cowdria ruminantium* in the vertebrate host, ticks and cultured endothelial cells. Onderstepoort J Vet Res. 1987;54 (3):193–6.

Purnell RE. Development of a prophylactic regime using Terramycin/LA to assist in the introduction of susceptible cattle into heartwater endemic areas of Africa. Onderstepoort J Vet Res. 1987;54(3):509–12.

Raliniaina M, Meyer DF, Pinarello V, Sheikboudou C, Emboulé L, Kandassamy Y, Adakal H, Stachurski F, Martinez D, Lefrançois T, Vachiéry N. Mining the genetic diversity of *Ehrlichia ruminantium* using map genes family. Vet Parasitol. 2010;167(2–4):187–95. https://doi.org/10.1016/j.vetpar.2009.09.020.

Rodrigues V, Fernandez B, Vercoutere A, Chamayou L, Andersen A, Vigy O, Demettre E, Seveno M, Aprelon R, Giraud-Girard K, Stachurski F, Loire E, Vachiéry N, Holzmuller P. Immunomodulatory effects of *Amblyomma variegatum* saliva on bovine cells: characterization of cellular responses and identification of molecular determinants. Front Cell Infect Microbiol. 2017;7:521. https://doi.org/10.3389/fcimb.2017.00521.

Roth JA, Richt JA, Morozov IA, editors. Vaccines and diagnostics for transboundary animal diseases. Proceedings of an international symposium, Ames, Iowa, September 2012. Developments in biologicals. vol 135. Basel: Karger; 2013. p. 226.

Sayler KA, Loftis AD, Mahan SM, Barbet AF. Development of a quantitative PCR assay for differentiating the agent of heartwater disease, *Ehrlichia ruminantium*, from the Panola Mountain Ehrlichia. Transbound Emerg Dis. 2016;63(6):e260–9. https://doi.org/10.1111/tbed.12339.

Schreuder BEC. A simple technique for the collection of brain samples for the diagnosis of heartwater. Trop Anim Health Prod. 1980;12(1):25–9. https://doi.org/10.1007/BF02242627.

Semu SM, Peter TF, Mukwedeya D, Barbet AF, Jongejan F, Mahan SM. Antibody responses to MAP 1B and other *Cowdria ruminantium* antigens are down regulated in cattle challenged with tick-transmitted heartwater. Clin Vaccine Immunol. 2001;8(2):388–96. https://doi.org/10.1128/CDLI.8.2.388-396.2001.

Stachurski F. Invasion of West African cattle by the tick *Amblyomma variegatum*. Med Vet Entomol. 2000;14:391–9.

Stachurski F, Musonge EN, Achu-Kwi MD, Saliki JT. Impact of natural infestation of *Amblyomma variegatum* on the liveweight gain of male Gudali cattle in Adamawa (Cameroon). Vet Parasitol. 1993;49:299–311.

Stewart CG. Specific immunity in farm animals to heartwater. Onderstepoort J Vet Res. 1987;54 (3):341–2.

Steyn HC, Pretorius A, McCrindle CME, Steinmann CML, Van Kleef M. A quantitative real-time PCR assay for *Ehrlichia ruminantium* using pCS20. Vet Microbiol. 2008;131(3–4):258–65. https://doi.org/10.1016/j.vetmic.2008.04.002.

Steyn HC, McCrindle CME, Du Toit D. Veterinary extension on sampling techniques related to heartwater research. J S Afr Vet Assoc. 2010;81(3):160–5.

Totté P, Blankaert D, Marique T, Kirkpatrick C, Van Vooren JP, Wérenne J. Bovine and human endothelial cell growth on collagen microspheres and their infection with the rickettsia *Cowdria ruminantium*: prospects for cells and vaccine production. Revue d'Elevage et de Médecine vétérinaire des Pays tropicaux. 1993;46(1–2):153–6.

Uilenberg G. Acquisitions nouvelles dans la connaissance du rôle vecteur de tiques du genre *Amblyomma* (Ixodidae). Revue d'Elevage et de Médecine vétérinaire des Pays tropicaux. 1983;36(1):61–6.

Uilenberg G. An historical account of tick and tick-borne disease control in East Africa with emphasis on tick-borne diseases. Implications for centres in West Africa. FAO seminar on African centres for tick and tick-borne disease control, Ouagadougou, 26–29 November 1984. FAO, Rome; 1984. p. 9.

Uilenberg G. International collaborative research: significance of tick-borne hemoparasitic diseases to world animal health. Vet Parasitol. 1995;57(1):19–41.

Vachiéry N, Jeffery H, Pegram RG, Aprelon R, Pinarello V, Lloyd R, Kandassamy Y, Raliniaina M, Molia S, Savage H, Alexander R, Frebling M, Martinez D, Lefrançois T. *Amblyomma variegatum* ticks and heartwater on three Caribbean Islands. Ann N Y Acad Sci. 2008;1149:191–5. https://doi.org/10.1196/annals.1428.081.

Van Amstel SR, Oberem PT. The treatment of heartwater. Onderstepoort J Vet Res. 1987;54 (3):475–9.

Van der Merwe L. The infection and treatment method of vaccination against heartwater. Onderstepoort J Vet Res. 1987;54(3):489–91.

Van Vliet AH, van der Zeijst BA, Camus E, Mahan SM, Martinez D, Jongejan F. Use of a specific immunogenic region on the *Cowdria ruminantium* MAP 1 protein in a serological assay. J Clin Microbiol. 1995;33(9):2405–10.

Walker JB, Olwage A. The tick vectors of *Cowdria ruminantium* (Ixodoidea, Ixodidae, genus *Amblyomma*) and their distribution. Onderstepoort J Vet Res. 1987;54:353–79.

Walker AR, Bouattour A, Camicas J-L, Estrada-Peña A, Horak I, Latif AA, Pegram RG, Preston PM. Ticks of domestic animals in Africa: a guide to identification of species. Bioscience reports, Edinburgh; 2003. p. 221.

Yonow T. The life-cycle of *Amblyomma variegatum* (Acari: Ixodidae): a literature synthesis with a view to modelling. Int J Parasitol. 1995;25(9):1023–60.

Zweygarth E, Josemans AI, Van Strijp MF, Lopez-Rebollar L, Van Kleef M, Allsopp BA. An attenuated *Ehrlichia ruminantium* (Welgevonden stock) vaccine protects small ruminants against virulent heartwater challenge. Vaccine. 2005;23(14):1695–702. https://doi.org/10.1016/j.vaccine.2004.09.030.